Advances in Mathematics and Applications

Carlile Lavor
Francisco A. M. Gomes
Editors

Advances in Mathematics and Applications

Celebrating 50 years of the Institute of Mathematics, Statistics and Scientific Computing, University of Campinas

Springer

Editors
Carlile Lavor
Institute of Mathematics
Statistics and Scientific Computing
University of Campinas
Campinas, SP, Brazil

Francisco A. M. Gomes
Institute of Mathematics
Statistics and Scientific Computing
University of Campinas
Campinas, SP, Brazil

ISBN 978-3-030-06772-4 ISBN 978-3-319-94015-1 (eBook)
https://doi.org/10.1007/978-3-319-94015-1

Mathematics Subject Classification: 22Exx, 37-XX, 62M10, 94Bxx, 01Axx, 01A73

This Springer imprint is published by the registered company Springer Nature Switzerland AG
The registered company address is: Gewerbestrasse 11, 6330 Cham, Switzerland

Preface

When the University of Bologna was founded, in 1088, Brazil had not been visited by the Portuguese navigators, which would only occur 412 years later, in 1500. An even greater gap was observed between the creation of the University of Coimbra, in 1290, the first in a Portuguese-speaking country, and the authorization of the Emperor Dom João VI for the installation of higher education courses in Brazil, in 1808, the year in which the Portuguese court arrives in the country, escaping from the pressure exerted by Napoleon Bonaparte.

Despite the imperial permission and the subsequent declaration of independence from Portugal, in 1822, universities, conceived as multidisciplinary institutions of higher education, were only created in Brazil in the twentieth century. The University of São Paulo (USP), for example, was founded in 1934, bringing together isolated colleges and schools, which was the model for higher education until then.

From 1940 to 1960, Brazilian higher education experienced a great advancement, multiplying by three the number of enrolled students. It is also from this period that the Brazilian Center for Research in Physics (CBPF), the National Council for Scientific and Technological Development (CNPq), the Coordination for the Improvement of Higher Education Personnel (CAPES), the Institute for Pure and Applied Mathematics (IMPA), and the São Paulo Research Foundation (FAPESP) were created, institutions that are of crucial relevance for the development of science in Brazil.

But it was the ebullient atmosphere of social and scientific changes that characterized the 1960s, in Brazil and around the world, which became a catalyst element for the experimentation of new ideas in all fields of sciences and higher education in the country. It is in this environment that the University of Campinas (Unicamp) appears, in 1966. Today, 52 years after being founded, there is no doubt that Unicamp was able to efficiently combine teaching activities with advanced research and outreach, which allowed it to approach universities of great international prestige, despite the several centuries of advantage that separate most of them from Unicamp.

Because it was conceived as a research university, Unicamp has adopted, from its very beginning, some innovative practices for the time from its founding, such as the creation of basic science institutes and the adoption of a common initial curriculum to be fulfilled by students of several programs, before they dedicated themselves to the specific courses of their specialties. This was how the Institute of Physics came into being in 1967, as well as the Institute of Mathematics, Statistics and Scientific Computation (IMECC) in 1968, months before the Brazilian Congress promulgated the so-called university reform, an act that officially instituted these practices in the rest of the country.

Naturally, the creation of an Institute of Mathematics in a city in the interior of the state of São Paulo, and its existence in a land that, 2 years previously, was just farmland, was a task worthy of being featured in an epic story. For the institute to fulfill this role, it was necessary to count upon the unlikely combination of the modernizing spirit of Zeferino Vaz, the first president of Unicamp, with the enthusiasm and dedication of the young researchers who accepted this task.

It was this creative and innovative drive of its founders that eventually became the trademark of IMECC. The institute was one of the first in the country to implement undergraduate programs in statistics, computer science, and applied mathematics. Moreover, its graduate program in mathematics was also one of the first to receive highest evaluation grade from CAPES, a federal agency that assesses the quality of graduate courses countrywide.

This volume starts with a description of the challenges faced in the initial years of the institute and a historical view with new opportunities for applied mathematics in Brazil, followed by research and survey articles of colleagues who lived the first years of existence of IMECC and, at the same time, stood out internationally in their research areas. Among them, we cannot but regret the passing away, just after finishing his contribution, of Prof. Waldyr Rodrigues Jr. We miss him and all other colleagues who are no longer among us and who helped build the IMECC.

It is with great pleasure and enthusiasm that, to celebrate the 50 years of IMECC history, we have gathered in this volume some articles that reflect a partial view of the institute's unique contributions to the development of mathematics, applied mathematics, and statistics in Brazil.

This date is not, however, an arrival in terms of IMECC's academic life. Rather, we consider it to be a starting point for our next 50 years – or more. This lays upon us, collectively, a serious responsibility: keep the high quality of our work as a full-fledged university institute and maintain our dedication to research, teaching, and outlook at a level of academic excellence.

Campinas, SP, Brazil — Carlile Lavor
Campinas, SP, Brazil — Francisco A. M. Gomes
October 2018

Contents

Contributors

Rodney Carlos Bassanezi Institute of Mathematics, Statistics and Scientific Computing, University of Campinas, Campinas, SP, Brazil

José Luiz Boldrini Institute of Mathematics, Statistics and Scientific Computing, University of Campinas, Campinas, SP, Brazil

Dan A. Agüero Cerna Institute of Mathematics, Statistics and Scientific Computing, University of Campinas, Campinas, SP, Brazil

Sueli I. R. Costa Institute of Mathematics, Statistics and Scientific Computing, University of Campinas, Campinas, SP, Brazil

João Frederico da Costa Azevedo Meyer Institute of Mathematics, Statistics and Scientific Computing, University of Campinas, Campinas, SP, Brazil

Djairo G. de Figueiredo Institute of Mathematics, Statistics and Scientific Computing, University of Campinas, Campinas, SP, Brazil

Sandro Gallo Center of Sciences, Federal University of São Carlos, São Carlos, SP, Brazil

Nancy L. Garcia Institute of Mathematics, Statistics and Scientific Computing, University of Campinas, Campinas, SP, Brazil

Otávio M. L. Gomide Institute of Mathematics, Statistics and Scientific Computing, University of Campinas, Campinas, SP, Brazil

Luiz Koodi Hotta Institute of Mathematics, Statistics and Scientific Computing, University of Campinas, Campinas, SP, Brazil

Luquesio P. M. Jorge Center of Sciences, Federal University of Ceará, Fortaleza, CE, Brazil

José Mario Martínez Institute of Mathematics, Statistics and Scientific Computing, University of Campinas, Campinas, SP, Brazil

Francesco Mercuri Institute of Mathematics, Statistics and Scientific Computing, University of Campinas, Campinas, SP, Brazil

Antonio Paques Institute of Mathematics, Federal University of Rio Grande do Sul, Porto Alegre, RS, Brazil

Alcibíades Rigas Institute of Mathematics, Statistics and Scientific Computing, University of Campinas, Campinas, SP, Brazil

Waldyr A. Rodrigues Jr. Institute of Mathematics, Statistics and Scientific Computing, University of Campinas, Campinas, SP, Brazil

Luiz A. B. San Martin Institute of Mathematics, Statistics and Scientific Computing, University of Campinas, Campinas, SP, Brazil

João E. Strapasson School of Applied Sciences, University of Campinas, Campinas, SP, Brazil

Marco Antonio Teixeira Institute of Mathematics, Statistics and Scientific Computing, University of Campinas, Campinas, SP, Brazil

Cristiano Torezzan School of Applied Sciences, University of Campinas, Campinas, SP, Brazil

Carlos Trucíos São Paulo School of Economics, Getúlio Vargas Foundation, São Paulo, SP, Brazil

Martin Tygel Institute of Mathematics, Statistics and Scientific Computing, University of Campinas, Campinas, SP, Brazil

Jayme Vaz Jr. Institute of Mathematics, Statistics and Scientific Computing, University of Campinas, Campinas, SP, Brazil

“And Now We’re in 2018. . . ”

João Frederico da Costa Azevedo Meyer

Abstract In this chapter, having been part of IMECC during its 50 years, I try to register, but not officially, of course, stories that make up the history of the Institute, as well as pointing out special moments, highlighting some of these. Some names are mentioned, but many other names do not appear since what I register is, mostly, the result of collective actions. The first years of IMECC are seen through the eyes (and feelings) of an undergraduate student and the remaining 46 years are described—and quite subjectively!—with the observation of a professor. The objective of this chapter, therefore, is to give an idea of where we came from and what made us who we are. . . .

The year was 1967. On paper our university—São Paulo State University at Campinas, in Portuguese “Universidade Estadual de Campinas”, or UNICAMP—had been formed in 1966, with all the necessary documents being signed and including the School of Medicine that had already existed in Campinas for some years. But, the first class, that of 1967, only began in April of that year. Brazilian society was living the first years of a violent coup d’état which had done away with a democratically elected government, closed Congress and done away with many individual and legal rights—the year was one of those which we identify today as the “Years of Lead.” On the other hand, literature, music, theater, and the change in social values were booming worldwide, and in Campinas too.

The University functioned as a whole in a single building close to the commercial center of Campinas, then a town with less than 370,000 inhabitants which already had strong industrial characteristics that existed alongside an intense agricultural production. This building had previously belonged to a technical school so it had the necessary conditions for installing classrooms, laboratories, bureaucratic offices, and storage, but it was right next to the municipal market and to the big open space

J. F. da Costa Azevedo Meyer (✉)
Institute of Mathematics, Statistics and Scientific Computing, University of Campinas, Campinas, SP, Brazil
e-mail: joni@ime.unicamp.br

C. Lavor, F. A. M. Gomes (eds.), *Advances in Mathematics and Applications*,
https://doi.org/10.1007/978-3-319-94015-1_1

to which came all the produce to be sold to intermediaries. Not surprisingly, all this commerce happened in an environment with several other activities: bars, early morning restaurants, pickpockets, street vendors .... The list is quite long. That is where we started. So, on a street called "The Cult of Science," next to this space of social vitality, with unheard of equipments and academic seriousness, both the University and its Mathematics Department were born.

In fact, this imagined University was the work of a very special person, Prof. Zeferino Vaz who had the courage (or the foolhardiness, or the vision) of accepting the challenge of creating something new as a university, something different from other schools, where new ideas from other countries could (and did!) influence our lives, and our academic and personal existences.

As in all Brazilian universities, there was an entrance exam, and those who passed thought of being engineers of some kind, at least one of which was not quite traditional on the Brazilian academic scenario. But Prof. Zeferino Vaz was not a man to dream or envision or imagine or create new things by himself. He quite literally seduced great researchers from other schools and towns (and other countries!), and talked them into coming to Campinas—and to share his dreams. And they came. The laboratories, the computer (which many called, in Portuguese, an "electronic brain"...) convinced them. Maybe. But, most of all, it was Zeferino Vaz's dream, and his promises that brought together a respectable team of professors.

So, for those of us who passed the exam to enter the university that year and in the subsequent years, an extraordinary privilege: we had classes with professors and researchers who had the courage to come after something new, challenging, motivating and, above all, the chance to come to a place where freedom and individual and civil rights were the rule.

In Mathematics, that was the background as well and many of us, having entered the university to become engineers, due to the contact with these scientists, some of them bright-eyed youngsters, some of them experienced professionals with histories of rebellion against imposed authority, was an opportunity to come in contact with something new, with new ideas, with new values, with a dream to be dreamt together.

In July of 1967, Ivam Resina (now Prof. Dr. Ivam Resina) and I were called to become monitors of the computer, a fantastic IBM 1130 with an unheard of capacity of 16K! Besides our good grades during the first semester (which I believe got us a small grant to be monitors), we were all involved in politics and sports: we were both part of the directing board of our student body—the existence of which the law forbade—Ivam played volleyball and football, whereas I was the swimmer in the free-style swimming as well as part of the relay group and a medium-distance runner....

At that time, Computer Science was something totally new, and was considered part of Applied Mathematics. And our teacher, the late Prof. Dr. Imre Simon was fantastic as a teacher and courageous as an academic—an example we had to look up to.

By this time, some of us were completely hooked by science, no more engineering for us. Many of our classmates, however, stayed in engineering and we all received this wonderful influence, our professors' enthusiasm was not wasted!

Mentioning names is a serious risk, since my memory will most certainly fail me, but some of the teachers who came from the State University at São Paulo included Angelo Barone, Mauro de Oliveira César, Paulo Boulos, and Ayrton Badelucci—and what an influence they had upon us!

With the arrival of the following year, 1968, a critical year for the whole country, a new class entered the university, and we saw the same enthusiasm overcome the participants of the class of 1968.... During this time, the political situation became more and more serious and dangerous and, finally, military authorities arrested ALL the students participating in the Annual Congress of the National Student Union in Brazil, called "UNE." Our three more outstanding leaders ended up in jail with so many others and our University President Zeferino Vaz was one of the very few who managed to visit them (and to take them cigarettes and chocolates, and messages from us, their colleagues).

That was when the Mathematics Institute was founded: the Institute for Mathematics, Statistics and Computer Science, later to become the Institute for Mathematics, Statistics and Scientific Computing. In 1968, it was born in the midst of political and social turmoil, but our teachers' enthusiasm was not diminished by what was happening in our midst as well as to the whole country. This was when Zeferino Vaz brought to UNICAMP the professors who had been fired by the military who had invaded the National University in Brasilia, our national capital. UNICAMP became a haven as well as the *locus* for serious science to be developed. Our Institute, IMECC came to be in difficult times but that was good: being forged in this scenario made it strong, sound, and very, very serious. Oh, it was not perfect, as always happened to university organizations, but it had a goal, a destiny, a task. And, our IMECC never left its pioneering activities, its innovative choice, its academic seriousness....

The following years, 1969 and 1970, were of an aggravated social situation, when many social, popular, and political leaders were arrested, underground movements were strengthened, and persecutions undertaken. So, in this situation, IMECC continued to thrive academically in no way separated from the national situation, but really as a part of this social turmoil! And, this was when the graduate studies appeared, to accompany the undergraduate efforts....

Yes, the political situation affected our efforts, yes, the risk of doubting and of speaking out was seriously dangerous, yes, many mathematicians were persecuted in Brazil and in Latin America: our graduate efforts were a place of resistance, and the seeds were set for an active political life for the IMECC!

During these initial years, two strong groups emerged: Logic and Differential Equations and, some years later, Mathematical Analysis and Algebra. During the academic semesters in 1970, fourth-year undergrads had classes together with those students who had come to our graduate programs, an opportunity to increase our experiences and our motivation.

By this time, we had already moved to the place where our Campinas campus is located, in a district north of Campinas called Barão Geraldo. There were buses in the morning from downtown to the campus and buses in the other sense at the end of the afternoon. If you missed these buses, the option was to hitch a ride to Barão

Geraldo and, from the entrance of the district, to walk more or less 3 miles to the classrooms and laboratories.

Research groups had grown, new professors had arrived, and there were a lot more undergrads and graduate students. Of course, we did not continue to live the initial honeymoon: discussions, some severe, involved many aspects and, although putting us through difficult moments, also helped forge a policy of understanding differences and diversities. So, in due time we had several different research groups working in our Institute, and we all received precious support from our dean, Prof. Dr. Ubiratan D'Ambrósio who succeeded Prof. Dr. Rubens Murillo Marques, who had laid the foundations for our becoming a nationally significant research center. Professor D'Ambrósio issued an international invitation, and a new group of brilliant youngsters attended this call coming from several countries. Among these came Alcibíades Rigas, Antônio Conde, and Francesco "Franco" Mercuri who joined the existing faculty, bringing with them further enthusiasm. Like them, IMECC received a group that came from Rio de Janeiro, a group that had been working with Professor Leopoldo Nachbin. In with this group came Mário Mattos, Carlos Mujica, and João Bosco Prolla. It was at this time that Algebra, Analysis, Geometry, Logic, Numerical Analysis, Measure Theory, Topology, Optimization, Operational Research, Computer Science, and Scientific Computing and Applications identified not only areas of work in Mathematics but groups of IMECC's faculty developing work, and teaching as well. By this time, our Graduate Programs had become a reality, a significant one in Brazil. But, it was important to add to these professors other young researchers who chose to come to Brazil to live and to work, helping to give IMECC what is at present a cosmopolitan faculty where most discussions are carried out with different accents, different cultures, different backgrounds, and histories: a healthy diversity that still marks our working (and personal!) coexistence....

In 1974, a special Program called the Programa para a Melhoria e a Expansão do Ensino (PREMEN)—the Program for Improvement and Expansion of Teaching—was created by the federal government, and the University President, Prof. Zeferino Vaz, invited prof. Ubiratan D'Ambrósio to be its Coordinator. Ubiratan immediately accepted, of course, he was a pioneer in most efforts to improve learning in all of its aspects. IMECC played a leading role in this national initiative. And, with the support IMECC received, new specialists were hired, and maybe one of them was paradigmatic: Prof. Henry "Hank" Wetzler who, after obtaining a doctorate in Differential Equations, dedicated himself to the area of Mathematical Education. Prof. Wetzler immediately proposed using television in the classroom and managed to build a small studio which, in the following years, began to work with Analytic Geometry and Linear Algebra (partially due to the high failure rates), two mass courses which became effective 2 years later with excellent results, influencing professors who preferred the traditional methods as well.... And, the small amateurish studio continued to grow and became what is today UNICAMP's Television.

The year of 1975 brought with it the choice of the Institute for hosting a graduate program totally supported by the OAS—Organization of American States—with students from the whole Latin American part of the continent. This was another experimental program that brought together professors from other countries (and universities) for working as students in chemistry, physics, and mathematics and to obtain a Master's Degree in a new as well as experimental transdisciplinary program. This was later reproduced in other Brazilian universities, in other countries as well as in Africa.

Not surprisingly, this environment affected IMECC's faculty: maybe one of the first Brazilian experiences in teaching Calculus in a computer-aided ambiance was undertaken at that time, in 1974 (using PASCAL and a mainframe computer lodged in an annex to IMECC's old building, a construction lent by the Physics Institute while IMECC's new facilities were being built).

But other innovative accomplishments were important, too! With financial and organizational support given by the main governmental agency for graduate studies, CAPES (CAPES literally means Coordination for the Improvement of Superior Education Teachers, and it is part of the Ministry of Education, responsible for evaluating Graduate Programs in all schools in the whole country), the Institute was the main agent in an experimental challenge: continued education in distance learning with state-of-the-art technology (back in 1975–1976): audio cassettes, Xerox, occasional phone calls, and snail-mail.... Some of those students seized the opportunity to forward their work in Mathematics and went on to obtain Doctorate Degrees!

These brief descriptions serve to illustrate what it was like to work in IMECC in those years, a reflection of what was happening in society worldwide. These were years in which IMECC's faculty were active in Mathematics, Statistics and Computer Science, of course, but also in other areas: Culture, University Professor's Union, Human Rights—and Sports!

All this academic ebullience had a consequence in IMECC's work in research: new groups began to work in Mathematical Logic in innovative aspects, Mathematical Modeling and Mathematical Education, Biomathematics, Artificial Intelligence, Computational Linguistics and, as a consequence, these became formal areas for graduate studies, and areas in which IMECC's pioneering work brought not only academic respect from other countries but helped create strong ties with many other universities in Brazil as well as in so many other countries.

And, another important factor in who we are at present was that IMECC created UNICAMP's first night course: a bachelor's for Mathematics Teachers, a step in which the whole university followed, making UNICAMP totally prepared to make effective a state law which obliged all state universities to have one third of their courses in the nocturnal period.

In 1990, the Computer Science Department formed an Institute of its own, a separation which had been ripening for some time, but the Institute maintained its same initials, changing its name to Institute for Mathematics, Statistics and Scientific Computing which, in Portuguese, keeps the same initials, a chance that was seized as a way of keeping, as well as these initials, the whole historic process which had formed IMECC until then.

At this time, a turning point came about with the arrival of one of the great Brazilian mathematicians: Professor Djairo Guedes de Figueiredo who promptly put together a group with professors, and graduate and undergraduate students and started to work with national projection.

Then, there was the creation of the Applied Mathematics Department, an initiative which began with Professors Miguel Taube and Vincent Buonomano. The Applied Mathematics Department received several faculty members from the Mathematics Department as well as other professors who felt motivated in working in this new mathematical environment. A special group migrated from Argentina and dedicated itself to serious research and responsible teaching—sometimes in very creative ways! The Applied Mathematics Department followed in the other departments' steps: it soon became a major force in the national academic scenario.

More recently, IMECC created a new aspect for students entering the university: when students take the entrance exam, they generally have to list their specific majors as options previously when enrolling for this entrance exam. So, the Mathematics and Physics Institutes created a joint entrance, enabling the students to choose between Physics, Mathematics, Applied Mathematics, and the day course for Mathematics Teachers 2 years after entering the university, guaranteeing precious time in making the right choice after the opportunity of contacting these themes....

And, even more recently, IMECC created a Professional Master's Degree (called "Mater's Program in Applied and Computational Mathematics"), which became an academic success due to the manner in which it permits graduate students with no grant and who continue to work, to study, obtain a Master's Degree, and greatly improve their mathematical education and with special emphasis on a transdisciplinary characteristic. This degree became IMECC's fourth graduate Program, alongside the Applied Mathematics', Mathematics', and Statistics' Programs—and all four have received high ratings in the constant evaluations undertaken by CAPES, a periodical evaluation which has placed IMECC as a major player in our national scenario, leading it's faculty to important positions in Academic Societies both nationally and internationally. But, this is also a challenge: maintaining these grades demands that we continue to work hard, keeping, nonetheless, the same bright-eyes enthusiasm of the Institute's founders.

In spite of this academic dedication, IMECC's academic community was able to jointly and morally expel a dean imposed upon the Institute during the military dictatorship.... This was in 1979 when the state governor cancelled the indication of all deans who had been chosen by the academic community in Schools in UNICAMP and had had their names confirmed by the governing board of each School as well as the University President. In IMECC, this act of expelling the interventor was done in a very patient, peaceful, and amusing way, and the interventor ended up by leading an enormous protest march all the way back to the University President's building (about a mile and a half...) from where he then left IMECC (and UNICAMP!) for good. The community was completely united in this act, in spite of all the academically natural differences which are a part of university lives! For all of us and many who came later, this was of paramount importance: the survival of this Institute with a thriving academic life, a history of facing challenges, a responsibility for innovation at the same time as differences are discussed and acted upon, and IMECC has had no shortage of these difficulties, like any other academic community. Yes, we agree and we disagree, yes, we sometimes bicker and discuss with quite a lot of energy (maybe this is an academic understatement...), yes we unite and separate as groups and as individuals due to our many different academic ideologies, but we have always acted jointly in a way in which we fought for a better Institute.

The purpose of this text was not that of a carefully documented historical report. Rather, it is a collection of personal memories—maybe not always precise, I must admit!—of a person who lived these years as part of IMECC's academic community. And the effort was to identify our Institute, from a participative point of view, as a *locus* where excellent Mathematical Applications, Mathematics and Statistics are developed, where excellent levels are maintained in research, in teaching, in cooperating nationally and internationally and, all in all, a place where it is fun to work and to study.

Acknowledgements The author thankfully acknowledges staff, colleagues, and students during these last 50 years for keeping alive facts (and their stories) and maintaining high hopes for our Institute of Mathematics, Statistics and Scientific Computing as well as their presence in facing the challenge of making a difference in our Latin American academic scenario.

Applied Mathematics in Brazil: Challenges and Opportunities

Martin Tygel

Abstract In this article I present an account and recollection of my experiences as a practitioner and promoter of Applied Mathematics, intensely and passionately exercised for more than 30 years at the Institute of Mathematics, Statistics and Scientific Computing (IMECC) at the University of Campinas (UNICAMP). Based on successes and, above all, failures along the road, I dare to share a few reflections about Applied Mathematics in Brazil, with special emphasis on its challenges and opportunities. In doing so, I take full responsibilities of the lack of impartiality, which seems unavoidable due to strong involvement. The style of the present text is the one usually employed by applied research articles: they start with an actual application and, after showing good results, proceed with the scientific formulation and arguments that justify the obtained results. In this article, I start with a brief description of the actual involvement and responses as an applied mathematician to the various challenges and opportunities faced along my career and then proceed with more general reflections and discussions about what can be learned from this process and it can contribute to make Applied Mathematics better and more useful.

1 Highlights of a Career in Applied Mathematics

Many applied mathematicians, including myself, started as undergraduates, not in Mathematics, but in other branches of Science and Engineering. My undergraduate was in Physics (starting at the Pontifical Catholic University of Rio de Janeiro (PUC-Rio) in 1964 and ending at the State University of Rio de Janeiro (UERJ) in 1969). Although the physics topics well addressed my desires for understanding nature, I felt unhappy with the somewhat excess of intuition and lack of rigor in the expositions. That pushed me into MSc studies in Mathematics at PUC-Rio (1971–1973) with disciplines and dissertation topic heavily based on Pure Mathematics

M. Tygel (✉)
Institute of Mathematics, Statistics and Scientific Computing, University of Campinas, Campinas, SP, Brazil

C. Lavor, F. A. M. Gomes (eds.), *Advances in Mathematics and Applications*,
https://doi.org/10.1007/978-3-319-94015-1_2

followed by a PhD in Mathematics at Stanford University (1974–1979) tending towards Applied Mathematics subjects.

Applied Mathematics and Geophysics Back to Brazil and eager to apply Mathematics to real-world problems, I was fortunate to have the opportunity of joining in 1981 the Research Center for Geophysics and Geology (CPGG) at the Federal University of Bahia (UFBa). My appointment was to be a Professor of Mathematics and to conduct research at the recently established Graduate Program in Geophysics of Petroleum Exploration. Under the leadership of Prof. Carlos Alberto Dias, and sponsored by the Brazilian Oil Company (Petrobras) and two other Government agencies, namely the Funding Authority for Studies and Projects (FINEP) and the National Council for Scientific and Technological Development (CNPq), the Program have been a pioneering, large-scale enterprise, designed to qualify human resources and undertake research and development projects in Petroleum Geophysics, to face the challenges of hydrocarbon exploration in Brazil. The Program was strongly based on internationally renowned experts of academy and industry that acted as MSc and PhD advisors on predefined and carefully designed topics related to the needs of the oil and gas sector, in particular the ones of interest of Petrobras. One of my main tasks was to design, organize, teach, and provide support on the mathematical courses of the Program. Those courses were tailored to the needs of the thesis and project works to be carried out by the students (many of them professionals from Petrobras) together with the visiting supervisors. It was a hard and fascinating work, which gave a pretty good perspective on how Mathematics could interact and contribute on a complex, real-world application. It also gave me the opportunity to interact with and conduct joint research with great scientists. One of them was Prof. Peter Hubral, at that time located at the Federal Institute for Geosciences and Natural Resources (BGR) in Hannover, Germany, with whom I started lifetime cooperation. That fruitful interaction helped me to understand how Mathematics training could be a valuable tool in an applied research area such as Petroleum Geophysics.

IMECC In 1984, I joined the University of Campinas (UNICAMP) as Professor of Applied Mathematics at IMECC. With a great support of the Department of Applied Mathematics (DMA), my aim was to continue and expand the Geophysics research started in Bahia. For this aim, a new research area, Mathematical Geophysics subsequently renamed Computational Geophysics, was established at the DMA.

In 1985–1987, IMECC granted me the generous permission to further continue and strengthen the cooperation with Peter Hubral at the BGR and later with the Geophysical Institute at the Karlsruhe Institute of Technology (GPI at KIT), where he has been appointed as a Professor. The stay in Germany was carried out under the framework of a fellowship of the prestigious Humboldt Foundation (AvH). The experience abroad provided a solid international cooperation network that was crucial to establish Geophysics (in particular Petroleum Geophysics) as a viable and attractive research area of Applied Mathematics at IMECC. Because that was a brand new area, not only at IMECC but in the Applied Mathematics community in Brazil, a first task was to organize its activities. That involved setting up a few courses and offer research topics attractive to DMA community.

Because of the complexities involved, most typically the multidisciplinary and interdisciplinary attitudes required by a specific and concrete application, that was a long process.

The first PhD in the area of Computation Geophysics was awarded in 1991 by Lucio Tunes dos Santos under my supervision. Lucio was already a Professor at DMA. Because at that time the Graduate Program at the DMA was not accredited for the Doctoral level, the PhD was granted by the School of Electrical and Computer Engineering (FEEC-UNICAMP). In 1995, Jörg Schleicher, who has recently fished his PhD in the research group at KIT lead by Prof. Hubral, arrived at the DMA as a post-doc on a joint program CNPq and the Alexander von Humboldt (AvH), under my supervision. Subsequently, Maria Amélia Novais Schleicher (1998) and Ricardo Biloti (2002) finished their PhD at the Graduate Program of the DMA, both under my supervision. Confirming its strong support to the Computational Geophysics research area, these three researchers became Professors at DMA in 1996, 2002, and 2005, respectively.

In 1997, under the leadership of Peter Hubral, the Wave Inversion Technology (WIT) Consortium was established. The WIT Consortium was a joint venture between the Universities of Karlsruhe, Hamburg, and Unicamp, designed to carry out research and development of seismic processing and imaging and funded by a pool of international oil companies. Until today, the WIT Consortium provides inspiration and funding support to tackle research and development challenges of interest of the sponsors carried out in close collaboration by the three universities.

In 2001, by means of a research and development project with the Brazilian National Agency of Petroleum, Natural Gas and Biofuels (ANP) and also with space provided by the DMA, the Laboratory of Computing Geophysics (LGC) has been established. Until today, the LGC provides the infrastructure and support for carrying out, besides Graduate works, projects with oil and gas industry (such as Petrobras) and also research funding agencies (such as CNPq).

In 2012, I considered that my aims and involvement with the Computing Geophysics Group and Lab have attained its objectives: The Group and Lab have been established as one possible model of the use of Applied Mathematics on a well-focused, well-funded, real-world problem.

The Center for Petroleum Studies (CEPETRO) Since my arrival at UNICAMP, and besides my duties at IMECC, I tried to get involved and contribute to any endeavor in Petroleum where my previous experiences could be useful. At that time, intense discussions concerning education, research, and development in key areas of the petroleum sector were taking place. As in the case of Bahia in 1981, Petrobras was a key partner and supporter. In 1987, the Graduate Program in Sciences and Petroleum Engineering (CEP) was established. From its beginning until today, I am attached to the CEP Program undertaking research, teaching and supervising activities. Later, the area Reservoir Geophysics was included at CEP under responsibility of the Computational Geophysics Group of IMECC.

Also in 1987, the Center for Petroleum Studies (Cepetro) was created, joining UNICAMP's several Centers and Nuclei devoted to multidisciplinary research in a wide range of areas of expertise. A strong relationship to Cepetro exists between IMECC and LGC, the former being a member of its scientific board and the latter a partner Lab. Cepetro is dedicated to research, development, and innovation in all petroleum-related areas. As stated in its homepage, "the main purpose of Cepetro is to provide a link between Unicamp and society, in the petroleum area, by offering support to courses, technological and scientific research projects and services." Cepetro has ten Laboratories in its premises and twenty five partner Laboratories in Institutes and Departments at Unicamp. It disposes also of an excellent administration that provides secretarial and accounting help to project executors. In 2010, Cepetro inaugurated its new headquarters. With a number of Laboratories and well-equipped infrastructure and office space, Cepetro significantly enhanced its capabilities as a key partner in research and development applied to the petroleum sector. In this promising environment, I welcomed the assignment as Associate Director of Cepetro. In view of the new responsibilities, I left (in good hands) the activities at the LGC with the aim of a full dedication to research and development projects in the framework of Cepetro.

The Laboratory of High-Performance Geophysics (HPG) Among the Labs envisaged in the new building, one of them was to be dedicated to geophysical studies well aligned with the aims of Cepetro. In 2013, under my coordination together with Edson Borin, Professor at the Institute of Computing (IC) at UNICAMP, the High-Performance Geophysics (HPG) was created to fulfill these purposes. The name HPG was derived from a decisive incorporation of Computing Science, most particularly high-performance computing (HPC) in the Lab activities. The synergy provided by Computer Science, Applied Mathematics and Geophysics enabled HPG to carry out more comprehensive projects of academic and industrial interest. A few distinguished characteristics of HPG are:

1. Located at Cepetro, HPG benefits from a vibrant multidisciplinary environment and in close relation of ongoing joint projects between academy and industry.
2. HPG benefits from the favorable administration of Cepetro, which maximizes the time devoted to the core (research and development) activities of projects.
3. Strong interaction with Computer Science, in particular in HPC, has a positive significant impact on the transfer of its technologies and deliveries (e.g., more professional coding and ready-to-use programs), especially in industry-oriented projects.
4. Within HPG, Geophysics and Computer Science are seen as truly synergetic partners, tackling problems that have actual or potential impact on both areas.
5. Long-run, industrial-funding projects enables HPG to maintain a multidisciplinary staff of researchers and professionals capable of undertaking complex challenges in the petroleum sector.
6. Many staff members of HPG have employment contracts paid with project money, so as to guarantee their full involvement with the project deliverables. Such employment is different from regular scholarships aimed primarily (and in many times exclusively) for Graduate purposes.

Since its establishment, HPG has significantly enlarged in volume and scope the geophysical activities carried out at Unicamp. A truly multidisciplinary staff in the areas of Geology, Geophysics, Applied Mathematics, Engineering and Computer Science are working together in comprehensive real-world problems. Under well-defined focus and concrete deliveries to attain, the HPG team manages to have a lively integrated work, where horizontal and vertical discussion can routinely take place. The three pillars of formulation, solution, and implementation of problems are systematically put in practice. In summary, the operation in HPG can be seen, in my evaluation, as one possible model for exercising Applied Mathematics in its full potential.

The Years to Come In 2017, at the age of 70 and after 46 years of dedicated work, I retired from Unicamp. Thanks to wise legislation at Unicamp, I am able to continue in the years to come to contribute to Cepetro and HPG as an Emeritus Collaborate Researcher. More specifically, my main interests are (1) to attract interest and funding, mainly from industry and government agencies of the petroleum sector, to research, development, and innovation projects within that sector; (2) use those funds to keep offering good opportunities for the young (in particular Applied Mathematicians) to be involved in relevant, real-world problems, and (3) be aware of open opportunities and activities where HPG can contribute. As present examples, Artificial Intelligence (AI) and Machine Learning (ML), in particular their application to seismic processing and imaging, are already ongoing research topics at HPG. Although not yet an activity area, Statistics is already in the HPG radar.

2 Some Historical Notes

In this section, a brief summary of the historical facts related to Applied Mathematics in Brazil is provided.

Mathematics and Pure Mathematics As a starting point of the considerations made here, I take the foundation of the Institute of Pure and Applied Mathematics (IMPA) in 1952. In my opinion, IMPA can be seen as the landmark of organized Mathematics in Brazil. In spite of word "Applied" attached to its name, the activities of IMPA were, until the beginning of the seventies, completely devoted to Pure Mathematics. The Brazilian Mathematical Society (SBM) that has been established in 1969 followed the same trend as in IMPA.

The view of Pure Mathematics as the legitimate representative of Mathematics was, since its early days, fully adopted by the Brazilian scientific community. Moreover, such view has strongly influenced the structure and content of the mathematical education in the country, throughout all of its levels.

Classical topics such as analysis, geometry, algebra, and dynamical systems were always the main focus of disciplines in undergraduate and graduate levels in Mathematics programs. In the same way, positions in Mathematics Departments were filled out mainly by specialists on such topics. Finally, official evaluation of education and research in Mathematics, as provided by the governmental agencies Coordination for the Improvement of Higher Education Personnel (CAPES) and the National Council for Scientific and Technological Development (CNPq), have been strongly based on performance in classical Pure Mathematics topics. A striking example is that CNPq has still today its Mathematical Committee composed by six sub-areas: four are topics of Pure Mathematics (Analysis, Geometry, Dynamical Systems, and Algebra) and the remaining two are Statistics and Probability and Applied Mathematics.

In that framework, studies devoted to the use of Mathematics to solve practical problems, in particular problems of areas other than Mathematics were tacitly considered of less relevance or away from the main stream of Mathematics. It is to be observed, in passing, that the above view encountered fertile ground in the Brazilian culture, in which theoretical work has been always valued higher than practical work. Furthermore, the incipient activity in high technology and innovation in Brazil did not favor the demand of scientists, in particular mathematicians, in industrial environments.

Applied Mathematics The early seventies can be traced as the time when Applied Mathematics was accepted as a legitimate player by the Brazilian Mathematical community. Acknowledged MSc/PhD Programs in Applied Mathematics appeared first at the Institute of Mathematics and Statistics of the University of São Paulo (IME-USP) in 1971 and second the University of Campinas (UNICAMP) in 1977. Later Applied Mathematics Programs, some of them within conventional Mathematics Programs, were installed, namely at: Department of Applied Mathematics at the Federal University of Rio de Janeiro (DMA-UFRJ) in 1986, Institute of Mathematics and Statistics at the Federal University of Rio Grande do Sul (IME-UFRGS) in 1995, Department of Mathematics at the Federal University of Paraná (UFPR) in 2002, and the Federal University of ABC (UFABC) in 2008. In 1978, the Brazilian Society of Applied and Computational Mathematics (SBMAC) (an Applied Mathematics counterpart of SBM) was founded and in 1980 the National Laboratory of Scientific Computing (LNCC) (an Applied Mathematics counterpart of IMPA) was established. Applied Mathematics is also strongly represented in undergraduate courses of Industrial Mathematics, as for example the one established in 2002 at the Federal University of Paraná (UFPR).

The rise of Applied Mathematics in the Brazilian scenario as a departure of the prevailing identification of Mathematics as Pure Mathematics came (roughly) about due to a combination of two reasons, one internal and another external:

(a) Internal: Within the universities, the mathematical community was exposed to the demands and opportunities of applying their expertise to a much wider spectrum of problems, both in academics and in industry. This contrasted with IMPA, an isolated institute solely devoted to (mainly Pure) Mathematics;
(b) External: The advances in science and technology achieved by the leading industrialized countries, most particularly in computer-related topics, strongly echoed in the Brazilian scientific community.

Recognition of Applied Mathematics as a new player in the well-established Brazilian (Pure) Mathematics environment was not at all a smooth process, especially when already limited budgets were disputed by ever more applicants. Typical arguments of representatives of Pure Mathematics were that Applied Mathematics results were second class in content and moreover did not follow acceptable practices of scientific rigor. Counter arguments of representatives of Applied Mathematics were that Pure Mathematics dealt with problems of no technologic, economic or social interest. Such quarrels are not uncommon in other parts of the world, leading to the separation of communities, departments, programs, and representative societies. In Brazil, scars and hard feelings are still present. As a newcomer that had to fight its way into an established system, Applied Mathematics, still today, suffers from a rebellion syndrome that, in extreme cases, manifests itself in a difficulty to accept general rules of academic hierarchy and tradition.

3 Problem-Oriented Science

The easy access and management of information provided by the internet introduced dramatic changes in the ways information and knowledge in all areas of interest can be acquired, produced, and disseminated. High-quality courses and lectures, until recently proprietary to renowned universities, are being offered online for free or very affordable prices. The market for such products is huge and potentially disruptive for conventional academic institutions. These institutions, even the well-established and renown ones, face organizational challenges, not only to keep their positions, but also to take advantages of the opened opportunities.

The internet made possible the most intense dialogue and cross fertilization of different areas of scientific activity leading researchers to benefit from contributions and collaborations in a much larger and broader scale. Such developments have been consolidated in what I call "problem-oriented science." As opposed to specific "disciplinary" conventional projects, problem-oriented projects are attached to multidisciplinary "big" scientific problems easily recognized as "meaningful" and "important." Big problems can be of a basic character (e.g., the origin of the universe) or of an applied character (e.g., search of clean and efficient energy sources). Problem-oriented projects make easier for government and private agencies to make decisions on where to allocate funds in ever tighter budgets.

As an important support to problem-oriented science, comprehensive computation literature, software (e.g., tool boxes, libraries, solvers, etc.) and hardware (e.g., cloud computing) designed for ready use for wide audience are easily available. As a result, "exclusive" or traditional areas of Applied Mathematics were "invaded" by outside fierce competitors. In other words, Applied Mathematics became pronouncedly global, as opposed to Pure Mathematics that still retains much of its original character.

The new trends of inter- and multidisciplinary character of modern scientific activity impose significant changes to education at all levels, with emphasis on formulation and solving problems. Isolated courses and programs do not help the young to find opportunities in science and technological areas. Applied Mathematics has a lot to contribute to overcome these difficulties.

A better understanding on how science and technology are carried out in a society can be gained by analyzing how these activities are funded. This we do below for the Brazilian reality. We divide our considerations into two parts, namely education of human resources and research and innovation.

4 Funding of Education of Human Resources

In Brazil, funding specifically devoted to human-resource scientific education is almost completely undertaken by governmental agencies in the form of scholarships. Typically, scholarships comprise Undergraduate (initial exposure to scientific activities) and Graduate (MSc, PhD, and Post-Doctoral) levels. Scholarships are provided on institutional or individual bases.

(a) **Institutional scholarships:** CNPq and CAPES make available scholarships in the undergraduate and graduate levels. In the undergraduate level, the so-called Scientific Initiation scholarships are provided to stimulate young students to be exposed to scientific problems and activities. In the graduate level, scholarships support students engaged in MSc and PhD programs.

Scholarships are provided to eligible institutions (universities and research institutes), which assign them to students according to their own criteria. To be eligible for institutional scholarships, programs have to be registered and regularly evaluated by a national (CAPES) system which establishes the number of scholarships assigned to each program.

(b) **CAPES evaluation system:** Graduate programs eligible for CAPES and CNPq institutional scholarships are regularly ranked by a comprehensive evaluation system elaborated by CAPES. Top-rank positions guaranties privileges such as larger number of scholarships and some fast-track processing. Fierce competition and the great complexity of the CAPES system represent a constant nightmare to Program coordinators, responsible for producing the report of Program activities to be analyzed by a special Committee composed by a few, elect representatives of the eligible programs. Evaluation is carried out every 4 years, being divided into several areas. Mathematics (which includes Applied Mathematics, Statistics, and Probability) comprise a single area. In spite of its legitimate intention of optimal use of public funding, the CAPES system has become an uncontrolled, number-crunching beast that transforms numerical indicators automatically extracted from the reports into consolidated ones (weighted sums of indicators) upon which the Program evaluation is produced. Indicators include number of publications (weighted by their importance to

Program areas of activity) divided by the number of Professors, average time for completing MSc and PhD studies, number of joint publications with students, student evasion, etc.

A backlash of the CAPES system is that the main preoccupation of the Program coordinators is to play with its indicators, so as to produce reports that attain the highest grades, many times at the expense of the actual quality of the Programs. The system also poses difficulties in the introduction of new areas or topics because of uncertainty to maintain indicators. Finally, University administrations, as dependent on the good performance of Graduate Programs, are also afraid of setting up new or experimental Programs because of the same uncertainties. The result is, invariably, more of the same.

(c) **Individual scholarships:** CAPES and CNPq also provide individual scholarships (i.e., applications are processed on a one-to-one basis, according to agency criteria/guidelines) for selective cases, such as studies e.g., Graduate of Post-Graduate studies at foreign universities. CNPq provides scholarships for researchers and professors employed at Brazilian academic institutions based on their research productivity. Individual scholarships for all levels are also granted by State funding agencies, the most important one being the Research Foundation of the State of São Paulo (FAPESP).

Comments The established formal education organization fails to provide the motivation and opportunities to fulfill the potential of Applied Mathematics as a contributing protagonist to solving real-world problems. The system reflects and reinforces the conservatism of the Brazilian academic community, which opposes the multidisciplinarity which is essential to the modern view of problem-oriented science. Programs are stimulated to strengthen their individual, vertical character, penalizing innovative, horizontal initiatives that would risk their status of "well-established," high-rank programs. Professors also refrain to explore new areas because of the risk of diminishing their productivity when departing from comfort-zone areas. Hiring new faculty is oriented to teaching needs of the same disciplines, rather than opening new directions, even ones able to attract new funding opportunities other than the usual ones from government sources.

As seen below, Research and Innovation agencies have a much different approach, with a far better alignment to problem-oriented science.

5 Funding for Research, Development and Innovation

As opposed to the difficulties faced by the Brazilian education system to modernize, a quiet and consistent revolution towards problem-oriented science is taking place in research and innovation in the country. Such revolution is being mainly pushed by a variety of State and Federal agencies, funds, and programs. Moreover, fiscal incentives stimulate the interest of industry to invest on activities of research, development, and innovation (R, D, & I) carried out within the country. A brief description of main public institutions engaged in such activities is provided below.

1. **The Research Foundation of the State of São Paulo (FAPESP):** In terms of budget and scope, FAPESP is Brazil's most powerful regional governmental research agency. Besides its well-recognized role of fostering fundamental research, FAPESP has fully embraced the concepts of problem-oriented science and applied research with careful attention to technology and innovation, as well as social and economic aspects. The foundation supports large research programs in Biodiversity, Bioenergy, Global Climate Change, and in e-Science. FAPESP maintains cooperation agreements with national and international research funding agencies, higher educational and research institutions, and business enterprises.
2. **National Agency for Petroleum, Natural Gas, and Biofuels (ANP)**: Created in 1997, it is the regulatory agency that oversees activities undertaken by the oil, natural gas, and biofuel industries in Brazil. Among its several duties, ANP is responsible for the management of the so-called RD&I Investment Clause, which sets forth that 1% of gross revenues from oil and natural gas exploration companies be invested in research, development, and innovation in the country. The Clause aims to stimulate research and adoption of new technologies for the sector. An appealing feature of the Clause is that it permits that the companies spend up to half of their obliged contribution (0.5% of gross revenues) in projects of their own choice, as long as certified by ANP. This means that companies can directly negotiate projects with Brazilian universities and research institutions accredited by ANP. Since the establishment of the Clause, billions of reals have been poured into the Brazilian Research, Development and Innovation system and that favorable situation is bound to remain in the long term.
3. **Funding Authority for Studies and Projects (FINEP):** FINEP is an organization of the Brazilian Federal Government under the Ministry of Science of Technology, devoted to funding of science and technology in the country. FINEP grants reimbursable and non-reimbursable funding to Brazilian research institutes and companies. FINEP's support encompasses every phase and dimension of the scientific and technological development cycle: basic research, applied research, product development and innovation, services and processes. FINEP also supports technology-based company incubation, installation of technological parks, structuring and consolidation of research processes, development and innovation for established companies, and market development. Furthermore, starting in 2012, FINEP also began to offer support for the implementation of first industrial units as well as acquisitions, mergers, and joint ventures.
4. **Brazilian Agency for Industrial Research and Innovation (EMBRAPII):** It is a Social Organization connected to the Ministry of Science, Technology, Innovation and Communications (MCTIC) and to the Ministry of Education (MEC). EMBRAPII's Management Contract was signed on December 2nd, 2013, and both federal Ministries share responsibility for its funding. EMBRAPII operates through cooperation with public or private technological and scientific

research institutions that are accredited as EMBRAPII Research Units. These Units focus on entrepreneurial demands and innovation projects that are in a pre-competitive stage.

5. **Serrapilheira Foundation:** Serrapilheira is a private nonprofit institution created to promote science and increase its visibility and impact in Brazil. In the words of its President, "We want to identify and support the best young researchers in Brazil, those who are posing the big questions in their fields. We have no preference when it comes to pure or applied sciences. Nor do we have any qualms about supporting risky research proposals, the sort where an audacious researcher may not always be successful." Presently, Serrapilheira supports research in the areas of chemistry of computer science, earth sciences, engineering, life sciences, mathematics, and physics.

Comments Research and Innovation agencies are very much aligned to the concept of problem-oriented science, definitely encouraging multidisciplinarity and interaction, not only between different academic groups, but most especially with industry partners. Such interaction obligatorily revolves over real-world problems of relevant social or economic interest. As has been done world-wide, aims of the funding agencies include breaking the separation obstacles between academy and industry, so as to unleash their potential.

6 New Trends and Scenarios

Problem-oriented science with decisive incorporation of technology and innovation is gaining significant momentum in Brazil. Pushed by government and industry funding, it privileges long-term multidisciplinary projects and partnerships with international institutions with a clear focus on economic and social returns of the investments. Academic-industry joint projects play a key role to guarantee a synergetic blend of science, technology, and innovation that is essential to solve real-world problems. Such trend constitutes a new driving force to move Brazilian science away from its apathy and to transform it into a living instrument for economic/social improvement. The new framework allows for long-needed comprehensive funding in which, besides scholarships strictly attached to MSc/PhD studies, also soft-money is available to recruit technical (professionals and researchers) and administrative personnel directly involved in carrying out the projects. It is to be observed that research projects, especially the ones with applied deliveries, pose specific demands (reports, computational codes, experiments, etc.) that conflict with the purpose of MSc/PhD studies in which attaining the degree is the prime interest. Provided by adequate projects, such qualified people can be recruited either by permanent employment by the university (generally very difficult) or by project soft money.

7 Challenges and Opportunities

In this section, I list a few challenges presently encountered by the established Applied Mathematics departments and programs in Brazil. Facing and eventually overcoming such challenges will hopefully contribute to enlarge the scope and role of Applied Mathematics, creating desirable opportunities both in the academic and productive sectors.

Applicable and Applied Mathematics As it is the general case of developing countries in which science and technology are not fully integrated within the economic, social, and cultural mainstream, academic programs and research activities officially recognized as Applied Mathematics in Brazil are mostly devoted to methods and proof-of-concept examples or illustrations (the so-called toy problems). In the late seventies, the name for that was "Applicable Mathematics." Strictly speaking, practically all Applied Mathematics undergraduate and graduate programs in Brazil should be more appropriately referred to as Applicable Mathematics programs. Real-world problems are always of a complex nature requiring not only mathematical expertise, but also active multi- and inter-disciplinary involvement. Contributions of Applied Mathematics to real problems have to be attached to specific areas of application (responsible for the motivation and also evaluation of the obtained results) and also computer science professionals (responsible of the effective data processing and algorithm implementation of the solutions).

In summary, present Applied Mathematics programs fail to deliver its promises of actual involvement and contributions to solve real problems. With some exaggeration, such programs advertise Applied Mathematics and deliver Applicable Mathematics. The result is that talented students interested in mathematical applications turn to opportunities in programs outside the established conventional ones in Applied Mathematics.

Applied Mathematics skills comprise not only mastering methods and tools to solve mathematical problems, but also abilities in problem formulations and solution implementations. Such skills can be essential in real-world multidisciplinary projects.

Applied Mathematics academic programs in Brazil are mainly focused on methods and tools only, leaving aside the equally important aspects of problem formulation (which relies on expertise of the specific area of application) and implementation (which relies on computing expertise).

In the same way as Mathematics is recognized as the language of Science and Technology, Applied Mathematics can be seen as an operational link between problem formulation and solution implementation of actual scientific and technological applications.

Multidisciplinary Centers and Laboratories One of the biggest challenges of academic programs in Applied Mathematics today is to provide education and training that excel in three aspects: methods, formulation, and implementation. This requires intense multidisciplinary interaction and also in-depth computational studies. It might well be that such challenge cannot be met within a single (Mathematics) Institute, but requires a broader Center which congregates associate researchers of different expertise. At UNICAMP, 21 Centers devoted to multidisciplinary research in a wide range of areas of expertise, already exist, being organized at the Coordination of Interdisciplinary Research Centers and Nuclei (COCEN). On a federal scale, the National Laboratory for Scientific Computing (LNCC) (albeit an isolated Institute such as IMPA) exists with the same purpose.

In this spirit, multidisciplinary Centers, not necessarily attached to a single Institute, can be of great use to awake the interest of Applied Mathematicians as partners to solve real-world problems. Good initiatives in this direction are the Research, Innovation and Dissemination Centers (RIDC), installed by FAPESP. Among them is the Center for Mathematics and Statistics Applied to Industry (CeMEAI), devoted to the systematic use of mathematical techniques to solve problems and to propose the construction of an infrastructure for this purpose.

On a more local level, Department Laboratories devoted to specific areas of interest can also be extremely useful. Such Laboratories help to provide not only infrastructure to perform academic tasks such as Dissertations and Theses, but also to carry out research and development projects, in particular with partners outside the University. Such projects could provide maintenance (e.g., equipment and staff) of the Laboratories and not constitute a burden to Department or University. A brief list of recommendations seems to be in order:

(a) Incorporate Applied Mathematics in the framework of problem-oriented science. Actions in this direction should include, among others: promoting participation in multidisciplinary projects both in academic and in industrial environments; stimulate undergraduate and graduate work with shared supervision of applied mathematicians and practitioners of other areas; special programs should be designed to be dedicated to applications of mathematics to other areas; stimulate contact with dynamic partners of the productive sector, such as startup companies.
(b) Introduce disciplines in Applied Mathematics programs that promote involvement with scientific and high-performance computing (HPC); stimulate activities related to coding, data processing, and computing applications, e.g., machine learning (ML).
(c) Introduce seminars and workshops with researchers and professionals of other areas to provide a horizontal view and awareness of the potentials of Applied Mathematics to be of use in its widest sense.
(d) Stimulate and provide opportunities for internships of students at companies or laboratories in which Applied Mathematics may contribute.
(e) Introduce disciplines on entrepreneurship, as well as seminars and workshops with industry, in particular startup companies.
(f) Promote involvement of Applied Mathematics in all levels of education (most particularly elementary education), implementing the philosophy of using mathematics to solve daily problems.

8 Summary and Conclusions

Applied Mathematics, as exercised in academic institutions in Brazil, has serious challenges to be a protagonist in real-world applications that would greatly benefit from its expertise. An analysis of difficulties and also some suggestions to overcome them have been presented.

As explained in the main text, a most import challenge is that activities presently labeled as Applied Mathematics evolve from their stage as Applicable Mathematics. A concrete example in this direction is the High-Performance Geophysics (HPG) at UNICAMP.

Acknowledgements I take the opportunity to thank all colleagues from IMECC for constructive criticism, inspiring suggestions, involvement, and support along all these years.

Nomenclature/Acronyms

ANP	National Agency of Petroleum, Natural Gas and Biofuels (http://www.anp.gov.br/)
AvH	Alexander von Humboldt Foundation (https://www.humboldt-foundation.de/web/home.html)
BGR	Federal Institute for Geosciences and Natural Resources (https://www.bgr.bund.de/EN/Home/homepage_node_en.html)
CAPES	Coordination for the Improvement of Higher Education Personnel (https://www.iie.org/Programs/CAPES)
CeMEAI	Center for Mathematics and Statistics Applied to Industry (http://www.cemeai.icmc.usp.br/)
CEP	Graduate Program in Sciences and Petroleum Engineering (http://www.cep.dep.fem.unicamp.br/?q=en/node/91)
CEPETRO	Center for PetroleumStudies (http://www.cepetro.unicamp.br/english/index.html)
CNPq	National Council for Scientific and Technological Development (CNPq; http://cnpq.br/)
CPGG	Research Center for Geophysics and Geology (http://www.cpgg.ufba.br/pesquisa/exploracao_petroleo-f.html)
COCEN	Coordination of Interdisciplinary Research Centers and Nuclei (http://www.cocen.unicamp.br/centros-e-nucleos)
DMA-UFPr	Department of Mathematics, Federal University of Paraná (http://www.mat.ufpr.br/)
DMA-UFRJ	Department of Applied Mathematics, Federal University of Rio de Janeiro (http://www.dma.im.ufrj.br/index.html)
EMBRAPII	Brazilian Agency for Industrial Research and Innovation (http://embrapii.org.br/en/categoria/institutional/aboutus/)

FAPESP	Research Foundation of the State of São Paulo (http://www.fapesp.br/en/about)
FINEP	Funding Authority for Studies and Projects (http://www.finep.gov.br/)
GPI at KIT	Geophysical Institute at the Karlsruhe Institute of Technology (https://www.gpi.kit.edu/english/GPIatKIT.php)
IMECC	Institute of Mathematics, Statistics and Scientific Computing (http://www.imecc50.ime.unicamp.br/historia)
IME-UFRGS	Institute of Mathematics and Statistics, Federal University of Rio Grande doSul (https://www.ufrgs.br/ime/institucional/historia/)
IME-USP	Institute of Mathematics and Statistics of the University of São Paulo (https://www.ime.usp.br/en)
IMPA	Institute of Pure and Applied Mathematics (https://impa.br/en_US/)
LNCC	National Laboratory of Scientific Computation (http://www.lncc.br/estrutura/default.php)
PUC-Rio	Pontifical Catholic University of Rio de Janeiro (http://www.puc-rio.br/english/aboutpuc/history.html)
RIDC	Research, Innovation and Dissemination Centers (http://www.fapesp.br/cepid/pasta_cepid_2013.pdf?t=1)
	Stanford University (https://www.stanford.edu/about/)
	School of Electrical and Computer Engineering (FEEC-UNICAMP) (http://www.internationaloffice.unicamp.br/english/teaching/graduate/school-electrical-computer-engineering/)
	Serrapilheira Foundation (https://serrapilheira.org/en/about-us/)
UERJ	Rio de Janeiro State University (http://www.uerj.br/idiomas.php#gb)
UFABC	Federal University of ABC (http://ufabc.edu.br/en/history/)
UFBa	Federal University of Bahia (https://www.ufba.br/)
UNICAMP	University of Campinas (http://www.unicamp.br/unicamp/node/64)
WIT	Wave Inversion Technology Consortium (http://www.wit-consortium.de/)

The Biomathematics in IMECC

A Historical Review

Rodney Carlos Bassanezi

Abstract The main motivation that led us to work with biomathematics is that we could understand some of the mechanisms of biological phenomena using techniques that came from mathematics. This existing interface between biology and mathematics, characterized by a great contact range, experiences a process of fast-track deepening nowadays. From this two-way process, not only basic biology issues have been solved, but also new lines of research in mathematics have arisen and taken on a life of their own. Moreover, it is important to notice the emerging new fields in applied mathematics, such as genetic algorithms, neural networks, sociobiological algorithms, fuzzy logic, etc., which we could call biological mathematics, as, in many cases, they owe their basic concepts to theoretical biology. It is hard to precisely say how biomathematics began as a research field at the Instituto de Matemática, Estatística e Computação Científica (IMECC). What I am about to tell you consists of some memories from the 1970s, when biomathematics was not talked about around here and the prey–predator system was just an example in the subject of differential equations, taught by Professor Torriani.

The Department of Applied Mathematics was created during Professor D'Ambrósio's term who had just returned from the USA and held the position as Director of the Institute when some professors were hired to strengthen the new department—others switched departments to work in the new one. In the early 1980s, research involving biological phenomena started with three dissertations oriented by Professor Alejandro Engel (L. Paraíba (83), S. Raimundo (86) and S. Bezerra (86)), whereas we would guide some outstanding students (Fenley, Moretti, Petrônio, Bia D'Ambrósio, Ibrain Saad and Andrea Hahn) in programs of scientific initiation.

R. C. Bassanezi (✉)
Institute of Mathematics, Statistics and Scientific Computing, University of Campinas, Campinas, SP, Brazil
e-mail: rodney@ime.unicamp.br

C. Lavor, F. A. M. Gomes (eds.), *Advances in Mathematics and Applications*,
https://doi.org/10.1007/978-3-319-94015-1_3

Contemporary biomathematics can be classified into three distinct branches concerning methods and approaches: the traditional interface offered by biophysical and biomechanical issues; the most recent one dedicated to the genomic analyses; and a third one called population dynamics. The field of greatest emphasis on the Instituto de Matemática, Estatística e Computação Científica (IMECC) graduate programs was population dynamics which, in its broader meaning, encompasses the study of the population of molecules, cells, microorganisms, higher organisms, diseases, and human societies. The synthesis and the foundation of this broad line of research proceed from a variety of mathematical models described by variational equations: ordinary and partial, continuous and discrete differential equations, and afterward, variational equations that envisage the subjectivity of parameters and state variables (fuzzy systems).

The use of mathematics in the formulation of biological laws is still in its initial stage if compared with its development and use in the physical sciences; however, in the past few years, along with the evolution of the computer sector, it has been demonstrated to be a crucial tool in cutting-edge research in several fields. The practical models that involve inter-relationships of a great number of variables are formulated through equation systems with countless parameters. In these cases, analytical treatment is usually impossible and the resolution of qualitative methods must be used, which favors the computational resolutions. The more complex or realistic the model is, the more difficult it will be to statistically show that it describes reality!

The first biomathematics paper at a Master's level with which we assisted took place in Londrina, where the IMECC acted as a partner in the "Fundagão Universidade Estadual de Londrina" postgraduate program—it was a dissertation about the dynamics of biodigesters (N. Martinhão/82: *"Exploração de Recursos Renováveis – Biodigestor" – Exploration of Renewable Resources – Biodigester*). This paper was expanded afterward and presented at the Congresso Nacional de Matemática Aplicada e Computacional (CNMAC) of Maringá (1983). Other dissertations followed at the IMECC, always involving ordinary differential equation (ODE) systems and population dynamics (C. Souza/85: *"Exploração de Recursos Renováveis: Otimização do Modelo de Beverton-Holt da pesca" – Exploration of Renewable resources: Optimization of Gordon's and Gould's models*; M. B. Custódio/86: *"Recursos Renováveis: a pesca – Comparação dos modelos de Gordon e Gould – Renewable resources: fishing – Comparison of Gordon's and Gould's models"*; A. T. Conceição/89: *"Modelos Compartimentais em Biomatemática – Compartment Models in Biomathematics"*; V. dos Santos/89: *"Sistema Presa Predador Generalizado – Generalized Prey-Predator System"*; A. P. Emérito/89: *"Modelo Matemático Determinístico em Doenças Transmissíveis – Deterministic Mathematical Model in Communicable Transmissible Diseases"*).

At the IMECC, biomathematics was only considered a field of research of applied mathematics as late as 1990. The first PhD thesis in this field, which we guided along with Professor Boldrini, was defended at the FEE-Unicamp by Prof. L. Vendite/88: *"Crescimento e Tratamento de Tumores Cancerígenos – Growth and Treatment of Cancerous Tumors"*.

With the recognition of biomathematics as an area of research, there was significant growth in the IMECC group with the admission of Professors Joni Meyer, Wilson C. Ferreira Jr, Laércio Vendite, Laécio Barros, and Silvio Pregnolatto—later the group became even stronger, with the hiring of Professor Hyan M. Yang (1993).

From 1991, the group started to publish the yearly BIOMATEMÁTICA journal containing the papers produced by its researchers and presented at the CNMAC. In the first issue, the diversity of interest and the subjects approached could be seen as a natural evolution consequence of the IMECC biomathematics group. The first papers presented can be gathered into two main topics: dynamic population systems (optimal control of tumors and bacteria, dengue evolution, resistance to fungicides and enzymatic kinetics), and the numerical analysis of parabolic–hyperbolic partial differential equations (PDE; river and sea pollution, hemodialysis, and potato drying). In the third issue of the BIOMATEMÁTICA journal, the following can be read at its presentation: "The difficulties remain present since it is a relatively new line of research at the IMECC both at local and national levels, where there are a few groups. Nevertheless, the growth of our group and its consolidation can be evaluated by what we present herein, besides the presence of the third mini course of this field taught at CNMAC—in that year, mathematical ecology." The BIOMATEMÁTICA journal, published for 25 years, rain or shine, and the mini courses in conferences, were crucial for the diffusion and consolidation of this field of research in the whole of Latin America. The graduate program started to receive a many students from other institutions who were interested in the field of biomathematics.

Professor Wilson C. Ferreira Jr defended his thesis at the IMECC (1993) in the program of applied mathematics: *"Mathematical models for the dynamics of populations distributed in aspect areas with non-local interaction: complexity paradigms"*. Also in 1993, we had the defense of Michel I. Silveira's thesis: *"Deterministic control of chemotherapy treatments"*, which we guided, along with Professor Boldrini. Other theses with papers reasoned on ODE systems and optimal control theory followed.

The biomathematics group began, then, to develop integrated research projects: "Growth and Treatment of Cancerous Tumors," (FAP UNICAMP) September/1987–August/1988; "Mathematical Modeling for Medicine Optimization in Cancerous Tumors,", CNPq August/1987–July/1989; "Mathematical Modeling in Biological Controls," CNPq September/1989–August/1991, "Mathematical Modeling of Dynamics and Control of Populations Subjected to Side Effects Due to Chemical Treatments", CNPq, October/1991–September/1993; "Mathematical Modeling of Interaction Between Two Epidemics: AIDS X Tuberculosis," CNPq, October/1993–September/1995 and "Mathematical Modeling of Epidemics Subjected to Dispersion and Migration Phenomena", CNPq, October/1995–September/1997.

We believe that the reason for the active continuity of this group is the result of modeling in biological processes supported by instruments resulting from the fuzzy theory. The study and research, using the fuzzy logic procedures, started in the Mathematics Department when we guided some Master's dissertations in 1988 about *"Fuzzy Measures"* (J. Gerônimo) and *"Fuzzy Integrals"* (J. Duarte). Next, we also had Heriberto Flores's thesis on *"Fuzzy Entropies"*. The first Biomathematical paper using fuzzy logic arguments occurred when we used, along with Heriberto, the structure of a foundation of fuzzy rules to study the process of medical diagnosis of childhood diseases.

In the Applied Mathematics program, the first dissertation which used the Fuzzy Logic concepts in formulating the models was the one presented by L. Barros/92: *"Deterministic Models with Subjective Parameters"*, in which a study of discrete models of population dynamics and its stability is carried out.

The use of the fuzzy concepts in models linked to biological phenomena was very well-accepted by the professional of the biological area and a great number of papers, joint work, arose from this union, up to the point Biomathematics would be confused with Fuzzy Logic and vice-versa. The elements of the Group, following their natural tendencies, began to dedicate themselves to specific matters. That's the reason why J. Meyer and W. Ferreira Jr. developed their activities in problems that use the process of diffusion with PDE; L. Vendite in processes of diagnoses and treatments; H. Yang in epidemiology and, L. Barros and R. Bassanezi in modeling that uses the subjectivity as a preponderant factor in biological processes.

The Group published some texts which became essential for those who intend to research in Biomathematics and/or Fuzzy Logic: *"Differential Equations with Applications"* (R. Bassanezi and W. C. Ferreira Jr.), Edit. Harbra, 1988 ; *"Theory of the fuzzy sets with applications"* (R. Jafelice, L. Barros and R. Bassanezi), SBMAC—Notas em Matemática Aplicada, Vol.17, 2005; *"Topics of Fuzzy Logic and Biomathematics"* (L. Barros and R. Bassanezi), Edit. Unicamp, 2006; *"Fuzzy Dynamic systems: Alternative modeling for biological systems"* (M. Cecconello, J. D. Mendes da Silva and R. Bassanezi), SBMAC—Notas em Matemática Aplicada, Vol.50, 2010. Recently, our book on fuzzy logic and biomathematics was expanded, translated into English, and published: *"The First Course in Fuzzy Logic, Fuzzy Dynamical Systems and Biomathematics – Theory and Applications"* (L. Barros, R. Bassanezi, and W. Lodaick), Springer, 2016.

The IMECC's Biomathematics Group organized and held, at Unicamp, the Congresso Latino Americano e Biomatemática – Latin America Biomathematics Conference (ALAB–ELAEM) on two separate occasions, in 2001 and 2010, and also organized the meeting's proceedings.

The activities of the group were intensified by the scientific cooperation with national research centers (EMBRAPA, ESALQ, CAISM, IB-Unicamp, Fundecitrus, etc.) and by the presence of foreign guest researchers who encouraged our research, several times even with effective collaboration.

In 1989, Professor Lee. A. Segel from the Weizmann Institute in Israel, one of the most renowned biomathematicians in the world, was present, as a guest of the group, and he lectured several times, emphasizing research into biomathematics. His motivation and power of enthusiasm leveraged our research and contributed to the paper developed in Professor Wilson's thesis (Fig. 1).

In 1994, we received Professor Alejandro Engel, from the University of Rochester, USA, working with "Implementation of fuzzy logic in artificial neural nets and applications to Biology"; Professor Gabriele Greco, from the University of

Fig. 1 Professor Segel visits the biomathematics group in 1989

Trento, has visited us several times, always bringing news concerning fuzzy logic. In the past few years, Professor Weldon Ludwick has honored us with his presence and cooperation.

Now, in a synthetic and simple inventory, we can say that the IMECC biomathematics group has accomplished its research and guidance project satisfactorily over these nearly 30 years of existence. The group has published more than a hundred articles in specialized journals and teaching and research books, it has taught several mini courses at conferences, and has guided 89 Master's and 44 doctorate theses in the field so far (as of 2016). Our former students are located throughout several regions of Brazil and Latin America, spreading knowledge in the field and encouraging other students to follow it.

It would be extremely complicated to summarize all the papers developed by the group. Therefore, we are going to present the ideas of some of the pioneering work that, in our opinion, have been used as an incentive for the continuity of biomathematics at the IMECC over these past 30 years.

Review of Some of the Papers Published by the Biomathematics Group of IMECC: Unicamp

After some Master's dissertations in the field, the first doctorate thesis came about. It was defended by Professor Vendite and the paper had been started in Italy and officially concluded at IEE-Unicamp.

1. **Mathematical Modeling for tumor growth and the problem of cellular resistance to antiblastic drugs** (Doctoral thesis of L. Vendite, 1988)

This work demonstrated the importance of pharmacoresistance from spontaneous mutations, as an intrinsic property of a tumor. The formal mathematical models show in this context different factors that can influence the efficacy of chemotherapy, such as tumor size, degree of cell resistance at the initiation of therapy, therapeutic program, the frequency of mutation-resistant cells, tumor kinetics, etc.

The results that were obtained suggest directions to be taken by therapists for the best choice of chemotherapy for its program, which is usually done empirically.

The proposed model initially considers C: tumor cells; S: sensitive cells; R_1: cells resistant to the first drug, and R_2: second drug-resistant cells:

$$\begin{cases} \dfrac{dS}{dt} = rS(1-kN) - \alpha_1 rS(1-kN) - \alpha_2 rS(1-kN) \\ \dfrac{dR_1}{dt} = rR_1(1-kN) + \alpha_1 rS(1-kN) - \alpha_2 rR_1(1-kN) \\ \dfrac{dR_2}{dt} = rR_2(1-kN) + \alpha_2 rS(1-kN) - \alpha_2 rR_2(1-kN) \\ \dfrac{dR_d}{dt} = rR_d(1-kN) + \alpha_2 rR_1(1-kN) - \alpha_1 rR_2(1-kN) \end{cases} \tag{1}$$

It is considered that the population R_d consists of a resistant part R_1 (sensitive to the second drug) that changes itself and the resistant part R_2 (sensitive to the first drug) that changes itself, in which, $N = S + R_1 + R_2 + R_d$ and α_i is the mutation from S to R_i and from $R_{j \neq i}$ to R_d.

As a result of this model, the factor of double resistance can be obtained, due to N and the rates α_i:

$$\frac{R_d}{N} = \left(1 - N^{-\alpha_1}\right) + \left(1 - N^{-\alpha_2}\right) + \left(1 - N^{-(\alpha_1+\alpha_2)}\right)$$

and also the percentage of resistant cells in a N-order:

$$\frac{R_1}{N} = N^{-\alpha_2} - N^{-(\alpha_1+\alpha_2)}$$

$$\frac{R_1}{N} = N^{-\alpha_1} - N^{-(\alpha_1+\alpha_2)}$$

Admitting that there is $\alpha \simeq \alpha_1 \simeq \alpha_2$ then,

$$\begin{array}{l} R_d = N\left(1 - N^{-\alpha}\right)^2 \\ R_1 = R_2 = N^{(1-\alpha)}\left(1 - N^{-\alpha}\right) \end{array}.$$

These relationships permit the resistant numbers to be calculated when the N-tumor mass and the mutation rate α are known.

In this paper, models with A and B alternative therapies having immediate effects and effects at fixed period intervals were also analyzed.

Thus, if the therapeutic program consists of two *Noncross-resistant* drugs, with period applications F_A and F_B interspersed, the model is changed to:

$$\begin{cases} \dfrac{dS}{dt} = rS(1-kN) - \alpha_1 rS(1-kN) - \alpha_2 rS(1-kN) - F(t)S \\ \dfrac{dR_1}{dt} = rR_1(1-kN) + \alpha_1 rS(1-kN) - \alpha_2 rR_1(1-kN) - F_B(t)R_1 \\ \dfrac{dR_2}{dt} = rR_2(1-kN) + \alpha_2 rS(1-kN) - \alpha_2 rR_2(1-kN) - F_A(t)R_2 \\ \dfrac{dRd}{dt} = rR_d(1-kN) + \alpha_2 rR_1(1-kN) - \alpha_1 rR_2(1-kN) \end{cases}.$$

Simulations carried out show the therapeutic advantage of using a program of alternate drugs over the mono-chemotherapy.

Some strategies of cancer control, formulated by Michel I. Silveira in his doctorate thesis at the Department of Applied Mathematics (1993), originated in a specialization course for high school teachers we held at Universidade de Guarapuava (University of Guarapuava) in 1986, in which the theme studied, with the modeling process, was bacteria control in papermaking. The tumor growth and cell resistance models were inspired by Vendite's thesis.

The following abstract is part of the paper developed in his doctorate thesis, which we guided with the cooperation of Professor Luis Boldrini.

2. **Optimal chemical control of populations developing drug resistance** (Michel I. da S. Costa, J. L. Boldrini and R. C. Bassanezi)—*IMA Journal of Mathematics Applied in Medicine & Biology, 1992.*

A system of differential equations for the control of tumor cell growth in cycle-nonspecific chemotherapy is presented. Drug resistance and toxicity are also taken into account. The aim of the control is to minimize the final tumor level and the toxicity. The analysis resorted to the optimal control theory and the results showed that maximum drug concentration featured in all treatments—in some cases it was the sole optimal strategy. Treatments dependent on tumor level were also optimal, whereas alternating maximum drug concentration and rest periods proved to be suboptimal, or an alternative strategy when there is no optimal solution.

Specifically, the model considered is given by the following systems of ordinary differential equations:

$$\begin{cases} \dfrac{dx}{dt} = xf(y) + \alpha f(y)(y-x) \\ \dfrac{dy}{dt} = yf(y) - u(t)g(y-x) \\ x(0) = x_0; \ y(0) = y_0 \end{cases}, \qquad (2)$$

where, y and x represents the total number of the population at time and drug-resistant individuals respectively, the resistance to drugs being acquired by spontaneous mutation, at a certain rate.

We are interested in solving the following free end-time optimal control problem associated with (2). The problem is to find a time $0 \leq t_f^* < +\infty$ and a bounded variation function $u^* : \left[0, t_f^*\right] \to \mathbb{R}$ with $0 \leq u^*(t) \leq u_{\max}$ almost everywhere in $\left[0, t_f^*\right]$ that will be the optimal drug administration treatment in the sense that

$$J_c(u^*(*), t_f^*) = \min \left\{ J_c(u, t_f) : u \varepsilon BV \left[0, t_f^*\right]; 0 \leq u(t) \leq u_{\max} \text{ a.e} \right\}, \tag{3}$$

where the functional J_c is defined by:

$$J_c(u, t_f) = y(t_f) + c \int_0^{t_f} u(t)dt, \tag{4}$$

with $c \geqslant 0$ a constant and $(x(t), y(t))$ a solution of (3). The pair $(u^*(*), t_f^*)$ is called an optimal strategy to the problem.

The term $y(t_f)$ is the number of tumor cells at the end of the treatment; the integral term of (4) is the total amount of the drug that reaches the tumor site during the treatment.

Using Pontryagin's minimum principle we prove that the optimal strategy is the *bang-off type*, we precisely prove the following:

Theorem 1 *Under the assumptions (2)–(4), and $x_0 \leqslant y_0 \leqslant y_m$, where y_m is the saturation level of the medium, the optimal strategy is given by $u(t) = u_{\max}$ with $0 \leqslant t \leqslant t_f$, where t_f is such that if $t_f > 0$, it is given by the condition $\frac{d}{dt}y(t_f) = -cu_{\max}$.*

This theorem is also applied in the case of cycle-nonspecific cancer chemotherapy and in control of bacteria populations in cellulose media.

Michel's thesis *"Controle determinístico de tratamentos quimioterápico"* (1993) promoted the publication of articles in scientific journals and encouraged the evolution of biomathematics at the IMECC. Among these works, in addition to those summarized here, we quote:

"Optimal chemotherapy: A Case study with drug resistance; saturation effect and toxicity" (with J. L. Boldrini and M. I. Costa); IMA J. Math. Applied in Medicine and Biology). Oxford University Press, 11, 1, pp. 45–59 (1994).

"Drug Kinetics and Drug Resistance in Optimal Chemotherapy" (with M. I. S. Costa and J. L. Boldrini); Math. Biosciences, 125, 2, pp. 191–209 (1995).

"Chemotherapeutic Treatments Involving Drug Resistance and Level of Normal Cells as a Criterion of Toxicity" (M. I. S. Costa, J. L. Boldrini, and R. C. Bassanezi); Math. Biosciences, V.125, 2, pp 211–218 (1995).

The research work in epidemiology began from the results obtained by Silvia Raimundo in her doctoral thesis: *"Uma Abordagem Deterministica da Interação de Doenças—AIDS e TB num Presídio"* in which we were her advisers in DMA, with the collaborations of Professor Drs Hyan Mo Yang and Alejandro Engel in 1996. The summary of the following publication is a part of her thesis.

3. **The Attracting Basins and the Assessment of the Transmission Coefficients for HIV and M. Tuberculosis Infections Among Women Inmates** (S. M. Raimundo, R. C. Bassanezi, H. M. Yang and M. Ferreira)—*Journal of Biological Systems, 2002; 10:61–83.*

It has been observed that in many cases, one infection can partially protect against another infection, or it may lead to a co-infection. For instance, the interaction between infections with different strains, such as dengue and malaria or tuberculosis and leprosy, induces cross immunity. On the other hand, individuals infected with HIV are much more susceptible to other infections, for instance, tuberculosis. We propose a compartmental model to describe the transmission of AIDS and tuberculosis in a closed community as an example of one infection activating the other one. When studying the dynamics of the interactions we obtain basins of attraction in which one infection prevails over the other and where both infections coalesce. Furthermore, we are taking into account an adaptation of the model to assess the transmission coefficients for HIV and *Mycobacterium tuberculosis* (MTB) infections among women inmates.

The present model describes the phenomenon of the interaction between HIV and MTB infections considering the populational dynamics theory. Taking into account the mathematical approach and the biological aspects of this phenomenon, we assess quantitatively the attracting basins for a single disease and both diseases. Initially, we present the biological features of the transmission of AIDS and TB diseases.

We developed a mathematical model to analyze the interaction between HIV and MTB infections. From our analyses, we obtained the attracting basins in which one infection prevails over the other one and where both infections coalesce, and based on these analyses, we also presented a simple epidemiological study. Let us take the developed countries as an epidemiological example of the interaction between MTB and HIV infections. As is well known, in many developed countries the MTB infection could be considered eradicated. In other words, in these countries, the value of the MTB transmission coefficient β_2 was lower than its critical value β_2^{th}, that is, $R_0^2 < 1$. In epidemiological terms this means that the only possible case of the existence of TB is among individuals with AIDS. That is, TB is an opportunistic infection in the course of AIDS disease.

The stated variables considered in the model are:

X_1 : susceptible individuals; X_2 : MTB-infected individuals; T_b : TB-diseased individuals; Y_1 : HIV-infected individuals; Y_2 : both HIV- and MTB-infected individuals; Y_{tb} : HIV-infected individuals with TB disease; A : individuals with AIDS disease; A_{tb} : individuals with both AIDS and TB diseases and $N = X_1 + X_2 + T_b + Y_{tb} + A + A_{tb} + Y_1 + Y_2$ is the total population size (assumed to be constant).

The mathematical model analyzes the interaction between AIDS and TB. As *Mycobacterium tuberculosis* is an airborne infection, and HIV is transmitted by contact with blood or products derived from blood, the quantitative descriptions for the insurgence of new cases of infections are different:

$$
\begin{cases}
\dfrac{dX_1}{dt} = \mu N + \theta(T_b + Y_{tb} + A_{tb}) + \alpha(A + A_{tb}) \\
\qquad -\frac{\beta_1}{N}(Y_1 + Y_2 + Y_{tb})X_1 - \beta_2(T_b + Y_{tb})X_1 - \mu X_1 \\
\dfrac{dX_2}{dt} = -\dfrac{\beta_1}{N}(Y_1 + Y_2 + Y_{tb})X_2 + \beta_2(T_b + Y_{tb})X_1 + \rho T_b - (\sigma + \mu)X_2 \\
\dfrac{dT_b}{dt} = \sigma X_2 - \dfrac{\beta_1}{N}(Y_1 + Y_2 + Y_{tb})T_b - (\rho + \theta + \mu)T_b \\
\dfrac{dY_1}{dt} = \dfrac{\beta_1}{N}(Y_1 + Y_2 + Y_{tb})X_1 - \beta_2(T_b + Y_{tb})Y_1 - (\omega + \mu)Y_1 \\
\dfrac{dY_2}{dt} = \dfrac{\beta_1}{N}(Y_1 + Y_2 + Y_{tb})X_2 - \beta_2(T_b + Y_{tb})X_1 - (\varpi + \sigma + \mu)Y_2 + \rho Y_{tb} \\
\dfrac{dY_{tb}}{dt} = \sigma Y_2 + \dfrac{\beta_1}{N}(Y_1 + Y_2 + Y_{tb})T_{tb} - (\rho + \xi + \theta + \mu)Y_{tb} \\
\dfrac{dA}{dt} = \omega Y_1 - \beta_2 A A_{tb} - (\alpha + \mu)A \\
\dfrac{dA_{tb}}{dt} = \varpi Y_2 + \xi Y_{tb} + \beta_2 A A_{tb} - (\alpha + \theta + \mu)A_{tb}
\end{cases}
$$

where the coefficients β_1 and β_2 are the transmission coefficients for HIV and MTB infections respectively; σ^{-1} and ρ^{-1} are the average re-activation and recovery periods of TB; ω^{-1} and ξ^{-1} are the average incubation periods of individuals who have AIDS disease and both AIDS and TB diseases respectively; μ is the natural mortality rate; α and θ are AIDS-related and TB-related mortality rates respectively; and ϖ^{-1}(assumed to be $\omega^{-1} + \xi^{-1}$) is the average incubation period of individuals with AIDS disease and MTB infection.

A study of the stability and critical points of the model is presented.

The paper of epidemiology that followed shows that we were on the right track with our research in biomathematics. A conjecture of Professor Yang was the theme of the thesis defended by Ma. Beatriz Leite in 1999. The result obtained in this research is essential in the process of the study of the persistence or not of diseases when the infectious individuals have different levels of infectiousness. An abstract of the paper, which contains its results, is presented hereinafter:

4. **The basic reproduction ratio for a model of directly transmitted infections considering the virus charge and the immunological response** (M. B. Leite, R. C. Bassanezi and H. M. Yang)—*IMA Journal of Mathematics Applied in Medicine and Biology (2000) 17, 15–31.*

First, we developed a mathematical model taking into account a heterogeneous infectivity based on a virus charge harbored by human hosts. From this model, we determined the formula for the basic reproduction ratio, when the infectious individuals were subdivided into k infective stages according to the interaction between the host's immunological response and the virus.

The mathematical model is

$$\begin{cases} \frac{d}{dt}S(t) = \mu + \delta R - \beta S \sum_{j=1}^{k} \varepsilon_j I_j - \mu S \\ \frac{d}{dt}E(t) = \beta S \sum_{j=1}^{k} \varepsilon_j I_j - (\mu + \sigma)E \\ \frac{d}{dt}I_1(t) = \sigma E - (\mu + \gamma_1)I_1 \\ \frac{d}{dt}I_j(t) = \gamma_{j-1}I_{j-1} - (\mu + \gamma_j)I_j, \text{ for } j = 2, \ldots, k \\ \frac{d}{dt}R(t) = \gamma_k I_k - (\mu + \delta)R \end{cases} . \tag{5}$$

The formula for the basic reproduction ratio was obtained by analyzing the stability of the trivial equilibrium point of system (5),

$$R_0 = \frac{\sigma}{\mu + \sigma} \sum_{j=1}^{k} P_{j-1} \frac{\beta \varepsilon_j}{\mu + \gamma_j}, \text{ where, } P_i = \begin{cases} \prod_{j=1}^{i} \frac{\gamma_j}{\mu + \gamma_j}, & \text{if } i = 1, 2, 3, \ldots, k \\ 1, & \text{if } i = 0 \end{cases},$$

and we demonstrated that the stability results can be assessed by analyzing the $\gamma-$independent term of the characteristic polynomial:

Theorem 2 *The trivial equilibrium point Q_0 is locally asymptotically stable (LAS) if the $\gamma-$independent term of the characteristic polynomial of (5) is strictly positive, and unstable if it is strictly negative.*

Theorem 3 *If the trivial equilibrium point Q_o is LAS then it is globally asymptotically stable.*

Modeling the Immunological Response When a susceptible individual has the first infective contact with a virus, this individual builds up an immunological response after a certain period of time. Also, this infective period is characterized by the abundance of the initial virus rising followed by it later decreasing because of the antibodies produced by the stimulated immunological system, which destroys completely (or at a very low level) the invading pathogens. In model (5) we did not take into account the heterogeneity among individuals. However, the genetic and nutritional aspects of the infected individual can affect both the infectious period and the virus charge.

A heterogeneous immunological response is obtained by dividing all the latent individuals into k different infection status classes according to their immunological response to the virus. The bilinear incidence model encompassing the heterogeneous immunological response and describing directly transmitted infection can now be set in terms of the fraction of individuals in each class as

$$\begin{cases} \frac{d}{dt}S(t) = \mu + \delta R - \beta S\sum_{j=1}^{k}\varepsilon_j I_j - \mu S \\ \frac{d}{dt}E(t) = \beta S\sum_{j=1}^{k}\varepsilon_j I_j - (\mu+\sigma)E \\ \frac{d}{dt}I_1(t) = \lambda_j \sigma E - (\mu+\gamma_j)I_j, \text{ for } j = 1, \ldots, k \\ \frac{d}{dt}R(t) = \sum_{j=1}^{k}\gamma_j I_j - (\mu+\delta)R \end{cases} . \tag{6}$$

Theorem 4 *The trivial equilibrium point of the model (6), Q_0 is LAS if the independent term a_n of the characteristic polynomial, given by the expression below, is strictly positive, and unstable if it is strictly negative, where*

$$a_n = (\mu+\sigma)\prod_{j=1}^{k}(\mu+\gamma_j) - \beta\sigma\sum_{j=1}^{k}\lambda_j\varepsilon_j \prod_{i=1, i\neq j}^{k}(\mu+\gamma_i).$$

Following the ideas of the works developed in the thesis by Michel, Renata Zotin obtained relevant results in her doctoral thesis, regarding the chemical treatment of pests in bean plants, using processes coming from optimal control. The following summary is part of her thesis paper:

5. **A Model For Optimal Chemical Control of Leaf Area Damaged by Fungi Population Parameter Dependence** (R. Zotin, R. C. Bassanezi, H. M. Yang and A. Adami)—*IMA Journal of Mathematics Applied in Medicine and Biology (2000) 17, 15–31.*

In the specific case of this study, we mathematically modeled an agricultural production situation subjected to the attacks of a susceptible and a resistant fungi population to maximize production (minimizing disease at the end of the harvest), controlling disease at a low cost.

We considered the dynamics of the damaged leaf area with the intraspecific competition between a susceptible and a resistant fungi population for the occupation of the leaf area. Such occupation is cumulative; that is, once the leaf area is damaged, it does not recover, which reduces plant productivity. The productivity is affected significantly if the damaged area exceeds some limit, which varies for each cultivar.

The control problem that we analyzed searched for the establishment of a minimum value for the damaged leaf area at the final time (crop) with the minimum cost of fungicides.

We present a model to study of a fungi population subjected to chemical control, incorporating the fungicide application directly into the model. From that, we obtain an optimal control strategy that minimizes both the fungicide application (cost) and the leaf area damaged by the fungi population during the interval between the moment when the disease is detected ($t = 0$) and the time of harvest ($t = t_f$): initially, the parameters of the model are considered constant. Later, we consider the apparent infection rate depending on the time (and the temperature) and perform some simulations to illustrate and compare with the constant case.

The model of the fungi population can be described by the following system of differential equations:

$$\begin{cases} \dfrac{dS}{dt} = r(t)N(1-N)(1-\beta u) + r(t)R\beta u(1-N) \\ \\ \dfrac{dR}{dt} = r(t)R(1-N) + \alpha r(t)(N-R)(1-N)(1-\beta u) \end{cases}, \tag{7}$$

where the fungi population in a given crop is represented by the total occupied area N and is subdivided into susceptible and resistant. We assign $S(t)$ and $R(t)$ to represent, at each time t, the proportion of the damaged leaf area for susceptible and resistant respectively.

The optimal control problem associated with the dynamic (7) consists of finding u that minimizes the functional

$$J(u) = N(t_f) + c_1 \int_0^{t_f} u(t)dt,$$

where c_1 is an adjustment constant and it is proportional to the cost of the fungicide.

The function u is the optimal control, which satisfies

$$J(u^*) = \min \left[N(t_f) + c_1 \int_0^{t_f} u(t)dt \right]. \tag{8}$$

Theorem 5 *The optimal control $u^*(t)$ that satisfies the Eqs. (8) with (7) is given by*

$$u(t) = \begin{cases} 0, & \text{if } g(t) > 0 \\ 1, & \text{if } g(t) < 0 \\ \text{indetermined}, & \text{if } g(t) = 0 \end{cases},$$

where $g(t) = c_1 - r(t)(1 - N)(N - R)\beta(\lambda_1 + \alpha\lambda_2)$ *and the constant equations*

$$\frac{d\lambda_1}{dt} = -\frac{\partial H}{\partial N}$$
$$\frac{d\lambda_2}{dt} = -\frac{\partial H}{\partial R}$$

are obtained from the Hamiltonian.

The final conditions for $\lambda_1^*$and λ_2^* are obtained when t_f is fixed and, $N(t_f)$ and $R(t_f)$ are free. The nonlinearity of the model hinders the analysis of the differential equations for λ_1 and λ_2 making it practically impossible to obtain analytic expressions for λ_1 and λ_2.

We initially consider the growth rate as being constant, that is, $r(t) = r$, $\forall t$.

Analytic Results In the constant case $r(t) = r$, we analyze which types of control are feasible. In practice, the parameter α is between 10^{-9} and 10^{-5} and it only has an effect after various crop cycles. As we analyze one cycle, we consider $\alpha \approx 0$.

Lemma 1 *If the change rate is null and the optimal control is zero* ($u(t) = 0$) *during some time interval* $[\bar{t}, t_f]$, *then* $u^*(t) = 0$, $\forall t \in [0, t_f]$.

From Lemma 1, we observe that the optimal control, if it is not always null, should end with $u(t) = 1$:

Lemma 2 *If the optimal control ends with* $u^*(t) = 1$ *and* $\alpha = 0$, *then it will have one commutation point at the most.*

Theorem 6 *For the control problems (7) and (8), with* $\alpha = 0$, *the optimal control* $u^*(t)$ *is given by*

$$u^*(t) = \begin{cases} 0, \; 0 < t < \bar{t} \\ 1, \; \bar{t} < t < t_f \end{cases},$$

with $\bar{t} \in [0; t_f]$.

We observed changes in the control strategy when we altered each of the parameters, as we did not obtain an explicit formula among $\bar{t}$ and the other parameters of the model.

The commutative point $\bar{t}$ as a function of growth rate r is observed in Fig. 2 and the optimal values of damaged leaf area of the final time t_f for each value fixed r are given in Fig. 3.

Professor Hyan has guided several theses involving epidemiological processes. Let us present here a summary of the work developed by Barrozo and Yang in 1999.

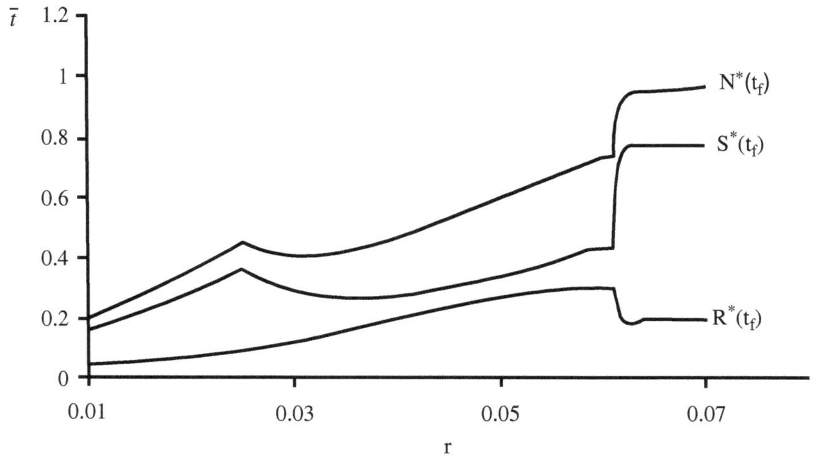

Fig. 2 The commutative point $\bar{t}$

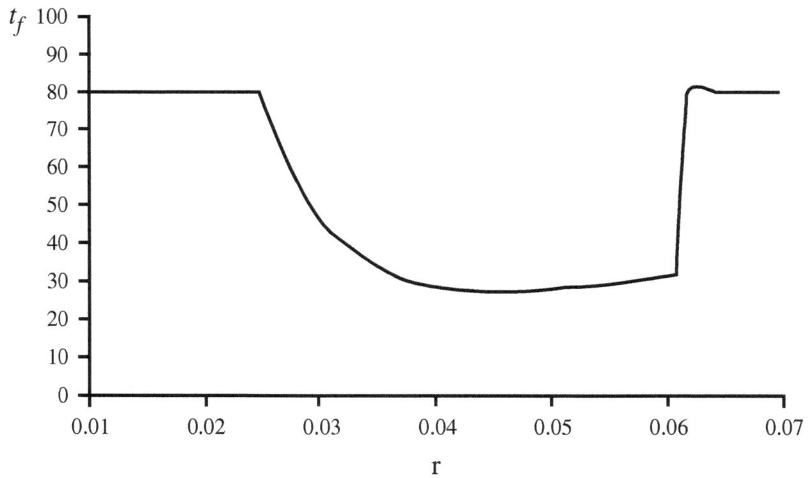

Fig. 3 $N^*(t_f)$, $S^*(t_f)$ and $R^*(t_f)$

6. Mathematical Modeling for Macroparasites with an emphasis on Schistosomiasis (S. Barrozo e H. M. Yang)—*Biomatemática 9 (1999), 73–89.*

Schistosomiasis is a parasitic disease caused by parasitic worms called *Schistosoma*, it occurs more frequently in tropical countries, especially in underdeveloped or developing ones. It can be a disease whose transmission takes place through the direct contact of the person with contaminated water, it is closely linked to the socioeconomic–cultural and educational conditions of each region. In Brazil, the species *Schistosoma mansoni*, which lives in the hepato-intestinal system of the infected individual, prevails. It is more common in the northeastern region of the country and in Minas Gerais, Espírito Santo, and Rio de Janeiro; however, in most Brazilian states, there are focal points with a low incidence.

The Schistosoma are parasites with an overly complex life cycle, involving two types of hosts (human and snail), two transmission stages (miracidia and cercaria) and several factors, such as environmental and immunological. In the definitive host (human), they undergo sexual reproduction, laying eggs that are eliminated in the feces or urine, depending on the species. These eggs hatch when in contact with water, freeing the miracidia, which are ciliate organisms with great mobility. When they find a specific snail (*Biomphalaria*), they infect it and reproduce asexually in hundreds of thousands of new organisms, called cercaria, which, after a few weeks, are eliminated in the water and constitute the form that infects humans. Once these cercaria find a human, they penetrate the skin with the help of special enzymes. Immediately after the infection, they become schistosomule and stay in the epidermis for just a few days. After that, they enter the blood circulation and migrate to the lungs and from there to the hepatoportal circulation, where they become sexually mature adults, they mate and migrate to the mesenteric veins, where they lay their eggs. They can remain in copulation and continue to lay eggs for many years. Each female lays about 300 eggs a day. These eggs fall into the circulation and are taken to several organs of the human organism. Some of them can transpose the natural barriers, move into the intestines, and are released into the environment with the feces to continue their cycle.

A deterministic model is presented here and it is aimed at describing the transmission of human schistosomiasis. To do so, it incorporates the concomitant immunity in the following way: it supposes that the susceptible individuals, when infected, begin to develop a certain immune response, which becomes effective after a certain period of time and it remains active only in the presence of parasites. If they lose the whole parasite load, after some time, they also lose immunity, continuing only with the immunological memory. The average parasite load per class of individuals infected in the population and the vital dynamics of the snails, which are intermediate hosts of the parasite, are also considered. Environmental and immunological factors are considered relevant in the transmission mechanism (see Fig. 4).

The compartments x_1, x_2, x_3, and x_4 represent, the fractions of susceptible population, developing immunity, immune, and with immunological memory respectively; the compartments w_2 and w_3 represent the average number of parasites in the population of the compartments x_1 and x_2 respectively, and the compartments y_1, y_2, and y_3 represent the fraction of susceptible, latent, and infectious snails in the community respectively.

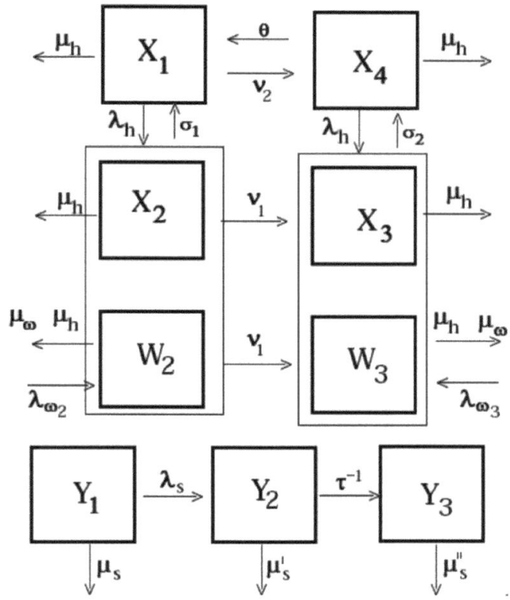

Fig. 4 Compartmental structure of the life cycle of the parasite

λ_h is the infection power of the susceptible individuals and those with immunological memory individuals λ_{ω_2} is the infection power of individuals developing immunity, λ_{ω_3} is the infection power of partially immune individuals, σ_1 is the rate at which the non-immune infected individuals who lose their parasites become susceptible again, ν_1 is the rate at which the infected ones become immune, σ_2 is the rate at which the partially immune, when losing their parasites, lose immunity and start having immunological memory, ν_2 is the rate at which the individuals with immunological memory become susceptible, θ is the rate of vaccination, which is applied to susceptible individuals, partially protecting them (making them have immunological memory), μ_h and μ_ω are the mortality rates of the individuals and parasites respectively, λ_c is the snail's basic reproduction ratio, τ is the latent period of infected snails, and μ_s, μ'_s, and μ''_s are the mortality rates of susceptible, latent, and infectious snails respectively.

We consider the total human population N_h and snails N_c to be homogeneously distributed and constant and do not consider age structure.

Applying the classical mass action law in a homogeneously distributed population and based on the transmission dynamics of the disease represented by the above scheme, we describe our model through the following system of differential equations:

$$\begin{cases} \dfrac{dx_1}{dt} = \mu_h + \sigma_1 x_2 + \nu_2 x_4 - [\lambda_h(t) + \mu_h + \theta]\, x_1 \\ \dfrac{dx_2}{dt} = \lambda_h(t) x_1 - [\sigma_1 + \nu_1 + \mu_h]\, x_2 \\ \dfrac{dx_3}{dt} = \nu_1 x_2 + \lambda_h(t) x_4 - [\sigma_2 + \mu_h]\, x_3 \\ \dfrac{dx_4}{dt} = \theta x_1 + \sigma_2 x_3 - [\lambda_h(t) + \mu_h + \nu_2]\, x_4 \\ \dfrac{dw_2}{dt} = \lambda_{\omega_2}(t) x_1 + \lambda_{\omega_2}(t) x_2 - [\nu_1 + \mu_h + \mu_\omega]\, w_2 \\ \dfrac{dw_3}{dt} = \lambda_{\omega_3}(t) x_3 + \lambda_{\omega_2}(t) x_4 + \nu_1 w_2 + - [\mu_h + \mu_\omega]\, w_3 \\ \dfrac{dy_2}{dt} = \lambda_s(t)\, [1 - y_2 - y_3] - \left[\tau^{-1} + \mu_s'\right] y_2 \\ \dfrac{dy_3}{dt} = \tau^{-1} y_2 - \mu_s'' y_3 \end{cases}.$$

In the study of this model, two situations were supposed: in the first, the basic reproduction ratio of the immune individuals depends both on the immunity and environmental factors; in the second one, those hypotheses were changed.

Concerning the diffusion phenomenon, several interesting papers were carried out by the group. We mention herein the one developed by S. Pregnolatto with the guidance of J. Meyer:

7. **A strategy for the numerical simulation of the evolutionary behavior of a descriptive PDE of the Trypanosoma evansi (surra) of capybaras—periodic infection rate** (João Frederico da C. A. Meyer and Sílvio de Alencastro Pregnolatto)—*BIOMATEMÁTICA 12 (2002), 01–18.*

In this paper, the aim is to fall back upon the numerical instrumental in type SIR or SIRS systems with spatial dissemination to which we add certain hypotheses related to a case study. In this example, we try to describe some of the hypotheses related to the evaluation of the *Trypanosoma evansi* in capybara populations, an epidemic with cyclic spatial dissemination in certain situations with a certain type of rodent in natural environments. The idea, then, would be to put an SIR or SIM model, but only with equations for Susceptible and Infected, as the removed ones are due to death (what SIM stands for), simulating the evolutionary behavior of the epidemic mentioned. This model would include the spatial dissemination of the infected capybaras, the contagion of the susceptible ones, and a population dynamics of the susceptible ones, as the infected ones hardly ever reproduce. Moreover, as the initial population of susceptible capybaras is homogeneous, there is no diffusibility at an early stage, but the spatial and temporal variation arises when, by contagion, part of the population of susceptible animals are infected and these, sometime later, are dead (or removed). The first model does not include *Trypanosoma equus* infection-carrying insects, which cause this endemic disease.

This first model is given, consequently, by:

$$\begin{cases} \dfrac{\partial S}{\partial t} - \alpha_s \Delta S + div(\overrightarrow{V} \cdot S) + \sigma_s S = \lambda S\left(1 - \dfrac{S+I}{K}\right) - \beta S I \\ \\ \dfrac{\partial I}{\partial t} + \alpha_I \Delta I + \sigma_I I = \beta S I - \gamma I \\ \\ \dfrac{\partial M}{\partial t} = \gamma I \end{cases}$$

$$S = S(x, y, t);\ I = I(x, y, t);\ M = M(x, y, t),$$

$$\text{with } (x, y) \in \Omega \subset \mathbb{R}^2 \text{ and } t \in (0, T].$$

A possible initial condition is given by $S(x, y; 0) = S_0$ steady in the whole domain, by $M(x, y; 0) = 0$ and by $I(x, y; 0) = I_0(x, y)$, where it is considered a population of healthy and susceptible animals, evenly distributed by the study domain, and a population of infected ones concentrated in some sub-region of the Ω domain. The boundary conditions are—in an initial exploratory modeling—of the homogeneous Dirichlet and Von Neumann types, indicating parts of the $\partial\Omega$ boundary in which there is an obstacle to the passage of individuals of the species studied (Γ_1) and where there are no individuals (Γ_0):

$$\begin{aligned} \frac{\partial S(x, y, t)}{\partial \eta}\Big|_{(x,y)\in\Gamma_1} &= 0 \text{ if } t \in (0, T] \\ S(x, y, t)\Big|_{(x,y)\in\Gamma_0} &= 0 \text{ if } t \in (0, T] \\ \frac{\partial I(x, y, t)}{\partial \eta}\Big|_{(x,y)\in\Gamma_1} &= 0 \text{ if } t \in (0, T] \\ I(x, y, t)\Big|_{(x,y)\in\Gamma_0} &= 0 \text{ if } t \in (0, T] \end{aligned}$$

One of the reasons that motivates the experimental use of this model is that a false steady state is obtained, given by:

$$\overline{S} = \frac{\sigma_I + \gamma}{\beta}$$

$$\overline{I} = \frac{\beta K(\sigma_s - \lambda) - \lambda(\sigma_I + \gamma)}{\beta(\lambda - \beta K)}$$

This pair of values, even if it remains (at least theoretically) steady in terms of the respective populations of susceptible and infected animals, corresponds to a linear variation in time for the population of dead individuals, counterbalancing the population growth through reproduction of the healthy ones.

An alternative model was studied incorporating qualitative traits with which it is possible to analyze the surra we intend to assess in an endemic population. Therefore, we approximated the infection rate, describing the contagion using a sine curve, which must show (in a first approximation) the periodic seasonal variation of the population of infection-carrying insects, that is, we associated the action of the infection carrier with its population density, replacing β with the expression $\beta + \delta \sin(\pi t/6)$.

We waived, of course, the exhibition of a simple expression for an analytical solution, and we dedicated ourselves, herein, to the quest and exploratory management of approximate solutions. The goal, much more than defending such solutions depicting possible realities, is to create a culture of models so that, as new traits of real and effective scenarios are incorporated into the model, an acquired intuition helps the assessment consideration of approximate solutions, allowing effective criticism of the use of the model and its results, going to the variational formulation of the system described above, choosing beforehand functional spaces suitable for this line of work. We obtained weak, or variational, formulation of the problem originally proposed, appropriate formulation for the discretization via the Galerkin method, with finite elements. Moreover, we fell back upon Crank–Nicolson in the discretization of the temporal variable.

With the insect dispersion process, we mention two papers oriented by Professor Wilson Ferreira Jr:

8. **Invasão de abelhas africanizadas: Dispersão não-local e taxia**—(*Invasion of Africanized bees: nonlocal dispersion and taxia*) (D. C. Mistro and W. C. Ferreira Jr)—*Memórias do IX Congreso Internacional de Biomatemática, p. 149–156, 1999.*

In this paper, a mathematical model, discrete in time and continuous in space, is built, which represents a dispersion of colonies of social insects that are particularly characterized by far reaching movements, strategical perception of the site (taxi) quality, and reproduction through fissions. The mathematical model is described using a nonlinear integral operator. The theoretical paper is illustrated by the dispersion phenomenon of Africanized bees, but the argument can be explained in the study of the collective behavior of other social insects concerning its macroscopic movement.

9. **Fitotaxia e agregação não-local na dispersão de insetos herbívoros**—(*Phyllotaxis and nonlocal clustering in herbivorous insect dispersion*) (L. A. D. Rodrigues and W. C. Ferreira Jr)—*Memórias do IX Congresso Internacional de Biomatemática, p. 199–206, 1999.*

In this paper, the mathematical model for the description of the dynamics of herbivore populations (growth and dispersion) in large textured crops under the circumstance of the oriented movement of herbivores concerning the quality of their food and including a nonlocal clustering behavior with regard to the conspecifics.

A discrete cellular automata-type model is used to analyze the nature of the variation of the quality of vegetation and the density of the herbivores in the space and time for parameter values that represent several situations of interest.

Effects such as the formation and spread of wave fronts and colonization patterns are observed on a macroscopic scale.

Fuzzy Theory and Biomathematics

The dynamics systems arose with the need to understand phenomena that evolve over time, following strict and predetermined rules. The deterministic variational models, formulated by differential equations or equations of differences, became essential tools for the knowledge and prediction of several situations, especially of a physical or biological nature. The essential trait of mathematical modeling of variational processes, using deterministic equations, is the *accuracy* obtained in the predictions of the phenomenon studied. Evidently, such predictions are always dependent on *accurate information*, which is inserted into the models using the mean values of the parameters involved.

The classic biomathematics models, particularly the models of population dynamics and epidemiology, are reasoned upon hypotheses, almost always deriving from physical–chemistry in which the reaction between two substances (variables in a state) is modeled by the product of their concentrations—the law of mass action. This same law is used in the Kermack–McKendrick epidemiology models. The parameter that represents the rate of predation or infection power of the epidemiological models are "mean" values simulated or obtained empirically.

In the models that deal with uncertainty, such as the stochastic, for instance, the solutions are stochastic processes whose means can be obtained after the event, when you have the distribution densities of the variables and/or of the parameters involved in the model referring to the phenomenon analyzed.

On the other hand, if in a population, besides quantifying its elements, we intend to take into account certain qualities of the individuals, the variables and/or parameters of the mathematical models must almost always be considered inaccurate or obtained with partial data. For example, in a population of prey of a certain species, to each prey we can take into account the ease with which it is predated, which may be related to its age, health condition, habitat, etc. Considerations of this kind (state variables with qualitative adjuncts) are very frequent in biological phenomena and often essential in the modeling and understanding of the phenomenon. In an evolutionary system, what seems to be insignificant can be of utmost importance in the future.

The several kinds of uncertainties that appear in the phenomena dealt by biomathematics can have well-varied modeling. When we opted for stochastic models, implicitly, we were supposing to know, a priori, the distribution of the probabilities of the parameters and the initial conditions of the phenomena studied.

This is the case of the Malthusian stochastic model studied by Pielou. The "realistic" models tend to also use stochastic equations in their formulations—much more complex and dependent on sophisticated computational models. Nevertheless, if, in the phenomenon at issue, we intend to take into account heterogeneity, such as gradualities that do not come from randomness, the use of tools coming from the fuzzy logic favor the variational modeling. Another more recent and maybe more practical and simple alternative, is to use fuzzy variational models, in which the variables and/or parameters are considered sets that display the pertinence level of their elements and, thus, the subjectivity is embedded in the concept of the state variable or the parameters themselves. Anyway, the deterministic models, while not correctly defining reality, can be markers of several stochastic or fuzzy models—when working with a large sample of individuals, it can be said that the process follows a deterministic course that represents the mean solution of isolated cases. In the fuzzy models, the solution is a fuzzy set of deterministic solutions that, when defuzzified, represent the means of the solutions. Still, of the deterministic solutions that form the fuzzy solution, the one that has the greatest pertinence or reliability level is the initial mean solution called "preferred."

The variational fuzzy models can encompass several uncertainty types (subjectivities or fuzziness) that can be attached in the parameters, initial conditions or the state variables themselves. If the subjectivity comes in the state variable or in the parameters of the models, we have demographic fuzziness or environmental fuzziness respectively. Therefore, when the state variables are modeled by means of the fuzzy sets, we have demographic fuzziness, and we have environmental fuzziness when just the parameters are considered fuzzy. In general, both kinds of fuzziness are present in biological phenomena.

This new way of modeling problems linked to the biological reality, in which not only the state variables, but also the parameters, are subjectivity holders, is gaining ground in the biomathematics field, with significant and very encouraging results. A great part of the research of the IMECC biomathematics group is linked to the use of variational fuzzy systems modeling biological phenomena.

The variational fuzzy equations have been studied using distinct methods. The first attempt to envisage nonrandom-type subjectivity in variational systems was with the use of the Hukuhara derivative, which was not very successful, the main cause being the fact that, with such a process, there was never solution stability—the uncertainty always increases with no limitation. Another way of envisaging the nonrandom subjectivity is through differential inclusions. Such a procedure, however, has been shown to be very complicated, even when applied in simple situations. An alternative method we have been using consists of fuzzifying the solutions of a deterministic model, using Zadeh's extension principle. We have recently shown that this relatively simple method, under certain circumstances, supplies the same solutions as the fuzzy differentials. We have also shown that these fuzzy systems, obtained via Zadeh's extension, behave in a similar manner to the associated deterministic models regarding the quality of their stationary points. In all the methods mentioned, the fuzzy process adopted for studying the variational systems is always derived from classic, deterministic or stochastic systems.

On the other hand, with the tools of fuzzy logic, we can study the dynamics of phenomena without the formal concepts of variations coming from the derivative or explicit differences or, then, from differential inclusions. The method consists of simply adopting the iterative process, considering variations obtained by means of a rule foundation and an operator (Mamdani's), which change them into "numbers." We call such systems purely fuzzy systems, or just p-fuzzy systems.

The solutions obtained from the fuzzy or p-fuzzy systems are apparently rougher and less precise than the deterministic ones; however, they are much safer. The fuzzy-system solutions are presented in interval ways in which each value has its reliability level as solution, at each instant. The p-fuzzy systems have the advantage of encompassing the subjectivity described by a specialist of the phenomenon studied.

After some satisfactory experience with research into fuzzy logic, which we had started in the 1980s, some interesting papers came about, not only from the perspective of mathematics, but also biology. Examples of these papers are: the thesis of Heriberto E. Roman Flores in 1989: "Sobre as Entropias Fuzzy—*About the Fuzzy Entropies"*, followed by the dissertations of Laécio C. Barros: "Modelos determinísticos com Parâmetros subjetivos—*Deterministic Models with subjective Parameters"* (1992) and Paulo Blinder: "Implementação da lógica fuzzy em redes neurais artificiais e aplicações à Biologia—*Fuzzy logic implementation in artificial neural nets and application to Biology"* in 1994. Laécio's doctorate thesis appeared straight after, in 1997, with Professor Pedro Tonelli's contribution: "Sobre Sistemas Dinâmicos Fuzzy—Teoria e Aplicações—*About Dynamic Fuzzy Systems—Theory and Applications"*.

From 2000, papers involving fuzzy logic in modeling processes of biological phenomena, especially in variational, discrete, and continuous models, began to emerge more frequently.

Next, we present some abstracts of papers that marked this period:

10. Fuzzy modelling in population dynamics (L. C. Barros, R. C. Bassanezi and P. A. Tonelli)—*Ecological Modelling, 128 (2000), 27–33.*

The aim of this paper is to analyze the behavior of models that describe the population dynamics, taking into account the subjectivity in the state variables or in the parameters. The models in this work have demographic and environmental fuzziness. The environmental fuzziness is presented using a life expectancy model where the fuzziness of the parameters is considered. The demographic fuzziness is presented using the continuous Malthus and logistic discrete models. An outstanding result in this case is the emergence of new fixed points and bifurcation values to the discrete logistic model with subjective state variables in the form of fuzzy sets. An interpretation is offered for this fact that differs from the deterministic one.

We consider the normalized logistic function

$$\begin{cases} x_{n+1} = ax_n(1 - x_n) = f(x_n) \\ 1 \leqslant x_n \leqslant 4; \;\; x_n \in \mathbb{R}^n \end{cases} \tag{9}$$

and the associated fuzzy system

$$\begin{cases} u_{n+1} = au_n(1-u_n) = \widehat{f}(u_n) \\ 1 \leqslant u_n \leqslant 4;\ u_n \in \mathscr{F}(\mathbb{R}^n) \end{cases}. \tag{10}$$

The fixed points of $\widehat{f}(u_n)$ are the critical points of (10), that is, the characteristic functions $\widehat{0}$ and $\widehat{x_a} = \widehat{(1-\frac{1}{a})}$. The fixed points of $\widehat{f}(u_n)$ that are different from the characteristic functions are:

- If $1 \leqslant a \leqslant 2$, the only fixed points are: $\widehat{0}$ and $\widehat{x_a}$ and $\overline{u_1}$ given by $[\overline{u_1}]^\alpha = [0, x_a], \forall\alpha \in [0, 1]$.
- If $2 < a \leqslant 3$, apart from $\widehat{0}$ and $\widehat{x_a}$, we have the fixed point $\overline{u_2}$ given by $[\overline{u_2}]^\alpha = \left[0, \frac{a}{4}\right], \forall\alpha \in [0, 1]$.
- If $3 < a \leqslant 1+\sqrt{5}$, excluding $\widehat{0}, \widehat{x_a}$ and $\overline{u_2}$, we also have the fixed point $\overline{u_3}$ given by $[\overline{u_3}]^\alpha = [x_1, x_2], \forall\alpha \in [0, 1]$, where $x_1, x_2 = \frac{a+1\pm\sqrt{(a-3)(a+1)}}{2a}$.
- If $1+\sqrt{5} < a < 4$, the fixed points are: $\widehat{0}, \widehat{x_a}, \overline{u_2}, \overline{u_3}, \overline{u_4}$ with $[\overline{u_4}]^\alpha = [f(a/4), a/4]$ and $\overline{u_5}$ given by

$$[\overline{u_5}] = \begin{cases} \left[0, f(\frac{a}{4})\right], \text{ if } \alpha \leqslant \overline{\alpha} \\ \\ \left[f(\frac{a}{4}), \frac{a}{4}\right], \text{ if } \alpha > \overline{\alpha} \end{cases}, \text{ for } \alpha \in [0, 1] \text{ and some } \overline{\alpha}.$$

- If $a = 4$, the only fixed points are $\widehat{0}, \widehat{x_a}$ and $\overline{u_6}$ with $[\overline{u_6}]^\alpha = [0, 1], \forall\alpha \in [0, 1]$.

The bifurcation diagram of the logistic function is presented. Although the deterministic branch presents fixed points and cycles, in the fuzzy branch we present only fixed points of f. The dashed and continuous lines indicate instability and stability respectively (Fig. 5).

11. **Attractors and asymptotic stability for fuzzy dynamical systems** (R. C. Bassanezi, L. C. de Barros and P. A. Tonelli)—*Fuzzy Sets and Systems 113 (2000), 473–483.*

We study the asymptotic properties of maps on fuzzy spaces that are extensions of maps on $\mathbb{R}$.

Definition 1 Let $f : T \times \mathbb{R}^n \longrightarrow \mathbb{R}^n$ be a map. For each fixed t we write $f_t(x) = f(t, x)$. Zadeh's extension of $f(t, x)$ is defined as:

$$\widehat{f}(t, u)(x) = \begin{cases} \sup_{f(t,z)=x} u(z), \text{ if } f_t^{-1}(x) \neq \phi \\ 0, \text{ if } f_t^{-1}(x) = \phi \end{cases}. \tag{11}$$

Definition 2 A discrete fuzzy dynamical system is an iterative system of the form

$$u_{n+1} = F(u_n), \tag{12}$$

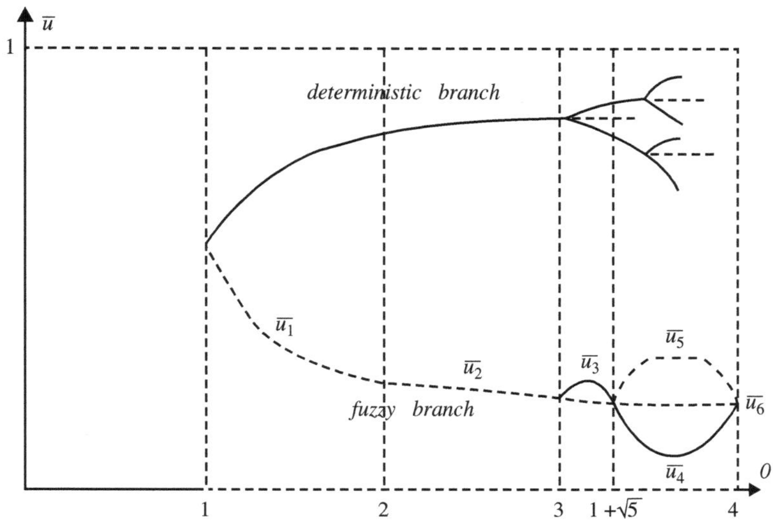

Fig. 5 Bifurcation diagram of the logistic fuzzy equation

where $F : \mathscr{F}(\mathbb{R}^n) \longrightarrow \mathscr{F}(\mathbb{R}^n)$ is a function. Given that $u_0 \in \mathscr{F}(\mathbb{R}^n)$, the sequence of elements $u_0, F(u_0), F(F(u_0)), \ldots$ is called the positive orbit of Eq. (12) from u_o and $F^n(u_o)$ denotes the n times composition of F.

$$x_{n+1} = \widehat{f}(u_n) \tag{13}$$

is the fuzzy system associated with the deterministic system

$$x_{n+1} = f(x_n). \tag{14}$$

Definition 3 Let $F : \mathscr{F}(\mathbb{R}^n) \longrightarrow \mathscr{F}(\mathbb{R}^n)$ a map. A point $\overline{u} \in \mathscr{F}(\mathbb{R}^n)$ is called a fixed point of F if $F(\overline{u}) = \overline{u}$.

Theorem 7 *Let $f : \mathbb{R}^n \longrightarrow \mathbb{R}^n$ be continuous with $f(\overline{x}) = \overline{x}$ and $\widehat{f}$ the Zadeh's extension of f. Then*

(a) $\chi_{\{\overline{x}\}}$ is stable for the system (13), if and only if, $\overline{x}$ is stable for the system (14),
(b) $\chi_{\{\overline{x}\}}$ is asymptotically stable for the system (13) if, and only if, $\overline{x}$ is asymptotically stable for the system (14).

Theorem 8 *Let $\widehat{f} : \mathscr{F}(\mathbb{R}^n) \longrightarrow \mathscr{F}(\mathbb{R}^n)$ Zadeh's extension of a continuous function f. If $\widehat{f}(\overline{u}) = \overline{u}$ and $\lim_{n\to\infty} D(\widehat{f}^n(u, \overline{u})) = 0$ for $D(u, \overline{u}) < r$, then the levels $[\overline{u}]^\alpha$ attract the levels $[u]^\alpha$ by f.*

Theorem 9 *If f and $\widehat{f}$ are as above, $\overline{u}$ is a fixed point of $\widehat{f}$, asymptotically stable with a diameter $diam\,[\overline{u}]^{\alpha} < r$, then $\overline{u}$ is the characteristic function of some point in $\mathbb{R}^n$; moreover, if $\overline{u}$ is globally asymptotically stable, then $\overline{u}$ is the characteristic function of some point in $\mathbb{R}^n$.*

We also study the stability of the new fixed points for Zadeh's extension logistic function $\widehat{f}(u_n) = au_n(1 - u_n)$.

The various types of uncertainties that appear in the evolutionary processes can have highly varied modeling. The fuzzy variational equations have been studied using distinct methods. A way of contemplating subjectivities of the nonrandom type in the state variables of a system is through problems of initial fuzzy values, denoted for

$$\begin{cases} \dfrac{dx}{dt} = F(x(t)) \\ \\ x(0) = x_o \in \mathcal{F}(U) \end{cases} .$$

The fuzzy continuous dynamic systems and the stability of their equilibrium points were analyzed in the thesis of M. Mizukoshi: "Sistemas dinâmicos fuzzy"; (2004) and M. Cecconello: "Sistemas dinâmicos em espaços métricos fuzzy—Aplicações em Biomatemática" (2010), whose main results are presented below in brief:

12. **Fuzzy differential equations and the extension principle** (M. Mizukoshi, L. Barros, Y. Chalco C., H. Roman F. and R.C. Bassanezi)— *Information Science (2007), p. 3627–3635*; *Stability of Fuzzy Dynamic Systems* (M. Mizukoshi, R. Bassanezi and L. Barros); *Int. Journal of Uncertainty, Fuzziness and Knowledge-Based Systems (2008), v. 17, n 1, p. 69–83. World Scientific Pu. Co.*; *On Topological Equivalence of Fuzzy Flows near Hyperbolic Equilibria* (M. T. Mizukoshi, L. C. Barros, R. C. Bassanezi and A. J. V. Brandão); *Fuzzy Sets and Systems, Volume 189, Issue 1, 16 (2012), p. 92–100*; *On the stability of fuzzy dynamical systems* (M. S. Cecconello, R. C. Bassanezi, A. J. V. Brandão and J. Leite); *Fuzzy Sets and Systems, DOI: 10.1016/j.fss.2013.12.009*; *About Projections of Solutions for Fuzzy Differential Equations* (M. S. Cecconello, J. Leite, R. C. Bassanezi and J. de Deus M. Silva); *Journal of Applied Mathematics 06/2013; 2013.*

The purpose of these works is to study fuzzy dynamical systems associated with deterministic systems. Zadeh's extension principle is used to obtain the fuzzy flow from the deterministic system. The Grobman–Hartman theorem states that, near hyperbolic equilibria, there exists a homeomorphism between the trajectories of the nonlinear system and those of the corresponding linearized system. That is, these systems are topologically equivalent. A similar theorem to Grobman–Hartman theorem to fuzzy flows is the main result in this article. It states that the fuzzy flows obtained from each system—the nonlinear and the linearized—are topologically equivalent.

We study the Cauchy problem for differential equations, considering its parameters and/or initial conditions given by fuzzy sets. These fuzzy differential equations are approached in two different ways: (a) by using a family of differential inclusions; and (b) Zadeh's extension principle for the solution of the model. We conclude that the solutions to the Cauchy problem obtained by both are the same.

Considering the initial value problem given by the autonomous equation

$$\begin{cases} \frac{dx}{dt} = f(t, x(t), w) \\ x_0 = x(0), \ \ w \in \mathbb{R}^n \end{cases}, \tag{15}$$

where $f : \mathbb{R}^n \longrightarrow \mathbb{R}^n$. It is such that the solution (15) exists and is unique.

The fuzzy associated system, denoted by

$$\begin{cases} \widehat{\frac{dx}{dt}} = \widehat{f}(t, \widehat{x}(t), \widehat{w}) \\ X_0 = \widehat{x_0}, \ \ \widehat{w} \in \mathscr{F}(\mathbb{R}^n) \end{cases}, \tag{16}$$

is defined and its solution $\widehat{x}(t)$ is given by Zadeh's extension of the solution $x(t)$ of the system (15).

Theorem 10 *Let $\overline{x}$ be a hyperbolic equilibrium point of (15) satisfying the hypotheses of the Grobman–Hartman theorem. Then, there exists a neighborhood B of $\chi_{\{\overline{x}\}}$ and a homeomorphism $\widehat{h} : B \longrightarrow B$ such that*

$$\widehat{h}(\widehat{\varphi}_t(X_0)) = \widehat{\psi}_t(\widehat{h}(X_o)),$$

where $\widehat{\varphi}_t(X_0)$ is the fuzzy flow of (16) and $\widehat{\psi}_t$ is the fuzzy flow of the associated linear system

$$\begin{cases} \frac{dy}{dt} = Df(\overline{x})y \\ y(0) = y_0 \end{cases}, \tag{17}$$

for all $t \geq 0$ with $Df(\overline{x})$, the Jacobian matrix of f around the equilibrium point $\overline{x}$.

Theorem 11 *Let x be a hyperbolic equilibrium point of the deterministic system (15). Then, $\chi_{\{\overline{x}\}}$is an equilibrium point for the fuzzy system associated with (16) and*

(i) If $\chi_{\{\overline{x}\}}$ is unstable for (15), then it is also unstable for (16);
(ii) If $\chi_{\{\overline{x}\}}$ is asymptotically stable for (15), then it is also asymptotically stable for (16).

Theorem 12 *Let $x_e : A \to U$ be continuous, $A \subseteq U$, $x_o \in F(U)$ with $[x_o]^0 \subset A$ and $x_e = \widehat{x}_e(x_o)$ (Figs. 6, 7, 8, 9 and 10).*

Under these conditions:

(i) If $\varphi_t(x_e(x)) = x_e(x)$ for every $x \in A$ then $\widehat{\varphi}_t(x_e) = x_e$ for every $t \geq 0$.

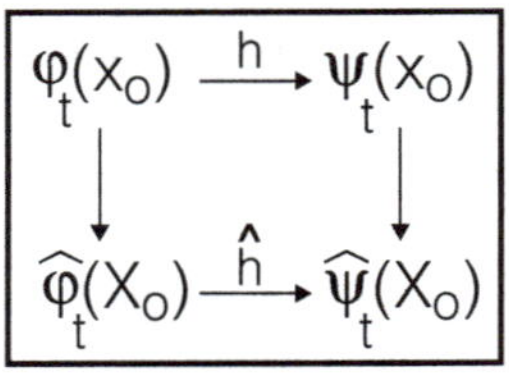

Fig. 6 Commutative diagram

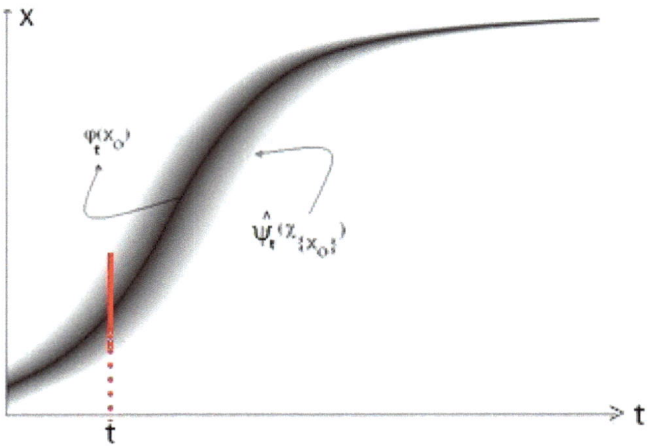

Fig. 7 Fuzzy and deterministic solutions

(ii) If $\varphi_t : U \to U$ *converges, uniformly in A, to* $x_e : A \to U$ *as* $t \to \infty$ *then* $\widehat{\varphi}_t(x_0)$ *converges to* x_e *and* $\widehat{\varphi}_t(x_e) = x_e$ *for every* $t \in R^+$.

In other words, the theorem affirms that if the deterministic solution $\varphi_t(x_o)$ converges uniformly to the function $x_e(x_o)$, so that Zadeh's extension $\widehat{\varphi}_t(x_0)$ converges to Zadeh's extension $\widehat{\varphi}_t(x_e)$.

If we consider that the subjectivity of a phenomenon, modeled by the system (15), appears in the w parameters of the function f, then we should apply Zadeh's extension to the flow of the deterministic system:

$$\begin{cases} \frac{dx}{dt} = f(x(t), w) \\ \frac{dw}{dt} = 0 \\ x_0 = x(0, w) \in \mathbb{R}^{n+m}, \quad w \in \mathbb{R}^m \end{cases} . \tag{18}$$

We study the projection of the fuzzy solution onto the phase space. By that projection, we may identify the behavior of each component of the fuzzy solution as a function of time.

Then, the existence, unicity, and stability are guaranteed by $n + m$ dimensional systems.

Example 1 Let the logistics be the fuzzy system in which the growth rate is a fuzzy number:

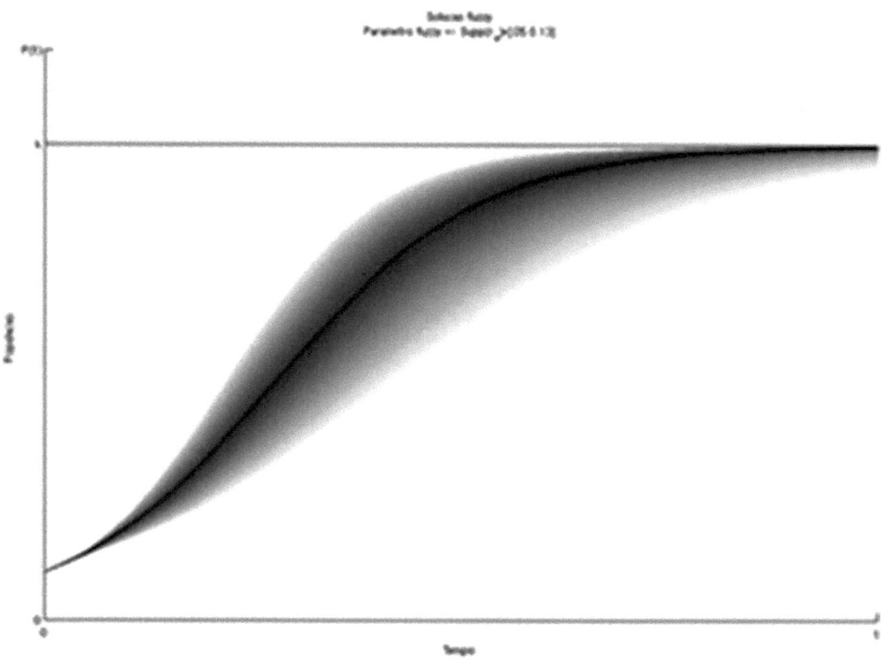

Fig. 8 Projection of the fuzzy solution of (19) on plane (t, x)

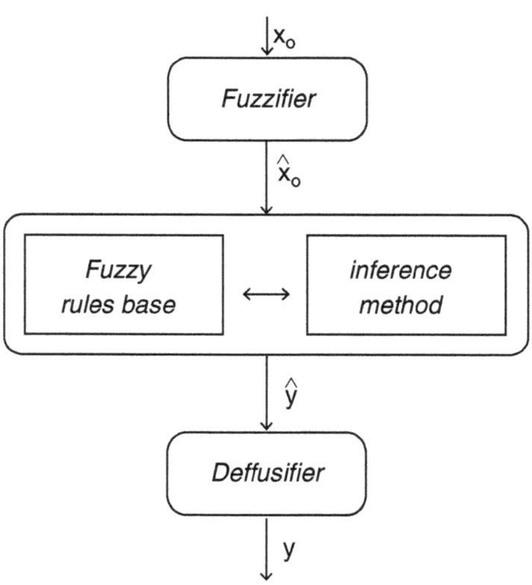

Fig. 9 Architecture of a fuzzy rule-based system

$$\begin{cases} \frac{dx}{dt} = rx(1-x) \\ x_0 \in \mathbb{R} \text{ and } r \in \mathscr{F}(\mathbb{R}) \end{cases} . \tag{19}$$

The solution of this equation is Zadeh's extension of the initial value problem associated with the deterministic two-dimensional system (20):

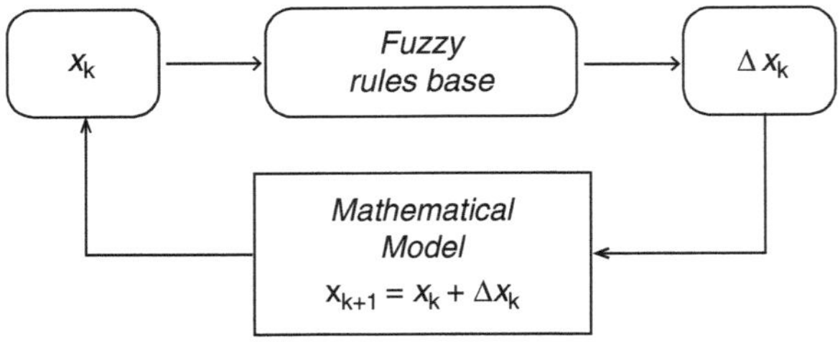

Fig. 10 Architecture of a p-fuzzy system

$$\begin{cases} \frac{dx}{dt} = rx(1-x) \\ \frac{dr}{dt} = 0 \\ X_0 = (x_0, r) \end{cases} . \tag{20}$$

When the convergence of a deterministic solution at the equilibrium point depends on the initial condition or parameters, it is possible to show that the fuzzy solution converges to a fuzzy equilibrium point, and we show how to determine the membership function of such an equilibrium point.

In summary, we have studied the case where the fuzzy systems are derived from deterministic systems, considering only the initial condition to be a fuzzy set. However, fuzzy autonomous systems with the parameters and uncertain initial condition can be treated as a system in which all the fuzziness is contained in the initial condition.

Theorem 13 *Let $A \subset U \times P$ be an open set, $y_e : A \to U \times P$ a continuous function and $y_o \in \mathscr{F}(U \times P)$ with $[y_o]^0 \subset A$. We are given that $\varphi_t(x_o, p_0) \to x_e(x_o, p_e)$ for every $(x_o, p_o) \in A$ as $t \to \infty$. If $x_e(x_o, p)$ is asymptotically stable, then $\widehat{\psi_t}(y_o)$ converges to the equilibrium point $y_e = \widehat{y_e}(y_o)$.*

The interpretation of fuzzy differential equations through Zadeh's extension principle captures the rich properties of stability.

We verified that the linearization captures the local dynamics of nonlinear systems, that is, if $\overline{x}$ is a crisp hyperbolic equilibrium point, there is a fuzzy neighborhood of $\chi_{\{x\}}$ in which the fuzzy systems (16) and (17) are topologically equivalent.

13. **Periodic orbits for fuzzy flows** (M. Cecconelo, R. Bassanezi, A. Brandão and J. Leite)—*Fuzzy Sets and Systems, 230, Nov/2013, p. 21–38. A special number of FSS: Differential Equations over Fuzzy Spaces—Theory, Applications and Algorithms.*

Periodic solutions are present in many of the various mathematical models that describe physical, chemical or biological phenomena. In this work, we investigate the existence of periodic solutions for problems of fuzzy initial values. We show that fuzzy solutions can present periodic points and develop tools of qualitative analysis for such solutions.

Let $\varphi_t : U \to U$ be the solution to the deterministic equation

$$\begin{cases} \frac{dx}{dt} = f(t, x(t), w) \\ x_0 = x(0), w \in \mathbb{R}^n \end{cases}, \tag{21}$$

and we suppose that $\varphi_t(x_o)$ is defined for all $x_o \in U$ and $t \in \mathbb{R}_+$. The solutions to autonomous differential equations, as they are considered here, satisfy the properties of a dynamic system on the open set $U \subset \mathbb{R}^n$, i.e., for all $x_o \in U$ and $t, s \in \mathbb{R}_+$ we have:

$$\varphi_0(x_o) = xo; \ \varphi_{t+s}(xo) = \varphi_t(\varphi_s(xo)).$$

For each $x_o \in U$, the subset of the phase space defined by $\gamma(xo) = \{\varphi_t(xo) \in U : t \geq 0\}$ is called the orbit or trajectory of the point x_o by the solution φ_t.

The $\omega -$ limit of a subset $B \subset U$, is defined by

$$\omega(B) = \cap_{s \geqslant 0} \overline{\cup_{t \geqslant s} \varphi_t(B)}.$$

We say that $p \in U$ is a *periodic point of period* τ, or, a $\tau -$ periodic point, for the flow φ_t when it exists $\tau > 0$ such that $\varphi_t = p$ and $\varphi_t \neq p$ for all $t < \tau$.

Theorem 14 *Let M be compact and invariant. Thus, M is asymptotically stable if, and only if, M is a uniform attractor.*

Corollary 1 *Let $\gamma \in U$ be a periodic orbit asymptotically stable and $A(\gamma) \subset U$ its attraction region. Therefore, given $\varepsilon > 0$ and a compact $K \subset A(\gamma)$ there exists $T(K, \varepsilon) > 0$ such that $dist(\varphi_t(x_o), \gamma) < \varepsilon$ for all $x_o \in K$ and $t > T(K, \varepsilon)$.*

The *fuzzy orbit* $\gamma(\mathbf{x}_o) \subset \mathscr{F}(U)$ of an initial state $\mathbf{x}_o \in \mathscr{F}(U)$ is defined as being a subset in the phase space $\mathscr{F}(U)$ defined by

$$\gamma(\mathbf{x}_o) = \cup_{t \in \mathbb{R}_+} \widehat{\varphi}_t(\mathbf{x}_o) = \{\widehat{\varphi}_t(\mathbf{x}_o) \in \mathscr{F}(U) : t \in \mathbb{R}_+\}.$$

For each $B \subset \mathscr{F}(U)$, the set $\omega -$ fuzzy limit is defined by

$$\omega(B) = \cap_{s \geqslant 0} \overline{\cup_{t \geqslant s} \widehat{\varphi}_t(B)}.$$

$\mathbf{p} \in \mathscr{F}(U)$ is a $\tau -$ periodic point for $\widehat{\varphi}_t$ when

$$\widehat{\varphi}_t(\mathbf{p}) = \mathbf{p} \text{ and } \widehat{\varphi}_t(\mathbf{p}) \neq \mathbf{p}, \ \text{ if } t \in (0, \tau).$$

We can also characterize the periodic points of the fuzzy flow $\widehat{\varphi}_t$ by its α-*levels*.

The following result merely ensures the existence of invariant sets for the fuzzy flow when the deterministic solution has a periodic orbit.

Theorem 15 *A point $p \in U$ is $\tau -$ periodic for φ_t if, and only if, $\chi_{\{p\}}$ is a periodic point of period τ for $\widehat{\varphi}_t$.*

Proposition 1 *If γ is a deterministic periodic orbit, so that the fuzzy periodic set*

$$\gamma = \left\{ \mathbf{x} \in \mathscr{F}(U) : [\mathbf{x}]^0 \subset \gamma \right\}$$

is closed, limited, and invariant because of the fuzzy flow.

Theorem 16 *Let γ be a periodic orbit for φ_t with period $\tau > 0$ and $\boldsymbol{\gamma}$ the fuzzy periodic set determined by γ.*

Thus:

(i) γ is stable for φ_t if, and only if, $\boldsymbol{\gamma}$ is stable for $\widehat{\varphi_t}$;
(ii) γ is asymptotically stable for φ_t if, and only if, $\boldsymbol{\gamma}$ is asymptotically stable for $\widehat{\varphi_t}$.

Theorem 17 *Let γ be a deterministic periodic orbit that is asymptotically stable and $x_0 \in \mathscr{F}(U)$. If $[\mathbf{x}_0]^0 \subset A$, then $\omega(\mathbf{x}_0) \subset \boldsymbol{\gamma}$ is a fuzzy periodic orbit.*

To establish an analogous result to the Poincaré–Bendixson theorem in the fuzzy metric spaces $E(\mathbb{R}^2)$, we present two theorems.

14. **Poincaré–Bendixson theorem in fuzzy metric space** $E(\mathbb{R}^2)$ (M. Diniz, R. C. Bassanezi and M. Cecconello)—Sent to Fuzzy Sets and Systems (2015).

Theorem 18 *Let $K \subset \mathbb{R}^2$ be a compact and invariant set, x_e is the unique singular point of K and $\mathbf{x}_0 \in E(\mathbb{R}^2)$. If x_e is unstable, then there exists a region $A \subset K$ such that, for $[\mathbf{x}_0]^0 \subset A$, $\widehat{\varphi_t}(x_o)$ converges to a fuzzy periodic orbit $\boldsymbol{\gamma}$.*

Theorem 19 *Let $K \subset \mathbb{R}^2$ be a compact and invariant set and $x_0 \in K$. If $\varphi_t(x_o)$ does not approach any equilibrium point, then there exists a region $A \subset K$ such that for $[\mathbf{x}_0]^0 \subset A$, $\widehat{\varphi_t}(x_o)$ is either a fuzzy periodic orbit or converges to one fuzzy periodic orbit.*

In both theorems it is necessary that the closure of the support of the fuzzy initial value is contained in a given region A, where A is the region of attraction of a periodic orbit that is asymptotically stable or a subset of a stable periodic orbit.

15. **Mathematical Modelling: medical diagnosis**

The basic idea for a medical diagnosis is to relate symptoms or patients' signs to possible diseases according to an expert's medical knowledge. This application can be summarized in an input–output system:

Input(Symptoms) $\rightarrow$ Knowledge-Based System $\rightarrow$ Output (Diagnosis).

Let us consider the following universal sets: U = set of patients; V = set of symptoms; W = set of diseases.

We want to obtain a fuzzy relation D such that $S \circ D = T$, where S and T are the matricial forms of the fuzzy relations of symptoms and patients defined in $U \times V$ and $U \times W$ respectively.

The knowledge base is composed of the fuzzy relations S and T where the matrix of the relation S is given by patients and their symptoms and T is the matrix related to diagnostic pattern.

The matrix of the relation $D = S^{-1} \otimes_g T$ gives the symptoms and diagnoses. Each element of the relation D indicates the degree of connection of each symptom with the diseases under consideration.

Using these arguments of the fuzzy logic some interesting works were realized:

"Relaciones fuzzy: optimizacion de diagnostico médico", Anais do Encontro Nacional de Ecologia (1989); R. C. Bassanezi and H. Roman-Flores.

"Construção e avaliação de um modelo matemático para predizer câncer de próstata e descrever seu crescimento utilizando a teoria dos conjuntos fuzzy", Tese de Doutorado (2005). M. J. P. Castanho.

"Software desenvolvido a partir de um modelo matemático fuzzy para predizer o estágio patológico do câncer de próstata", Biomatemática 18 (2008), 27–36. G. P. Silveira, L. L. Vendite, and L. C. Barros.

"Postoperative vomiting in pediatric oncologic patients: prediction by a fuzzy logic model", Pediatric Anesthesia, 2012. B. S. B. Bassanezi1, A. G. de Oliveira-Filho, R. S. M. Jafelice, J. Bustorff-Silva, and A. Uldesmann.

"Diagnosis of Incidence Risk of Cardiovascular Diseases", Master's Dissertation, 2015. L. Bassani.

The variational systems can also be given by means of a rule-based system, in which case the systems are called *p-fuzzy*. Such rules are usually provided by a specialist when modeling of a particular phenomenon is desired.

Conditions of existence of p-fuzzy systems were obtained in the thesis of João de Deus M. Silva. In the next work, we present new concepts and techniques for modeling using systems based on fuzzy rules. We enunciate and prove theorems that ensure the existence of a stationary point for each equilibrium viable set of the p-fuzzy system.

16. Stationary points: I. One-dimensional P-fuzzy Dynamic Systems (J. de Deus M. Silva, J. Leite, R. C. Bassanezi, and M. S. Cecconello)— *Journal of Applied Mathematics 09/2013, 2013.*

P-Fuzzy dynamical systems are variational systems whose dynamic is obtained by means of a Mamdani-type fuzzy rule-based system. In this paper, we show the $1-$dimensional p-fuzzy dynamical systems and present theorems that establish the conditions of existence and uniqueness of stationary points. Besides the analytical results obtained, we present examples that illustrate and confirm the mathematical results.

The efficiency of a deterministic model depends on knowledge of the relationships between variables and their variations. Moreover, in many situations, such relations are only partially known; therefore, the modeling with deterministic variational systems, or even with stochastic ones, may not be adequate. In addition, fuzzy systems derived from deterministic models, which have subjectivity regarding some parameters, are not appropriate when we have only incomplete information of the phenomenon being analyzed. Thus, the use of a rule-based system can be adopted as an alternative to modeling partially known phenomena or those carried out with imprecision. Fuzzy rule-based systems have been used with success in some areas such as control, decision-taking, recognition systems, etc. This success is because of its simplicity and interrelation with human ways of reasoning. Fuzzy rule-based systems are conceptually simple. Such systems are basically threefold: an input (fuzzifier), an inference mechanism composed of a base of fuzzy rules together with an inference method, and, finally, an output (defuzzifier) stage.

Formally, a p-fuzzy system in $\mathbb{R}^n$ is a discrete dynamic system:

$$\begin{cases} x_{k+1} = F(x_k) \\ x_0 \text{ given and } x_k \in \mathbb{R}^n \end{cases}$$

where the F function is given by $F(x_k) = x_k + \Delta x_k$ and $\Delta x_k \in \mathbb{R}^n$ is obtained by means of a fuzzy rule-based system; that is, Δx_k is the defuzzification value of the rule-based system.

Theorem 20 (Existence) *Let S be a p-fuzzy system and A^* an equilibrium viable set of S of the type $(A_i, A_{i+1}) \to (C, B)$. Then, there is at least one stationary point of S in A^*. That is, $\exists x^* \in A^*$ such that $\Delta x^* = 0$.*

Theorem 21 (Uniqueness) *Let S be a p-fuzzy system and A^* an equilibrium viable set of S of the type $(A_i, A_{i+1}) \to (C, B)$.*

If the pertinence functions μ_{A_i} and $\mu_{A_{i+1}}$ are piecewise monotone and $A^* \subset [z_1, z_2]$, then there exists only one stationary point in A^*.

Theorem 22 (Uniqueness) *Let S be a p-fuzzy system and A^* an equilibrium viable set of S of the type $(A_i, A_{i+1}) \to (C, B)$.*

If the pertinence functions μ_{A_i}, $\mu_{A_{i+1}}$, μ_B and μ_C are continuously differentiable, μ_{A_i}, $\mu_{A_{i+1}}$ are piecewise monotone and μ_C are strictly monotone, such that:

(i) $\mu_C(t) \leqslant \mu_B(-t), \forall t \in (0, a)$.

(ii) $\dfrac{\mu'_C(q)}{\mu'_B(p)} > \left(\dfrac{p}{q}\right)^3, \forall p \in supp(B), \forall q \in supp(C)$ *and* $\mu_B(p) < \mu_C(q)$; $\mu_{A_i}(x) \neq \mu_{A_{i+1}}(x)$.

(iii) $\dfrac{\mu'_{A_i}(x)}{\mu'_{A_{i+1}}(x)} \leqslant 0, \forall x \in (c_1, z_0)$.

Then, S has only one stationary point, x^* in A^* and $x^* \in (c_1, z_0)$ (Fig. 11).

Corollary 2 *Let S be a p-fuzzy system and A^* an equilibrium viable set of S of the type $(A_i, A_{i+1}) \to (C, B)$.*

If the pertinence functions μ_{A_i}, $\mu_{A_{i+1}}$, μ_B and μ_C are triangular fuzzy numbers, then S has only one stationary point in A^*.

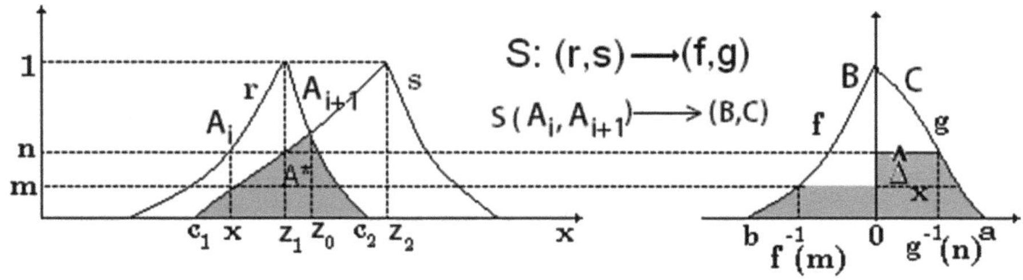

Fig. 11 Mamdani's inference process

Just as we did for the one-dimensional case, we apply the theorem 21 to ensure the uniqueness of the solution. To ensure the uniqueness of the solution, it is important to know what the sufficient conditions are for a p-fuzzy system to be Lipschitzian. Such conditions are described in the theorem 23.

Theorem 23 *Let* $Fp : U \subset \mathbb{R}^2 \rightarrow V \subset \mathbb{R}^2$, *the output function of a two dimensional p-fuzzy system S, with well-posed rules. If all the membership functions of the input fuzzy sets are Lipschitzian and all the membership functions of the output fuzzy sets are Lipschitzian and have inverse Lipschitzian, then the p-fuzzy system is Lipschitzian.*

Predator-Prey p-Fuzzy Model In this example, we present a p-fuzzy model type predator–prey. We can represent the rules by the diagram of arrows in Fig. 12. In this system, there are two input variables, and each of these variables can take five different fuzzy state values. Using similar reasoning to the previous example, it is easy to see that this p-fuzzy system is a well-posed two-dimensional p-fuzzy system (Fig. 13).

Other recent and unpublished results were obtained in relation to the existence and uniqueness of the solutions of p-fuzzy systems *Existence and uniqueness for continuous p-fuzzy systems* (2014) (M. M. Diniz and R.C. Bassanezi).

The p-fuzzy systems have also been used to model the diffusion process using a rule base:

17. **P-fuzzy diffusion equation using rules base** (J. Leite, R. C. Bassanezi, Jaqueline Leite and M. Cecconello)—*Journal of Applied Mathematics, Vol. 2014, Article ID 478241.*

The p-fuzzy systems incorporate subjective information in both variables as the variations and their relationships with the variables and is therefore a very useful tool for modeling phenomena whose behavior is partially known. The fuzzy systems are generally the result of a generalization of the classical systems, i.e., in this approach the uncertain concepts are incorporated into these systems. A central feature of fuzzy systems is that they are based on the concept of fuzzy partition information. The use of fuzzy sets allows a generalization of information that is associated with the introduction of imprecision ignoring the phenomena. In essence,

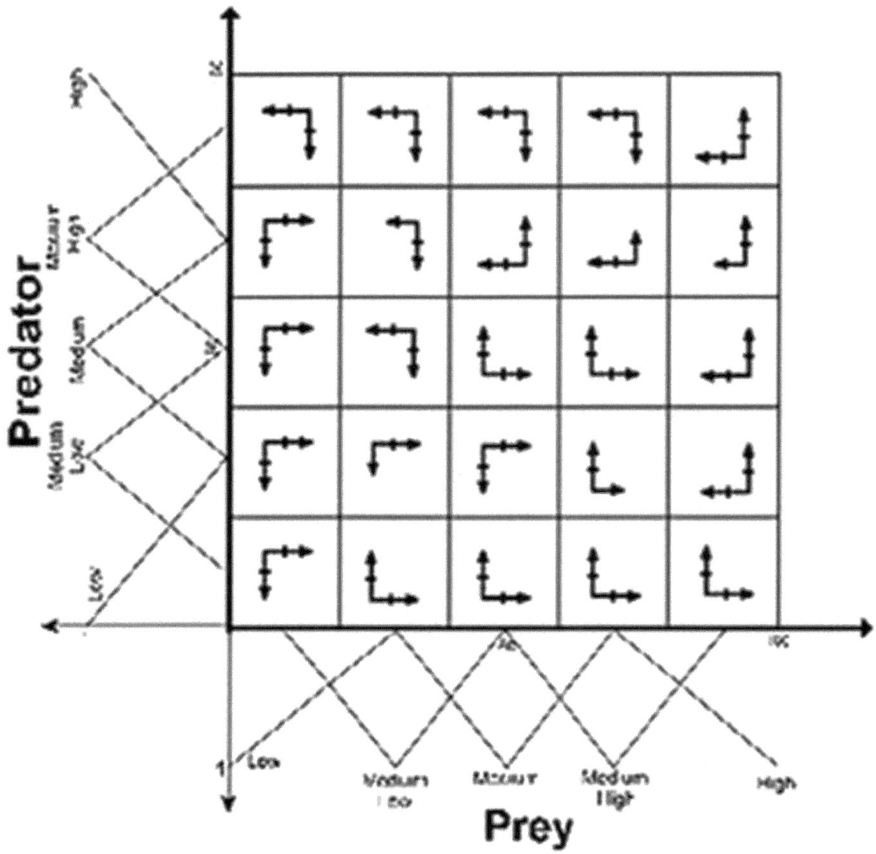

Fig. 12 Diagram of the vector fields of a two-dimensional p-fuzzy system

the representation of information in fuzzy systems tries to imitate the process of human reasoning, considering heuristic knowledge and information across the disconnected principle. In this work, we describe a diffusive process without the use of their analytical solution, using p-fuzzy dynamical systems and given a rule base. It is worth noting that the results obtained in terms of solutions are very similar to the deterministic case.

The rule base is a set consisting of fuzzy rules that relate the linguistic terms of the input variables and output variables. The rule base is considered an element, a member of the fuzzy controller core. Each rule base satisfies the following structure:

$$\text{IF } a \text{ is in } A_i \text{ THEN } b \text{ is in } B_i,$$

where A_i and B_i are fuzzy sets that represent linguistic terms for input variables and output variables respectively.

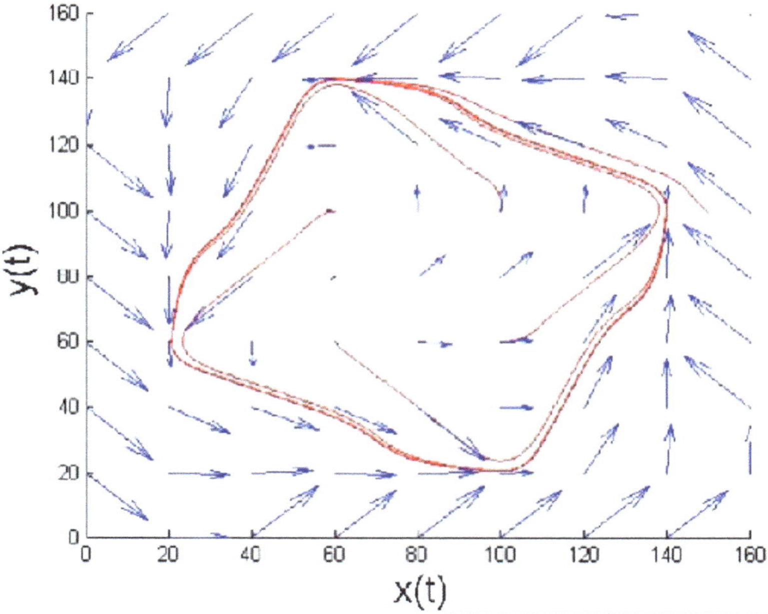

Fig. 13 Trajectory in phase plane

Thus, consider how linguistic variables for the position of the population (distance to origin): *low positive* (Bp), *mean positive* (Mp), *mean high positive* (MAp), *positive high* (Ap), *low negative* (Bn), *mean negative* (Mn), *mean high negative* (MAn), and *negative high* (An), where the positive terms or negative mean distance from the origin to the right or left respectively.

Considering the known results about the diffusion process, consider the following basis for fuzzy rules:

(a) If the position of the individuals is *low positive* Bp, then the variation of the population is *low positive* Bp.
(b) If the position of the individuals is *positive average* Mp, then the variation of the population is *positive average* Mp.
(c) If the position of the individuals is *average high positive* MAp, then the variation of the population is *average high positive* MAp.
(d) If the position of the individuals is *high positive* Ap, then the variation in the population is *high positive* Ap.
(e) If the position of the individuals is *low negative* Bn, then the variation in the population is *low negative* Bn.
(f) If the position of the individuals is *average negative* Mn, then the variation of the population is *average negative* Mn.
(g) If the position of the individuals is *average high negative* MAn, then the variation of the population is *average high negative* MAn.

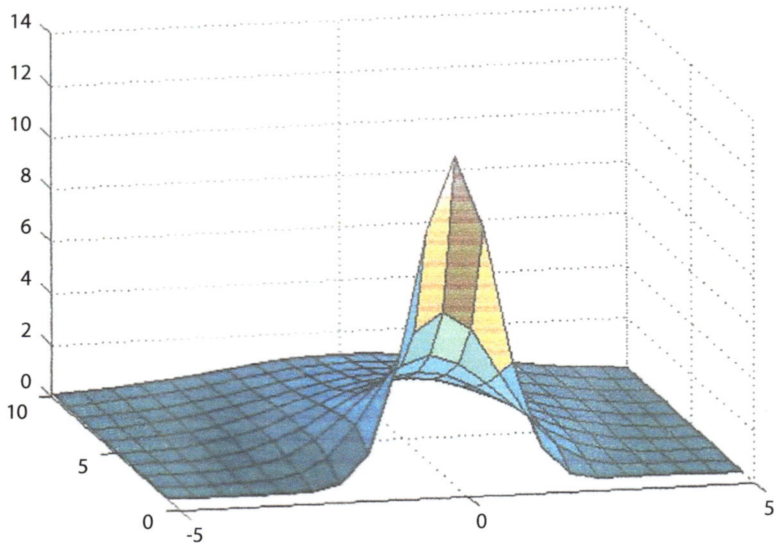

Fig. 14 Union of p-fuzzy solutions

(h) If the position of the individuals is *high negative An*, then the variation of population is *high negative An* (Fig. 14).

18. On Fuzzy Solutions for Diffusion Equation (J. Leite, M. Cecconello, Jac. Leite, and R. C. Bassanezi)—*Journal of Applied Mathematics Volume 2015, Article ID 874931, 10 pages.*

Our main objective in this paper is to explore properties such as uniqueness and stability of the fuzzy solution of a fuzzy differential equation associated with a classical advection–diffusion–reaction equation using Zadeh's extension.

Diffusion models have been extensively employed to investigate dispersal and have yielded considerable insight into the dynamics of animal movement in space and time. Diffusion models can be written in the simplest form as:

$$\frac{\partial u}{\partial t} = D\nabla^2 u + f\left(u, \frac{\partial u}{\partial x}\right) \tag{22}$$

where the operator ∇ denotes the spatial gradient, t is time, $u(x, y, t)$ is the local population density in the spatial variables x and y; D is the coefficient of diffusion, and $f(u, \frac{\partial u}{\partial x})$ is the reaction–advection term describing the net population change due to birth, death, and direction of travel. Most of the phenomena involving diffusion are described by models that suggest a dynamic in $\mathbb{R}$ and $\mathbb{R}^2$ is what will be treated below. Consider the initial value problem given by

$$\begin{cases} \frac{\partial u}{\partial t}(x,t) = D\Delta(u(x,t)), \ x \in \mathbb{R}, \ t \geqslant 0 \\ u(x,0) = N\delta(x), \ x \in \mathbb{R}, \ t = 0 \end{cases} \tag{23}$$

where, $u : U \subset \mathbb{R} \to \mathbb{R}$ is a real function.

The classical solution to problem (23) exists, is unique for values of x in bounded domains $U \subset \mathbb{R}$, and is given by

$$u(x,t) = \frac{N_o}{\sqrt{4\pi Dt}} e^{-\frac{x^2}{4Dt}}.$$

If incorporating into (23), the parameters of reaction and advection have a problem of the form

$$\frac{\partial u}{\partial t}(x,t) = D\Delta(u(x,t)) + a\frac{\partial u}{\partial x}(x,t) + bu(x,t) \tag{24}$$

whose solution is given by

$$u(x,t) = \frac{u_o}{\sqrt{4\pi Dt}} e^{-\frac{(x-at)^2}{4Dt+bt}}.$$

If we know that the phenomenon occurs by diffusion, but its initial condition is not well determined, we can consider the initial condition as a fuzzy number. Thereafter. the principle of Zadeh's extension in the initial condition applies. Thus, we have the solution to the fuzzy initial value problems:

$$\widehat{u}(x,t) = \frac{\widehat{u}_o}{\sqrt{4\pi Dt}} e^{-\frac{(x-at)^2}{4Dt+bt}}.$$

We also develop an alternative method for treatment parameters, which may involve some uncertainty; for this, the fuzzy initial condition and setting conditions for the fuzzy solution are unique and ensure that these solutions are stable (Fig. 15).

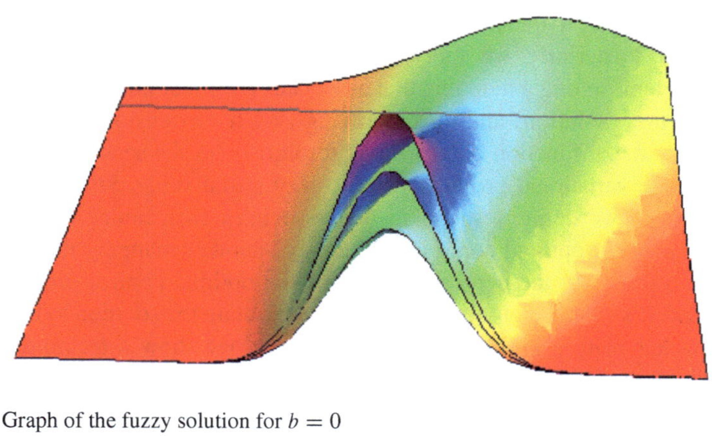

Fig. 15 Graph of the fuzzy solution for $b = 0$

Still using the Zadeh extension principle in EDP other interesting works were developed as: *"On Fuzzy Solutions for Partial Differential Equations," Fuzzy Sets and Systems, 2012*; A. M. Bertone, R. M. Jafelice, L. Barros, and R. C. Bassanezi.

The fuzzy logic provided alternative forms of mathematical modeling. Systems-based rules, cellular automata, and fuzzy sets of type 2 were some of these arguments applied in biomathematics models:

"On Fuzzy control of Soybean Aphid" (2016); (M. Peixoto, R. C. Bassanezi, L. C. Barros, and O. A. Fernandes);

"A Fuzzy delay approach for HIV dynamics using a cellular automaton". Journal of Applied Mathematics, V. 25, ID 378753 (2015); (R. Motta Jafelice, C. A. F. Silva, L. C. Barros, and R. C. Bassanezi);

"A Study on Subjectivities of Type 1 and 2 in Parameters of Differential Equations". TEMA—Tendências em Matemática Aplicada e Computacional, 16, N. 1 (2015); (R. M. Jafelice, A. M. Bertone and R. C. Bassanezi).

The most important problems in Biomathematics arise from situations that require some type of control. Classical mathematics provides tools for control decisions, especially optimal control when one has a series of information that allows deterministic modeling. However, the uncertainties inherent in biological phenomena stimulate thinking in the sense of seeking some instrumentation to obtain a more adequate *optimal control* as in the thesis of Michael Diniz:

19. Optimization of functions, functional, and fuzzy control, IMECC (2016) (Michael Diniz)

In this thesis, we study an optimization process that allows optimal control to be established when subjectivity or a lack of information is present in the variables or in the experimental data. We study optimization of real functions with fuzzy parameters, variational problems with fuzzy boundary values, and optimal control with fuzzy initial values. Using ordinal optimization, we study the minimization of real functions with fuzzy parameters. We verify that, under some hypotheses, it is possible to establish mapping that associates each parameter value to the minimizer of the function; thus, we show that Zadeh's extension of this mapping, in fact, is a minimizer, according to the notion of the smallest element on the established partial order relation. Posteriorly, we apply a similar reasoning for the fuzzy variational calculus problem, setting a mapping that associates each real initial value with each optimal function (solution) and thus, applying Zadeh's extension to this map, we prove that the fuzzy function obtained is the solution (in the sense of the smallest element) of the variational problem with a fuzzy initial value. Still following the same reasoning, we define a mapping that associates each initial value with the state solution of the optimal control problem, and another mapping that associates each initial value with the control solution of the optimal control problem; thus, applying Zadeh's extension to these mappings, we obtain the solution to the optimal control problem with a fuzzy initial value. Finally, we show a technique to find an approximate solution for the control problem whose states, controls, and time variables are given by fuzzy numbers. For this, we build a fuzzy grid and apply a dynamic programming algorithm to find the rules. As a result, we obtained a fuzzy rule-based system whose inputs are the states and the time and the output is the best control (decision) to be applied.

We focus on cases where the objective *functional* has as an image the set of fuzzy numbers, and therefore we use the concept of a smaller (greater) element, to define the minimizer/maximizer of the functional being evaluated.

Definition 4 (The Biggest Element) Let P be a partially ordered set. An element $g \in P$ is the largest element of P, if $\forall a \in P, a \preceq g$. In an analogous way, we can set the smallest element.

Definition 5 Let (M, d) be a metric space and P be a partially ordered set according to the order relationship $\preceq_P$. Consider a function $f : M \to P$ and $x^* \in \Omega \subset M$. We say that x^* is a local minimizer of f in Ω when there exists a $\delta > 0$, such that $f(x^*) \preceq_P f(x)$, for all $x \in B(x^*, \delta)$; x^* is called the global minimizer of f in Ω when $f(x^*) \preceq_P f(x)$, for all $x \in \Omega$.

Consider the following optimal control problem:

$$\begin{aligned} &\min_{u \in U} J(u) = \int_{t_0}^{t_f} F(t, x, u) dt \\ &\widehat{x}(t_0) = \widehat{x}_0 \in \mathscr{F}\left(\left[x_0^L, x_0^R\right]\right), \; x(t_f) = x_f \end{aligned} \quad . \tag{25}$$

We define the solution of the optimal control problem with an initial fuzzy condition as Zadeh's extension of the solution in relation to the initial condition $x(t_0)$. In addition, we study conditions in the optimal control problem that ensure that the objective functional image is a fuzzy number when the initial condition is a fuzzy number.

Theorem 24 *Let the optimal control problem given in (25), such that $F(t, x, u)$ is continuous with respect to all variables, f is of class C^1, and the (global) solution of the classical problem, $x^*(t, x_0) \in u^*(t, x_0)$, be continuous in relation to $x_0 \in \left[x_0^L, x_0^R\right]$. Then, $\widehat{x^*}(t, x_0) \in \widehat{u^*}(t, \widehat{x}_0)$ form the (global) solution to the problem (25) at $B_F^*(\widehat{u^*}(t, x_0), \in)$ for all $\widehat{x}_0 \in \mathscr{F}\left(\left[x_0^L, x_0^R\right]\right)$.*

Finally, we present an approximate solution to the optimal control problem whose state, control, and time variables are defined by fuzzy numbers. For this, we construct a fuzzy net and apply the algorithm of dynamic programming, or define the rules of a Mamdani-type controller. As a result, we obtain a system based on fuzzy rules that, at each established state and time, assigns the best decision (control) to be applied.

Phase Field: A Methodology to Model Complex Material Behavior

José Luiz Boldrini

Abstract This work was done in commemoration of the 50th anniversary of the inauguration of the Institute of Mathematics, Statistics and Scientific Computation of the University of Campinas, Brazil (Instituto de Matemática, Estatística e Computação Científica da Universidade Estadual de Campinas). Our objective is just to give a rather fast introduction to some important modeling aspects of the phase field approach to model complex material behavior; we aim at students of mathematics who have almost no previous background in continuum thermomechanics. Thus, we briefly recall some of its main concepts and explain the main approaches used to derive the governing equations including the phase field variables (diffusification, energetic variational, and entropy approaches); we comment on some of their limitations and relationships, and briefly describe a few simple applications.

1 Introduction

Commemorating the 50th anniversary of the inauguration of the Institute of Mathematics, Statistics and Scientific Computation of the University of Campinas, Brazil (Instituto de Matemática, Estatística e Computação Científica da Universidade Estadual de Campinas), we present here a rather fast introduction to some modeling aspects of the important phase field methodology when used to derive the equations governing complex material behavior. Specifically, we consider situations where structures and interfaces may appear and evolve in time in a material.

We stress that modeling and analyzing such situations are not easy tasks since such structures and interfaces may interact in a complex and nonlinear way with the material properties; moreover, their appearances, shapes, and positions are not a priori known and must be determined together with the other physically relevant variables.

J. L. Boldrini (✉)
Institute of Mathematics, Statistics and Scientific Computing, University of Campinas, Campinas, SP, Brazil
e-mail: boldrini@ime.unicamp.br

C. Lavor, F. A. M. Gomes (eds.), *Advances in Mathematics and Applications*,
https://doi.org/10.1007/978-3-319-94015-1_4

In this work, we want to clarify the role of the phase field approach in modeling situations as just delineated. For this, we start by describing the more traditional sharp-interface methodology, briefly explaining some of its difficulties. Next, we describe in general terms the diffuse-interface (phase field) methodology and contrast it with the sharp-interface approach, explaining how phase fields deal with the pointed difficulties.

The Sharp-Interface Methodology We exemplify this approach by considering an old and famous problem studied in the late nineteenth century by J. Stefan. He analyzed the temperature distribution and freezing-front history of a solidifying slab of water, having as a basic assumption that the freezing-front was sharp; that is, it was a regular surface (actually a planar surface in the original Stefan's problem) with solid water at one of its side and liquid water at the other side. Along the time, Stefan's assumption has been applied to more general situations and different problems, leading to mathematical problems nowadays called sharp-interface models. For instance, we consider the following slight generalization of the original Stefan's problem, taken from Rubinstein [104], see also Alexiades and Solomon [1] where the reader can find more details. Consider a material that may assume either of two phases, e.g., solid or liquid, and occupies a spatial region $\Omega \subset \mathbf{R}^n$ separated at an instant t by an interface $\Gamma(t)$. Let $T_m \in \mathbf{R}$ be the melting temperature at equilibrium, i.e., the temperature at which both phases may coexist in equilibrium separated by just an interface assumed to be planar for simplicity. The temperature $T(\mathbf{x}, t)$ must then satisfy a heat diffusion equation in each side of the interface:

$$\rho C_v T_t = \operatorname{div}(K \nabla T) \qquad \text{in} \quad \Omega \setminus \Gamma(t). \tag{1}$$

Here, C_v is the specific heat, K is the thermal conductivity, and ρ is the mass density. For simplicity of exposition, we assume that either $C_v = C_v^s > 0$ or $C_v = C_v^l > 0$, respectively, on the solid and liquid part of $\Omega \setminus \Gamma(t)$ with constant $C_v^s 0$ and $C_v^l 0$; similarly, $K = K^s > 0$ or $K = C_v^l > 0$, respectively, on the solid and liquid part of $\Omega \setminus \Gamma(t)$, with constant K^s and K^l; $\rho > 0$ is the same constant for both liquid and solid phases.

Moreover, the interface must be at the melting temperature, and the rate of change of the latent heat equals the amount by which the heat flux jumps across the interface. These lead to the following conditions at the interface:

$$\begin{aligned} T &= T_m && \text{on } \Gamma(t), \\ \ell \mathbf{v} &= -\left[K \nabla T \cdot \mathbf{n}\right]_-^+ && \text{on } \Gamma(t), \end{aligned} \tag{2}$$

where ℓ is the latent heat, $\mathbf{v}$ is the (normal) velocity to the interface $\Gamma(t)$, $\mathbf{n}$ is the unit normal at $\Gamma(t)$, and $[\cdot]_-^+$ denotes the jump in the quantity as one crosses the interface from solid to liquid. Thus, the sharp-interface problem is stated as finding T and Γ subject to (1), (2) and suitable initial and boundary conditions.

This sharp-interface approach can be used in many other physical situations, leading, as we can see from the previous example, to free-boundary problems. We remark that such problems are in general very difficult to analyze, both from the theoretical and numerical point of views, for the reasons we explain in the following.

By thinking a little about the fact that the equation for the motion of the interface (2) is a key ingredient of sharp-interface models, one quickly sees some complications.

First, from the physical point of view, it is not in general easy to incorporate the effects of several physical phenomena that may be relevant to realistic analysis (for instance, supercooling and superheating effects, surface tension effects, and so on); even when this is done, it is not clear whether it was done in a thermodynamically consistent way.

Second, from the geometrical point of view, the very formulation of the equation for the motion of the interface requires the existence of the normal $\mathbf{n}$ to the surface (see (2)); thus, this approach requires at least a certain regularity of the interface, preventing the possibility of directly describing the formation of kinks, cusps, branching, contact, coalescence, dendrites, and other complex geometric behaviors that may occur during the evolution of such interfaces. In the sharp-interface methodology, therefore, these possibilities must be approached in an ad hoc and sometimes unclear way.

Third, and again from the physical point of view, in several situations the basic hypothesis of this methodology, that is, that transitions are abrupt, is not correct. For instance, in problems involving solidification/melting, there is the possibility of occurrence of extended transitions (mushy) zones between pure solid and pure liquid phases, where a mixture of solid and liquid materials predominate.

Due to all these difficulties, rigorous mathematical analyses of sharp-interface models are in general very difficult to perform; see, for instance, Rubinstein [104], Cannon et al. [28, 29], DiBenedetto and Friedman [43], and DiBenedetto and O'Leary [44]. Moreover, the geometrical difficulties of sharp-interface models translate into similar ones found in numerical simulations, requiring the numerical tracking of possible complex evolving interfaces (front-tracking), which is a very demanding and difficult computational task.

The Diffuse-Interface (Phase Field) Methodology The previously described complications motivated the introduction of another modeling methodology, in which sharp interfaces are replaced by continuous variations that are measured by a new auxiliary variable (sometimes more than one new variable). This new variable is called either a phase field or an order parameter or a kinetic descriptor, depending on the context of the problem being considered; in the present work, we just use the generic name phase field. The key idea in this approach is that the interfaces are in fact diffuse transitions layers instead of sharp fronts and that the position of such layers is specified by the level sets of the phase fields considered in the problem. Due to these characteristics, this approach is also called the diffuse-interface methodology.

To illustrate these ideas, we mention two historical articles. The first phase field model was originally developed in 1958 by Cahn and Hilliard in [27] to describe the process of phase separation of two fluids. For this, those authors developed a fourth-order nonlinear partial differential equation, presently known as the classical Cahn–Hilliard equation, for a variable $u(x, t)$ (the phase field) related the continuous concentration function of one of the fluids in their mixture. Such variable had the range of their values given by the interval $-1 \leq u(x, t) \leq 1$, and the region where $u = 1$ indicated the region occupied by one of the fluids, while the region where $u = -1$ indicated the region occupied by the other fluid; the fluids were then separated by a transition region defined by a diffuse interface associated to the region where $-1 < u < 1$. In 1972, Allen and Cahn in [2] developed a second-order nonlinear partial differential equation, which is presently known as the classical Allen–Cahn equation, describing the phase separation in iron alloys. Both of those articles used the Ginzburg–Landau free-energy functional; however, as we will explain later on, the Cahn–Hilliard equation is conservative, while the Allen–Cahn equation is the nonconservative.

The phase field approach has several advantages over the sharp-interface approach as we explain in the following.

First, from the physical point of view, although it is not so for every phase field model one can find in the literature, by using the entropy approach to be explained in detail in Sect. 5 and following a rather standard argumentation scheme, one can derive phase field models that are automatically thermodynamically consistent even in complex situations. To explain this claim is the objective of the present work, but we advance here the main argumentation steps. In a first step, one chooses the physical fields that are relevant to the problem under consideration and also the phase fields to be used to describe the possible structures and interfaces (transition layers); at this point, one also chooses whether each phase field will be considered as an internal or a dynamical variable (we will give details on these aspects later on). In a second step, one obtains the general forms of the dynamical equations (the equations governing the time evolution of the physical fields and the phase fields that were considered as physical variables); for this, one uses the standard balance laws of mass, momentum, and energy (one uses the principle of virtual powers instead of the balance of momentum when there are dynamical phase fields), and also other physical laws (like Maxwell's equation, and so on) as required by the physical variables. In a third step, one uses the concepts of free-energy and of pseudo-potential of dissipation, the principle of entropy, and general dynamical equations obtained in the previous step to get the general forms of the constitutive relations in terms of free-energy and of pseudo-potential of dissipation. Finally, in a fourth step, one chooses the specific forms of the free-energy and of the pseudo-potential of dissipation that are adequate to the situation and material at consideration; once this is done, the mathematical model is determined and automatically thermodynamically consistent. Obviously, it is not easy to complete this argumentation scheme in proper way, and there are points that require careful studies of the particular situation in order to choose in a physically sound way the free-energy and of the pseudo-potential of dissipation. However, at least there is a general approach to obtain consistent models; in contrast, the inclusion of complex phenomena in a physically consistent way is much more difficult in sharp-interface models.

Second, from the geometrical point of view, since transition layers are localized by specific level sets of the phase field, they may present kinks, cups, intersections, coalescences, and so on; thus, they are suitable for describing very complex evolving geometries. Moreover, the evolution of such complex geometries is automatically done in a physically sound way since the equations obtained with the phase field methodology hold even for these complex geometries; this is in contrast with the sharp-interface methodology where the introduction of ad hoc (and unclear) criteria are necessary to proceed with the evolution of complex geometries.

Third, by its very concept, the phase field methodology can easily handle extended transition layers.

Phase fields are thus key ingredients of a successful modeling strategy for situations involving appearance and evolutions of several kinds of interfaces and may be used to model the appearance, evolution, and interaction of structures in macro-, meso-, and microscales in problems involving phase transitions, membranes, damage in materials, bubbles, growth of tissues, and so on. Moreover, from the point of view of numerical simulations, phase field methods can be thought as physically consistent level sets methods, and the evolution of complex interfaces geometries can be obtained rather easily. Interesting examples of this successful approach can be seen, for instance, in the numerical simulations of the growth of dendritic patterns in solidification processes in Kobayashi [76], Karma and Rappel [72], and Nestler et al. [87]. Thus, it is safe to say that nowadays the phase field method has emerged as a powerful tool in the task of understanding material behavior.

On the other hand, we must also draw the reader's attention to the fact that the use of a particular phase field model in a practical situation requires realistic values for the physical parameters appearing in its description; however, these values are not easy to achieve, requiring suitable laboratory tests and other kinds of analyses. Obviously, the practical use of a sharp-interface model also requires the knowledge of the values of its own parameters; but, since this is an older and traditional modeling approach, presently there is more laboratory technology and data to estimate these parameters. Nonetheless, by using asymptotic analyses, it is possible to associate phase field models to corresponding sharp-interface models, relating in this way also their respective parameters; such relations can then be used to estimate the phase field parameters from the known associated sharp-interface ones. Therefore, the study of the relationship between phase field and sharp-interface models via asymptotic analyses is an important subject that has been considered along the years; examples of articles on this subject are Caginalp [23], Caginalp and Xie [26], and Colli and Sprekels [36, 37].

Finally, despite their importance, we stress that in this work we do not comment on rigorous mathematical or numerical analyses of phase field models, neither do we comment on practical aspects of their numerical simulations; the references mentioned in the next section deal with these aspects, and the interested reader may consult them and their bibliographies. As we have already said, our objective here is just to give a rather fast introduction to important modeling aspects of the phase field approach; this could serve to mathematical students who have almost no previous background in continuum thermomechanics but are interested in this field of study.

For this, we recall some basic physical concepts and explain the main approaches used to derive the governing equations (diffusification, energetic variational, and entropy approaches), commenting on some of their limitations and relationships.

The outline of this work is as follows. Section 2 gives some references for more information on the aspects that we left out; Sect. 3 very briefly comments on the diffusification approach; Sect. 4 deals with the main ideas used in the energetic variational approach; and Sect. 5 explains the entropy approach.

2 Some Representative References

Due to its flexibility and usefulness, presently there are many hundreds of scientific articles dealing with the phase field methodology; thus, it is impossible to present here all the relevant works concerning this approach and comment on their results. Therefore, we drastically reduce our scope, mentioning only a few references that, we think, may represent some aspects of the approach. We leave to the reader the task of consulting their bibliographies for much further information. Some references we mention are concerned with the physical derivation of the mathematical models; others strive for rigorous mathematical analyses of such models and deal with the questions of existence or qualitative properties of solutions of the model equations; others else propose and analyze numerical methods for the approximation of such solutions; some others are more concerned with the practical implementation and numerical simulations or model validation. Since in the present work we are just focusing in the modeling aspects, we do not explicit comment on the results of each of those references but just group them by their main application areas.

First of all, we mention that very interesting general references are Provatas and Elder [96], Fremond [56, 57], Emmerich [51], and Gomez and van der Zee [66].

Turning to references concerned with specific application areas, since the original work of Cahn and Hilliard [27], many authors considered the interaction among different fluids using phase fields. Some articles dealing with this topic are Anderson et al. [5], Kim [75], Liu and Shen [80], Feireisl et al. [54], Cao and Gal [31], Vasconcelos et al. [110], Eleuteri et al. [49, 50], and Dai et al. [42].

Many other articles also used the phase field approach to study solidification/melting of materials or crystal growth processes. Fix [55] seems to be the first one to do this; many other authors followed, studying several phase transitions problems: some of them are Collins and Levine [40], Caginalp [22, 24], Kobayashi [76], Caginalp and Jones [25], Karma and Rappel [72], Nestler et al. [87, 88], and Provatas et al. [97]. Among the many papers considering solidification of alloys, we mention Warren and Boettinger [115], Wheeler et al. [117], Boldrini and Planas [16], Frémond and Rocca [59] Vaz and Boldrini [111], Boldrini et al. [11], and Calsavara Caretta and Boldrini [30]. Some articles also included in the model the influence of the macroscopic motions of the material (in particular, the convection in the melt); a few of them are Blanc et al. [9], Beckermann et al. [8], Diepers et al. [45], Rappaz and Scheid [98], Boldrini and Vaz [18], Scheid [105], Planas and Boldrini [94, 95], Boldrini and Planas [17], and Rocca and Rossi [99]. We mention that a particular approach that has been used to model phase transitions employ the thermodynamic potential known as enthalpy (H-method); this can be seen as a particular case of phase field since its values determine the material phases. Some articles using this particular approach are Voller and Prakash [112], Voller et al. [113], Peicleous et al. [91], O'Leary [90], and Boldrini et al. [15].

Another kind of fluid–structure interaction may be found in articles studying the motion of membranes (vesicles) immersed in fluids; see, for instance, Du et al. [46, 47], and Entringer and Boldrini [52].

The phase field approach has also been used to study the interplay among elasticity, plasticity, phase change, damage, fatigue, and fracture of materials. Examples of references doing this are Frémond and Nedjar [58], Frémond [56, 57], Nedjar [86], Rocca and Rossi [100], Heinemann and Kraus [67], Heinemann and Rocca [68], Duda et al. [48], Bonetti et al. [20], Miehe et al. [81, 82], Ambati et al. [3, 4], Boldrini et al. [10], and Nguyen et al. [89].

Phase field models taking in consideration the second principle of thermodynamics can be found, for example, in Penrose and Fife [92, 93], Zheng [119], Wang et al. [114], Sprekels and Zheng [107], Laurençot [77], Colli and Laurençot [34], Colli and Sprekels [39], Kenmochi and Kubo [74], Ito and Kenmochi [70], Ito et al. [71], Fabrizio et al. [53], Assunção and Boldrini [7], and Boldrini et al. [10].

Besides the important question concerning the relationship between phase field and sharp-interface model, which was already mentioned in the Introduction, there are many other interesting aspects that must be considered. For instance, asymptotic properties are studied in Kenmochi and Niezgódka [73], Miranville and Zelik [84], Rocca and Schimperna [101, 102], Gal et al. [61], da Silva and Boldrini [41], and Gal and Grasselli [62, 63]. Memory effects are important in some situations; some articles considering this aspect are Colli and Sprekels [38], Colli et al. [32], Bonetti et al. [19], and Frémond [56, 57]. Control problems related to phase field models can also be considered; for instance, Hoffman and Jiang [69], Boldrini et al. [12], Rocca and Sprekels [103], Colli et al. [33, 35], Frigeri et al. [60], and Araruna et al. [6]. Finally, besides those already mentioned articles, we also refer to the following interesting ones: Moroşanu [85], Miranville and Quintanilla [83], Wells et al. [116], Gomez and Hughes [65], and Guillén-González and Tierra [64].

3 Diffusification Approach

In the beginning of their historical development, diffuse-interface (phase field) models were usually thought not as physical models per se, but just as convenient approximations (regularizations) of sharp-interface models, to be used just as way to avoid the difficulties with the front-tracking of the sharp interfaces in numerical simulations.

Following this idea of regularization, a diffuse-interface model is then derived from a previously given sharp-interface model by introducing a smooth field, the phase field $\varphi(\mathbf{x}, t)$, where $\mathbf{x}$ denotes the points in the spatial domain Ω and t, the time. This field is seen as a regularization of the jump appearing at a sharp interface by a smooth profile; the commonly used profile is the one given by the hyperbolic tangent, that is, φ is taken as $\varphi(\mathbf{x}, t) := \tanh(d_t(\mathbf{x})/\sqrt{2\epsilon})$, where $d_t(\mathbf{x})$ denotes the signed distance from $\mathbf{x}$ to the sharp interface, and $\epsilon > 0$ is a parameter related to the thickness of the corresponding approximate diffuse interface. We stress that this field is designed to attribute value $\varphi = -1$ to one of the phases, value $\varphi = 1$ to the other (for instance, respectively, liquid and solid phases in the example described in the Introduction); the intermediate value $0 < \varphi < 1$ is related to the transition region between the two pure phases.

The next step in the arguments is to look for expressions for the geometrical entities appearing in the equation for the motion of the interface in terms of φ. More precisely, by using the case of Eq. (2) to exemplify these ideas, $\mathbf{v}$ and $\mathbf{n}$ must be written in terms of φ and maybe its temporal and spatial derivatives (we remark that, if we had not taken a planar interface in that example, the curvature of the sharp interface would also appear in (2) and the curvature should also be written in terms of φ and its temporal and spatial derivatives). Once these expressions are found, they are substituted back in the equation for the motion of the interface, (2) in our example; this leads to a partial differential equation for φ that is assumed to be the equation governing the evolution of the phase field not just near the interface, but also in all the domain Ω. This equation replaces (2) in the associated diffuse-interface model; we do not write this equation here and refer to Gomez and van der Zee [66] where a more general situation is discussed.

The next step is to obtain a unique equation in the diffuse-interface model that corresponds to Eq. (1); such equation should depend also on φ and hold on all the domain Ω; this contrasts with the sharp-interface model in which we have different equations each one holding in one side of the sharp interface. To obtain the required equation, one possibility is to take one with the form as in (1), but with coefficients C_v and K defined on all the domain Ω and with values smoothly varying from one phase to the other; for instance, by taking the averages $C_v = \frac{(1-\varphi)}{2} C_v^l + \frac{(1+\varphi)}{2} C_v^s$ and $K = \frac{(1-\varphi)}{2} K^l + \frac{(1+\varphi)}{2} K^s$.

Once the previous steps are accomplished, one gets a system of equations coupling the temperature and the phase field. However, as one can observe, this approach is difficult to generalize to more complex situations, and it requires certain choices that sometimes are difficult to justify; moreover, there is no systematic way to verify the thermodynamical consistency of the diffuse-interface models derived by this method. Therefore, we do not give here more details of this approach and refer to the very interesting article by Gomez and van der Zee [66] for further discussion on it.

On the other hand, some of the ideas used in the diffusification approach are relevant for the two approaches to be described in the next sections. In fact, both of them use the key concept of free-energy density, which must be expressed in terms of the phase field. In particular, certain terms of free-energies densities, like surfaces energies, that are rather well-known in the case of the sharp-interface models, are rewritten in terms of the phase field variable with the help of some of the previous ideas.

4 The Energetic Variational Approach

4.1 Phase Field Equations (Isothermal Processes)

We firstly recall certain basic ideas of the continuum mechanics, which here are described in **Eulerian coordinates**. Suppose that a certain physical process occurs in a domain $\Omega \subset R^n$, $n = 1, 2, 3$, on an interval of time $[0, T]$; for simplicity, we focus in just a scalar physical variable, for instance, mass, electric charge, or energy, which is distributed in Ω. To describe the changing of such variable, three basic concepts are required: the **density of the physical variable** of interest, $\rho : \Omega \times [0, T] \rightarrow R$; the **density of sources or sinks of the physical variable**, $g : \Omega \times [0, T] \rightarrow R$; and the **flux of that physical variable**, $\mathbf{F} : \Omega \times [0, T] \rightarrow R^n$.

Many of the equations from classical continuum mechanics are derived by relating the previous three concepts by using the physical principle known as **law of balance**; this says that at any subregion $\mathcal{V}$ of Ω, the rate of change of the total amount of the physical variable in $\mathcal{V}$ is equal to sum of the amount generated (or consumed) in $\mathcal{V}$ by the sources and sinks and the amount left in $\mathcal{V}$ by the flux that crosses its boundary. In mathematical terms, under suitable smoothness conditions on the previous fields and the use of the divergence theorem, the law of balance is written as the **general (scalar) balance equation in integral form:**

$$\frac{d}{dt} \int_{\mathcal{V}} \rho \, d\mathbf{x} = - \int_{\partial \mathcal{V}} \mathbf{F} \cdot \mathbf{n} \, dS + \int_{\mathcal{V}} g \, d\mathbf{x}, \tag{3}$$

for any (suitable) $\mathcal{V} \subset \Omega$. Here, $\mathbf{n}$ denotes the external unitary normal field on $\partial\Omega$, and dS denotes the area element. By assuming enough regularity to use the divergence theorem and using the fact that $\mathcal{V} \subset \Omega$ is "arbitrary," we get the **general (scalar) balance equation in differential (local) form:**

$$\partial_t \rho + \operatorname{div} \mathbf{F} = g. \tag{4}$$

Thus, in any particular physical situation, to complete the derivation of the equation governing the phenomenon of interest, we must find the right expressions for g and $\mathbf{F}$. Examples of fluxes are the advection flux ($\mathbf{F} = \rho\mathbf{v}$, where $\mathbf{v}$ is a velocity field of the material) and the diffusion flux ($\mathbf{F} = -k\nabla\rho$, where $k \geq 0$ is a diffusion coefficient).

Inspired in the previous arguments, the **energetic variational approach** proposes the following modified form of the balance equation (4) (in Eulerian coordinates) as the evolution equation for a phase field φ:

$$\partial_t \varphi + \operatorname{div}(\varphi \mathbf{v}) = L(\varphi) + g. \tag{5}$$

Here, $\mathbf{v}$ is the macroscopic velocity of the material, and thus an advection flux is included in the equation because it is natural to assume that the structures determined by φ may be advected by the velocity flow; g is a source term, whose expression depends on the situation being considered; and $L(\varphi)$ denotes a maybe nonlinear operator that must be determined by further arguments. For simplicity, we will look for an expression for $L(\varphi)$ in a situation without (macroscopic) motion of the material ($\mathbf{v} = 0$), without sources or sinks ($g = 0$) (recall that all other physical variables, including temperature, are kept constant); in this situation, the phase field equation is reduced to

$$\partial_t \varphi = L(\varphi).$$

The basis of the energetic variational approach to determine suitable expressions for the operator $L(\varphi)$ is the assumption that the evolution in time of φ must occur such way that the **total free-energy** of the physical system under investigation does not increase with time. Let us initially apply this idea in the case that the total free-energy is expressed in the following form, which depends only on the derivatives up to first order of the phase field variable:

$$\mathscr{E} = \int_\Omega E(\varphi, \nabla\varphi) d\mathbf{x}. \tag{6}$$

By denoting $p_i = \partial_i \varphi$ and writing: $E(\varphi, \nabla\varphi) = E(\varphi, p_1, \ldots, p_n)$, and also assuming enough smoothness, we obtain $\frac{d\mathscr{E}}{dt} = \int_\Omega \partial_\varphi E(\cdot)\, \partial_t \varphi + \partial_{p_i} E(\cdot)\, \partial_t \partial_i \varphi \; d\mathbf{x}$, where we used the usual Einstein's index notation that repeated index must be added up. By using integration by parts with suitable boundary condition on φ (either $\varphi = 0$ or $\partial\varphi/\partial\nu = 0$ on $\partial\Omega$), we get

$$\frac{d\mathscr{E}}{dt} = \int_\Omega \frac{\delta E}{\delta \varphi}\, \partial_t \varphi \; d\mathbf{x} = \int_\Omega \frac{\delta E}{\delta \varphi} L(\varphi)\; d\mathbf{x}. \tag{7}$$

Here, $\frac{\delta E}{\delta \varphi}$ is called the **variational derivative** and is given by

$$\frac{\delta E}{\delta \varphi} = \partial_\varphi E - \partial_i \partial_{p_i} E. \tag{8}$$

The previous arguments can be easily generalized for functionals depending on higher order derivatives of φ. For instance, suppose that the free energy has the following form:

$$\mathscr{E} = \int_\Omega E(\varphi, \nabla\varphi, \Delta\varphi) d\mathbf{x}. \tag{9}$$

By proceeding exactly as before, with suitable boundary conditions, we get again expression (7), but now with the variational derivative given by

$$\frac{\delta E}{\delta \varphi}(\cdot) = \partial_\varphi E(\cdot) - \partial_i \partial_{p_i} E(\cdot) + \Delta \partial_w E(\cdot), \tag{10}$$

where we used the same notations as before and also $w = \Delta\varphi$.

Expression (7) suggests that a first possibility to guarantee that the free-energy does not increase in time is to take

$$L(\varphi) = -\lambda(\cdot)\frac{\delta E}{\delta \varphi},$$

where $\lambda > 0$ is a coefficient called the **relaxation factor**. In fact, in this case, we have the **dissipative energy law** expressed by $\frac{d\mathcal{E}}{dt} = -\int_\Omega \lambda(\frac{\delta E}{\delta\varphi})^2 d\mathbf{x} \leq 0$. Thus, under the stated conditions, the total free-energy is automatically a Lyapunov functional, and we expect that as the time t goes to infinity $\varphi(\cdot, t)$ approaches an equilibrium state given by the equation: $\frac{\delta E}{\delta\varphi} = 0$, which is exactly the **Euler–Lagrange equation** for the critical points of the total free-energy functional $\mathcal{E}$.

Another possibility to guarantee the decay of the total free-energy is to take

$$L(\varphi) = \mathrm{div}\left(M\nabla\frac{\delta E}{\delta \varphi}\right),$$

where $M > 0$ is now a coefficient called **mobility**. By using this in (7), with the help of integration by parts and the use of suitable boundary conditions (either $\frac{\delta E}{\delta\varphi} = 0$ or $M\frac{\partial}{\partial\nu}(\frac{\delta E}{\delta\varphi}) = 0$ on $\partial\Omega$), we obtain another **dissipative energy law** expressed by $\frac{d\mathcal{E}}{dt} = -\int_\Omega M|\nabla\frac{\delta E}{\delta\varphi}|^2 \; d\mathbf{x} \leq 0$. Thus, similarly as before, the total free-energy is automatically a Lyapunov functional.

By using these previous expressions for $L(\varphi)$ in the general situation (5), that is, when $\mathbf{v}$ and g are not necessarily null, we obtain the following possibilities for the equation governing the evolution of the phase field:

$$\textbf{Allen-Cahn}: \; \partial_t\varphi + \mathrm{div}(\varphi\mathbf{v}) = -\lambda\frac{\delta E}{\delta\varphi} + g, \tag{11}$$

$$\textbf{Cahn-Hilliard}: \; \partial_t\varphi + \mathrm{div}(\varphi\mathbf{v}) = \mathrm{div}(M\nabla\frac{\delta E}{\delta\varphi}) + g. \tag{12}$$

Remarks

(i) In the Cahn–Hilliard equation, $\mu(\varphi) = \frac{\delta E}{\delta\varphi}(\varphi)$ is called **chemical potential**.

(ii) The Cahn–Hilliard equation is said to be **conservative** because, with the boundary conditions $\mathbf{v} = 0$ and $M\frac{\partial}{\partial\nu}(\frac{\delta E}{\delta\varphi}) = 0$ on $\partial\Omega$ and source term $g = 0$, by integration on Ω we formally obtain that the **"total mass"** $\int_\Omega \varphi(\cdot, t) \; d\mathbf{x}$ is constant in time. This does not hold for the previous Allen–Cahn equation;

so it is said to be **nonconservative**. However, there is a conservative modified form of the Allen–Cahn: $\partial_t\varphi + \mathrm{div}(\varphi\mathbf{v}) = -\lambda\frac{\delta E}{\delta\varphi} + g + \frac{1}{|\Omega|}(\int_\Omega(\lambda\frac{\delta E}{\delta\varphi} - g)d\mathbf{x})$ ($|\Omega|$ denotes the volume of Ω), with the boundary conditions $\mathbf{v} = 0$ on $\partial\Omega$; since its numerical treatment is simpler than that of the Cahn–Hilliard equation, which involves fourth-order differential operators, some authors prefer to use this modified Allen–Cahn; see, for instance, Yang et al. [118].

An Example: Solidification/Melting at a Given Temperature

We consider a solidification/melting isothermic process of a pure material, assuming that a given θ constant temperature, that macroscopic velocity is null ($\mathbf{v} = 0$), and that there are no heat sources ($g = 0$). Now, we consider the phase field φ variable (an order parameter) such that associates values $\varphi \leq -1$ to pure solid state, $\varphi \geq 1$ to pure liquid states, and $-1 < \varphi < 1$ to the transition layers between solid and liquid state.

We assume that volumetric density of free-energy is of the form

$$E(\varphi, \nabla\varphi) = \frac{\gamma}{2}|\nabla\varphi|^2 + \frac{1}{\gamma}\mathscr{H}(\varphi) - \varphi\ell(\theta - \theta_m).$$

Here, the first term is related to the interfacial energy (it attributes more energy to regions where the gradient of φ is larger) and the $\gamma > 0$ is a constant related to the width of the transitions layers; the second term $\mathscr{H}(\varphi) = (1/4)[(\varphi^2 - 1)]^2$ is the classical two-well potential; thus, the first two terms in the last expression correspond to the classical Ginzburg–Landau free-energy. In the third term, $-\varphi\ell(\theta - \theta_m)$, the coefficient $\ell >$ is related to the latent heat of the material, while θ_m is the given melting temperature; this term $-\varphi\ell(\theta - \theta_m)$ expresses qualitative changes in the free-energy according the temperature. In fact, for $\theta = \theta_m$, the total bulk potential density $\mathscr{H}_\theta(\varphi) = \frac{1}{\gamma}\mathscr{H}(\varphi) - \varphi\ell(\theta - \theta_m)$ has two absolute minimum points at $\varphi_{m1} = -1$ (pure solid state) and at $\varphi_{m2} = 1$ (pure liquid state); for $\theta > \theta_m$, $\mathscr{H}_\theta$ has a single absolute minimum point at $\varphi_m \geq 1$ (pure liquid state), and for $\theta < \theta_m$, $\mathscr{H}_\theta$ has a single absolute minimum point at $\varphi_m \leq -1$ (pure solid state). See more physical details in Caginalp [22].

Under these conditions, by using (8), the Allen–Cahn equation (11) becomes

$$\partial_t\varphi = \lambda\gamma\Delta\varphi + \frac{\lambda}{\gamma}(\varphi - \varphi^3) + \lambda\ell(\theta - \theta_m).$$

4.2 *Phase Field Equation Coupled with the Equation for the Macroscopic Motion (Isothermal Processes)*

The question now is how to couple in proper way the phase field equations with the dynamical equations governing the motion of that same material.

To answer this, we need to recall the concept of **balance of linear momentum**. The important ideas are the following: (a) **linear momentum** is a vectorial physical variable whose density is given by the expression ρu, where ρ is the mass density and $\mathbf{v} = (v_1, \ldots, v_n)$, $n = 1, 2$, or 3, is the velocity; (b) each component ρv_i, $i = 1, \ldots, n$, of the linear momentum is advected by the velocity flow; that is, there is an advection flux of the form $\rho v_i u$ (this special case is called convection); and (c) the sources and sinks of linear momentum are the forces acting on the body.

Thus, applying the balance law, Eq. (3), to each i-th component of the linear momentum in an "arbitrary" subregion $\mathscr{V} \subset \Omega$, one obtains that $\frac{d}{dt}\int_{\mathscr{V}} \rho v_i dx = -\int_{\mathscr{V}} \rho v_i \mathbf{v} \cdot \mathbf{n} dS + \int_{\mathscr{V}} f_i dx$, where as before $\mathbf{n}$ denotes the unitary external normal at the boundary of $\mathscr{V}$, and f_i denotes the volumetric density i-the component of the total force $\int_{\mathscr{V}} \mathbf{f} dx$. In vectorial terms, we get:

$$\frac{d}{dt}\int_{\mathscr{V}} \rho \mathbf{v} d\mathbf{x} = -\int_{\partial\mathscr{V}} (\rho \mathbf{v} \otimes \mathbf{v}) \cdot \mathbf{n} dS + \int_{\mathscr{V}} \mathbf{f} d\mathbf{x},$$

where $\otimes$ denotes the tensorial product; in the present case, $\mathbf{v} \otimes \mathbf{v}$ is an $n \times n$ matrix whose (i, j)-element is given by $v_i v_j$.

The total force $\int_{\mathscr{V}} \mathbf{f} d\mathbf{x}$ is the sum of body forces, contact forces, and microscopic forces. **Body forces** are forces like gravity; when their volumetric density is given by a volumetric density field $\mathbf{f}_b$, the total body force acting on $\mathscr{V}$ is given by $\int_{\mathscr{V}} \mathbf{f}_b d\mathbf{x}$. **Contact forces** are forces that one part of the body acts on the other parts through their common boundary; they are obtained by using the concept of **Cauchy stress tensor** $\mathbf{T}_0 = [\mathbf{T}_{0,ij}]_{n\times n}$ of the material; the balance of angular moments requires that $\mathbf{T}_0$ be a symmetric tensor. The total contact force that the part $\Omega - \mathscr{V}$ of the body acts on $\mathscr{V}$ is known to be given by (Cauchy's Theorem) $\int_{\partial\Omega} \mathbf{T}_0 \cdot \mathbf{n} dS$. **Microscopic forces** are forces due to internal structures, in case that they exist. We assume that such forces are given by a volumetric density field $\mathbf{f}_{\text{micro}}$, whose expression will be related later on to the phase field variable that is used to describe such structures, and the total microscopic force acting on $\mathscr{V}$ is then given by $\int_{\mathscr{V}} \mathbf{f}_{\text{micro}} d\mathbf{x}$. Thus, $\int_{\mathscr{V}} \mathbf{f} d\mathbf{x} = \int_{\mathscr{V}} \mathbf{f}_b d\mathbf{x} + \int_{\partial\Omega} \mathbf{T}_0 \cdot \mathbf{n} dS + \int_{\mathscr{V}} \mathbf{f}_{\text{micro}} d\mathbf{x}$. By substituting this in the balance of linear momentum, using the divergence theorem and the fact that $\mathscr{V}$ is arbitrary, we obtain the differential form for the balance of linear momentum:

$$\partial_t(\rho \mathbf{v}) + \operatorname{div}(\rho \mathbf{v} \otimes \mathbf{v}) = \operatorname{div} \mathbf{T}_0 + \mathbf{f}_b + \mathbf{f}_{\text{micro}}. \tag{13}$$

We recall that the stress tensor $\mathbf{T}_0$ determines many of the main properties of the material, and that an expression of $\mathbf{T}_0$ in terms of other variables of the physical problem is called a **constitutive relation**. In Sect. 5, we describe a thermodynamical argument that gives general expressions for σ in terms of the free-energy and the pseudo-potential of dissipation.

Microscopic Forces in Terms of the Phase Field To find an expression for the microscopic forces $\mathbf{f}_{\text{micro}}$, we will use the following form of the Principle of Virtual Power which is adequate for the energetic variational approach that we are considering in this section. It says that at any time the power of the forces acting on any part of a material body subjected to any **virtual displacement** (and thus with corresponding **virtual velocity**) must equal the rate of variation of the total energy along the same virtual displacements. We also recall that virtual displacements are arbitrary displacements with the only requirement that they satisfy all restrictions that one might have for the motion (examples: rigid walls, incompressibility, etc.).

We remark that this principle in particular implies the balance law for the linear momentum. Another important remark is that, since the expression for the total energy may have several parts, by the application of Principle of Virtual Power, from each of these parts one gets a particular type of force. In particular, one of these parts of this total energy is the part of free-energy associated to the energy contribution due to the structure determined by φ. By assuming the simplifying hypothesis that the phase field φ does not appear in the other terms of the total energy, and since in our equation for the balance of linear momentum we already know the expression for the forces, with the exception of $\mathbf{f}_{\text{micro}}$, we can apply a simplified form of the Principle of Virtual Power by observing that $\mathbf{f}_{\text{micro}}$ will come from the rate of variation of the free energy along virtual displacements.

We apply the previous arguments in the case of a viscous fluid in a still domain Ω (and thus, we have the restriction: $u|_{\partial\Omega} = 0$, since the fluid sticks to the walls), in which there is an evolving structure determined by a phase field φ. Moreover, since the process of obtaining the expression for the microscopic forces is simpler in the case without further restrictions on the virtual displacements, in the following we explain how to do that under the extra hypothesis that the fluid is incompressible (and thus, we have the restriction: div $\mathbf{v} = 0$). Additionally, we assume that the free-energy depends only on φ; that is, the other thermodynamics variables are kept constant.

To construct virtual displacements satisfying our restrictions, we consider the vector fields in the set

$$\mathscr{V}(\Omega) = \{\hat{\mathbf{v}} \in (C_0(\Omega))^n : \text{ div } \hat{\mathbf{v}} = 0\}, \tag{14}$$

Then, take any $\mathbf{v} \in \mathscr{V}(\Omega)$ and at any fixed time t and for each $\mathbf{x} \in \Omega$ consider the displacements given by solving the auxiliary family of systems of ordinary differential equations:

$$\begin{cases} \dfrac{d\mathbf{z}}{d\tau} = \hat{\mathbf{v}}(\mathbf{z}), \\ \mathbf{z}|_{\tau=0} = \mathbf{x}. \end{cases}$$

The solutions $\mathbf{z} = \mathbf{z}(\mathbf{x}, \tau)$ are the virtual displacements that we will use.

Thus, by using our previous notations, the chain rule, and integration by parts, the previous formulation of the Principle of Virtual Power gives us that:

$$\int_\Omega \mathbf{f}_{\text{micro}} \cdot \mathbf{v}\, d\mathbf{x} = \frac{d}{d\tau}\mathscr{E}(\varphi(\mathbf{z}(\mathbf{x}, \tau), t))|_{\tau=0} = \int_\Omega \frac{\delta E}{\delta\varphi}\nabla\varphi \cdot \hat{\mathbf{v}}\, d\mathbf{x}, \quad \text{for all } \hat{\mathbf{v}} \in \mathscr{V}(\Omega).$$

Thus, $\int_{\Omega}(\mathbf{f}_{\text{micro}} - \frac{\delta E}{\delta\varphi}\nabla\varphi)\cdot\hat{\mathbf{v}}\,d\mathbf{x} = 0$, $\forall\mathbf{v}\in\mathscr{V}(\Omega)$, that is, $\mathbf{f}_{\text{micro}} - \frac{\delta E}{\delta\varphi}\nabla\varphi$ is orthogonal to $\mathscr{V}(\Omega)$ in $L^2(\Omega)$. Thus, Theorem 1.4, p. 11, in Temam [108] implies that there is q such that $\mathbf{f}_{\text{micro}} = -\nabla q + \frac{\delta E}{\delta\varphi}\nabla\varphi$.

Thus, substituting back this last expression in the balance of linear momentum equations (13), we get the following equations governing the motion of a material with an evolving structure determined by a phase field:

$$\begin{cases} \partial_t(\rho\mathbf{v}) + \operatorname{div}(\rho\mathbf{v}\otimes\mathbf{v}) = \operatorname{div}\mathbf{T}_0 - \nabla q + \mathbf{f}_b + \dfrac{\delta E}{\delta\varphi}\nabla\varphi, \\ \operatorname{div}\mathbf{v} = 0. \end{cases} \tag{15}$$

Remark We stress that the previous arguments assumed that the Cauchy stress tensor $\mathbf{T}_0$ was exactly that of the virgin material (that is, the material disregarding the presence of the structure associated to the phase field). Thus, in this model the interaction between the structure and the rest of the material is not realized through contact forces but just through the microforces $\mathbf{f}_{\text{micro}}$ which were considered part of the body forces. However, in Sect. 5, we show that thermodynamical consistency in general requires suitable modification of the Cauchy stress tensor and contact interaction forces between the structure and the rest of material do appear.

Equations (15) must be coupled with an equation for the phase field; this may be an Allen–Cahn or Cahn–Hilliard equation according to the kind of structure immersed in the fluid. Moreover, a free-energy must be specified. Next, we illustrate this procedure.

Example: Motion of Vesicles in Fluids Du et al. [46] (see also Du et al. [47]) consider a phase field model for the motion of a vesicle immersed in a homogeneous incompressible Newtonian viscous fluid in a domain $\Omega\subset R^n$, $n = 2$ or 3. They assume that the same fluid was in the exterior and in the interior of the vesicle, that membrane density is comparable (equal, actually) to the fluid density, and that there are no external forces and no sources of the membrane material. A phase field variable φ is used to describe the relative vesicle position: at time t, the **interior** of the vesicle is given by $\{\mathbf{x}\in\Omega : \varphi(\mathbf{x},t) > 0\}$; the **exterior** of the vesicle is given by $\{\mathbf{x}\in\omega : \varphi(\mathbf{x},t) < 0\}$; and the **membrane** of the vesicle is at $\{\mathbf{x}\in\Omega : \varphi(\mathbf{x},t) = 0\}$.

By supposing a homogeneous incompressible Newtonian viscous fluid, with constant density $\rho = 1$, for simplicity of exposition, the Cauchy stress tensor is $\mathbf{T}_0 = -pI + \mu_0\frac{1}{2}(\nabla\mathbf{v} + (\nabla\mathbf{v})^t)$, where p is the hydrostatic pressure; thus, by incorporating q into the hydrostatic pressure p and calling $\tilde{p} = p + q$, Eq. (15) simplify. By putting together the equations for the fluid motion and the phase field equation (an Allen–Cahn type in this case), we obtain

$$\begin{cases} \partial_t\mathbf{v} + \mathbf{v}\cdot\nabla\mathbf{v} - \mu_o\Delta\mathbf{v} + \nabla p = f + \dfrac{\delta E}{\delta\varphi}\nabla\varphi, \\ \operatorname{div}\mathbf{v} = 0, \\ \partial_t\varphi + \mathbf{v}\cdot\nabla\varphi = -\lambda\dfrac{\delta E}{\delta\varphi}, \end{cases} \tag{16}$$

where the expression for $\frac{\delta E}{\delta \varphi}(\cdot)$ is given by (10) since, as we will explain, the free-energy functional associated to the vesicle is a functional of form (9).

To find the expression for the total free-energy $\mathscr{E}$ in terms of the phase field, the authors of [46] firstly consider that in a sharp front approach the corresponding free energy should be what is known as the **Helfrich bending energy**; this is then rewritten in terms of the phase field (which includes a small positive parameter ϵ and recovers the bending energy of the sharp front model as ϵ approaches zero). Moreover, under the conditions considered in [46], **two constraints** appear: the volume of the vesicle and the surface area of vesicle membrane should be fixed constants. The authors treat these volumetric and area constraints by penalization, including them in a final total free-energy functional with two penalization parameters. Then, the free energy considered in [46] is the sum of the following three terms:

$$\mathscr{E} = \mathscr{E}_{bending} + \mathscr{E}_{volume} + \mathscr{E}_{area}.$$

The first term is $\mathscr{E}_{bending} = \frac{k}{2\epsilon} \int_\Omega |e(\varphi)|^2 d\mathbf{x}$, which is a simplified form of elastic bending energy for the phase field. Here, $\epsilon > 0$ is a small parameter related to the width of the transition layer; k is the bending modulus; c_0 is the spontaneous curvature of the vesicle, and $e(\varphi) = \epsilon \Delta \varphi + (\frac{1}{\epsilon}\varphi + c_0\sqrt{2})(1 - \varphi^2)$. The second term is $\mathscr{E}_{volume} = \frac{1}{2} M_1 (A(\varphi) - \alpha)^2$, which is a penalization term for the volume of the vesicle. Here, $M_1 > 0$ is a large penalization term; α is related to the required volume, and $A(\varphi) = \int_\Omega \varphi d\mathbf{x}$. The third term is $\mathscr{E}_{area} = \frac{1}{2} M_2 (B(\varphi) - \beta)^2$, which is a penalization term for the surface area. Here, M_2 is a large penalization term; β is related to the required surface area, and $B(\varphi) = \int_\Omega \frac{\epsilon}{2}|\nabla\varphi|^2 + \frac{1}{4\epsilon}(\varphi^2 - 1)^2 d\mathbf{x}$.

Next, for simplicity of exposition, take $c_0 = 0$. Then, a direct computation using (10) shows that $\frac{\delta E}{\delta \varphi}(\varphi) = kg(\varphi) + M_1(A(\varphi) - \alpha) + M_2(B(\varphi) - \beta)e(\varphi)$, where $g(\varphi) = -\Delta e(\varphi) + \frac{1}{\epsilon}(3\varphi^2 - 1)e(\varphi)$.

Finally, the equations governing the interaction between the membrane and the fluid are given by (16), where, by using (10) and $c_0 = 0$, the expression of the variational derivative is $\frac{\delta E}{\delta \varphi} = kg(\varphi) + M_1(A(\varphi) - \alpha) + M_2(B(\varphi) - \beta)e(\varphi)$.

Dissipative Energy Laws; Further Short Commentaries The just described problem formally satisfies a dissipative energy law of form

$$\frac{d}{dt} \int_\Omega \frac{1}{2}|\mathbf{v}|^2 dx + \mathscr{E}(\varphi) = -\mu_0 \int_\Omega |\nabla \mathbf{v}|^2 dx - \lambda \int_\Omega \left|\frac{\delta E}{\delta \varphi}\right|^2 dx.$$

The first term in the left-hand side of the last inequality is the time derivative of the kinetic energy $\mathscr{K}(\mathbf{v}) = \int_\Omega \frac{1}{2}|\mathbf{v}|^2 dx$ (recall that for simplicity the density was taken to be one), while $\mathscr{E}(\varphi) = \int_\Omega E(\varphi) dx$ is the total free-energy. This dissipative energy law can be obtained by the following formal computations: multiply the first equation in (16) by $\mathbf{v}$ and the third equation by $\frac{\delta E}{\delta \varphi}$; integrate on Ω, using standard integration by parts, the second equation (div $\mathbf{v} = 0$). By adding the corresponding results, observing that the term coming from the last one in the first equation and the term coming from the second one in the third equation cancel each other, we obtain the stated dissipative energy law. Although we do not have space to comment this aspect as it deserves, many phase field models, derived by using the variational energy approach, do satisfy suitable dissipative energy laws. For this reason, such models are popular and convenient, specially from the mathematical and numerical point of view.

However, it is not clear in general whether they are thermodynamically consistent (i.e., satisfy the entropy principle), specially in non-isothermal situations. Some authors argue that in order to satisfy the entropy principle, the principle of nonincreasing of the total free-energy, which was used for determining the phase field equations, should be replaced by to the requirement of nonincreasing in time of the following modified free-energy functional: $\mathscr{E} = \int_\Omega \frac{1}{\theta} E d\mathbf{x}$, where E is as before and $\theta > 0$ is the absolute temperature. Although our impression is that these arguments are a bit confusing, some truth must be in them since, as we will see in the next section, where we describe a thermodynamically consistent approach, at least one term of the derived equation satisfies this last claim for phase fields considered as internal variables.

5 The Entropy Approach

Some of the arguments presented in this section are generalizations of the ones in Boldrini et al. [10] for the special case of a phase field model for damage and fatigue in materials.

We describe here a physically sound approach to obtain phase field models, in the sense that the standard physical principles, including the second principle of thermodynamics (entropy condition), are required to hold. Such methodology is called the entropy approach, and to explain how it works, we assume that all the stated variables and other mathematical entities that follow have enough regularity for the required computations hold. We start by considering a body that at time t occupies a domain denoted by $\Omega_t \subset \mathbf{R}^3$ described by **Eulerian (spatial) coordinates** $\mathbf{x}$ (we will briefly comment on the use of **Lagrangian (reference) coordinates** in the last section); $\mathscr{D}_t$ denotes arbitrary regular subdomains of Ω_t moving with the body. The variables characterizing the thermodynamical state of the body are the following. A **mass density** ρ that must satisfy the standard conservation of mass; the **displacement** and **velocity** vector fields, denoted, respectively, by $\mathbf{u}$ and $\mathbf{v}$, are dynamical variables, and the governing equation for $\mathbf{v}$ will be obtained by applying the Principle of Virtual Power (PVP) (see, for instance, Frémond [56]); the **specific density of the internal energy** e (density by unit of mass) whose governing equation will be obtained by applying the first principle of thermodynamics, that is, the balance of energy.

Since we want to exemplify the application of the entropy approach in a rather general setting, we consider **two phase fields of different types** as we will explain. At this point of the arguments, we do not attribute any physical meaning to those phase fields because we want just to distinguish them by the way their respective governing equations are obtained; later on, we will consider an example where specific physical meanings will be attributed to those phase fields. We assume:

- A first phase field φ that is considered a **dynamical variable** in the sense that its corresponding governing equation will also be obtained by applying the Principle of Virtual Power (PVP);
- A second phase field $\mathscr{F}$ that is considered an **internal variable**; we assume that its governing equation is a constitutive differential equation to be determined by the second principle of thermodynamics, that is, such that a suitable form of the entropy inequality be satisfied.

Concerning notation, $\dot{g} = g_t + \mathbf{v} \cdot \nabla g$ denotes the material derivative of any given variable $g(\mathbf{x}, t)$ (in particular, $\dot{\mathbf{v}}$ is the acceleration) and $\nabla^S \mathbf{w} = \text{sym}\,(\nabla \mathbf{w})$ denotes the symmetric part of the gradient of any given vector field $\mathbf{w}$. In particular, $\mathbf{E} = \nabla^S \mathbf{u}$ and $\mathbf{D} = \nabla^S \mathbf{v}$ are, respectively, the infinitesimal strain tensor and the rate of strain tensor fields.

In the present context, we use the expression **macroscopic velocity** to refer to the standard (classical) velocity, that is, the time rate of change of the displacement, $\mathbf{v}$; we use the term **microscopic velocity** to refer to the time rate of change of the dynamical phase field φ, that is, $\dot{\varphi}$, which is denoted here by c. Moreover, for the application of the Principle of Virtual Power (PVP), we denote by $\hat{\mathbf{v}}$ any **admissible virtual macroscopic velocity** and by $\hat{c}$ any **admissible virtual microscopic velocity**. The term **admissible** means that such velocities must satisfy any possible physical or geometrical restrictions. For instance, irreversibility, incompressibility, or nonpenetrability of rigid walls, and so on; we recall that in the simplified application of the Principle of Virtual Power done in the previous section, in the arguments to find an expression for $\mathbf{f}_{\text{micro}}$, we had the requirement that admissible virtual motions should be incompressible, and thus the associated virtual macroscopic velocities should have null divergence, that is, we had to require $\hat{\mathbf{v}} \in \mathscr{V}(\Omega)$, which is defined in (14). However, to simplify the presentation of the arguments, we do not consider in this section any restriction and take for any fixed time t the following admissible virtual velocities sets:

$$\hat{\mathbf{v}} \in \mathscr{V}_{\text{macro}}(\Omega_t) = (C_0(\Omega_t))^n, \quad \hat{c} \in \mathscr{V}_{\text{micro}}(\Omega_t) = C_0(\Omega_t). \tag{17}$$

At the end of this section, we briefly comment on other possibilities.

5.1 General Governing Equations

The first physical law to be satisfied is the **conservation of mass**, which is expressed by the **continuity equation** for the **material density** ρ:

$$\dot{\rho} + \rho\, \text{div}\mathbf{v} = 0. \tag{18}$$

Next, to obtain the dynamic equations, we closely follow the arguments in Frémond [56]. For this, we consider the virtual powers of several kinds of forces.

The **virtual power of the interior forces** is given for any $(\mathscr{D}_t, \hat{\mathbf{v}}, \hat{c})$ by:

$$\mathscr{P}_i(\mathscr{D}_t, \hat{\mathbf{v}}, \hat{c}) = -\int_{\mathscr{D}_t} \mathbf{T} : \hat{\mathbf{D}} d\mathbf{x} - \int_{\mathscr{D}_t} (b\hat{c} + \mathbf{h} \cdot \nabla \hat{c}) d\mathbf{x}. \tag{19}$$

Here, $\mathbf{T}$ is the **Cauchy stress tensor**, b is the **volumetric density of energy exchanged** by variation of a unit of the time rate of φ; and $\mathbf{h}$ is the **flux of energy** associated to the spatial variation of a unit of the time rate of φ. The first term in the right-hand side of the previous equation is the **classical stress power**. The next two other terms are the **powers of generalized interior forces associated to microscopic motions** described, respectively, by the phase fields φ and $\mathscr{F}$.

The **virtual power of the exterior forces** is given for any $(\mathscr{D}_t, \hat{\mathbf{v}}, \hat{c})$ by:

$$\mathscr{P}_e(\mathscr{D}_t, \hat{\mathbf{v}}, \hat{c}) = \int_{\mathscr{D}_t} \rho \mathbf{f} \cdot \hat{\mathbf{v}}\, d\mathbf{x} + \int_{\mathscr{D}_t} \rho a \hat{c}\, d\mathbf{x} + \int_{\partial \mathscr{D}_t} \mathbf{t} \cdot \hat{\mathbf{v}}\, dS + \int_{\partial \mathscr{D}_t} t_h \hat{c}\, dS. \tag{20}$$

In this last expression, $\mathbf{f}$ is the **body force vector per unit of mass**, a is the **specific (by unit of mass) density of energy supplied from the exterior to the evolving structures** (for example, if the phase fields are used to describe material damage, a could be energies supplied by external irradiation or electrical or chemical resulting from external actions modifying the microscopic bounds), $\mathbf{t}$ is the **macroscopic contact force** and t_h is the **superficial density of energy supplied to the material by the flux h**. The first two integrals in (20) are **virtual powers of actions at distance**; the last two integrals in (20) are **virtual powers of contact forces.**

The **virtual power of the inertia (acceleration) forces** is expressed for any $(\mathscr{D}_t, \hat{\mathbf{v}}, \hat{c})$ as follows:

$$\mathscr{P}_a(\mathscr{D}_t, \hat{\mathbf{v}}, \hat{c}) = \int_{\mathscr{D}_t} \rho \dot{\mathbf{v}} \cdot \hat{\mathbf{v}} d\mathbf{x}. \tag{21}$$

Remark In (21), the acceleration forces associated to the phase field φ are assumed to be null; so there is no virtual power associated to them. This is a usual hypothesis, which implies in a purely dissipative evolution for the structures described by φ. However, as is pointed out by Frémond [56, p. 5], in certain specific situation it is necessary to take into account also the acceleration forces of the microscopic motions. In such cases, we must add the term $\int_{\mathscr{D}_t} \hat{\rho}\, \dot{c}\, \hat{c}\, d\mathbf{x}$ to $\mathscr{P}_a(\mathscr{D}_t, \hat{\mathbf{v}}, \hat{c})$, where $\hat{\rho}$ is a parameter associated to the "inertia" of the evolving structure (related, for instance, to the mass of the bonds in certain damage modeling; see Frémond [56, Section 12.2], Frémond and Nedjar [58], and Nedjar [86]), $\dot{c} = \ddot{\varphi}$ is the acceleration of φ, that is, the material derivative of the microscopic velocity c, and $\hat{c}$ is a virtual microscopic velocity.

The **Principle of Virtual Power** (PVP) is stated as follows: for any $(\mathscr{D}_t, \hat{\mathbf{v}}, \hat{c})$,

$$\mathscr{P}_a(\mathscr{D}_t, \hat{\mathbf{v}}, \hat{c}) = \mathscr{P}_i(\mathscr{D}_t, \hat{\mathbf{v}}, \hat{c}) + \mathscr{P}_e(\mathscr{D}_t, \hat{\mathbf{v}}, \hat{c}). \tag{22}$$

From Eqs. (19) to (22), with $\hat{c} \equiv 0$, using the fact that the virtual velocities satisfy (17) and standard arguments, we obtain:

$$\begin{aligned} \rho\dot{\mathbf{v}} &= \operatorname{div}\mathbf{T} + \rho\mathbf{f} \quad \text{in } \mathscr{D}_t, \\ \mathbf{T}\mathbf{n} &= \mathbf{t} \quad \text{in } \partial\mathscr{D}_t. \end{aligned} \tag{23}$$

Similarly, taking $\hat{\mathbf{v}} \equiv \mathbf{0}$ in (22), we also have:

$$\begin{aligned} 0 \quad &= \operatorname{div}\mathbf{h} - b + \rho a \quad \text{in } \mathscr{D}_t, \\ \mathbf{h}\cdot\mathbf{n} &= t_h \quad \text{in } \partial\mathscr{D}_t. \end{aligned} \tag{24}$$

To the previous dynamical equations, we must add another one governing the evolution of **F**; since this phase field is considered an internal variable, we assume that it satisfies a **constitutive differential relation** as follows:

$$\dot{\mathscr{F}} = F. \tag{25}$$

The expression of F will be determined later on by using the entropy condition.

Next, we must impose the **first principle of thermodynamics**, that is, the **balance of energy** in the system:

$$\frac{d}{dt}\int_{\mathscr{D}_t} \rho e \, d\mathbf{x} + \frac{d}{dt}K(\mathscr{D}_t, \mathbf{v}) = \mathscr{P}_e(\mathscr{D}_t, \mathbf{v}, \dot{\varphi}) + \int_{\mathscr{D}_t} \rho r \, d\mathbf{x} - \int_{\partial\mathscr{D}_t} \mathbf{q}\cdot\mathbf{n}\, dS,$$

where

$$K(\mathscr{D}_t, \mathbf{v}) = \int_{\mathscr{D}_t} \frac{1}{2}\rho\mathbf{v}\cdot\mathbf{v}\, d\mathbf{x}$$

is the **macroscopic kinetic energy**, r is the **specific heat source density**, e is the **specific internal energy density**, and $\mathbf{q}$ is the **heat flux**.

Remark When the acceleration forces associated to the phase field φ are not null (see Remark just after (21)), the kinetic energy must be modified to $K(\mathscr{D}_t, \mathbf{v}, c) = \int_{\mathscr{D}_t} \frac{1}{2}\rho\mathbf{v}\cdot\mathbf{v}\, d\mathbf{x} + \int_{\mathscr{D}_t} \frac{1}{2}\hat{\rho}|c|^2\, d\mathbf{x}$.

The previous expression of the balance of energy, combined with the balance of mechanical work, which is obtained from (22) by taking $\hat{\mathbf{v}} = \mathbf{v}$ and $\hat{c} = \dot{\varphi}$, gives the reduced form of the balance of energy in the integral form as:

$$\frac{d}{dt}\int_{\mathscr{D}_t} \rho e \, d\mathbf{x} = -\mathscr{P}_i(\mathscr{D}_t, \mathbf{v}, \dot{\varphi}) + \int_{\mathscr{D}_t} \rho r \, d\mathbf{x} - \int_{\partial\mathscr{D}_t} \mathbf{q}\cdot\mathbf{n}\, dS.$$

Due to the conservation of mass (18), we have $\frac{d}{dt}\int_{\mathscr{D}_t} \rho e \, d\mathbf{x} = \int_{\mathscr{D}_t} \rho \dot{e} \, d\mathbf{x}$, and thus, from the last integral identity we obtain the following local form:

$$\rho \dot{e} = -\mathrm{div}\, \mathbf{q} + \rho r + \mathbf{T} : \mathbf{D} + b\dot{\varphi} + \mathbf{h} \cdot \nabla \dot{\varphi} \quad \text{in } \mathscr{D}_t. \tag{26}$$

Finally, we must also impose the **second principle of thermodynamics**, that is, the **entropy inequality**. For this, since we have two phase fields: φ, which is considered a dynamical variable, and $\mathscr{F}$, which is considered an internal variable, we will combine arguments similar to the ones in Frémond [56] and Fabrizio et al. [53], but with a more general form of the second principle of thermodynamics:

$$\rho \dot{\eta} \geq \mathrm{div}\, \mathbf{F} + \rho s \quad \text{in } \mathscr{D}_t. \tag{27}$$

Here, η, $\mathbf{F}$, and s are, respectively, the **specific entropy density**, the **entropy flux**, and the **specific entropy production term**.

The entropy flux is assumed to be of form

$$\mathbf{F} = \frac{\mathbf{q}}{\theta} + \mathbf{k},$$

where, as before, $\mathbf{q}$ is the **heat flux**, $\theta > 0$ is the absolute temperature (from now on, we assume that θ is always positive); we observe that $\mathbf{q}/\theta$ is the classical entropy flux, while $\mathbf{k}$ is an entropy flux correction due to the physical processes associated to the evolution of the structures described by the phase fields.

Similarly, the specific entropy production term is of form

$$s = \frac{r}{\theta} + \omega,$$

where r/θ is the classical specific entropy production due to heat generation, and ω is an entropy production correction again due to the evolution of the structures described by the phase fields.

Suitable expressions for $\mathbf{k}$ and ω will be obtained in the next subsection; however, we firstly observe that certain restrictions are natural. We assume that there is no flux of entropy due to microstructure evolution through the body's boundary, that is, the entropy production correction must satisfy

$$\mathbf{k} \cdot \mathbf{n} = 0 \quad \text{on } \partial \Omega_t. \tag{28}$$

Also, although local decreasing of entropy due to microscopic evolution is acceptable, this has to be compensated by corresponding increase in other parts of the boundary in such way that the total entropy production due to microscopic evolution in the body cannot decrease; that is, we must have

$$\int_{\Omega} \rho \omega \, d\mathbf{x} \geq 0. \tag{29}$$

We observe that the under such restrictions, the second law of thermodynamics assumes its standard form for the whole body.

Therefore, with the previous conditions, the entropy inequality (27) becomes:

$$\rho\dot{\eta} \geq -\mathrm{div}\left(\frac{\mathbf{q}}{\theta} + \mathbf{k}\right) + \frac{\rho r}{\theta} + \rho\omega. \tag{30}$$

Collecting the previous results, the basic governing equations are the following:

$$\begin{cases} \dot{\rho} + \rho\,\mathrm{div}\,\mathbf{v} = 0, \\ \dot{\mathbf{u}} = \mathbf{v}, \\ \rho\dot{\mathbf{v}} = \ \mathrm{div}\ \mathbf{T} + \rho\mathbf{f}, \\ 0 = \ \mathrm{div}\,\mathbf{h} - b + \rho a, \\ \rho\dot{e} = -\ \mathrm{div}\,\mathbf{q} + \rho r + \mathbf{T} : \mathbf{D} + b\dot{\varphi} + \mathbf{h}\cdot\nabla\dot{\varphi}, \\ \dot{\mathscr{F}} = F. \end{cases} \tag{31}$$

The previous expressions together with (30), (28), (29) constitute the general equations for the models considered in this work. As usual, the constitutive relations must be found in such way that the entropy inequality (30) is satisfied for all possible admissible processes.

5.2 *Constitutive Relations*

We recall that we are using Eulerian (spatial) coordinates; also, for simplicity, in the following arguments we assume that the body under consideration is under the hypothesis of small strains; we will briefly comment on what must be changed when this is not so, that is, when the body is subjected to large strains.

To obtain thermodynamically consistent expressions for the constitutive relations, we follow arguments similar to the ones introduced by Truesdell and Noll [109]. We start by assuming that the constitutive properties are expressed in terms of specific the free-energy density

$$\psi = e - \theta\eta \tag{32}$$

and that $\psi = \psi(\Gamma)$, that is, it is a function of the following variables:

$$\Gamma = (\rho, \theta, \varphi, \mathscr{F}, \nabla\rho, \nabla\theta, \nabla\varphi, \nabla\mathscr{F}, \mathbf{E}), \tag{33}$$

By rewriting (30) in terms of the specific free-energy with the help of the equation for the balance of energy (31) (iv), we obtain:

$$-\rho(\dot{\psi} + \eta\dot{\theta}) + \mathbf{T} : \mathbf{D} + b\dot{\varphi} + \mathbf{h}\cdot\nabla\dot{\psi} - \frac{1}{\theta}\mathbf{q}\cdot\nabla\theta + \theta\,\mathrm{div}\,\mathbf{k} - \rho\omega \geq 0. \tag{34}$$

As in Frémond [56], **T**, b, **h**, and **q** are split in their reversible (non-dissipative) and irreversible (dissipative) parts, which are indicated, respectively, by the superscripts $^{(r)}$ and $^{(ir)}$:

$$\begin{aligned} \mathbf{T} &= \mathbf{T}^{(r)} + \mathbf{T}^{(ir)},\ b = b^{(r)} + b^{(ir)}, \\ \mathbf{h} &= \mathbf{h}^{(r)} + \mathbf{h}^{(ir)},\ \ \mathbf{q} = \mathbf{q}^{(r)} + \mathbf{q}^{(ir)}. \end{aligned} \tag{35}$$

Here, $\mathbf{T}^{(r)}$ and $\mathbf{T}^{(ir)}$ are symmetric tensors; the non-dissipative (reversible) parts may in general depend on the variables Γ (see (33)); the dissipative (irreversible) parts may in general depend on the variables in Γ and on some of their derivatives in time or space. The following arguments will lead to specific dependences on such derivatives.

For simplicity of arguments, again as in Frémond [56, p. 27], we assume that dissipation (irreversibility) appears only due to $\dot{\varphi}$ and $\nabla\theta$ in (34), that is,

$$\mathbf{h}^{(ir)} \equiv 0, \tag{36}$$

and also that the heat flux is purely dissipative (irreversible); that is,

$$\mathbf{q}^{(r)} \equiv 0, \tag{37}$$

The expressions in (35) must be found such that the entropy condition is satisfied for any admissible process. To do that, we recall that for any sufficiently smooth field $g(\mathbf{x}, t)$ depending on the spatial position $\mathbf{x}$ and time t, the following holds (see, for instance, Lemma 1, p. 146, in Fabrizio et al. [53]):

$$\dot{\overline{\nabla g}} = \nabla \dot{g} - (\nabla \mathbf{v})^T \nabla g. \tag{38}$$

Next, we use the chain rule for ψ and Eq. (38) with ρ, $\mathscr{F}$, and $\mathbf{E}$ in place of g. From (25), (31), and the entropy condition (34) (written in terms of the free-energy) and the fact that $\mathbf{T}$ and $\partial_{\mathbf{E}}\psi$ are symmetric tensors, after some manipulation, collecting similar terms, and rearranging, we obtain:

$$\begin{aligned} &-\rho(\eta + \partial_\theta \psi)\dot{\theta} + (-\rho\partial_\varphi\psi + b^{(r)} + b^{(ir)})\dot{\varphi} + \rho^2 \partial_{\nabla\rho}\psi\nabla(\operatorname{div}\mathbf{v}) \\ &\quad + \rho\partial_{\nabla\rho}\psi\left((\operatorname{div}\mathbf{v})\,\mathbf{I} + (\nabla\mathbf{v})^T\right)\nabla\rho - \rho\partial_{\nabla\theta}\psi\dot{\overline{\nabla\theta}} - (\rho\partial_{\nabla\varphi}\psi - \mathbf{h}^{(r)})\dot{\overline{\nabla\varphi}} \\ &\quad + (\mathbf{T}^{(r)} + \mathbf{T}^{(ir)} - \rho\,\partial_{\mathbf{E}}\psi + \rho^2\partial_\rho\psi\mathbf{I} + \rho\nabla\mathscr{F}\otimes\partial_{\nabla\mathscr{F}}\psi + \nabla\varphi\otimes\mathbf{h}) : \nabla\mathbf{v} \\ &\quad - \rho\partial_{\mathscr{F}}\psi F - \rho\partial_{\nabla\mathscr{F}}\psi\cdot\nabla F - \frac{1}{\theta}\mathbf{q}^{(ir)}\cdot\nabla\theta + \theta\operatorname{div}\mathbf{k} - \rho\omega \\ &\quad + \frac{1}{2}\rho\,\partial_{\mathbf{E}}\psi : [(\nabla\mathbf{v})^T\nabla\mathbf{u} + (\nabla\mathbf{u})^T\nabla\mathbf{v}] \geq 0. \end{aligned} \tag{39}$$

Since we are considering only the case of small strains, the last term in the previous inequality can be disregarded, and we are left with:

$$
\begin{aligned}
&-\rho(\eta+\partial_\theta\psi)\dot{\theta}+(-\rho\partial_\varphi\psi+b^{(r)})\dot{\varphi}+\rho^2\partial_{\nabla\rho}\psi\nabla(\operatorname{div}\mathbf{v})\\
&\quad-\rho\partial_{\nabla\theta}\psi\overset{\cdot}{\overline{\nabla\theta}}-(\rho\partial_{\nabla\varphi}\psi-\mathbf{h}^{(r)})\overset{\cdot}{\overline{\nabla\varphi}}\\
&\quad+(\mathbf{T}^{(r)}-\rho\,\partial_{\mathbf{E}}\psi+\rho^2\partial_\rho\psi\mathbf{I}+\rho\nabla\mathscr{F}\otimes\partial_{\nabla\mathscr{F}}\psi+\nabla\varphi\otimes\mathbf{h}^{(r)}):\nabla\mathbf{v}\\
&\quad+\rho\partial_{\nabla\rho}\psi\left((\operatorname{div}\mathbf{v})\,\mathbf{I}+(\nabla\mathbf{v})^T\right)\nabla\rho+b^{(ir)}\dot{\varphi}+\mathbf{T}^{(ir)}:\nabla\mathbf{v}\\
&\quad-\rho\partial_{\mathscr{F}}\psi F-\rho\partial_{\nabla\mathscr{F}}\psi\cdot\nabla F-\frac{1}{\theta}\mathbf{q}^{(ir)}\cdot\nabla\theta+\theta\operatorname{div}\mathbf{k}-\rho\omega\geq 0.
\end{aligned}
\tag{40}
$$

Next, we choose the reversible terms of the last inequality such that they do contribute to the increase in the entropy for any admissible process, that is,

$$
\begin{aligned}
&-\rho(\eta+\partial_\theta\psi)\dot{\theta}+(-\rho\partial_\varphi\psi+b^{(r)})\dot{\varphi}\\
&\quad+\rho^2\partial_{\nabla\rho}\psi\nabla(\operatorname{div}\mathbf{v})-\rho\partial_{\nabla\theta}\psi\overset{\cdot}{\overline{\nabla\theta}}-(\rho\partial_{\nabla\varphi}\psi-\mathbf{h}^{(r)})\overset{\cdot}{\overline{\nabla\varphi}}\\
&\quad+(\mathbf{T}^{(r)}-\rho\,\partial_{\mathbf{E}}\psi+\rho^2\partial_\rho\psi\mathbf{I}+\rho\nabla\mathscr{F}\otimes\partial_{\nabla\mathscr{F}}\psi+\nabla\varphi\otimes\mathbf{h}^{(r)}):\nabla\mathbf{v}=0.
\end{aligned}
\tag{41}
$$

Since in (41) the dependence on $\dot{\theta}$, $\nabla\mathbf{v}$, $\nabla(\operatorname{div}\mathbf{v})$, and $\overset{\cdot}{\overline{\nabla\theta}}$ are linear and, at any point $\mathbf{x}$ and time t, such quantities can assume arbitrary values (due to the possibility of choosing in suitable ways the forcing terms $\mathbf{f}$ and r), their respective coefficients must be zero. Thus, we must have

$$\partial_{\nabla\rho}\psi=0,\qquad \partial_{\nabla\theta}\psi=0, \tag{42}$$

$$\eta=-\partial_\theta\psi. \tag{43}$$

In addition, by taking the reversible parts of b, $\mathbf{h}$, and $\mathbf{T}$, respectively, as

$$b^{(r)}=\rho\partial_\varphi\psi, \tag{44}$$

$$\mathbf{h}^{(r)}=\rho\partial_{\nabla\varphi}\psi. \tag{45}$$

$$\mathbf{T}^{(r)}=\rho\,\partial_{\mathbf{E}}\psi-\rho^2\partial_\rho\psi\mathbf{I}-\rho\nabla\mathscr{F}\otimes\partial_{\nabla\mathscr{F}}\psi-\nabla\varphi\otimes\mathbf{h}^{(r)}, \tag{46}$$

identity (41) is automatically satisfied. Then, from (37) and (45), we get

$$\mathbf{h}=\mathbf{h}^{(r)}=\rho\partial_{\nabla\varphi}\psi. \tag{47}$$

From (46), using that $\mathbf{T}$, $\partial_{\mathbf{E}}\psi$, and $\partial_\rho\psi\mathbf{I}$ are symmetric tensors, we then obtain

$$\mathbf{T}^{(r)}=\rho\,\partial_{\mathbf{E}}\psi-\rho^2\partial_\rho\psi\mathbf{I}-\rho\operatorname{sym}(\nabla\mathscr{F}\otimes\partial_{\nabla\mathscr{F}}\psi+\nabla\varphi\otimes\partial_{\nabla\varphi}\psi), \tag{48}$$

together with the following restriction:

$$\operatorname{skw}(\nabla\varphi\otimes\partial_{\nabla\varphi}\psi)\equiv 0\quad\text{and}\quad\operatorname{skw}(\nabla\mathscr{F}\otimes\partial_{\nabla\mathscr{F}}\psi)\equiv 0. \tag{49}$$

By using the previous results and that $\mathbf{T}^{(ir)}$ is a symmetric tensor, (40) is then reduced to

$$b^{(ir)}\dot{\varphi} + \mathbf{T}^{(ir)} : \mathbf{D} - \rho\partial_{\mathscr{F}}\psi F - \rho\partial_{\nabla\mathscr{F}}\psi \cdot \nabla F - \frac{\mathbf{q}^{(ir)}}{\theta} \cdot \nabla\theta + \theta\,\mathrm{div}\,\mathbf{k} - \rho\omega \geq 0. \tag{50}$$

Let us now look for constitutive relations for $\mathbf{k}$ and $\mathbf{q}$ guaranteeing thermodynamic consistency. For this purpose, we use the identities $\theta\,\mathrm{div}\,\mathbf{k} = \mathrm{div}\,(\theta\mathbf{k}) - \mathbf{k}\cdot\nabla\theta$ and $\rho\partial_{\nabla\mathscr{F}}\psi \cdot \nabla F = \mathrm{div}\,(\rho\partial_{\nabla\mathscr{F}}\psi F) - \mathrm{div}\,(\rho\partial_{\nabla\mathscr{F}}\psi)F$ in (50); then, after some manipulation, it can be rewritten as

$$\begin{aligned} &b^{(ir)}\dot{\varphi} + \mathbf{T}^{(ir)} : \mathbf{D} - \frac{\mathbf{q}^{(ir)}}{\theta} \cdot \nabla\theta - (\rho\partial_{\mathscr{F}}\psi - \mathrm{div}\,(\rho\partial_{\nabla\mathscr{F}}\psi))F \\ &-\mathrm{div}\,(\rho\partial_{\nabla\mathscr{F}}\psi F - \theta\mathbf{k}) - \mathbf{k}\cdot\nabla\theta - \rho\omega \geq 0. \end{aligned} \tag{51}$$

Allen–Cahn Type Systems

To simplify expression (51), we choose $\mathbf{k}$ exactly as in [53]:

$$\mathbf{k} = \frac{\rho}{\theta}\partial_{\nabla\mathscr{F}}\psi F. \tag{52}$$

We also take the correction term for the entropy production due to microscopic evolution, ω, to be null (then (29) is automatically satisfied), that is,

$$\omega = 0.$$

By using these last two expressions, respectively, in the fifth and seventh terms of (51), after dividing by θ and some manipulation, the inequality reduces to

$$\frac{b^{(ir)}}{\theta}\dot{\varphi} + \frac{\mathbf{T}^{(ir)}}{\theta} : \mathbf{D} - \frac{\mathbf{q}^{(ir)}}{\theta^2} \cdot \nabla\theta - F\xi \geq 0, \tag{53}$$

where we denoted

$$\xi = \frac{\rho}{\theta}\partial_{\mathscr{F}}\psi - \mathrm{div}\left(\frac{\rho}{\theta}\partial_{\nabla\mathscr{F}}\psi\right). \tag{54}$$

The next main idea is to automatically satisfy expression (53) by using the concept of **pseudo-potential of dissipation**. In the case we are discussing, this is a functional

$$\psi_d = \psi_d(\dot{\varphi}, \mathbf{D}, \nabla\theta, \xi, \tilde{\Gamma}), \tag{55}$$

where $\tilde{\Gamma} = (\rho, \theta, \varphi, \mathscr{F}, \nabla\varphi, \nabla\mathscr{F}, \mathbf{E})$ (we took in consideration (33) and (42)), satisfying: $\psi_d(\dot{\varphi}, \mathbf{D}, \nabla\theta, \xi, \tilde{\Gamma}) \geq 0$ for all $(\dot{\varphi}, \mathbf{D}, \nabla\theta, \xi, \tilde{\Gamma})$, $\varphi(0, 0, 0, 0, \tilde{\Gamma}) = 0$ and to be continuous and convex with respect to the variables $\dot{\varphi}$, $\mathbf{D}$, $\nabla\theta$, ξ.

To obtain (53), it is enough to take $[b^{(ir)}/\theta, \mathbf{T}^{(ir)}/\theta, -\mathbf{q}^{(ir)}/\theta^2, -F]$ as the gradient of $\psi_d(\cdot)$ with respect to $[\dot\varphi, \mathbf{D}, \nabla\theta, \xi]$ (recall that for simplicity of exposition we assumed that $\psi_d(\cdot)$ is differentiable with respect to variables $[\dot\varphi, \mathbf{D}, \nabla\theta, \xi]$). In fact, the convexity of $\psi_d(\cdot)$ implies that $0 = \psi_d(0, 0, 0, 0, \tilde{\Gamma}) \geq \psi_d(\dot\varphi, \mathbf{D}, \nabla\theta, \xi, \tilde{\Gamma}) + \partial_{\dot\varphi}\psi_d(\dot\varphi, \mathbf{D}, \nabla\theta, \xi)\,(0 - \dot\varphi) + \partial_{\mathbf{D}}\psi_d(\dot\varphi, \mathbf{D}, \nabla\theta, \xi) : (0 - \mathbf{D}) + \partial_{\nabla\theta}\psi_d(\dot\varphi, \mathbf{D}, \nabla\theta, \xi)\cdot(0 - \nabla\theta) + \partial_\xi\psi_d(\dot\varphi, \mathbf{D}, \nabla\theta, \xi)(0 - \xi)$. Since $\psi_d(\dot\varphi, \mathbf{D}, \nabla\theta, \xi, \tilde{\Gamma}) \geq 0$, we obtain the inequality $\partial_{\dot\varphi}\psi_d(\dot\varphi, \mathbf{D}, \nabla\theta, \xi)\dot\varphi + \partial_{\mathbf{D}}\psi_d(\dot\varphi, \mathbf{D}, \nabla\theta, \xi) : \mathbf{D} + \partial_{\nabla\theta}\psi_d(\dot\varphi, \mathbf{D}, \nabla\theta, \xi)\cdot\nabla\theta + \partial_\xi\psi_d(\dot\varphi, \mathbf{D}, \nabla\theta, \xi)\xi \geq 0$. Therefore, in order to have (53) satisfied, it is enough to take

$$
\begin{aligned}
\frac{b^{(ir)}}{\theta} &= \partial_{\dot\varphi}\psi_d, \quad \frac{\mathbf{T}^{(ir)}}{\theta} = \partial_{\mathbf{D}}\psi_d, \\
-\frac{\mathbf{q}^{(ir)}}{\theta^2} &= \partial_{\nabla\theta}\psi_d, \; -F = \partial_\xi\psi_d.
\end{aligned}
\tag{56}
$$

Remark When ψ_d is not differentiable, the results are similar, but with the partial derivatives replaced by the corresponding subdifferentials and the equalities replaced by inclusions since subdifferentials are not necessarily single valued.

Thus, using (56) and all the previous results, we obtain

$$
\begin{aligned}
b &= \rho\partial_\varphi\psi + \theta\partial_{\dot\varphi}\psi_d, \\
\mathbf{T} &= \rho\,\partial_{\mathbf{E}}\psi - \rho^2\partial_\rho\psi\mathbf{I} - \rho\,\mathrm{sym}\,(\nabla\mathscr{F}\otimes\partial_{\nabla\mathscr{F}}\psi + \nabla\varphi\otimes\partial_{\nabla\varphi}\psi) + \theta\partial_{\mathbf{D}}\psi_d, \\
\mathbf{q} &= -\theta^2\partial_{\nabla\theta}\psi_d, \\
F &= -\partial_\xi\psi_d
\end{aligned}
\tag{57}
$$

By collecting all the previous results and recalling that $e = \psi + \theta\eta = \psi - \theta\partial_\theta\psi$, we finally rewrite the governing equations (31) as

$$
\begin{cases}
\dot\rho + \rho\,\mathrm{div}\,\mathbf{v} = 0, \\
\dot{\mathbf{u}} = \mathbf{v}, \\
\rho\dot{\mathbf{v}} = \mathrm{div}\,\mathbf{T} + \rho\,\mathbf{f}, \\
\theta\partial_{\dot\varphi}\psi_d = \mathrm{div}\,(\rho\partial_{\nabla\varphi}\psi) - \rho\partial_\varphi\psi + \rho a, \\
\rho\dot e = \mathrm{div}\,\left(\theta^2\partial_{\nabla\theta}\psi_d\right) + \mathbf{T} : \mathbf{D} + (\rho\partial_\varphi\psi + \theta\partial_{\dot\varphi}\psi_d)\dot\varphi + \rho\partial_{\nabla\varphi}\psi\cdot\nabla\dot\varphi + \rho r, \\
\dot{\mathscr{F}} = -\partial_\xi\psi_d. \\
\mathbf{T} = \rho\,\partial_{\mathbf{E}}\psi - \rho^2\partial_\rho\psi\mathbf{I} - \rho\,\mathrm{sym}\,(\nabla\mathscr{F}\otimes\partial_{\nabla\mathscr{F}}\psi + \nabla\varphi\otimes\partial_{\nabla\varphi}\psi) + \theta\partial_{\mathbf{D}}\psi_d, \\
e = \psi - \theta\partial_\theta\psi.
\end{cases}
\tag{58}
$$

We observe that the fifth equation in the previous system is usually written in terms of the temperature θ; moreover, suitable initial and boundary conditions must be added to the system to complete the evolution problem.

We stress that the sixth equation $\dot{\mathscr{F}} = -\partial_\xi\psi_d$ in system (58) can be thought as a **generalized Allen–Cahn type equation**. In fact, let us consider, for instance, the mathematically simplest case: a quadratic pseudo-potential given by

$$\psi_d(\dot{\varphi}, \mathbf{D}, \nabla\theta, \xi, \tilde{\Gamma}) = \frac{\tilde{\lambda}}{2}|\dot{\varphi}|^2 + \frac{\tilde{b}}{2}|\mathbf{D}|^2 + \frac{\tilde{c}}{2}|\nabla\theta|^2 + \frac{\tilde{F}}{2}|\xi|^2,$$

where the coefficients are nonnegative and may depend on $\tilde{\Gamma}$. Then, we obtain $F = -\partial_\xi \psi_d = -\tilde{F}\xi = -\tilde{F}[\frac{\rho}{\theta}\partial_{\mathscr{F}}\psi - \operatorname{div}(\frac{\rho}{\theta}\psi_{\nabla\mathscr{F}})]$, by recalling the definition (54) of ξ, and we arrive at the rather standard **thermo-modified Allen–Cahn equation**:

$$\dot{\mathscr{F}} = \tilde{F}\left[\operatorname{div}\left(\frac{\rho}{\theta}\partial_{\nabla\mathscr{F}}\psi\right) - \frac{\rho}{\theta}\partial_{\mathscr{F}}\psi\right]. \tag{59}$$

From (52), condition (28) is satisfied if we assume either of the following boundary conditions: $\frac{\rho}{\theta}\partial_{\mathscr{F}}\psi - \operatorname{div}\left(\frac{\rho}{\theta}\partial_{\nabla\mathscr{F}}\psi\right) = 0$ or $\partial_{\nabla\mathscr{F}}\psi \cdot \mathbf{n} = 0 \quad \text{on } \partial\Omega$.

Cahn–Hilliard Type Systems

There are other possibilities to the expression of F giving the differential constitutive relation for the phase field system. For instance, assume that

$$F = \operatorname{div}\mathbf{H}, \tag{60}$$

where $\mathbf{H}$ has to be found. Then, we rewrite inequality in (51) in terms of $\mathbf{H}$ and ξ (see (54)) as

$$\begin{aligned} & b^{(ir)}\dot{\varphi} + \mathbf{T}^{(ir)} : \mathbf{D} - \frac{\mathbf{q}^{(ir)}}{\theta} \cdot \nabla\theta - \tilde{\xi}\operatorname{div}\mathbf{H} \\ & -\operatorname{div}(\rho\partial_{\nabla\mathscr{F}}\psi \operatorname{div}\mathbf{H} - \theta\mathbf{k}) - \mathbf{k}\cdot\nabla\theta - \rho\omega \geq 0, \end{aligned}$$

where we denoted

$$\tilde{\xi} = \rho\partial_{\mathscr{F}}\psi - \operatorname{div}(\rho\partial_{\nabla\mathscr{F}}\psi). \tag{61}$$

By taking

$$\mathbf{k} = \frac{\rho}{\theta}\partial_{\nabla\mathscr{F}}\psi \operatorname{div}\mathbf{H} \tag{62}$$

and observing that $\tilde{\xi}\operatorname{div}\mathbf{H} = \operatorname{div}(\tilde{\xi}\mathbf{H}) - \mathbf{H}\cdot\nabla\tilde{\xi}$, the last inequality becomes

$$b^{(ir)}\dot{\varphi} + \mathbf{T}^{(ir)} : \mathbf{D} - \frac{\mathbf{q}^{(ir)}}{\theta} \cdot \nabla\theta - \operatorname{div}(\tilde{\xi}\mathbf{H}) + \mathbf{H}\cdot\nabla\tilde{\xi} - \mathbf{k}\cdot\nabla\theta - \rho\omega \geq 0.$$

Next, by taking the correction term for the entropy production as

$$\omega = \frac{1}{\rho}\operatorname{div}(\xi\mathbf{H}), \tag{63}$$

we finally get

$$b^{(ir)}\dot{\varphi} + \mathbf{T}^{(ir)} : \mathbf{D} - \left(\frac{\mathbf{q}^{(ir)}}{\theta} + \mathbf{k}\right) \cdot \nabla\theta + \mathbf{H} \cdot \nabla\tilde{\xi} \geq 0. \tag{64}$$

Similarly as before, expression (64) can be satisfied with the help of a pseudo-potential, but now of form

$$\psi_d = \psi_d(\dot{\varphi}, \mathbf{D}, \nabla\theta, \nabla\tilde{\xi}, \tilde{\Gamma}), \tag{65}$$

where $\tilde{\Gamma}$ is as before, and ψ_d is such that $\psi_d(\dot{\varphi}, \mathbf{D}, \nabla\theta, \nabla\tilde{\xi}, \tilde{\Gamma}) \geq 0$ for all $(\dot{\varphi}, \mathbf{D}, \nabla\theta, \nabla\tilde{\xi}, \tilde{\Gamma})$, $\psi_d(0, 0, 0, 0, \tilde{\Gamma}) = 0$, and it is continuous and convex with respect to the variables $\dot{\varphi}$, $\mathbf{D}$, $\nabla\theta$, $\nabla\tilde{\xi}$. As before, (64) is satisfied if we take

$$\begin{aligned} b^{(ir)} &= \partial_{\dot{\varphi}}\psi_d, \quad \mathbf{T}^{(ir)} = \partial_{\mathbf{D}}\psi_d, \\ -\frac{\mathbf{q}^{(ir)}}{\theta} - \mathbf{k} &= \partial_{\nabla\theta}\psi_d, \ \mathbf{H} = \partial_{\nabla\tilde{\xi}}\psi_d. \end{aligned} \tag{66}$$

Thus, using the previous results, we obtain

$$\begin{aligned} b &= \rho\partial_{\varphi}\psi + \partial_{\dot{\varphi}}\psi_d, \\ \mathbf{T} &= \rho\,\partial_{\mathbf{E}}\psi - \rho^2\partial_{\rho}\psi\mathbf{I} - \rho\,\mathrm{sym}\,(\nabla\mathscr{F} \otimes \partial_{\nabla\mathscr{F}}\psi + \nabla\varphi \otimes \partial_{\nabla\varphi}\psi) + \partial_{\mathbf{D}}\psi_d, \\ \mathbf{q} &= -\theta\partial_{\nabla\theta}\psi_d - \rho\partial_{\nabla\mathscr{F}}\psi\ \mathrm{div}(\partial_{\nabla\tilde{\xi}}\psi_d), \\ F &= \mathrm{div}(\partial_{\nabla\tilde{\xi}}\psi_d). \end{aligned} \tag{67}$$

Remark As before, when φ is not differentiable with respect to $[\dot{\varphi}, \mathbf{D}, \nabla\theta, \nabla\tilde{\xi}]$, we have similar expressions but with the partial derivatives replaced by subdifferentials and the equalities replaced by inclusions since subdifferentials are not necessarily single valued operators.

By collecting all the previous results and recalling that $e = \psi + \theta\eta = \psi - \theta\partial_{\theta}\psi$, we finally rewrite the governing equations (31) as:

$$\begin{cases} \dot{\rho} + \rho\,\mathrm{div}\,\mathbf{v} = 0, \\ \dot{\mathbf{u}} = \mathbf{v}, \\ \rho\dot{\mathbf{v}} = \ \mathrm{div}\,\mathbf{T} + \rho\,\mathbf{f}, \\ \partial_{\dot{\varphi}}\psi_d = \ \mathrm{div}\,(\rho\partial_{\nabla\varphi}\psi +) - \rho\partial_{\varphi}\psi + \rho a, \\ \rho\dot{e} = \mathrm{div}\,(\theta\partial_{\nabla\theta}\psi_d + \rho\partial_{\nabla\mathscr{F}}\psi\ \mathrm{div}(\partial_{\nabla\tilde{\xi}}\psi_d)) + \mathbf{T} : \mathbf{D} + (\rho\partial_{\varphi}\psi + \partial_{\dot{\varphi}}\psi_d)\dot{\varphi} \\ \qquad + \rho\partial_{\nabla\varphi}\psi \cdot \nabla\dot{\varphi} + \rho r, \\ \dot{\mathscr{F}} = \mathrm{div}(\partial_{\nabla\tilde{\xi}}\psi_d) \\ \mathbf{T} = \rho\,\partial_{\mathbf{E}}\psi - \rho^2\partial_{\rho}\psi\mathbf{I} - \rho\,\mathrm{sym}\,(\nabla\mathscr{F} \otimes \partial_{\nabla\mathscr{F}}\psi + \nabla\varphi \otimes \partial_{\nabla\varphi}\psi) + \partial_{\mathbf{D}}\psi_d, \\ e = \psi - \theta\partial_{\theta}\psi. \end{cases} \tag{68}$$

We stress that the sixth equation in (68), which is the differential constitutive equation for the phase field $\mathscr{F}$, is in **conservative form** and can be thought as a **generalized Cahn–Hilliard type equation**. In fact, as before, let us consider, for instance, the mathematically simplest case of a quadratic pseudo-potential:

$$\psi_d(\dot{\varphi}, \mathbf{D}, \nabla\theta, \nabla\xi, \tilde{\Gamma}) = \frac{\tilde{\lambda}}{2}|\dot{\varphi}|^2 + \frac{\tilde{b}}{2}|\mathbf{D}|^2 + \frac{\tilde{c}}{2}|\nabla\theta|^2 + \frac{\tilde{M}}{2}|\nabla\tilde{\xi}|^2,$$

where the coefficients are nonnegative and may depend on $\tilde{\Gamma}$. Then, we obtain $\mathbf{H} = \partial_{\nabla\tilde{\xi}}\psi_d = \tilde{M}\nabla\tilde{\xi} = \tilde{M}\nabla\left(\rho\partial_{\mathscr{F}}\psi - \text{div}\left(\rho\psi_{\nabla\mathscr{F}}\right)\right)$. By recalling again the definition (61) of $\tilde{\xi}$, we arrive at the rather standard **Cahn–Hilliard equation**:

$$\dot{\mathscr{F}} = \text{div}\left[\tilde{M}\nabla\left(\text{div}\left(\rho\partial_{\nabla\mathscr{F}}\psi\right) - \rho\partial_{\mathscr{F}}\psi\right)\right], \tag{69}$$

where $\tilde{M}$ functions as the mobility.

We also observe that condition (29) is satisfied when one imposes the boundary condition $\tilde{\xi} = \rho\partial_{\mathscr{F}}\psi - \text{div}\left(\rho\partial_{\nabla\mathscr{F}}\psi\right) = 0$ on $\partial\Omega$; in fact, in this case we have $\int_\Omega \rho\omega = \int_\Omega \text{div}(\tilde{\xi}\mathbf{H}) = \int_{\partial\Omega}\tilde{\xi}\mathbf{H} = 0$.

Example: Constitutive Relations for Solid Materials Under Damage and Fatigue

A particular case of the previously described situation was presented in Boldrini et al. [10]. In that article, the authors develop a phase field model for the evolution of fatigue and damage in materials, leading eventually to fracture, under non-isothermal processes. Moreover, two phase field variables, also denoted by φ and $\mathscr{F}$, were used to give, respectively, the level of damage and fatigue in the material; the variable φ was the volumetric fraction of damaged material φ (and so $0 \leq \varphi \leq 1$; virgin material when $\varphi = 0$; fractured material when $\varphi = 1$) and was considered a dynamical variable; the variable associated to fatigue $\mathscr{F}$ was considered an internal variable. The model equations were similar to the ones in (58); in the particular case used by the authors for their numerical simulations, a nearly incompressible approximation was taken (density approximately constant given by ρ_0, see the commentaries in Sect. 5.3), and the volumetric free-energy density had the following form:

$$\rho_0\psi = (1-\varphi)^2\frac{1}{2}\mathbf{E}^T\mathscr{C}\mathbf{E} - c_V\,\theta\ln\theta + g_c\left(\frac{\gamma}{2}|\nabla\varphi|^2 + \frac{1}{\gamma}\mathscr{H}(\varphi)\right) + \frac{1}{\gamma}\mathscr{F}\mathscr{H}_f(\varphi). \tag{70}$$

Here, $\mathscr{C}$ is the symmetric fourth-order elasticity tensor whose coefficients give the elastic properties of the virgin material; g_c is the critical Griffith-type fracture energy parameter and for simplicity is assumed to be a positive constant; $\gamma > 0$ is related to the width of the fracture layers and again for simplicity is assumed a positive constant; and $\mathscr{H}(\varphi)$ and $\mathscr{H}_f(\varphi)$ are the following potentials:

$$\mathcal{H}(\varphi) = \begin{cases} \frac{1}{2}\varphi^2 \text{ for } 0 \leq \varphi \leq 1, \\ \frac{1}{2} \quad \text{ for } \varphi > 1, \\ 0 \quad \text{ for } \varphi < 0. \end{cases} \quad \text{and} \quad \mathcal{H}_f(\varphi) = \begin{cases} -\varphi \text{ for } 0 \leq \varphi \leq 1, \\ -1 \text{ for } \varphi > 1, \\ \ \ 0 \text{ for } \varphi < 0, \end{cases}$$

The pseudo-potential of dissipation in that particular case was

$$\psi_d = \frac{\lambda}{2}|\dot{\varphi}|^2 + \frac{b}{2}|\mathbf{D}|^2 + \frac{c}{2}|\nabla\theta|^2 + \frac{\tilde{F}}{2}|\xi|^2,$$

where ξ is the expression defined in (54), and the coefficients cannot depend on $\dot{\varphi}$, $\mathbf{D}$, ∇, θ, and ξ. More details and justifications can be found in Boldrini et al. [10], where several numerical simulations were also presented to show the potentiality of this kind of phase field models.

5.3 Further Commentaries

Incompressibility Systems (58) and (68) give the governing equations compressible materials; the term $-\rho^2\partial_\rho\psi\mathbf{I}$ in the expression of $\mathbf{T}$ is the thermodynamic pressure. When the material is incompressible (isochoric), the null divergence of velocity is required ($\operatorname{div}\mathbf{v} = 0$), and the first admissible virtual velocities space in (17) must be replaced by $\mathcal{V}_{\text{macro}}(\Omega) = \{\hat{\mathbf{v}} \in (C_0(\Omega))^n : \operatorname{div}\hat{\mathbf{v}} = 0\}$. The arguments then lead to the addition of extra term to the stress tensor $\mathbf{T}$; this is related to a hydrostatic-type pressure p, and $\mathbf{T}$ now becomes

$$\mathbf{T} = -p\mathbf{I} + \rho\,\partial_{\mathbf{E}}\psi - \rho^2\partial_\rho\psi\mathbf{I} - \rho\,\text{sym}\,(\nabla\mathcal{F}\otimes\partial_{\nabla\mathcal{F}}\psi + \nabla\varphi\otimes\partial_{\nabla\varphi}\psi) + \theta\partial_{\mathbf{D}}\psi_d.$$

Nearly Incompressible Processes Besides the small strains hypothesis, another rather common simplifying assumption is the nearly incompressibility of solid materials. In such approximation, the material density is assumed to be a known constant ρ_0; the first equation in previously obtained system is disregarded, and the density ρ is replaced by ρ_0 in the other governing equations; in this approximation, the stress tensor has no additional pressure term.

Quasi-Static Processes Another simplifying hypothesis, frequently used in conjunction to the nearly incompressibility, assumes that the equilibrium of forces and damage (fracture) occur at a much faster timescale than the equilibrium of thermal energy and fatigue. This is a quasi-static situation, and the previous systems are simplified by taking the approximations $\dot{\mathbf{v}} \equiv 0$, $\dot{\varphi} \equiv 0$.

Irreversible Phase Fields In some physical situations, the physical consequences described by a phase field ϕ are irreversible. Examples are solidification of several polymers (the white of eggs that cannot naturally turn back to nonsolid state after being fried) and several kinds of damage (fracturing) in materials (no healing after having occurred). These situations translate in the mathematical requirement that for admissible processes we must have $\dot{\varphi} \geq 0$. One possibility to deal with this is to replace the second admissible velocity space in (17) by the admissible virtual microvelocity set $\mathscr{V}_{\text{micro}}(\Omega) = \{\hat{c} \in C_0(\Omega) : \hat{c} \geq 0\}$; this leads to imposition of Kuhn–Tucker type conditions, similar, for instance, as in Simo and Hughes [106] (in the plasticity context). Another possibility is to impose the irreversibility by modifying the pseudo-potential of dissipation by the addition of the extra term $I_-(\dot{\varphi})$, where $I_-(z)$ denotes the potential defined by $I_-(z) = 0$ for $z \geq 0$ and $I_-(z) = +\infty$ for $z < 0$. This forces $\dot{\varphi} \geq 0$ at the expense that now $\partial_{\dot{\varphi}}\psi_d$ must be understood as a subdifferential and that, in the equations where this term appears, the equalities must be replaced by inclusions; see, for instance, Bonfanti et al. [21], Laurençot et al. [78], Luterotti et al. [79], and Boldrini et al. [13, 14]. In particular situations, some authors consider the irreversibility of phase fields using alternative approaches; see, for instance, Miehe et al. [81, 82] in the context of phase field modeling of damage and fracture of materials.

Anisotropy Material anisotropy can be included by suitably changing the part depending on the gradient of the phase field in the free-energy density. For instance, in the example described in the last subsection, the term $|\nabla\varphi|^2$ in (70) could be replaced by $\langle\nabla\varphi, A\nabla\varphi\rangle$, where $\langle\cdot,\cdot\rangle$ denotes the canonical inner product in R^n and A is a positive definite matrix associated to the anisotropy.

Energy Inequalities The phase field models derived in this section automatically satisfy an energy identity (and thus corresponding inequalities). This is because they were derived in a physically consistent way, but one can also see this directly by formally proceeding as follows: first, we integrate on Ω the fifth equation in (58), using the information given by conservation of mass (the first equation); second, we take the scalar product of the third equation in (58) by the velocity $\mathbf{v}$ and integrate on Ω, using again the information given by conservation of mass (the first equation), integration by parts and the fact that the Cauchy tensor $\mathbf{T}$ is symmetric; third, we multiply the fourth equation in (58) by $\dot{\varphi}$ and integrate on Ω and use integration by parts; fourth, we add the resulting identities obtained in the previous three steps to obtain the following **conservation of energy**:

$$\frac{d}{dt}\int_\Omega \rho e\, d\mathbf{x} + \frac{d}{dt}\int_\Omega \frac{\rho|\mathbf{v}|^2}{2}\, d\mathbf{x} = \int_\Omega \rho r\, d\mathbf{x} + \int_\Omega \rho\mathbf{f}\cdot\mathbf{v}\, d\mathbf{x} + \int_\Omega \rho a\dot{\varphi}\, d\mathbf{x}$$

we observe that the sixth equation in (58), the equation for the evolution of the phase field $\mathscr{F}$, was not used to obtain this identity, and $\mathscr{F}$ appears only implicitly in it. This situation is consistent with the choice of this phase field as an internal variable. However, given the specific free-energy and the pseudo-potential of dissipation densities, one can try to obtain modified "energy" inequalities explicitly involving $\mathscr{F}$. For this, one could multiply the sixth equation in (58) by $\dot{\mathscr{F}}$, for instance, and proceed as usual, trying to combine the result with the other equations. Exactly, the same observations hold for system (68).

Energetic Variational Approach and Thermodynamical Consistency By using the notation of (27), when the total entropy entering the region Ω from the exterior is null ($\int_{\partial\Omega} \mathbf{F} \cdot \mathbf{n} dS = 0$) and the total internal production is positive ($\int_{\Omega} \rho s \; d\mathbf{x} \geq 0$), the **total entropy** $\int_{\Omega} \rho\eta d\mathbf{x}$ must be nondecreasing. In fact, from (27) and the previous conditions, using the conservation of mass (18), we must have $\frac{d}{dt}\int_{\Omega} \rho\eta d\mathbf{x} = \int_{\Omega} \rho\dot{\eta} d\mathbf{x} \geq \int_{\partial\Omega} \varphi dS + \int_{\Omega} \rho s \geq 0$. The classical Allen–Cahn (11) and Cahn–Hilliard (12) phase field equations obtained by the energetic variational approach (see Sect. 4) automatically satisfy this requirement when applied to nearly incompressible isothermal processes (constant mass density and temperature) for materials with the internal energy e depending only on the temperature. In fact, from the definition of the specific free-energy (32), we have $\rho\eta = \rho e/\theta - \rho\psi/\theta = \rho e/\theta - E/\theta$ (recall that E is the volumetric free-energy density); thus, $\frac{d}{dt}\int_{\Omega} \rho\eta \; d\mathbf{x} = -\frac{1}{\theta}\frac{d}{dt}\int_{\Omega} E \; d\mathbf{x} \geq 0$. However, as we had already mentioned, it is not a priori clear that more general phase field models obtained by the energetic variational approach are thermodynamically consistent in the sense of satisfying an entropy inequality.

Large Strains The last term in (39) appeared because Eulerian (spatial) coordinates were used; it does not appear in Lagrangian (reference) coordinates; moreover, in such coordinates the other terms in the expression corresponding to (39) appear in simpler forms. This means that in Lagrangian coordinates no approximation is required at that point of the arguments since it is not necessary to pass from (39) to (40), and thus, for large strains, it is more convenient to follow the previous arguments and derivations using Lagrangian coordinates; by doing this, one gets expressions similar to the just obtained ones (but in Lagrangian coordinates). However, there is a "mathematical price" to be paid in any specific situation. To explain this, we just observe that, since everything must be written in Lagrangian coordinates, in particular, the same is so for the free-energy density. The difficulties appear in the cases with gradient terms in the free-energy, as in the situation we just described; in fact, by their physical origin, such gradients are naturally gradients with respect to Eulerian (spatial) coordinates since they correspond to fluxes or diffusions occurring in the spatial (deformed) configuration. Thus, to apply a theory written in Lagrangian coordinates, one must firstly rewrite the free-energy density in such coordinates, which brings nonlinearities involving also the deformation gradient and results in more complicated mathematical expressions.

References

1. V. Alexiades, A. D. Solomon, *Mathematical Modeling of Melting and Freezing Processes*, Hemisphere Publishing Corporation, Washington, 1993.
2. S. M. Allen, J. W. Cahn, Ground state structures in ordered binary alloys with second neighbor interactions. *Acta Metallurgica*, 20 (3), 423–433, 1972.
3. M. Ambati, T. Gerasimov, L. De Lorenzis, Phase-field modeling of ductile fracture. *Comput Mech*, 55 (5), 1017–1040, 2015.

4. M. Ambati, R. Kruse, L. De Lorenzis, A phase-field model for ductile fracture at finite strains and its experimental verification. *Comput Mech*, 57, 149–167, 2016.
5. D. M. Anderson, G. B. McFadden and A. A. Wheeler, Diffusive-interface methods in fluid mechanics. *Ann. Rev. Fluid Mech.* 30, 139–165, 1998.
6. F. D. Araruna, J. L. Boldrini, B. M. R. Calsavara, Optimal Control and Controllability of a Phase Field System with One Control Force. *Appl. Math. Optim.*, 70(3), 539–563, 2014.
7. W.V. Assunção, J.L. Boldrini, Analysis of a system related to a model for phase transitions in dissipative isochoric thermovisco elastic materials. *Math. Methods Appl. Sci.*, 40 (10), 3504–3527, 2017.
8. C. Beckermann, H.-J. Diepers, I. Steinbach, A. Karma, X. Tong, Modeling melt convection in phase-field simulations of solidification. *J. Comput. Phys.* 154, 468–496. 1999.
9. Ph. Blanc, L. Gasser, J. Rappaz, Existence for a stationary model of binary alloy solidification. *Math. Mod. Num. Anal.*, 29 (06), 687–699, 1995.
10. J.L. Boldrini, E.A. Barros de Moraes, L.R. Chiarelli, F.G. Fumes, M.L. Bittencourt, A non-isothermal thermodynamically consistent phase field framework for structural damage and fatigue. *Comput Methods Appl Mech Eng.*, 312, 395–427, 2016.
11. J.L. Boldrini, B.M. Calsavara Caretta, E. Fernändez-Cara, Analysis of a two-phase field model for the solidification of an alloy. *J. Math. Anal. Appl.*, 357, 25–44, 2009.
12. J.L. Boldrini, B.M. Calsavara Caretta, E. Fernändez-Cara, Some optimal control problems for a two-phase field model of solidification. *Rev. Mat. Complut.*, 23, 49–75, 2010.
13. J.L. Boldrini, L.H. de Miranda, G. Planas, On singular Navier-Stokes equations and irreversible phase-transitions. *Commun. Pure Appl. Anal.*, 11, 2055–2078, 2012.
14. J.L. Boldrini, L.H. de Miranda, G. Planas, A Mathematical analysis of fluid motion in irreversible phase transitions. *Z. Angew. Math. Phys.*, 66(3), 785–817, 2015.
15. J.L. Boldrini, S. Lorca P., H. Soto, Stationary solutions of a singular Navier-Stokes enthalpy-heat conduction system. *Differ. Integral Equ.*, 27(5–6), 511–530, 2014.
16. J.L. Boldrini, G. Planas, Weak Solutions of a Phase-Field Model for Phase Change of an Alloy with Thermal Properties. *Math. Meth. Appl. Sci.*, 25(14), 1177–1193, 2002.
17. J.L. Boldrini, G. Planas, A tridimensional phase-field Model with convection for phase change of an alloy. *Discrete Contin. Dyn. Syst.*, 13(2), 429–250, 2005.
18. J. L. Boldrini, C. L. D. Vaz, Existence and regularity of solutions of a phase field model for solidification with convection of pure materials in two dimensions. *Electron. J. Differential Equations*, 109, 1–25, 2003.
19. E. Bonetti, M. Frémond, E. Rocca, A new dual approach for a class of phase transitions with memory: existence and long-time behaviour of solutions. *J. Math. Pures Appl.*, 88, 455–481, 2007.
20. E. Bonetti, E. Rocca, R. Rossi, M. Thomas, A rate-independent gradient system in damage coupled with plasticity via structured strains, *ESAIM: Proceedings and Surveys*, 54, 54–69, 2016.
21. G. Bonfanti, M. Frémond, F. Luterotti, Global solution to a nonlinear system for irreversible phase changes. *Adv. Math. Sci. Appl.* 10, 1–24, 2000.
22. G. Caginalp, An Analysis of phase field model of a free boundary, *Arch. Rat. Mech. Anal.* 92, 205–245, 1986.
23. G. Caginalp, Stefan and Hele-Shaw type models as asymptotic limits of the phase-field equations. *Phys. Rev. A*, 39(11), 5887–5896, 1989.
24. G. Caginalp, Phase field computations of single-needle crystals, crystal growth and motion by mean curvature. *SIAM J. Sci. Comput.*, 15(1), 106–126, 1994.
25. G. Caginalp, J. Jones, A derivation and analysis of phase field models of thermal alloys. *Annal. Phys.*, 237, 66–107, 1995.
26. G. Caginalp, W. Xie, Phase-field and sharp-interface alloy models. *Phys. Rev. E*, 48(03), 1897–1999, 1993.
27. J. W. Cahn, J. E. Hilliard, Free energy of a nonuniform system. i. interfacial free energy, *J. Chem. Phys.* 28 (2), 258–267, 1958.

28. J. R. Cannon, E. DiBenedetto, G.H. Knightly, The steady state Stefan problem with convection. *Arch. Rat. Mech. Anal.*, 73, 79–97, 1980.
29. J. R. Cannon, E. DiBenedetto, G. H. Knigthly, The bidimensional Stefan problem with convection time dependent case. *Comm. Partial. Diff. Eqs.*, 8(14), 1549–1604, 1983.
30. B.M. Calsavara Caretta, J.L. Boldrini, Three-dimensional solidification with two possible crystallization states: Existence of solutions with flow in the melt. *Math. Methods Appl. Sci.*, 33 (5), 655–675, 2010.
31. C. Cao, C.G. Gal, Global solutions for the 2D NS–CH model for a two-phase flow of viscous, incompressible fluids with mixed partial viscosity and mobility. *Nonlinearity*, 25(11), 2012.
32. P. Colli, M. Frémond, E. Rocca, K. Shirakawa, Attractors for the 3D Fremond model of shape memory alloys, *Chin. Ann. Math. Ser. B*, 27, 683–700, 2006.
33. P. Colli, G. Gilardi, G. Marinoschi, E. Rocca, Optimal control for a phase field system with a possibly singular potential. *Math. Control Relat. Fields*, 6, 95–112, 2016.
34. P. Colli, Ph. Laurençot, Weak solution to the Penrose-Fife phase field model for a class of admissible heat flux laws. *Phys. D*, 111, 311–334, 1998.
35. P. Colli, G. Marinoschi, E. Rocca, Sharp interface control in a Penrose-Fife model. *ESAIM: COCV*, 22, 473–499, 2016.
36. P. Colli, J. Sprekels, On a Penrose-Fife model with zero interfacial energy leading to a phase-field system of relaxed Stefan type. *Ann. Mat. Pura Appl.*, 169(4), 269–289, 1995.
37. P. Colli, J. Sprekels, Stefan problems and the Penrose-Fife phase field model. *Adv. Math. Sci. Appl.*, 7(2), 911–934, 1997.
38. P. Colli, J. Sprekels, Weak solution to some Penrose-Fife phase-field systems with temperature-dependent memory. *J. Diff. Eq.*, 142(1), 54–77, 1998.
39. P. Colli, J. Sprekels, Global solution to the Penrose-Fife phase-field model with zero interfacial energy and Fourier law. *Adv. Math. Sci. Appl.*, 9(1), 83–391, 1999.
40. J. B. Collins, H. Levine, Diffuse interface model of diffusion-limited crystal growth. *Phys. Rev. B* , 31, 6119–6122, 1985.
41. P.N. da Silva, J.L. Boldrini, Maximal attractor for an Ostwald ripening model. *J. Math. Anal. Appl.*, 351, 107–119, 2009.
42. M. Dai, E. Feireisl, E. Rocca, G. Schimperna, M. Schonbek, On asymptotic isotropy for a hydrodynamic model of liquid crystals. *Asymptot. Anal.*, 97, 189–210, 2016.
43. E. DiBenedetto, A. Friedman, Conduction-convection problems with change of phase. *J. Diff. Eqs.* 62, 129–185, 1986.
44. E. DiBenedetto, M. O'Leary, Three-dimensional conduction-convection problems with change of phase. *Arch. Rat. Mech. Anal.*, 123, 99–116, 1993.
45. H.-J. Diepers, C. Beckermann, I. Steinbach, Simulation of convection and ripening in a binary alloy mush using the phase-field method. *Acta. Mater.*, 47(13), 3663–3678, 1999.
46. Q. Du, M. Li, C. Liu. Analysis of a phase field Navier-Stokes vesicle-fluid interaction model. *Discrete Continuous Dyn. Syst. Ser. B*, Vol. 8, No. 3, pp. 539–556, 2007.
47. Q. Du, C. Liu, R. Ryham, X. Wang, Energetic variational approaches in modeling vesicle and fluid interactions, *Phys. D*, 238, 923–930, 2009.
48. F.P. Duda, A. Ciarbonetti, P.J. Sánchez, A.E. Huespe, A phase-field/gradient damage model for brittle fracture in elastic-plastic solids. *Int. J. Plast.*, 65, 269–296, 2015.
49. M. Eleuteri, E. Rocca, G. Schimperna, On a non-isothermal diffuse interface model for two-phase flows of incompressible fluids. *Discrete Contin. Dyn. Syst.*, 35, 2497–2522, 2015.
50. M. Eleuteri, E. Rocca, G. Schimperna, Existence of solutions to a two-dimensional model for nonisothermal two-phase flows of incompressible fluids. *Ann. Inst. H. Poincaré Anal. Non Linéaire*, 33, 1431–1454, 2016.
51. H. Emmerich, *The Diffuse Interface Approach in Material Science: Thermodynamic Concepts and Applications of Phase-Field Models*. Lecture Notes in Physics 73, Springer Verlag, Berlin, 2003.
52. A.P. Entringer, J.L. Boldrini, A phase field α-Navier-Stokes vesicle-fluid interaction model: existence and uniqueness of solutions. *Discrete Continuous Dyn. Syst. Ser. B*, 20(2), 397–422, 2015.

53. M. Fabrizio, C. Giorgi, A. Morro, A thermodynamic approach to non-isothermal phase-field evolution in continuum mechanics. *Phys. D*, 214, 144–156, 2006.
54. E. Feireisl, H. Petzeltova, E. Rocca, G. Schimperna, Analysis of a phase-field model for two-phase compressible fluids. *Math. Models Methods Appl. Sci.*, 20(7), 2010.
55. G. J. Fix, Phase field methods for free boundary problems, in A. Fasano, M. Primicerio (Eds), *Free Boundary Problems: Theory and Applications*, Pitman, Boston, 580–589, 1983.
56. M. Frémond, *Non-Smooth Thermo-Mechanics*, Springer Verlag, Berlin Heidelberg, 2010.
57. M. Frémond, *Phase Change in Mechanics*, Springer Verlag, Berlin Heidelberg, 2012.
58. M. Frémond, B. Nedjar, Damage, gradient of damage and principle of virtual power. *Int. J. Solids Structures*, 33(8), 1083–1103, 1996.
59. M. Frémond, E. Rocca, Solid liquid phase changes with different densities. *Quart. Appl. Math.*, 66, 609–632, 2008.
60. S. Frigeri, E. Rocca, J. Sprekels, Optimal distributed control of a nonlocal Cahn-Hilliard/Navier-Stokes system in 2D. *SIAM J. Control Optim.*, 54, 221–250, 2016.
61. C.G. Gal, M. Grasselli, A. Miranville, Robust Exponential Attractors for Singularly Perturbed Phase-Field Equations with Dynamic Boundary Conditions. *Nonlinear Differ. Equ. Appl.*, 15, 535–556, 2008.
62. C.G. Gal, M. Grasselli, Asymptotic behavior of a Cahn-Hilliard-Navier-Stokes system in 2d. *Ann. Inst. H. Poincaré Anal. Non Linéaire*, 27 (1), 401–436, 2010.
63. C.G. Gal, M. Grasselli, Instability of two-phase flows: A lower bound on the dimension of the global attractor of the Cahn-Hilliard-Navier-Stokes system. *Phys. D*, 240(7), 629–635, 2011.
64. F. Guillén-González, G. Tierra, On linear schemes for a Cahn-Hilliard diffuse interface model. *J. Comput. Phys.*, 234: 140–171, 2013.
65. H. Gomez, T. Hughes, Provably unconditionally stable, second-order time-accurate, mixed variational methods for phase-field models. *J. Comput. Phys.*, 230(13), 5310–5327, 2011.
66. H. Gomez, K.G. van der Zee, Computational phase-field modeling. In: *Encyclopedia of Computational Mechanics, Second Edition*, Ewing Stein, René de Borst and Thomas J.R. Hughes, Eds. John Wiley & Sons, Ltd., 2017.
67. C. Heinemann, C. Kraus. *Phase Separation Coupled with Damage Processes. Analysis of Phase Field Models in Elastic Media*, Springer Spektrum, Wiesbaden, 2014.
68. C. Heinemann, E. Rocca, Damage processes in thermoviscoelastic materials with damage-dependent thermal expansion coefficients. *MMAS*, 38, 4587–4612, 2015.
69. K. H. Hoffman, L. Jiang, Optimal control a phase field model for solidification. *Numer. Funct. Anal. Optimiz.*, 13(1 & 2), 11–27, 1992.
70. A. Ito, N. Kenmochi, Inertial set for a phase transition model of Penrose-Fife type. *Adv. Math. Sci. Appl.*, 10(1), 353–374, 2000.
71. A. Ito, N. Kenmochi, M. Kubo, Non-isothermal phase transition models with Neumann boundary conditions. *Nonlinear Anal.* 53(7–8), 977–996, 2003.
72. A. Karma, W.-J. Rappel, Quantitative phase-field modeling of dendritic growth in two and three dimensions. *Phys. Rev. E*, 57(4), 4323–4349, 1998.
73. N. Kenmochi, M. Niezgódka, Systems of nonlinear parabolic equations for phase change problems. *Adv. Math. Sci. Appl.*, 3, 89–117, 1993/94.
74. N. Kenmochi, M. Kubo, Weak solutions of nonlinear systems for non-isothermal phase transitions. *Adv. Math. Sci. Appl.*, 9(1), 499–521, 1999.
75. J. Kim, Phase-field models for multi-component fluid flows. *Commun. Comput. Phys.*, 12(03), 613–661, 2012.
76. R. Kobayashi, Modeling and numerical simulation of dentritic crystal growth. *Phys. D*, 63, 410–479, 1993.
77. Ph. Laurençot, Weak solutions to a phase-field model with non-constant thermal conductivity. *Quart. Appl. Math.*, 15(4), 739–760, 1997.
78. Ph. Laurençot, G. Schimperma, U. Stefanelli, Global existence of a strong solution to the one-dimensional full model for irreversible phase transitions. *J. Math. Anal. Appl.* 271, 426–442, 2002.

79. F. Luterotti, G. Schimperma, U. Stefanelli, Global solution to a phase field model with irreversible and constrained phase evolution. *Q. Appl. Math.* 60, 301–316, 2002.
80. C. Liu, J. Shen, A phase field model for the mixture of two incompressible fluids and its approximation by a Fourier-spectral method. *Phys. D*, 179(3), 211–228, 2003.
81. C. Miehe, F. Welschinger, M. Hofacker, Thermodynamically consistent phase-field models of fracture: variational principles and multi-field FE implementations. *Int. J. Numer. Meth. Eng.*. 83 (10), 1273–1311, 2010.
82. C. Miehe, M. Hofacker, L.-M. Schänzel, F. Aldakheel, Phase field modeling of fracture in multi-physics problems. Part II. Coupled brittle- to-ductile failure criteria and crack propagation in thermo-elastic-plastic solids. *Comput. Methods Appl. Mech. Eng.* 294, 486–522, 2015.
83. A. Miranville, R. Quintanilla, A generalization of the Caginalp phase-field system based on the Cattaneo law. *Nonlinear Anal. Theory Methods Appl.*, 71(5–6), 2278–2290, 2009.
84. A. Miranville, S. Zelik, Robust exponential attractors for singularly perturbed phase-field type equations. *Electron. J. Differential Equations*, Vol. 2002(2002), No. 63, 1–28, 2002.
85. C. Moroşanu, D. Motreanu, A generalized phase-field system. *J. Math. Anal. Appl.* 237, 515–540, 1999.
86. B. Nedjar, Damage and gradient of damage in transient dynamics. In: *IUTAM Symposium Variations de domaines et frontières libres en mécaniques*, Kluwer, Amsterdam, 1998.
87. B. Nestler, D. Danilov, P. Galenko, Crystal growth of pure substances: phase-field simulations in comparison with analytical and experimental results. *J. Comput. Phys.* 207(1), 221–239, 2005.
88. B. Nestler, H. Garcke and B. Stinner, Multicomponent alloy solidification: phase-field modeling and simulations. *Phys. Rev. E*, 71, 041609–P2005
89. T.-T. Nguyen, J. Réthoré, J. Yvonnet, M.-C. Baietto, Multi-phase-field modeling of anisotropic crack propagation for polycrystaline materials. *Comp. Mech.* 60, 289–314, 2017.
90. M. O'Leray, Analysis of the mushy region in conduction-convection problems with change of phase. *Elect. Journal. Diff. Eqs.* 1997(4), 1–14, 1997.
91. K. A. Pericleous, M. Cross, G. Moran, P. Chow, K.S. Chan, Free surface Navier-Stokes flows with simultaneous heat transfer and solidification/melting. *Adv. Comput. Math.*, 6, 295–308, 1996.
92. O. Penrose and P.C. Fife. Thermodynamically consistent models of phase-field type for the kinetic phase transitions. *Phys. D*, 43, 44–62, 1990.
93. O. Penrose, P.C. Fife, On the relation between the standard phase-field model and a thermodynamically consistent phase-field model. *Phys. D*, 69, 107–113, 1993.
94. G. Planas, J.L. Boldrini, Weak solutions of a phase-field model with convection for solidification of an alloy. *Comm. Appl. Anal.*, 8(4), 503–532, 2004.
95. G. Planas, J.L. Boldrini, A bidimensional phase-field model with convection for change phase of an alloy. *J. Math. Anal. Appl.*, 303(2), 669–687, 2005.
96. N. Provatas, K. Elder, *Phase-Field Methods in Material Science and Engineering*. Wiley-VCH, Weinheim, 2010.
97. N. Provatas, M. Greenwood, B. Athereya, N. Goldenfeld, J. Dantzig, Multiscale modeling of solidification: Phase-Field methods to adaptive mesh refinement. *Internat. J. Modern Phys. B*, 19(31), 4525–4565, 2005.
98. J. Rappaz, J. F. Scheid, Existence of solutions to a Phase-field model for the isothermal solidification process of a binary alloy. *Math. Meth. Appl. Sci.*, 23, 491–512, 2000.
99. E. Rocca, R. Rossi, Analysis of a nonlinear degenerating PDE system for phase transitions in thermoviscoelastic materials. *J. Differential Equations*, 245, 3327–3375, 2008.
100. E. Rocca, R. Rossi, A degenerating PDE system for phase transitions and damage. *Math. Models Methods Appl. Sci.*, 24, 1265–1341, 2014.
101. E. Rocca, G. Schimperna, Universal attractor for some singular phase transition systems. *Phys. D*, 192(3-4), 279–307, 2004.
102. E. Rocca, G. Schimperna, Global attractor for a parabolic-hyperbolic Penrose-Fife phase field system. *Discrete Contin. Dyn. Syst.*, 15(4), 1192–1214, 2006.

103. E. Rocca, J. Sprekels, Optimal distributed control of a nonlocal convective Cahn-Hilliard equation by the velocity in 3D, *SIAM J. Control Optim.*, 53, 1654–1680, 2015.
104. L. I. Rubinstein, *The Stefan Problem*, Am. Math. Soc. Transl. 27, AMS, Providence, 1971.
105. J.-F Scheid, Global solutions to a degenerate solutal phase-field model for the solidification of a binary alloy. *Nonlinear Anal. Real World Appl.* 5 (1), 207–217, 2004.
106. J.C. Simo, T.H.R. Hughes, *Computational Inelasticity*, Springer Verlag, New York, 1998.
107. J. Sprekels, S. M. Zheng, Global smooth solutions to a thermodynamically consistent model of phase-field type in higher space dimensions. *J. Math. Anal. Appl.* 176(1), 200–223, 1993.
108. R. Temam, *Navier-Stokes Equations, Theory and Numerical Analysis*, AMS Chelsea Publishing, American Mathematical Society, Providence, RI, 2001.
109. C. Truesdell, W. Noll, *The Non-Linear Field Theories of Mechanics*. Springer Verlag, Heidelberg, 1965.
110. D. Vasconcelos, A. Rossa, A. Coutinho, A residual-based Allen–Cahn phase field model for the mixture of incompressible fluid flows. *Int. J. Numer. Methods Fluids*, 75(9), 645–667, 2014.
111. C.L.D. Vaz, J.L. Boldrini, A mathematical analysis of a nonisothermal Allen-Cahn type system, *Math. Methods Appl. Sci.*, 35(12), 1392–1405, 2012.
112. V. R. Voller, C. Prakash, A fixed grid numerical modelling methodology for convection-diffusion mushy region phase-change problems. *Int. J. Mass. Tranfer*, 30(8), 1709–1719, 1987.
113. V. R. Voller, M. Cross, N. C. Markatos, An enthalpy method for convection/diffusion phase field models of solidification. *Int. J. Num. Methods. Eng.*, 24(1), 271–284, 1987.
114. S.-L. Wang, R.F. Sekerka, A.A. Wheeler, B.T. Murray, S.R. Coriel, R.J. Braun, G.B. McFadden, Thermodynamically-consistent phase-field models for solidification. *Phys. D*, 69, 189–200, 1993.
115. J. A. Warren, W. J. Boettinger, Prediction of dendritic growth and microsegregation patterns in a binary alloy using the phase-field method. *Acta Metall. Mater.*, 43(2), 689–703, 1995.
116. G.N. Wells, E. Kuhl, K. Garikipati, A discontinuous Galerkin method for the Cahn-Hilliard equation. *J. Comput. Phys.*, 218(2), 860–877, 2006.
117. A. A. Wheeler, W. J. Boettinger and G. B. McFadden, Phase-field model for isothermal phase transitions in binary alloys. *Phys. Rev. A*, 45, 7424–7439, 1992.
118. X. Yang, J.J. Feng, C. Liu and J. Shen, Numerical simulations of jet pinching-off and drop formation using an energetic variational phase-field method. *J. Comput. Phys.*, 218, 417–428, 2006.
119. S. M. Zheng, Global existence for a thermodynamically consistent model of phase field type. *Differ. Integral Equ.*, 5(2), 241–253, 1992.

Spherical Codes from Lattices

Sueli I. R. Costa, João E. Strapasson, and Cristiano Torezzan

Abstract Lattices are homogeneous discrete sets in the n-dimensional space that have been used in different applications in communication areas such as coding for Gaussian or fading channels and cryptography. This chapter approaches the connection between quotient of lattices and spherical codes, presenting a survey on contributions to this topic mainly based on Costa et al. (Flat tori, lattices and spherical codes. In: 2013 Information Theory and Applications Workshop (ITA), February 2013, pp 1–8), Siqueira and Costa (Des Codes Cryptogr 49(1–3):307–321, 2008), Torezzan et al. (IEEE Trans Inf Theory 59(10):6655–6663, 2013), Costa et al. (Lattice applied to codding for reliable and secure communications. Springer, 2017), and Torezzan et al. (Des Codes Cryptogr 74(2):379–394, 2015).

1 Introduction

Lattices are discrete sets in $\mathbb{R}^n$ given as integer linear combinations of a set of independent vectors. Problems on the geometry of lattices and their packings have already been approached since the seventeenth century in works by J. Kepler, I. Newton, F. Gauss, J-L. Lagrange and H. Minkowski, and this is still a very active field of research with applications in different areas [44]. In communication and coding theory lattices have been used for coding for Gaussian and fading channels [45] and also in cryptography [28, 29].

Group codes as introduced by Slepian in [31], and developed in [3, 4, 24, 27], are defined as finite sets on a sphere in the n-dimensional space generated by a group of orthogonal matrices. These spherical codes are geometrically uniform codes [25]

S. I. R. Costa (✉)
Institute of Mathematics, Statistics and Scientific Computing, University of Campinas, Campinas, SP, Brazil
e-mail: sueli@ime.unicamp.br

J. E. Strapasson · C. Torezzan
School of Applied Sciences, University of Campinas, Campinas, SP, Brazil
e-mail: joao.strapasson@fca.unicamp.br; cristiano.torezzan@fca.unicamp.br

C. Lavor, F. A. M. Gomes (eds.), *Advances in Mathematics and Applications*,
https://doi.org/10.1007/978-3-319-94015-1_5

capturing the highly desirable properties that come from homogeneity such as same distance profile and congruent Voronoi region for each codeword.

Lattices in $\mathbb{R}^L$ with orthogonal sublattices can be used to construct spherical codes in $\mathbb{R}^{2L}$, generated by a finite commutative group of orthogonal: matrices, which are contained in a flat torus. By considering the unit sphere in $\mathbb{R}^{2L}$ as foliated by flat tori, which are L-dimensional surfaces of zero curvature we can also construct quasi-commutative group codes on layers of flat tori. Lattices and flat tori can also be used to construct homogeneous spherical curves for transmitting a continuous alphabet source over an AWGN (additive white Gaussian noise) channel. In both cases the performance is related to the packing density of the involved lattices. In the continuous case the packing density of curves (fat-strut problem) relies on the search for projection lattices with good packing density.

We present here a survey on this topic, mainly based in [13, 16, 30, 39], and summarize most of the contributions of our research group up to now. This research group has been formed, since 1997, through an interaction between faculty members, graduated and undergraduated students and post-doctoral researchers of the Institute of Mathematics, Statistics and Computer Science (IMECC) with the Faculty of Electrical and Computer Engineering (FEEC) and the Institute of Computing (IC) of the University of Campinas. An important support for this group has been given by the FAPESP foundation through four consecutive thematic projects.

This chapter is organized as follows. A very brief introduction to lattices and to spherical codes is presented in Sects. 2 and 3. In Sect. 4 flat tori in the sphere are described and related bounds for distances are derived. The connection between lattices and commutative group codes in even dimensions as well as the constructions of codes using this connection through different approaches is presented in Sects. 5 and 6. Quasi-commutative group codes on layers of flat tori are described in Sect. 7 and in Sect. 8 the structure of flat tori and lattices comes together again in the proposal of homogeneous spherical curves for transmitting a continuous alphabet source over an AWGN channel. An application to the wiretap channel is also included.

2 Lattices

In this section we briefly introduce the concept of lattice and its main proprieties to be used hereafter. A general reference for lattices is the classical book [10] and applications to communication areas can also be found in [45] and [13].

A *lattice* $\Lambda \subset \mathbb{R}^n$ is a discrete set of vectors composed by all integer linear combinations of a subset of independent vectors $\beta = \{\boldsymbol{b}_1, \boldsymbol{b}_2, \ldots, \boldsymbol{b}_m\}$:

$$\Lambda = \Lambda_\beta = \{u_1\boldsymbol{b}_1 + \cdots + u_m\boldsymbol{b}_m; u_1, \ldots, u_m \in \mathbb{Z}\}. \quad (1)$$

The set β is said to be a *basis* and m is the *rank* of Λ. If $m = n$, we say that Λ is *full-rank*. In this chapter we only consider full rank lattices and vectors described

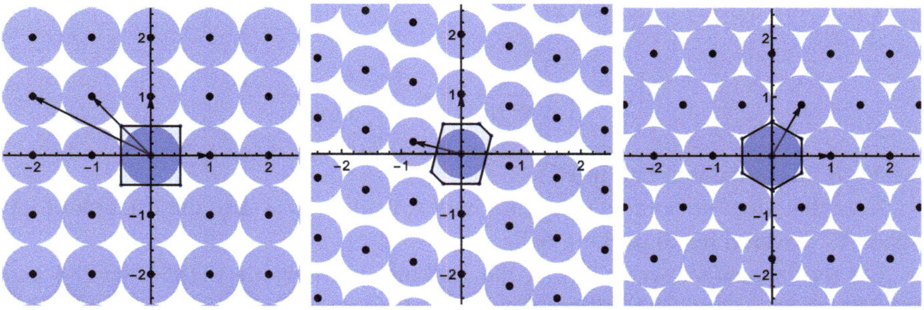

Fig. 1 Three lattices in $\mathbb{R}^2$ and their packings

in column form. A *generator matrix* B for a lattice Λ is a matrix whose columns are a basis for it, i.e. $\Lambda = \{\mathrm{B}\boldsymbol{u};\ \boldsymbol{u} \in \mathbb{Z}^m\}$, where $\mathbb{Z}^m$ denotes the set of integers m-uples. Note that, for full rank lattices, a generator matrix $B_{n\times m}$ is invertible and any lattice Λ admits an infinite number of basis, since any matrix $\mathrm{H} = \mathrm{B}\,\mathrm{U}$, where U is unimodular (U has integer elements and $\det(\mathrm{U}) = \pm 1$), is also a generator matrix for Λ [13, Ch. 2].

Example 1 Three different lattices in $\mathbb{R}^2$ are shown in Fig. 1, which also illustrate their packing densities, to be introduced next. The lattices are represented by black dots and the bases are $B_1 = \{(1,0), (0,1)\}$, $B_2 = \{(-4/5, 1/5), (0,1)\}$ $B_3 = \left\{(1,0), (1/2, \sqrt{3}/2)\right\}$ for the lattices on the left, at the centre and on the right, respectively.

An example of full rank lattice in $\mathbb{R}^n$ is given by the checkerboard lattice D_n, defined as $D_n = \{(x_1, \dots x_n) \in \mathbb{Z}^n;\ x_1 + \cdots + x_n \text{ is even}\}$.

Since two generator matrices B and A for the same full rank lattice satisfy $\mathrm{A} = \mathrm{B}\,\mathrm{U}$, for some unimodular matrix U, we must have $|\det(\mathrm{A})| = |\det(\mathrm{B})|$. This number $|\det(\mathrm{B})|$, which is basis invariant, is also the n-dimensional *volume* of the Voronoi region of the lattice at the origin. Such region is defined as the set of points of $\mathbb{R}^n$ which are closer to the origin than to any other lattice point. The Voronoi region of a lattice including its boundary provides a tilling of $\mathbb{R}^n$ through translations by lattice vectors.

The *packing radius* of a lattice, i.e. the largest radius of congruent open balls centred at lattice points that can be packed without overlapping, is given by $d/2$, where d is the minimum distance between any two distinct lattice points. The *packing density*, $\bar{\Delta}(\Lambda)$, of lattice is the ratio between the volume of an n-dimensional ball $B^n(d/2)$ of the packing radius $d/2$ and the volume $\mathrm{vol}(\Lambda) = |\det(\mathrm{B})|$ of the Voronoi region,

$$\bar{\Delta}(\Lambda) = \frac{\mathrm{vol}(B^n(d/2))}{|\det(\mathrm{B})|}.$$

The centre density is defined as

$$\Delta(\Lambda) = \frac{\bar{\Delta}(\Lambda)}{\text{vol}(B(1))} = \frac{d^n}{2^n|\det(B)|}, \qquad (2)$$

where $B^n(1)$ is the unit ball in $\mathbb{R}^n$.

Note that, due the homogeneity of a lattice, its density is also the fraction of the space $\mathbb{R}^n$ covered by all packing balls.

For the lattices shown in Fig. 1, the packing densities are 0.7854, 0.6283 and 0.9069, respectively. The lattice Λ generated by B_3 presented on the right side is called the hexagonal lattice, and it has the largest possible packing density in dimension 2 while the lattice D_n has the best packing density in dimensions 3, 4 and 5.

The *dual* Λ^* of a full rank lattice $\Lambda \subset \mathbb{R}^n$ is a lattice defined as

$$\Lambda^* = \{\boldsymbol{y} \in \mathbb{R}^n; \boldsymbol{x} \cdot \boldsymbol{y} \in \mathbb{Z}, \forall \boldsymbol{x} \in \Lambda\}$$

and it plays an important role in applications involving projection of lattices, as it will be discussed in Sect. 8.

2.1 Quotient of Lattices and q-Ary Codes

A lattice Λ' is said to be a sublattice of Λ if $\Lambda' \subset \Lambda$. Note that if Λ' and Λ have generator matrices B$'$ and B, Λ' is sublattice of Λ, if only if B$'$ = BH, where H is a matrix of integers.

Since the lattices Λ and Λ' are commutative groups, if $\Lambda' \subset \Lambda$, the *quotient of lattices* represented by $\frac{\Lambda}{\Lambda'}$ is also a group and, for full rank lattices, the number of elements in this group is given by

$$\left|\frac{\Lambda}{\Lambda'}\right| = \frac{\text{vol}(\Lambda')}{\text{vol}(\Lambda)} = |\det(\text{H})|.$$

Any integer square matrix H, of order n, can be decomposed into the so-called Smith normal form, H = UDV, where U and V are unimodular matrices, D is a diagonal matrix with diagonal elements $d_j \in \mathbb{N}$ and $d_i|d_{i+1}$. [9, Sec 2.4]

The Smith normal form can be used to obtain special bases for a pair of nested full rank lattices and also to identify the quotient group Λ/Λ' as well as its generators. This will be used to describe spherical codes in the following sections. Given a nested pair of full rank lattices $\Lambda' \subset \Lambda$, there are special bases $\{\mathbf{w}_1, \ldots \mathbf{w}_n\}$ of Λ' and $\{\mathbf{v}_1, \ldots, \mathbf{v}_n\}$ of Λ such that $\mathbf{w_i} = k_i\mathbf{v_i}$, for $i = 1, \ldots, n$; $k_i \in \mathbb{N}$.

Let B be a generator matrix of Λ and BH a generator matrix of Λ'. Consider the Smith decomposition, H = UDV. The matrix A = BHV^{-1} = (BU)D is also a generator matrix of Λ' and BU is a generator matrix of Λ, since U and V^{-1} are unimodular matrices. If we consider $\{\mathbf{w}_1, \ldots, \mathbf{w}_n\}$ and $\{\mathbf{v}_1, \ldots, \mathbf{v}_n\}$ as the columns of matrices A and BU, respectively, we get $\mathbf{w}_j = d_j\mathbf{v}_j$. This implies that the group Λ/Λ' is isomorphic to $\mathbb{Z}_{d_k} \oplus \mathbb{Z}_{d_{k+1}} \oplus \cdots \oplus \mathbb{Z}_{d_n}$, where $k = \min\{i; d_i \neq 1\}$. The generators of this quotient are the classes $\bar{\mathbf{v}}_j$, $j = k, \ldots, n$.

This approach of quotient lattices allows to describe the orthogonal group of matrices which generate the spherical code in the next sections.

Quotient of special lattices are also associated with q-ary codes what provides a connection between the spherical code construction presented here and these codes.

For a lattice Λ with integer coordinates which contains $\Lambda' = q\mathbb{Z}^n$ as a sublattice, $\Lambda/\Lambda' \subseteq \mathbb{Z}_q^n$ is identified with a linear code over a q-ary alphabet (a *q-ary code*). A natural distance to be considered for error correction in these codes is the Lee distance, which is induced from the metric l_1 (or graph distance) in $\mathbb{Z}^n$.

For example, the code $\mathscr{C}_1 = \{k(4, 3) \mod 25; 0 \leqslant k \leqslant 24\}$ in $\mathbb{Z}_{25}^2$ is perfect regarding the Lee metric, in the sense that balls of packing radius 3 centred at its points cover $\mathbb{Z}_{25}^2$. $\mathscr{C}_2 = \{k(1, 10) \mod 25; 0 \leqslant k \leqslant 24\}$ has packing radius 2 and is not perfect (Fig. 4 illustrates these codes). Perfect q-ary codes on flat tori were approached in [14]. There is a long-standing conjecture stated by Golomb-Weech in 1970 asserting that for dimension $\geqslant 3$ the only perfect codes in $\mathbb{Z}^n$ regarding the graph metric have packing radius one. This will imply the same assertion for q-ary codes with $q \geqslant 2R + 1$. Quasi-perfect codes in the Lee-metric and perfect codes in the l_p metric were considered in [5, 35].

If we consider the (cyclic) code $\mathscr{C}_1$ ordered by k, we have a circulant graph where each point has four neighbours (at graph distance 7). This geometric view through quotient of lattices [14, 33] may provide tools to analyse circulant and Cayley graphs which are used in parallel computing schemes.

The search for good codes regarding the Lee metric (l_1 metric) relies on analysing balls of maximum radius centred at the code points that do not intersect each other (packing balls) that offer the maximum possible covering of the full space $\mathbb{Z}_q^n$. This approach is considered in [12, 14, 18] and also in [5, 35] regarding the l_p metric.

3 Spherical Codes

A *spherical code* is a finite set of M points on a sphere of radius a, $S^{n-1}(a) = \{\mathbf{x} \in \mathbb{R}^n; \|\mathbf{x}\| = a\}$. Usually we consider only spherical codes on the sphere of radius one, $S^{n-1} = S^{n-1}(1)$ and all the conclusions will be extended by similarity to a sphere of radius a. Two dual optimization (packing) problems regarding spherical codes, which have several applications such as the ones in physics, chemistry, architecture and signal processing, can be stated as:

(i) Given a number M to find a spherical code with M points and maximum possible minimum distance between them.

(ii) Given a minimum distance d to find a spherical code with the largest number M of points such that each two of them are at a distance at least d.

Codes which are solutions for one of these problems are called *optimal spherical codes*. Optimal spherical codes are known only in a few cases. In dimension 2, codes which are vertices of regular polygons inscribed in the circle S^1 provide solutions for both problems. The solution of (i) in dimension 3 is, up to now, only known for $1 \leqslant M \leqslant 12$ and for $M = 24$ [19, 26]. For $M = 2, 3$ and 4 the optimal spherical codes in $\mathbb{R}^3$ are two antipodal points, the vertices of an equilateral triangle inscribed on an equator and the vertices of an inscribed regular tetrahedron in $S^2 \subset \mathbb{R}^3$, respectively. For $M = 8$, the optimal spherical code in $\mathbb{R}^3$ is given by the vertices not of a cube as one should possibly expected, but of a regular anti-prism with eight vertices and with same length edges (see Fig. 3). Spherical codes which are known to be optimum in any dimension n are the antipodal code ($M = 2$), the simplex code ($M = n + 1$) and the bi-orthogonal code ($M = 2n$) given by permutation of vectors $(\pm 1, 0, 0, \ldots, 0)$ [19].

A group code $\mathscr{C}$ is a spherical code given as an orbit of a finite multiplicative group of orthogonal matrices $\mathscr{G} = \{\mathrm{G}_i, i = 1, \ldots, M\}$, that is $\mathscr{C} = \mathscr{G}\boldsymbol{u} = \{\mathrm{G}_i\boldsymbol{u}, \mathrm{G}_i \in \mathscr{G}\}$. $\boldsymbol{u} \in S^{n-1}$ is called the initial vector of the group code. A group code may not provide an optimum code but it has more structure, can be easily generated and is quite homogeneous. Its minimum distance can be given as:

$$d := \min_{\substack{\boldsymbol{x},\,\boldsymbol{y} \in \mathscr{C} \\ \boldsymbol{x} \neq \boldsymbol{y}}} \|\boldsymbol{x} - \boldsymbol{y}\| = \min_{\substack{\mathrm{G}_i \in \mathscr{G} \\ \mathrm{G}_i \neq I_n}} \|\mathrm{G}_i\,\boldsymbol{u} - \boldsymbol{u}\|,$$

where $\|.\|$ and I_n denote the standard Euclidean norm and the identity matrix of order n, respectively.

We remark that the minimum distance of a group code depends on the generator group and on the chosen initial vector. Isomorphic groups may also present different minimum distances for the same initial vector, as it will be seen in the following sections.

4 Flat Tori

The unit sphere $S^{2L-1} \subset \mathbb{R}^{2L}$ can be foliated by flat tori (Clifford Tori) as follows. For each unit vector $\boldsymbol{c} = (c_1, c_2, \ldots, c_L) \in \mathbb{R}^L$, $c_i > 0$, $\sum_{i=1}^{L} c_i^2 = 1$, and $\boldsymbol{u} = (u_1, u_2, \ldots, u_L) \in \mathbb{R}^L$, let $\varphi_{\boldsymbol{c}} : \mathbb{R}^L \to \mathbb{R}^{2L}$ be defined as

$$\varphi_{\boldsymbol{c}}(\boldsymbol{u}) = \left(c_1 \cos\left(\frac{u_1}{c_1}\right), c_1 \sin\left(\frac{u_1}{c_1}\right), \ldots, c_L \cos\left(\frac{u_L}{c_L}\right), c_L \sin\left(\frac{u_L}{c_L}\right)\right). \qquad (3)$$

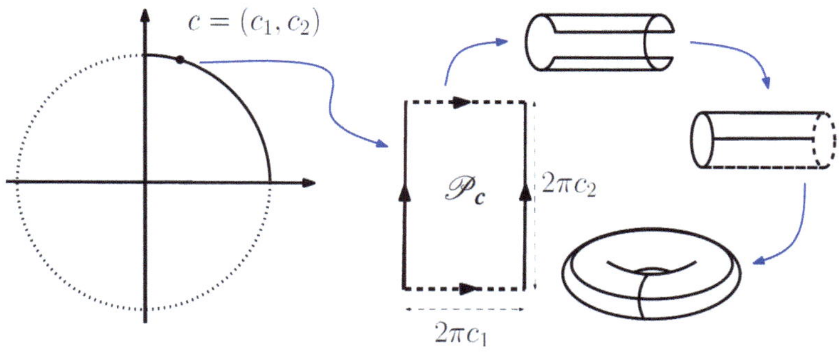

Fig. 2 Illustration of a torus layer T_c

The image of this periodic mapping φ_c is the torus T_c, a flat L-dimensional surface in the unit sphere S^{2L-1}. T_c is also the image of an L-dimensional box $\mathscr{P}_c$,

$$\mathscr{P}_c = \{u \in \mathbb{R}^L; 0 \leqslant u_i < 2\pi c_i\}, \quad 1 \leqslant i \leqslant L, \tag{4}$$

in which $\mathscr{P}_c$ is injective (see Fig. 2).

For $c \in S^{L-1}$ and $c_i \geqslant 0$, if $c_i = 0$ for some $1 \leqslant i \leqslant L$, we may replace in (3) both coordinates related to c_i by 0 and obtain a degenerated flat torus T_c, which is an embedding of a $(L-k)$-dimensional box in $\mathbb{R}^{2L}$, where k is the number of zero coordinates of $\boldsymbol{c}$.

The Gaussian curvature of a torus T_c is zero and T_c can be cut and flattened into the box, $\mathscr{P}_c$, just as a cylinder in $\mathbb{R}^3$ can be cut and flattened into a 2-dimensional rectangle. The mapping φ_c is a local isometry, what means that any measure of length, area and volume up to dimension $L-k$ on T_c is the same of the corresponding pre-image in the box $\mathscr{P}_c$.

The family of flat tori T_c and their degenerations, with $\boldsymbol{c} = (c_1, c_2, \ldots, c_L)$, $\|\boldsymbol{c}\| = 1$, $c_i \geqslant 0$, defined above is a foliation of the unit sphere of $S^{2L-1} \subset \mathbb{R}^{2L}$. This means that any vector of S^{2L-1} belongs to one and only one of these flat tori.

The following results [39, 41] allow to relate the distances between two points in $\mathbb{R}^L$ and their spherical image on a flat torus in $\mathbb{R}^{2L}$ and will be used in the construction of spherical codes.

Proposition 1 *Let T_b and T_c be two flat tori, defined by unit vectors $\boldsymbol{b}$ and $\boldsymbol{c}$ with non-negative coordinates. The minimum distance $d(T_c, T_b)$ between two points on these flat tori is*

$$d(T_c, T_b) = \|\boldsymbol{c} - \boldsymbol{b}\| = \left(\sum_{i=1}^{L} (c_i - b_i)^2\right)^{1/2}. \tag{5}$$

The distance between two points $\varphi_{\boldsymbol{c}}(\boldsymbol{u})$ and $\varphi(\boldsymbol{v})$ on the same torus $T_{\boldsymbol{c}}$, defined by a vector $\boldsymbol{c} = (c_1, \ldots, c_L)$, is given by

$$\|\varphi_{\boldsymbol{c}}(\boldsymbol{u}), \varphi_{\boldsymbol{c}}(\boldsymbol{v})\| = 2\sqrt{\sum c_i^2 \sin^2\left(\frac{u_i - v_i}{2c_i}\right)} \tag{6}$$

and it is bounded according to the next proposition.

Proposition 2 *Let* $\boldsymbol{c} = (c_1, c_2, \ldots, c_L) \in S^{2L-1}$, $c_i > 0$, $c_\xi = \min_{1 \leqslant i \leqslant L} c_i \neq 0$, $\Delta = \|\boldsymbol{u} - \boldsymbol{v}\|$, *for* $\boldsymbol{u}, \boldsymbol{v} \in P_{\boldsymbol{c}}$. *Suppose* $0 < \Delta \leq \pi c_\xi$, *then*

$$\frac{2\Delta}{\pi} \leq \sin\left(\frac{\Delta}{2c_\xi}\right) 2c_\xi \leq \|\varphi_{\boldsymbol{c}}(\boldsymbol{u}) - \varphi_{\boldsymbol{c}}(\boldsymbol{v})\| \leq \frac{\sin\frac{\Delta}{2}}{2} \leq \Delta$$

In the next section we show that a commutative group codes must lie on a flat torus, and the above proposition allows to derive bounds on the minimum distance of these codes.

5 Commutative Group Codes

Consider a unit vector $\boldsymbol{u}$ in the n-dimensional Euclidean space $\mathbb{R}^n$ and a finite commutative group of orthogonal matrices, $\mathcal{G} = \{G_1, G_2, \ldots, G_M\}$. The orbit of $\boldsymbol{u}$ under the action of $\mathcal{G}$ is the set of points $\mathcal{C} = \{G\,\boldsymbol{u}, \forall\, G \in \mathcal{G}\}$. Due to orthogonality, the points in $\mathcal{C}$ belong to the surface of the unit sphere $S^{n-1} \subset \mathbb{R}^n$, and this spherical code is called a *commutative group code.*

Commutative group codes belong to the family of *Slepian* group codes, introduced in [31] or, in a more general sense, geometrically uniform codes [25]. Such codes have been widely applied in communication theory, for instance, to match signal sets to groups [27].

There is a well-known representation of a finite commutative group of orthogonal matrices $\mathcal{G}$, as stated in the next proposition:

Proposition 3 ([21] Theorem 12.1) *Every finite commutative group* $\mathcal{G}$ *of orthogonal matrices,* $n \times n$, *can be carried by the same real orthogonal transformation* Q *into a block-diagonal form:*

$$QG_iQ^t = [R(a_{i\,1}), \ldots, R(a_{i\,q}), \mu(i)_{2q+1}, \ldots, \mu(i)_n]_{n\times n},$$

$R(a_{i\,j})$ *is the rotation matrix,*

$$R(a_{i\,j}) = \begin{bmatrix} \cos(a_{i\,j}) & -\sin(a_{i\,j}) \\ \sin(a_{i\,j}) & \cos(a_{i\,j}) \end{bmatrix}, \tag{7}$$

where $a_{i\,j} = \frac{2\pi b_{i\,j}}{M}$, $b_{i\,j} \in \mathbb{Z}$, $0 \leqslant b_{i\,j} \leqslant M$ *and* $\mu(i)_l = \pm 1$, $l = 2q + 1, \ldots, n$, $j = 1, \ldots, q$, $\forall G_i \in \mathcal{G}$.

Any commutative group $\mathcal{G}$ is isomorphic to the group $\mathbb{Z}_{m_1} \oplus \cdots \oplus \mathbb{Z}_{m_k}$, where m_i divides m_{i+1}. If $\mathcal{G}$ is a commutative group of orthogonal matrices of order M, any $G \in \mathcal{G}$ can be given a product of powers of generator matrices [9, 21]:

$$\mathcal{G} = \{\mathrm{G}_{j_1}^{i_1} \ldots \mathrm{G}_{j_k}^{i_k}; 0 \leqslant i_1 \leqslant m_1 - 1, \ldots, 0 \leqslant i_k \leqslant m_k - 1 \text{ and } m_{i_1} \ldots m_{i_k} = M\}.$$

We will then denote $\mathcal{G} = \langle \mathrm{G}_{j_1}, \ldots \mathrm{G}_{j_k} \rangle$.

The above result provides a characterization of the geometric locus, [30], of a commutative group code.

Proposition 4 (Geometric Locus of Commutative Group Codes) *Every commutative group code of order M is congruent to a commutative group code X whose initial vector is $\boldsymbol{u} = (u_1, \ldots, u_n)$ and its points have the form*

$$\Big(\mathrm{R}(a_{i\,1})(u_1, u_2), \ldots, \mathrm{R}(a_{i\,q})(u_{2q-1}, u_{2q}), \mu_{2q+1}(i)u_{2q+1}, \ldots, \mu_n(i)u_n\Big), \qquad (8)$$

where $a_{i\,j} = \dfrac{2\pi b_{i\,j}}{M}$. Moreover,

1. *If $n = 2L$, X is contained in the flat torus T_δ, where $\delta = (\delta_1, \ldots, \delta_L)$ satisfies $\delta_i^2 = u_{2i-1}^2 + u_{2i}^2$.*
2. *If $n = 2L + 1$ and X is not contained in a proper subspace, $X = X_1 \cup X_2$, where X_i is contained in the hyperplane $\Pi = \{(x_1, \ldots, x_{2L+1}) \in \mathbb{R}^{2L+1}; x_{2L+1} = (-1)^i u_n\}$. Also, X_i is contained in the torus T_δ of a sphere in $\Pi \equiv \mathbb{R}^{2L}$ with radius $(1 - u_n^2)^{1/2}$, where $\delta_i^2 = u_{2i-1}^2 + u_{2i}^2$.*

Remark 1 By applying again Proposition 3 we can see that in the above Proposition 4 the initial vector in $\mathbb{R}^{2L}$ can always be considered as $\boldsymbol{u} = (\delta_1, 0, \ldots, \delta_L, 0)$.

Examples of commutative group codes, using the notation of Proposition 3:

1. The spherical code in $\mathbb{R}^4$, $\mathcal{G}_1\boldsymbol{u}$, where $\mathcal{G}_1 = \langle \mathrm{G}_1 \rangle$, $\mathrm{G}_1 = \big[\mathrm{R}\big(\frac{4\pi}{25}\big), \mathrm{R}\big(\frac{3\pi}{25}\big)\big]$, and initial vector $\boldsymbol{u} = (\sqrt{2}/2, 0, \sqrt{2}/2, 0)$ is a cyclic group code of order $M = 25$.
2. The spherical code in $\mathbb{R}^4$, $\mathcal{G}_2\boldsymbol{u}$, where $\mathcal{G}_2 = \langle \mathrm{G}_1, \mathrm{G}_2 \rangle$, $\mathrm{G}_1 = \big[\mathrm{R}\big(\frac{\pi}{5}\big), \mathrm{I}_2\big]$, $\mathrm{G}_2 = \big[\mathrm{I}_2, \mathrm{R}\big(\frac{\pi}{5}\big)\big]$ and initial vector $\boldsymbol{u}$ as above. Note that $\mathcal{G}_2 \approx \mathbb{Z}_5 \times \mathbb{Z}_5$ is not cyclic.
3. For $M = 8$ the regular anti-prism in $\mathbb{R}^3$. In can be described as $\mathcal{G}_3$, where $\mathcal{G}_3 = \langle \big[\mathrm{R}\big(\frac{\pi}{4}\big), -1\big] \rangle$ and $\boldsymbol{v} = (0.859533, 0, 0.511081)$ (Fig. 3).

The following proposition [11, 15, 30] gives a characterization for commutative group codes in even dimensions as a torus mapping (3) image of lattices in half of the dimension.

Proposition 5 *Let $\mathcal{G}\boldsymbol{u}$ be a commutative group code of order M in even dimension $n = 2L$, where $\boldsymbol{u} = (\delta_1, 0, \ldots, \delta_L, 0)$. If $L = q$ in (7), i.e. if the elements of $\mathcal{G}$ are free from 2×2 reflection blocks, $\bar{\boldsymbol{u}} = (\delta_1, \ldots, \delta_L)$, then the inverse image $\varphi_{\bar{\boldsymbol{u}}}^{-1}(\mathcal{G}\boldsymbol{u})$, through the torus mapping (3), is the full rank lattice $\Lambda_{\mathcal{G}\boldsymbol{u}}$ generated by the set*

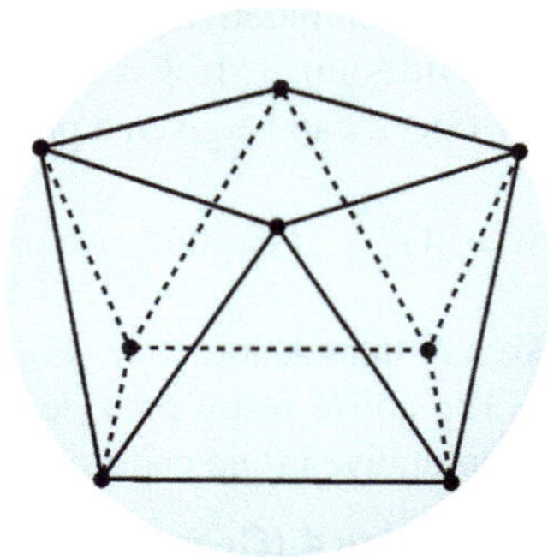

Fig. 3 The optimum commutative group code with $M = 8$ in $\mathbb{R}^3$: The anti-prism generated by $\mathscr{G} = \langle [R\frac{\pi}{4}, -1] \rangle$ and initial vector (0.859533, 0, 0.511081)

$$\left\{ \boldsymbol{v}_i;\ \boldsymbol{v}_i = \left(\frac{2\pi b_{i\,1}\delta_i}{M}, \ldots, \frac{2\pi b_{i\,L}\delta_L}{M} \right) \right\},$$

which has the orthogonal lattice $\Lambda' = \prod_{j=1}^{L}(2\pi\delta_j)\mathbb{Z}$ as a sublattice. The group $\mathscr{G}$ is isomorphic to $\Lambda_{\mathscr{G}\boldsymbol{u}}/\Lambda'$.

As a consequence of the last proposition the following bound is derived in [30].

Proposition 6 (Bounds for Commutative Group Codes) *Bounds for the number of points M of a commutative group code free from reflection blocks in $\mathbb{R}^{2L}$, with minimum distance at least d, are given by*

$$M \leqslant \frac{\pi^L \Pi_{j=1}^{L} \delta_j\ \Delta(\Lambda)}{\left(\arcsin \frac{d}{4}\right)^L} \leqslant \frac{\pi^L \Delta_L}{\left(\arcsin \frac{d}{4}\right)^L L^{L/2}}, \tag{9}$$

where $\Delta(\Lambda)$ is the centre density (2) *of the associated lattice described in the last proposition and Δ_L is the maximal centre density of a lattice in $\mathbb{R}^L$ and $\boldsymbol{u}$ $= (\delta_1, 0, \delta_2, 0, \ldots, \delta_L, 0)$ is the initial vector.*

As we can see from the above proposition, the search for good spherical codes in dimension $2L$ generated by a commutative group of orthogonal matrices is related to lattices with good packing density in dimension L.

Consider the examples of the spherical codes $\mathscr{C}_1 = \mathscr{G}_1\boldsymbol{u}$, $\mathscr{C}_2 = \mathscr{G}_2\boldsymbol{v}$ in $\mathbb{R}^4$, where $\mathscr{G}_1$ is generated by $G_1 = \left[R\left(4\frac{2\pi}{25}\right), R\left(3\frac{2\pi}{25}\right)\right]$ ($\mathscr{G}_1 = \langle G_1 \rangle$) and $\boldsymbol{u} = \left(\frac{\sqrt{2}}{2}, 0, \frac{\sqrt{2}}{2}, 0\right)$, $\mathscr{G}_2 = \left\langle \left[R\left(\frac{2\pi}{25}\right), R\left(10\frac{2\pi}{25}\right)\right]\right\rangle$, $\boldsymbol{v} = (0.741048, 0, 0.671452, 0)$. The associated lattices $\Lambda_{\mathscr{G}_1\boldsymbol{u}}$ and $\Lambda_{\mathscr{G}_2\boldsymbol{v}}$ are illustrated in Fig. 4 on the left and on the right, respectively. $\mathscr{C}_2$ is proved to be the commutative group code of order 25 in $\mathbb{R}^4$ with the greatest minimum distance. We can observe that the lattice on the right has greater density than the one on the left (it "approaches" the hexagonal lattice).

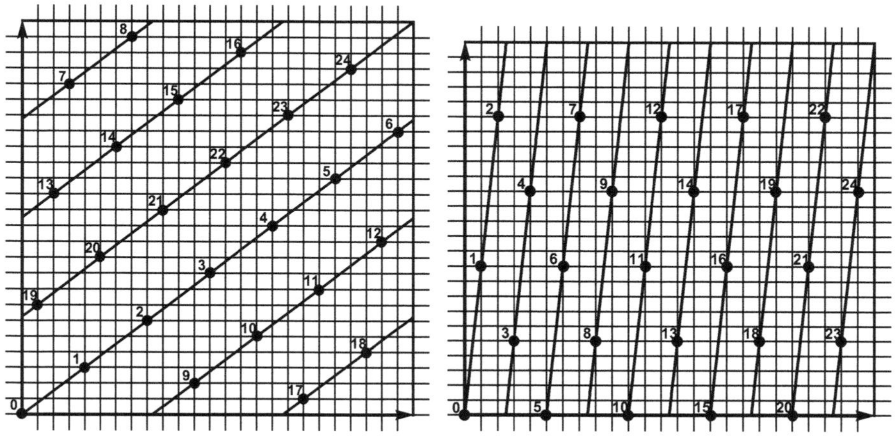

Fig. 4 Lattices which are pre-images of cyclic group codes through torus mappings from group codes $\mathscr{C}_1 = \mathscr{G}_1 \boldsymbol{u}$(left) and $\mathscr{C}_2 = \mathscr{G}_2 \boldsymbol{v}$ (right), where $\mathscr{G}_1 = \left\langle \left[\mathrm{R}\left(4\frac{2\pi}{25}\right), \mathrm{R}\left(3\frac{2\pi}{25}\right) \right] \right\rangle$ and $\boldsymbol{u} = \left(\frac{\sqrt{2}}{2}, 0, \frac{\sqrt{2}}{2}, 0\right)$, $\mathscr{G}_2 = \left\langle \left[\mathrm{R}\left(\frac{2\pi}{25}\right), \mathrm{R}\left(10\frac{2\pi}{25}\right) \right] \right\rangle$ and $\boldsymbol{v} = (0.741048, 0, 0.671452, 0)$

6 Constructive Spherical Codes from Lattices

In this section we present three different approaches for the problem of finding good commutative group codes in even dimensions.

6.1 Commutative Group Codes Obtained from Quotient of Lattices with Good Packing Density

Considering the bound established in Proposition 6, one strategy to construct good commutative group codes is to search for orthogonal sublattices of lattices with good packing density. This is the approach explored in [2] and [34]. For small distances d or big M, good commutative codes can be derived.

Consider $\alpha = \{\mathbf{v}_1, \mathbf{v}_2, \ldots, \mathbf{v}_n\}$ and $\beta = \{\mathbf{w}_1, \mathbf{w}_2, \ldots, \mathbf{w}_n\}$ bases of lattices Λ_α and Λ_β, $\Lambda_\beta \subset \Lambda_\alpha$, and the associated generator matrices B_α, B_β. Then $\mathrm{B}_\beta = \mathrm{B}_\alpha \mathrm{H}$, where H is an integer matrix. Suppose that β is composed by orthogonal vectors (Λ_β is a suborthogonal lattice). The next proposition [13, Ch. 5] describes the spherical code in $\mathbb{R}^{2L}$ attached to the nested pair of lattices $\Lambda_\beta \subset \Lambda_\alpha \subset \mathbb{R}^L$, Λ_β orthogonal, using the Smith normal form (Sect. 2.1).

Proposition 7 *In the notation above, let* $p_i = \|\mathbf{w}_i\|$, $p = \left(\sum_{j=1}^n \|\mathbf{w}_j\|^2\right)^{\frac{1}{2}}$, $c_i = \frac{p_i}{p}$, $\mathbf{c} = (c_1, c_2, \ldots, c_L)$ *and* $\varphi_{\mathbf{c}}$ *be the torus mapping regarding this frame* β. *Then to the normalized nested pair* $(1/p)\Lambda_\beta \subset (1/p)\Lambda_\alpha$ *of lattices it is associated*

Table 1 Examples of commutative group codes in $\mathbb{R}^n$, $n = 4, 6, 8, 16$, constructed through the quotient of A_2, D_3, D_4, E_8 by "orthogonal" sublattices

n	M	$d_{\min}$	Upper bound	Group
4	141,180	0.012706	0.0127061	$\mathbb{Z}_{141180}$
4	423,540	0.00733585	0.00733588	$\mathbb{Z}_{423540}$
6	32	1.1547	1.26069	$\mathbb{Z}_2 \oplus \mathbb{Z}_4^2$
6	2048	0.318581	0.320294	$\mathbb{Z}_8 \oplus \mathbb{Z}_{16}^2$
8	648	0.707107	0.736258	$\mathbb{Z}_3 \oplus \mathbb{Z}_6^3$
8	10,368	0.366025	0.369712	$\mathbb{Z}_6 \oplus \mathbb{Z}_{12}^3$
16	65,536	0.707107	0.780361	$\mathbb{Z}_2 \oplus \mathbb{Z}_4^6 \oplus \mathbb{Z}_8$
16	16,777,216	0.382683	0.392069	$\mathbb{Z}_4 \oplus \mathbb{Z}_8^6 \oplus \mathbb{Z}_{16}$

Their minimum distances approach the upper bound (9)

a spherical code in $\mathbb{R}^{2L}$ *with initial vector* $(c_1, 0, c_2, 0, \ldots, c_L, 0)$ *and generator group of matrices determined by the Smith normal decomposition of* H, *as described in Sect. 2.1.*

Starting from this result it is studied in [2] the existence of orthogonal sublattices of A_2, D_3, D_4, E_8, which are the densest lattices in dimensions 2, 3, 4 and 8, respectively. It is then obtained spherical codes in the double of these dimensions which approaches the bound of Proposition 6, particularly when M increases (Table 1).

This kind of spherical code construction was also recently approached in [34] by using dual lattices in the search for orthogonal sublattices.

Consider a full rank lattice $\Lambda \subset \mathbb{R}^n$, with generator matrix B such that $\mathrm{B}^* = (\mathrm{B}^T)^{-1}$ (generator matrix of the dual lattice) has integer entries. Then we can assert that $\mathrm{A} = \mathrm{BB}^*$ is a generator matrix of an orthogonal sublattice Λ' of Λ.

A good commutative group code can be asymptotically reached through the following proposition [34].

Proposition 8 *Let* Λ *be a lattice with generator matrix* B *such that* $\mathrm{B}^* = (\mathrm{B}^T)^{-1}$ *has integer entries and* $\Lambda^\star_{w,\mathrm{P}}$ *a lattice with generator matrix* $\mathrm{B}^\star_{w,\mathrm{P}} = w\mathrm{B}^* + \mathrm{P}$, *where* P *has integer entries and* w *is integer. Then the lattice* $\Lambda_{w,\mathrm{P}}$ *with generator matrix* $\mathrm{adj}(\mathrm{B}^\star_{w,\mathrm{P}})$ *has* $\Lambda'_{w,\mathrm{P}} = \det(\Lambda^\star_{w,\mathrm{P}})\mathbb{Z}^L$ *as an orthogonal sublattice. Moreover*

$$\frac{1}{w}\Lambda^\star_{w,P} \longrightarrow \Lambda^*(w \to \infty)$$

and by continuity of the matrix inversion process, $\frac{1}{\det(\frac{1}{w}\mathrm{B}^\star_{w,\mathrm{P}})}\Lambda_{w,\mathrm{P}} \longrightarrow \Lambda(w \to \infty)$.

In [34], it is discussed the conditions to be imposed on the matrix P to obtain faster convergence. For a lattice such that the matrix of its dual does not have integer entries we can consider $\mathrm{B}^\star_{w,P} = w\mathrm{B}^* + \mathrm{P}$, where $\mathrm{P} = \lfloor w\mathrm{B}^* \rceil - w\mathrm{B}^*$.

Table 2 Performance of spherical commutative group codes in dimension 48 using Proposition 8

	$P_{24,1}$		$P_{24,2}$	
w	$\log_{10} M$	Distance	$\log_{10} M$	Distance
7	27.6113	0.177774	31.1194	0.128473
8	28.9791	0.156625	32.5112	0.112635
9	30.1901	0.139890	**33.7389**	**0.100256**
10	31.2763	0.126336	34.8371	0.0903175
11	32.2609	0.115147	35.8305	0.0821655
12	33.1610	0.105760	36.7374	0.0753593

In particular, if B is upper triangular and $P = C_n = (c_{i\,j})$ where $c_{i\,j} = 1$ if $i = j+1$ and $c_{i\,j} = 0$ otherwise (cyclic perturbation), the spherical code associated with $\frac{\Lambda_{w,P}}{\Lambda'_{w,P}}$ is a cyclic group although the convergence is not a fast one.

Example 2 The Leech lattice was considered in [34] as a sublattice of the lattice $E_8 \times E_8 \times E_8$ to which it was associated with two perturbation matrices (associated with different representations of E_8), ($P_{24,1} = \begin{bmatrix} P_{8,1} & 0 & 0 \\ 0 & P_{8,1} & 0 \\ 0 & 0 & P_{8,1} \end{bmatrix}$), where

$$P_{8,1} = \begin{bmatrix} 0 & 1 & 1 & 0 & 0 & 0 & 0 & 0 \\ -1 & 0 & 0 & 0 & 0 & 0 & 0 & 0 \\ 0 & 1 & 0 & -1 & 0 & 0 & -1 & 0 \\ 0 & 0 & 0 & 0 & 0 & 0 & 0 & 0 \\ 0 & -1 & -1 & 0 & 0 & 0 & 1 & -1 \\ 0 & 0 & 1 & 1 & 1 & 0 & 1 & 0 \\ -1 & -1 & 0 & 0 & 0 & -1 & 0 & 1 \\ 1 & 1 & 0 & -1 & 0 & 0 & -1 & 0 \end{bmatrix},$$

and the null perturbation, $P_{24,2}$.

Table 2 illustrates the minimum distance versus the number of points of the associated spherical codes in $\mathbb{R}^{48}$ using the perturbation matrices $P_{24,1}$ and $P_{24,2}$. It is interesting to note that the code with $M = 5.48151 \cdot 10^{33}$ points and minimum distance equal to 0.100256 is comparable with the torus layer code in the same dimension described in Sect. 7, which is a not commutative group code.

6.2 *Optimum Commutative Group Codes*

The minimum distance of a commutative group code $\mathscr{C}$, generated by a finite group of orthogonal matrices $\mathscr{G} = \{G_1, G_2, \ldots, G_M\}$, depends on the choice of the initial vector $\boldsymbol{u}$. So for each $\mathscr{G}$ we search for an "optimal" initial vector (initial vector problem—IVP). It should be also remarked that isomorphic groups considered with their optimum initial vectors may provide non-congruent spherical codes, as it is the case presented in Fig. 4, where both groups $\mathscr{G}_1$ and $\mathscr{G}_2$ are isomorphic to cyclic group $\mathbb{Z}_{25}$. For $\mathscr{G}_1 = \left\langle \left[\mathrm{R}_1(4\frac{2\pi}{25}), \mathrm{R}_2(3\frac{2\pi}{25}) \right] \right\rangle$ the best initial vector is $(\sqrt{2}/2, 0, \sqrt{2}/2, 0)$ and minimum distance is 0.857441, whereas considering $\mathscr{G}_2 = \left\langle \left[\mathrm{R}_1(1\frac{2\pi}{25}), \mathrm{R}_2(10\frac{2\pi}{25}) \right] \right\rangle$ (see Proposition 3) the best initial vector is $(0.741048, 0, 0.671452, 0)$ and the minimum distance is 0.871154.

Therefore, the search for an optimum (with largest minimum distance) n-dimensional commutative group code with M points requires the consideration of all order M groups of orthogonal matrices and the solution of the initial vector problem for each of these groups. Thus, an efficient search for an optimum commutative group code requires to classify the non-congruent commutative groups with M points. This is a crucial aspect to be considered.

In [3] it is proposed a first approach to find optimal cyclic group codes by discarding isometric cases. Even discarding many isometries, the authors estimated that the number of cases to be tested is of order $\binom{M/2}{n/2}$, which can be computationally infeasible for codes with a large number of points. In [40] it is proposed a more general characterization for non-congruent commutative group codes through a special triangular basis (*Hermite* basis) combined with the elimination of isometries by permutation and also considering the ones related to the Ádám's condition. The main result is summarized in the next proposition, which allows a significant reduction of number cases now estimated to be of order $\dfrac{M^L}{2^L \varphi(M)}$ where φ is the Euler Phi function.

Proposition 9 *Every commutative group code $\mathscr{C} \subset S^{2L-1}$ generated by a finite group $\mathscr{G}$ of orthogonal matrices free of 2×2 reflection blocks (Proposition 5) is isometric to a code obtained as image of a lattice $\Lambda_{\mathscr{G}\boldsymbol{u}}$, $\boldsymbol{u} = (\delta_1, 0, \ldots, \delta_L, 0)$, where $\Lambda_{\mathscr{G}\boldsymbol{u}}$ is up to scaling factors $\frac{2\pi\delta_i}{M} (i = 1, \ldots, L)$, a lattice Λ with a generator matrix* T *satisfying the following conditions:*

1. T *is lower triangular such that:*

   a. $0 < \mathrm{T}(i,i) \leqslant \mathrm{T}(i+1,i+1),\ \forall\ 1 \leqslant i \leqslant L-1;$
   b. $0 \leqslant \mathrm{T}(i,[1:i-1]) < \mathrm{T}(i,i),\ \forall\ 2 \leqslant i \leqslant L;$
   c. $\mathrm{T}(i,i) \leqslant \gcd(\mathrm{T}(i,[j:i])),\ \ \forall\ 1 \leqslant j < i \leqslant L;$

   where $\mathrm{T}(r,[p:q])$ *are the elements in the columns p to q of the rth row of* T.

2. $\det(\mathrm{T}) = M^{L-1};$

3. There is a matrix H, *with integer elements satisfying* $\mathrm{T}\mathrm{H} = M\,\mathrm{I}_L$, *where* I_L *is the $L \times L$ identity matrix;*

4. The elements of the diagonal of T *satisfy* $\mathrm{T}(i,i) = \dfrac{M}{a_i}$ *where a_i is a divisor of M (what implies from 2. that $(a_i)^i \cdot (a_{i+1} \ldots a_l) \leqslant M$, $\forall i = 1, \ldots, L$).*

5. $\mathscr{G} \approx \Lambda/(M\mathbb{Z}^L)$ *and the classification of this group and its generators are obtained from Smith normal form of H (Sect. 2.1).*

Table 3 Some optimum commutative group codes of order M in $\mathbb{R}^4$

M	$d_{\min}$	δ_1	δ_2	Group	Gen. (b_{ij})	Bound
10	1.224	0.707	0.707	$\mathbb{Z}_{10}$	(1 3)	1.474
20	0.959	0.678	0.734	$\mathbb{Z}_{20}$	(3 4)	1.054
30	0.831	0.707	0.707	$\mathbb{Z}_{30}$	(3,5)	0.864
40	0.714	0.607	0.794	$\mathbb{Z}_{40}$	(4 5)	0.750
50	0.628	0.707	0.706	$\mathbb{Z}_{50}$	(7 2)	0.672
100	0.468	0.757	0.653	$\mathbb{Z}_5 \oplus \mathbb{Z}_{20}$	(0 20), (5 10)	0.476
200	0.330	0.750	0.660	$\mathbb{Z}_{200}$	(93 1)	0.337
300	0.273	0.656	0.754	$\mathbb{Z}_5 \oplus \mathbb{Z}_{60}$	(60 120), (10 15)	0.275
400	0.237	0.686	0.727	$\mathbb{Z}_{400}$	(189 1)	0.238
500	0.211	0.674	0.738	$\mathbb{Z}_{500}$	(13 20)	0.213
600	0.193	0.676	0.736	$\mathbb{Z}_{600}$	(191 198)	0.194
700	0.180	0.718	0.695	$\mathbb{Z}_{700}$	(14 25)	0.180
800	0.168	0.670	0.742	$\mathbb{Z}_{800}$	(16 25)	0.168
900	0.158	0.704	0.709	$\mathbb{Z}_{900}$	(197 2)	0.159
1000	0.149	0.716	0.697	$\mathbb{Z}_{1000}$	(33 4)	0.150

Based on this proposition, it is derived in [40] a two-step algorithm which searches for an optimum commutative group code $\mathscr{C}$ of order M in an even dimension. The first step consists of storing all matrices T according to Proposition 9 and using the Ádám's relation to discard isometric groups. For each one of these matrices T it is established a linear programming problem to determine the initial vector $\boldsymbol{u}$ which maximizes the minimum distance of the group code $\varphi_{\bar{\mathbf{u}}}(\Lambda_{\mathscr{G}_{\mathbf{u}}})$ (3) in $\mathbb{R}^{2L}$. The algorithm is summarized as a pseudo code in [40]. In Table 3 [40] some optimum commutative group codes in dimension 4 are displayed.

6.3 A Heuristic Method for Large Number of Points and Higher Dimensions

Although the approach presented in [40], discussed in the previous section, provides a significant reduction in the number of non-congruent cases, it still demands a brute-force search, since $\frac{M^L}{2^L\varphi(M)}$ grows fastly with M. For instance, for $(2L, M) = (16, 1024)$, which corresponds to a commutative group code in $S^{15} \subset \mathbb{R}^{16}$, with $M = 1024$ points, the approach proposed in [40] to find the optimal code demands a full search to a set with about 2^{63} elements. In addition, it also means solving 2^{63} linear programming problems to find the best initial vector for each case.

To overcome this difficulty, a heuristic approach for designing cyclic group codes in dimension $n = 2L$ is presented in [36]. The idea is to restrict the search to a subset of candidates that are likely to contain codes with good minimum distances, what is done by checking only a special set of cyclic group codes.

According to Proposition 5, a vector $\boldsymbol{v} = (v_1, \ldots, v_L) \in \mathbb{Z}^L$ defines a generator (orthogonal matrix) G of a cyclic group $\mathscr{G}$, if and only if, $\gcd(v_1, \ldots, v_L, M) = 1$. We can trivially guarantee this condition by choosing $v_1 = 1$ and, of course, reducing the set of candidates (eventually loosing optimality). This is the first restriction proposed in [36].

The main reduction in the set of candidates comes from rewriting the bound given in Proposition 6 in terms of d,

$$d \leqslant 4 \sin \left(\frac{\pi^L \Delta_L}{M L^{L/2}} \right)^{1/L} . \tag{10}$$

It means that the minimum distance of a cyclic group code $\mathscr{C}(M, 2L)$ is bounded by $\check{d} := 4 \sin \left(\frac{\pi^L \Delta_L}{M L^{L/2}} \right)^{1/L}$. By considering the underling lattice of a cyclic group code, the authors propose to restrict the search for generator vectors $\boldsymbol{v} \in \mathbb{Z}^L$, such that

$$||\boldsymbol{v}|| \simeq \frac{2M\sqrt{k}}{\pi} \arcsin(\check{d}/4) \tag{11}$$

This approach reduces significantly the number of cases to be checked and allows to find good codes with large number of points in high dimensions. Table 4 presents some results from [36] for 6-dimensional cyclic group codes of order M. The columns "Exact" and "Heuristic" display the number of cases tested in the full search approach of [40] and by the heuristic method, respectively. It is possible to

Table 4 Minimal distance of 6-dimensional cyclic group codes for several values of M found by the heuristic approach compared with the respective upper bound and optimum commutative group code of the same order

M	Exact	Heuristic	Bound	Optimum	Heuristic
10	31	25	1.820	1.414	1.345
20	125	25	1.465	1.240	1.190
30	422	25	1.287	1.133	1.056
40	500	25	1.173	1.044	1.007
50	781	25	1.091	0.976	0.946
100	3125	25	0.870	0.804	0.786
200	12,500	500	0.692	0.673	0.633
300	42,188	500	0.605	0.585	0.568
400	50,000	500	0.550	0.540	0.525
500	78,125	500	0.511	0.504	0.479
600	168,750	500	0.481	0.472	0.458
700	178,646	500	0.457	0.445	0.439
800	200,000	500	0.437	0.427	0.415
900	379,688	500	0.420	0.413	0.403
1000	312,500	500	0.406	0.397	0.394

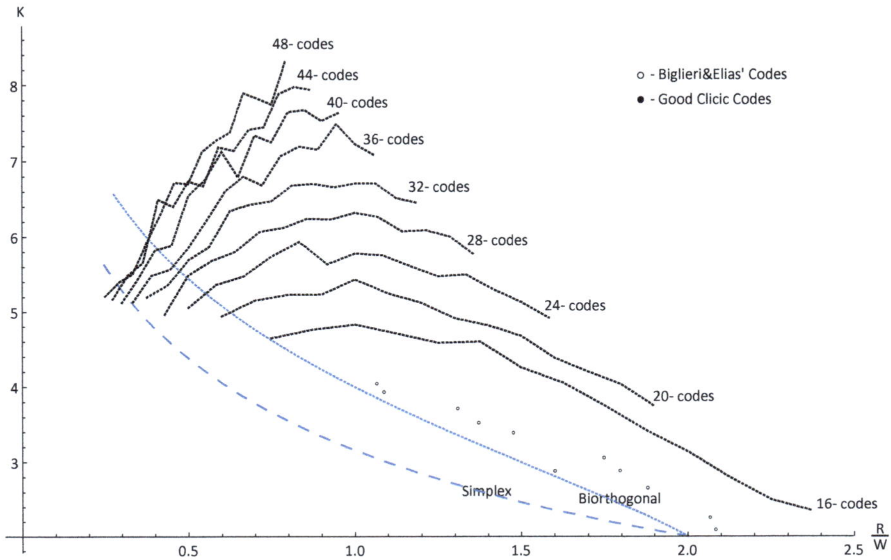

Fig. 5 Efficiency comparison of cyclic group codes found using the heuristic method and some commutative group codes found in the literature

see that the codes found by the heuristic approach have minimum distance close to the correspondent best codes, even testing a few number of candidates.

The codes presented in Fig. 5 were obtained by the heuristic approach considering at most 512 candidates, what is a very small portion of all the combinatorial possibilities. For instance, for the smallest case presented in Fig. 5, $(M, 2L) = (2^6, 16)$, an exact method would need to check 2^{35} candidates and for $(M, 2L) = (2^{19}, 48)$ the number of candidates is of order 2^{414}. Both cases are computationally prohibitive because for each candidate we must solve a linear programming problem for the IVP.

7 Spherical Codes on Torus Layers

The connection between flat tori and commutative group codes, explored in the previous sections, can also be extended to construct a more general family of spherical codes, using not only a single flat torus, but a pile of them, designed carefully in order to achieve a large number M of points, given a target minimum distance. This idea was originally presented in [38] and then expanded in [39], where a formal construction is presented.

The construction of a $2L$-dimensional spherical code in layers of flat tori ($TLSC$ codes) starts from a set $SC(L,d)_+ = \{\boldsymbol{c} \in SC(L,d), c_i \geq 0, \;\; 1 \leq i \leq L\}$ of L-dimensional unit vectors, with non-negative coordinates, such that the distance between any two vectors in this set is at least a given number d. Thus, each point $\boldsymbol{c}$ in $SC(L, d)_+$ defines a flat torus and the distance between two of these flat tori in $\mathbb{R}^{2L}$ at least d (Proposition 1). Each box $\mathscr{P}_{\boldsymbol{c}}$ is then filled with a suitable set of lattice points (or any other set of points), say $Y_{T_c} \subset \mathscr{P}_{\boldsymbol{c}}$, in such a way that the distance, after embedding in the $2L$ dimension by the mapping defined in Eq. (3), is not smaller than a target value d, i.e.

$$\|\varphi_c(\boldsymbol{y}) - \varphi_c(\boldsymbol{x})\| \geq d \;\; \forall \boldsymbol{x}, \boldsymbol{y} \in Y_{T_c}. \tag{12}$$

As an example, we reproduce here the construction of a 4-dimensional spherical code in layers of tori. Let us start from a set of 2-dimensional unit vectors defined by

$$SC(2, d)_+ = \left\{(\cos(\alpha_{\pm j}), \sin(\alpha_{\pm j})), 0 \leq \alpha_{\pm j} \leq \frac{\pi}{2}\right\},$$

where $\alpha_{\pm j} = \frac{\pi}{4} \pm (2j-1)\arcsin\left(\frac{d}{2}\right)$ and $1 \leq j \leq \left\lfloor \frac{\pi - 2\arcsin(d/2)}{8\arcsin(d/2)} \right\rfloor$. These points belong to the positive quadrant of $\mathbb{R}^2$ and the minimum distance between them is at least d, as illustrated in Fig. 6.

Note that each point in $SC(2, d)_+$ defines a rectangle with sides of length $2\pi\cos(\alpha_{\pm j})$ and $2\pi\sin(\alpha_{\pm j})$, and hence a flat torus on the surface of the 4-dimensional unit sphere, according to the Eq. (3). Moreover, since the distance

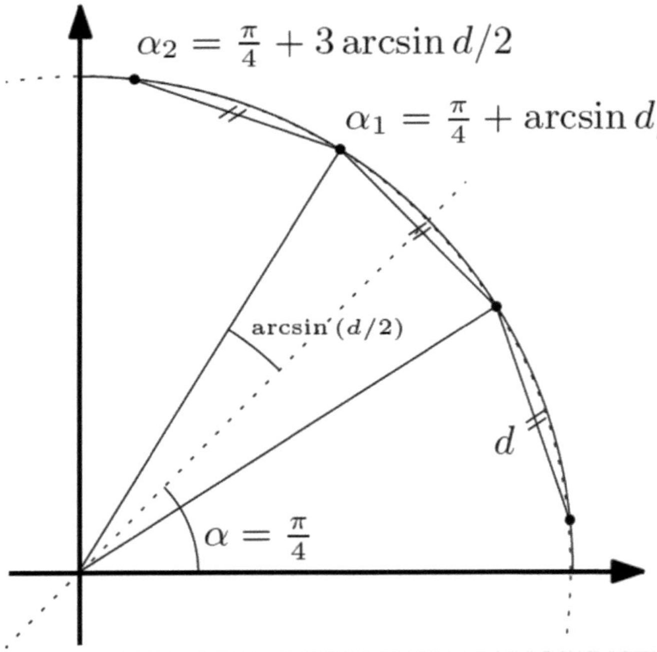

Fig. 6 $SC(2, d)_+$ symmetric in relation to $y = x$

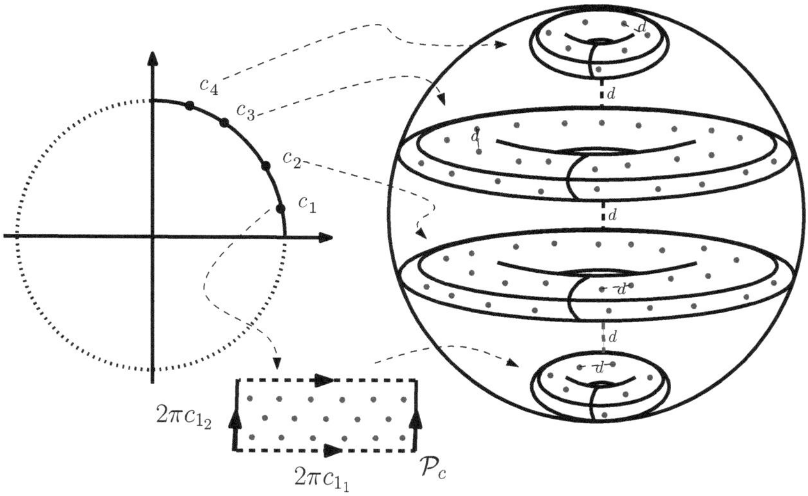

Fig. 7 Construction of a 4th-dimensional spherical code in layers of flat tori

between any two points in $SC(2, d)_+$ is at least d, the distance between any two points on different associated tori will be also d.

After setting the layers, the construction of the spherical code is completed by choosing a suitable set of points inside each one of the rectangles (flattened tori) defined by $SC(2, d)_+$, providing that the minimum distance given by the Eq. (12) is satisfied. At this stage, a good option is to fit a dense lattice inside the box and so to come up with a commutative group code in each layer. The whole process is illustrated in Fig. 7.

In [39], starting from a orthogonal sublattice of the Leech lattice it is presented a $TLSC$ in dimension 48 with more than $2^{113} = 1.03846 \cdot 10^{34}$ points placed in 24 layers of flat tori with minimum distance 0.1. This code is generated by using only 12 matrices.

Codes constructed on layers of flat tori have been proved to be very effective when compared to other state-of-the-art techniques as the well-known wrapped spherical codes [22], laminated spherical codes [23] and apple-peeling codes [20], especially for not asymptotically small distances. In addition, such codes present advantages in the coding/decoding processes inherited from their homogeneous structure and the underlying lattice codebook in the half of the code dimension.

The ideas behind the construction of torus layer codes have also been used in other applications, such as coding for continuous alphabet sources, presented in the next section.

8 Continuous Constructions

The homogeneous structure of flat tori and lattices can be used in a proposal for transmitting a continuous alphabet source over an AWGN channel, as discussed in [41]. Consider the problem illustrated in Fig. 8. Suppose that a value x from a source with probability density function (pdf) having support $[0, 1)$ has to be transmitted over a Gaussian (AWGN) channel of dimension N. The encoder will use a function $\boldsymbol{s} : [0, 1) \to \mathbb{R}^N$ and then transmit the encoded value $\boldsymbol{s}(x)$ over the channel, such that the receiver will observe a noisy vector $\boldsymbol{y} = \boldsymbol{s}(x) + \boldsymbol{z}$. The objective is to recover an estimate $\boldsymbol{x}$ of the sent value, attempting to minimize the mean square error (mse) $E[(X - \hat{X})^2]$. If the N-dimensional Gaussian channel has power P and variance σ^2, then the average of the transmitted value should be no greater than P, i.e. $\boldsymbol{s}$ should satisfy the constraint $E[\|\boldsymbol{s}(x)\|^2] \leq P$. This essentially means that the image of $[0, 1)$ by the mapping $\boldsymbol{s}$ needs to be contained within a sphere of radius P. On the other hand, it can be shown that for low noise, the mse of the scheme represented in Fig. 8 is given by

$$E[(X - \hat{X})^2] \approx \sigma^2 \int_0^1 f(x) \|\dot{\boldsymbol{s}}(x)\|^{-2} dx := E_{\text{low}}[(X - \hat{X})^2], \tag{13}$$

The quantity $\mathscr{S}(x) := \|\dot{\boldsymbol{s}}(x)\|$ is called the *stretch* of the curve. Summarizing, we want to stretch the function as much as possible, but since $\boldsymbol{s}(x)$ is contained in a sphere, it has to be twisted or folded. If the distance between the folds of the curve become too small, large errors will occur frequently when the noise is high, and the mse will approach Eq. (13) only if the noise is sufficiently low. This is called the threshold effect.

Constant curvature curves on a flat torus [17] in dimension $N = 2L$ are easily described as being homogeneous and allow to control the distance between its folds ("fat strut" problem Fig. 9).

It was shown in [42] that curves on the flat torus determined by the vector $\boldsymbol{c} = (1/\sqrt{L})(1, \ldots, 1)$ can be used to obtain the proper scaling of the mse with the signal-to-noise ratio (SNR). Given a vector $\boldsymbol{a} \in \mathbb{Z}^L$, the curves considered in [42] are of the type

$$\boldsymbol{s}(x) = \varphi_{\boldsymbol{c}}\left(\frac{2\pi}{\sqrt{L}}\boldsymbol{a}x (\text{mod } 1)\right), \tag{14}$$

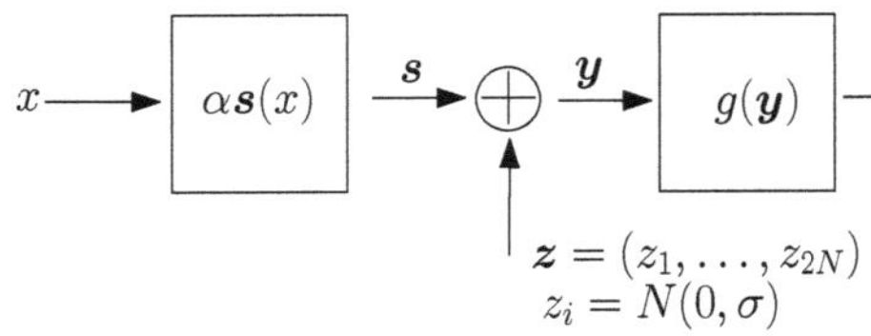

Fig. 8 Continuous encoder

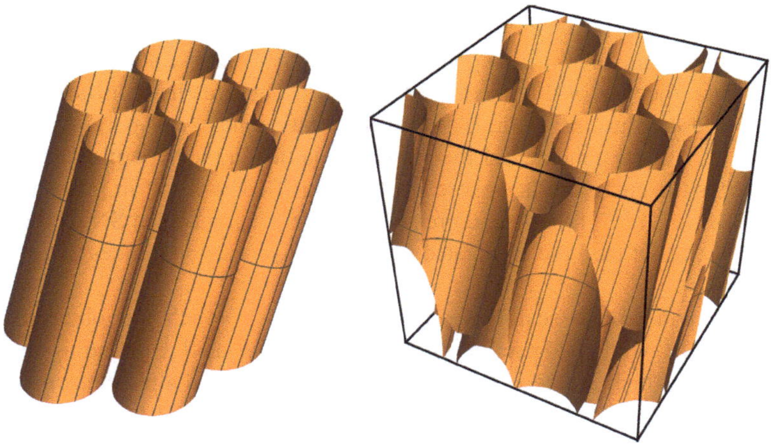

Fig. 9 Packing of a curve in a torus of $\mathbb{R}^6$, represented in a $3D$ box

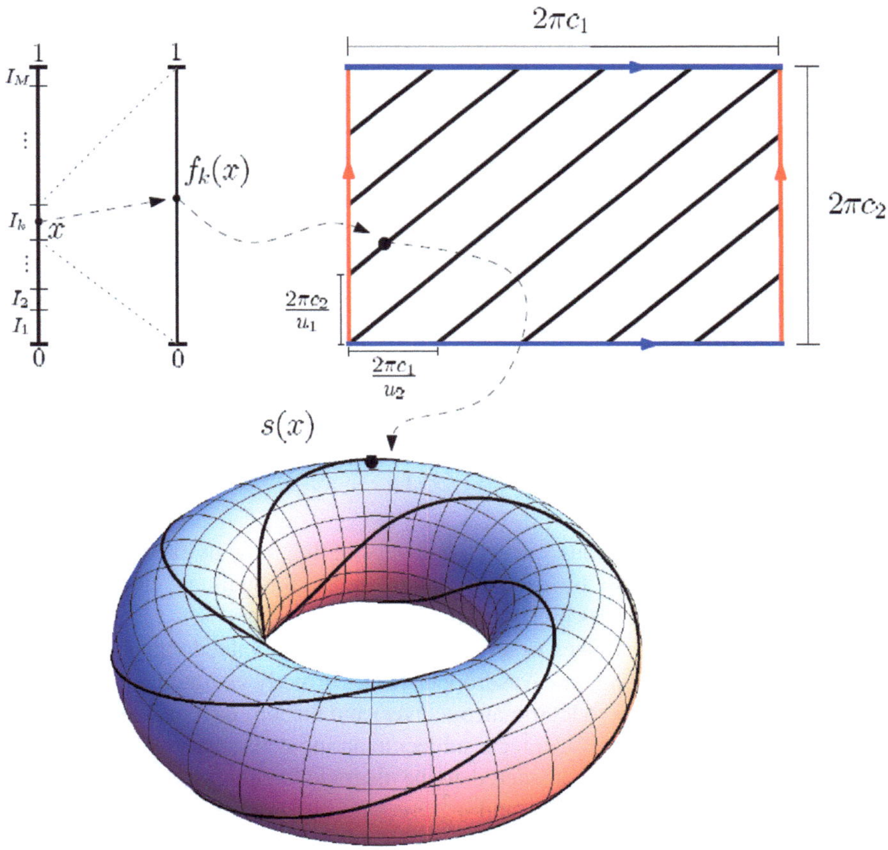

Fig. 10 Encoding process—a flat torus curve in $\mathbb{R}^4$, represented in $\mathbb{R}^3$, $\boldsymbol{a} = (1, 5/4)$

where $\boldsymbol{a}x \pmod 1 = \boldsymbol{a}x - \lfloor \boldsymbol{a}x \rceil$ (Fig. 10), the length of this curve is $2\pi \|\boldsymbol{a}\|/\sqrt{L}$ and the small ball radius of $\boldsymbol{s}(x)$ is approximately the minimum distance of the lattice which is the projection of $\mathbb{Z}^L$ onto the subspace $\boldsymbol{a}^\perp$. The stretch is constant and equal to $2\pi/\sqrt{L}\,\|\boldsymbol{a}\|^2$. Thus, projections of the cubic lattice play an important role in the design of such curves: the denser the projection lattice in the hyperplane $\boldsymbol{a}^\perp \subset \mathbb{R}^L$, the denser is the curve on the sphere $\mathcal{S}^{2L-1}$ (Fig. 9). The result in [42] states that if $\boldsymbol{a} = (1, a, \ldots, a^{L-1})$ then the correct scaling between mse and SNR is achieved when $a \to \infty$, but then the associated sequence of projection lattices converges to $\mathbb{Z}^{L-1}$.

The problem of the search for lattices in $\mathbb{R}^{L-1}$ which are obtained from projection of $\mathbb{Z}^L$ with good packing density is then associated with finding spherical curves in $\mathbb{R}^{2L}$ which are good for transmitting a continuous alphabet source. This problem was approached in [6, 32] (lifting construction) and also projections from higher dimensions are considered in [7].

Discrete sets of points selected on a continuous closed curve on a flat torus, as described in this section, have been used in [37] to approach good commutative group codes which are cyclic.

By using layers of tori it is possible to generalize the construction in [42] as it is presented in [8]. In this context, a good collection of tori (i.e. such that each of them is separated at least a certain distance from each other) can be obtained by designing a suitable spherical code for a given minimum distance. On the other hand, finding good curves in each torus is equivalent to finding good projections of the orthogonal lattice $c_1\mathbb{Z} \oplus \cdots \oplus c_L\mathbb{Z}$. In this case, it is possible to generalize the lifting construction and exhibit sequences of projections of $c_1\mathbb{Z} \oplus \cdots \oplus c_L\mathbb{Z}$ converging to any $(L-1)$-dimensional lattice. The construction of curves on torus layers meaningfully increase the total length of the curves, while keeping a good distance between their laps, hence enhancing the performance of the codes proposed in [42] which compare favorably with other previous constructions [43].

8.1 Continuous Curves and Secrecy

Schemes based on continuous curves on layers of flat tori, as described in the previous section, can also be used to design codes for wiretap channels with continuous input alphabets as presented in [1].

In this case, a sender wishes to reliably transmit a real valued signal $x \in \mathbb{R}$ to a receiver, while preventing an eavesdropper from correctly estimating x. Both the main channel (from sender to legitimate receiver) and the wiretap channel (sender to the eavesdropper) are AWGN channels subject to an input average power constraint P. The wiretap channel is considered to be degraded with respect to the main channel, i.e. $\sigma_w^2 > \sigma_m^2$, where σ_m^2 and σ_w^2 are the noise variances associated with the main and wiretap channels.

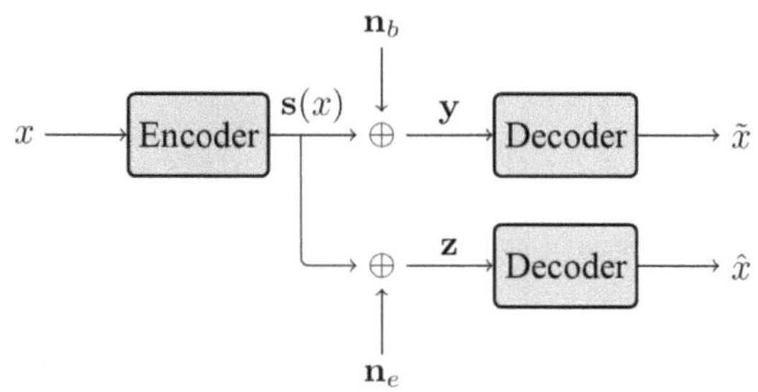

Fig. 11 The AWGN wiretap channel model

To transmit the source value x, the sender employs a spherical code as described in Sect. 8, i.e. he employs an encoder that maps x onto a codeword $\mathbf{s}(x) \in \mathbb{R}^{2L}$. The codeword $\mathbf{s}(x)$ is then transmitted to the destination over the main channel and corrupted by the additive noise vector $\mathbf{n}_b$, where $\mathbf{n}_b = (n_{b,1}, \ldots, n_{b,2L})$, with $n_{b,i} \sim \mathcal{N}(0, \sigma_m^2)$. Similarly, the eavesdropper observes the transmission of $\mathbf{s}(x)$ over the wiretap channel, which is corrupted by the noise vector $\mathbf{n}_e$, where $\mathbf{n}_e = (n_{e,1}, \ldots, n_{e,2N})$, $n_{e,i} \sim \mathcal{N}(0, \sigma_w^2)$. The legitimate receiver obtains the (main) channel output sequence $\mathbf{y} = \mathbf{s}(x) + \mathbf{n}_b$, while the eavesdropper obtains the (wiretap) channel output sequence $\mathbf{z} = \mathbf{s}(x) + \mathbf{n}_e$ (Fig. 11). Then both receivers estimate the source message using some decoder that tries to minimize the mean square error.

As it was shown in [1] a careful parametrization of these codes, which takes into account their geometrical properties, enables legitimate users to communicate under a small distortion, while forcing the eavesdropper to operate at large distortions. Moreover, the proposed construction is tunable, as it provides a simple mechanism to trade-off reliability and secrecy.

Acknowledgements The authors acknowledge all the support from IMECC-Unicamp and from FAPESP (2013/25977-7) and CNPq (312926/2013-8; 400441/2014-4) foundations. The first author wishes to thank her present and past students for their contributions on the subject discussed in this chapter.

References

1. J. Almeida, C. Torezzan, and J. Barros. Spherical codes for the gaussian wiretap channel with continuous input alphabets. In *IEEE 14th Workshop on Signal Processing Advances in Wireless Communications*, 2013.
2. C. Alves and S. I. R. Costa. Commutative group codes in $\mathbb{R}^4$, $\mathbb{R}^6$, $\mathbb{R}^8$ and $\mathbb{R}^{16}$ approaching the bound. *Discrete Mathematics*, 313(16):1677–1687, 2013.
3. E. Biglieri and M. Elia. Cyclic-group codes for the gaussian channel. *IEEE Transaction on Information Theory*, 22(5):624–629, Sept. 1976.
4. G. Caire and E. Biglieri. Linear block codes over cyclic groups. *IEEE Transaction on Information Theory*, 41(5):1246–1256, Sept. 1995.
5. A. Campello, G. C. Jorge, J. E. Strapasson, and S. I. R. Costa. Perfect codes in the ℓ_p metric. *European Journal of Combinatorics*, 53:72–85, 2016.

6. A. Campello and J. Strapasson. On sequences of projections of the cubic lattice. *Computational and Applied Mathematics*, 32(1):57–69, Apr. 2013.
7. A. Campello, J. Strapasson, and S. I. R. Costa. On projections of arbitrary lattices. *Linear Algebra and its Applications*, 439(9):2577–2583, 2013.
8. A. Campello, C. Torezzan, and S. I. R. Costa. Curves on flat tori and analog source-channel codes. *IEEE Transactions on Information Theory*, 59(10):6646–6654, Oct. 2013.
9. H. Cohen. *A Course in Computational Algebraic Number Theory*. Springer Berlin Heidelberg, 1993.
10. J. H. Conway and N. J. A. Sloane. *Sphere packings, lattices and groups*, volume 290 of *Grundlehren der Mathematischen Wissenschaften [Fundamental Principles of Mathematical Sciences]*. Springer-Verlag, New York, third edition, 1999. With additional contributions by E. Bannai, R. E. Borcherds, J. Leech, S. P. Norton, A. M. Odlyzko, R. A. Parker, L. Queen and B. B. Venkov.
11. S. I. R. Costa, E. Agustini, M. Muniz, and R. Palazzo. Slepian-type codes on a flat torus. In *Proc. IEEE International Symposium on Information Theory*, page 58, 25–30 June 2000.
12. S. I. R. Costa, M. Muniz, E. Agustini, and R. Palazzo. Graphs, tessellations, and perfect codes on flat tori. *IEEE Transactions on Information Theory*, 50(10):2363–2377, Oct. 2004.
13. S. I. R. Costa, F. Oggier, A. Campello, J-C. Befiore, and E. Viterbo. *Lattice Applied to Codding for Reliable and Secure Communications*. Springer, 2017.
14. S. I. R. Costa, J. E. Strapasson, M. M. S. Alves, and Carlos T. B. Circulant graphs and tesselations on flat tori. *Linear Algebra and Its Applications*, 432:369–382, 2010.
15. S. I. R. Costa, J. E. Strapasson, R. M. Siqueira, and M. Muniz. Circulant graphs, lattices and spherical codes. *International Journal of Applied Mathematics.*, 20:581–594, 2007.
16. S. I. R. Costa, C. Torezzan, A. Campello, and V. A. Vaishampayan. Flat tori, lattices and spherical codes. In *2013 Information Theory and Applications Workshop (ITA)*, pages 1–8, Feb. 2013.
17. S. I. R. Costa. On closed twisted curves. *Proceedings of the American Mathematical Society*, 109(1):205–214, 1990.
18. A. C. de A. Campello, G. C. Jorge, and S. I. R. Costa. Decoding q-ary lattices in the lee metric. In *2011 IEEE Information Theory Workshop*, pages 220–224, Oct. 2011.
19. T. Ericson and V. Zinoviev. *Codes on Euclidean Spheres*. North-Holland Mathematical Library, 2001.
20. A. A. El Gammal, L. A. Hemachandra, I. Shperling and V. K. Wei. Using simulated annealing to design good codes. *IEEE Trans. Inf. Theor*, 1987.
21. F. R. Gantmacher. *The theory of matrices. Vol. 1. Transl. from the Russian by K. A. Hirsch. Reprint of the 1959 translation.* Providence, RI: AMS Chelsea Publishing, 1998.
22. J. Hamkins and K. Zeger. Asymptotically dense spherical codes. i. wrapped spherical codes. *IEEE Transactions on Information Theory*, 43(6):1774–1785, Nov. 1997.
23. J. Hamkins and K. Zeger. Asymptotically dense spherical codes .ii. laminated spherical codes. *IEEE Transactions on Information Theory*, 43(6):1786–1798, Nov. 1997.
24. I. Ingemarsson. Group codes for the gaussian channel. In G. Einarsson, T. Ericson, I. Ingemarsson, R. Johannesson, K. Zigangirov, and C.-E. Sundberg, editors, *Topics in Coding Theory*, volume 128 of *Lecture Notes in Control and Information Sciences*, pages 73–108. Springer Berlin Heidelberg, 1989.
25. G. D. Forney Jr. Geometrically uniform codes. *IEEE Transactions on Information Theory*, 37(5):1241–1260, 1991.
26. C. C. Lavor, M. M. S. Alves, R. M. Siqueira, and S. I. R. Costa. *Uma intrdução à teoria de códigos*. Number 21 in Notas em Matemática Aplicada. SBMAC, 2006.
27. H.-A. Loeliger. Signal sets matched to groups. *IEEE Transactions on Information Theory*, 37(6):1675 –1682, Nov. 1991.
28. D. Micciancio and O. Regev. *Lattice-based Cryptography*, pages 147–191. Springer Berlin Heidelberg, Berlin, Heidelberg, 2009.
29. C. Peikert. A decade of lattice cryptography. *Foundations and Trends® in Theoretical Computer Science*, 10(4):283–424, 2016.

30. R. Siqueira and S. I. Costa. Flat tori, lattices and bounds for commutative group codes. *Designs Codes Cryptography*, 49(1-3):307–321, December 2008.
31. D. Slepian. Group codes for the gaussian channel. *The Bell System Technical Journal*, 47: 575–602, 1968.
32. N. J. A. Sloane, V. A. Vaishampayan, and S. I. R. Costa. A note on projecting the cubic lattice. *Discrete Computational Geometry*, 46(3):472–478, October 2011.
33. J. Strapasson, S. Costa, and M. Muniz. A note on quadrangular embedding of abelian cayley graphs. *Trends in Applied and Computational Mathematics*, 17(3):331, 2016.
34. J. E. Strapasson. A note on sub-orthogonal lattices. *Linear Algebra and its Applications*, 543(15):31–41, 2018.
35. J. E. Strapasson, G. C. Jorge, A. Campello, and S. I. R. Costa. Quasi-perfect codes in the ℓ_p metric. *Computational and Applied Mathematics*, Aug 2016.
36. J. E. Strapasson and C. Torezzan. A heuristic approach for designing cyclic group codes. *International Transactions in Operational Research*, 23(5):883–896, 2016.
37. R. M. Taylor, L. Mili, and A. Zaghloul. Structured spherical codes with asymptotically optimal distance distributions. In *2017 IEEE International Symposium on Information Theory (ISIT)*, pages 2188–2192, June 2017.
38. C. Torezzan, S. I. R. Costa, and V. A. Vaishampayan. Spherical codes on torus layers. *International Symposium on Information Theory.* 2009.
39. C. Torezzan, S. I. R. Costa, and V. A. Vaishampayan. Constructive spherical codes on layers of flat tori. *IEEE Transactions on Information Theory*, 59(10):6655–6663, Oct. 2013.
40. C. Torezzan, J. E. Strapasson, S. I. R. Costa, and Rogerio M. Siqueira. Optimum commutative group codes. *Designs, Codes and Cryptography*, 74(2):379–394, Feb 2015.
41. V. A. Vaishampayan and S. I. R. Costa. Curves on a sphere, shift-map dynamics, and error control for continuous alphabet sources. *IEEE Transactions on Information Theory*, 49(7):1658–1672, 2003.
42. V. A. Vaishampayan, N. J. A. Sloane, and S. I. R. Costa. Dynamical systems, curves and coding for continuous alphabet sources. In *Proceedings International Telecommunications Symposium.* ITW2002, Bangalore, 2002.
43. N. Wernersson, M. Skoglund, and T. Ramstad. Polynomial based analog source-channel codes. *IEEE Transactions on Communications*, 57(9):2600–2606, Sept. 2009.
44. R. Zamir. Lattices are everywhere. In *2009 Information Theory and Applications Workshop*, pages 392–421, Feb 2009.
45. R. Zamir. *Lattice Coding for Signals and Networks: A Structured Coding Approach to Quantization, Modulation, and Multiuser Information Theory*. Cambridge University Press, 2014.

Nonvariational Semilinear Elliptic Systems

Djairo G. de Figueiredo

Abstract In this paper we survey questions regarding the existence of solutions of the Dirichlet problem for systems of semilinear elliptic equations of the type

$$-\Delta u = f(x, u, v, \nabla u, \nabla v), \ -\Delta v = g(x, u, v, \nabla u, \nabla v) \text{ in } \Omega, \tag{1}$$

where Ω is a bounded subset of $\mathbb{R}^N$, $N \geq 3$. The existence of solutions is discussed here using Topological Methods. In order to use this method, the main point is the proof of a priori bounds for the eventual solutions of the systems above. These bounds will be obtained by three different methods, namely Hardy-type inequalities, Moving Planes techniques, and Blow-up. This last method leads to interesting questions about Liouville problems for systems.

1 Introduction

In this paper we survey some of our results on the solvability of the following system of semilinear elliptic equations

$$-\Delta u = f(x, u, v), \ -\Delta v = g(x, u, v) \text{ in } \Omega, \tag{2}$$

and the more general one, where the nonlinearities depend also on the gradients, namely

$$-\Delta u = f(x, u, v, \nabla u, \nabla v), \ -\Delta v = g(x, u, v, \nabla u, \nabla v) \text{ in } \Omega. \tag{3}$$

On the above equations u and v are real-valued functions $u, v : \overline{\Omega} \to \mathbb{R}$, where Ω is some domain in $\mathbb{R}^N$, $N \geq 3$, and $\overline{\Omega}$ its closure. The regularity of the solutions will

D. G. de Figueiredo (✉)
Institute of Mathematics, Statistics and Scientific Computing, University of Campinas, Campinas, SP, Brazil
e-mail: djairo@ime.unicamp.br

C. Lavor, F. A. M. Gomes (eds.), *Advances in Mathematics and Applications*,
https://doi.org/10.1007/978-3-319-94015-1_6

be detailed later. Although we concentrate in the case of the Laplacian differential operator $\Delta = \sum_{j=1}^{N} \frac{\partial^2}{\partial x_j^2}$, many results stated here can be extended to general second order elliptic operators. The nonlinearity of the problems appears only in the real-valued functions $f, g : \overline{\Omega} \times \mathbb{R} \times \mathbb{R} \times \mathbb{R}^N \times \mathbb{R}^N \to \mathbb{R}$.

Viewing the existence of solutions for the Dirichlet problem for the above systems, we shall discuss mainly the following question pertaining to the above systems:

- A quick review of the special systems that can be treated by Variational Methods in Sect. 2
- Our treatment of Nonvariational Systems via Topological Methods is done in Sect. 3. In order to use this method, we need a priori bounds for the eventual solutions of the systems above, in order to use Leray-Schauder degree theory. These bounds will be obtained by three different methods, namely
  - Hardy-type inequalities, Sect. 3.1,
  - Moving Planes techniques, Sect. 3.2,
  - Using Blow-up, Sect. 3.3.

Remark There has been recently an ever-increasing interest in systems of nonlinear elliptic equations. Many aspects of this recent research are not touched here. Our objective in this paper is to survey some of our results on Nonvariational Semilinear Elliptic Systems.

2 On Variational Methods

In this section we study two special classes of systems, the Gradient Systems and the Hamiltonian Systems, which can be treated by Variational Methods. We say that the system (3) above is of the **Gradient type** if there exists a function $F : \overline{\Omega} \times \mathbb{R} \times \mathbb{R} \to \mathbb{R}$ of class C^1 such that

$$\frac{\partial F}{\partial u} = f, \ \frac{\partial F}{\partial v} = g,$$

and it is said to be of the **Hamiltonian type** if there exists a function $H : \overline{\Omega} \times \mathbb{R} \times \mathbb{R} \to \mathbb{R}$ of class C^1 such that

$$\frac{\partial H}{\partial v} = f, \ \frac{\partial H}{\partial u} = g.$$

Associated with Gradient Systems we have the functional

$$\Phi(u, v) = \frac{1}{2} \int_{\Omega} |\nabla u|^2 + \frac{1}{2} \int_{\Omega} |\nabla v|^2 - \int_{\Omega} F(x, u, v), \tag{4}$$

which is defined in the Cartesian product $H_0^1(\Omega) \times H_0^1(\Omega)$ provided

$$|F(x,u,v)| \leq |u|^p + |v|^q, \ \forall x \in \Omega, \ u, v \in \mathbb{R}$$

with $p, q \leq \frac{2N}{N-2}$, if the dimension $N \geq 3$. Associated with Hamiltonian systems we will first consider the functional

$$\Phi(u,v) = \int_\Omega \nabla u.\nabla v - \int_\Omega H(x,u,v), \tag{5}$$

which is defined in the Cartesian product $H_0^1(\Omega) \times H_0^1(\Omega)$ provided again that

$$|H(x,u,v)| \leq |u|^p + |v|^q, \forall x \in \Omega, \ u, v \in \mathbb{R}$$

with $p, q \leq \frac{2N}{N-2}$, if the dimension $N \geq 3$. However, it has been observed that the restriction on the powers of u and v as above is too restrictive in the case of Hamiltonian systems. We can allow different values of p, q. As a matter of fact the same values of p, q that have appeared in our work with P. Clement and E. Mitidieri of equation (3) studied there by Topological Methods, cf. [24] (see also [62, 76].) The interesting fact is that p, q can take any values $1 < p, q < \infty$ with the restriction that they are below the so-called critical hyperbola

$$\frac{1}{p+1} + \frac{1}{q+1} = 1 - \frac{2}{N}$$

In this more general situation, one has to use fractional Sobolev spaces instead of $H_0^1(\Omega)$. A reference to this is our paper with P. Felmer [36].

3 Nonvariational Elliptic Systems

In this section we study systems of the general form (2) that do not fall in the categories of the systems studied in the previous section. Since they are not variational systems, we will treat them by Topological Methods. This method also is applied to system (3). The main tool used to prove the existence of a solution is a result due to Krasnoselskii [64] stated below. The difficulty when one uses this result is obtaining a priori bounds for the solutions. There are several methods to tackle this question. We will comment three of them, including the use of Hardy-type inequalities and Moving Planes. However, the most successful one in our framework seems to be the Blow-up Method. This method leads naturally to Liouville type theorems, that is, theorems asserting that certain systems have no non-trivial solution in the whole space $\mathbb{R}^N$ or in a half-space $\mathbb{R}^N_+$. In Sect. 4, we present some results on Liouville theorems for systems.

Theorem 3.1 (Krasnoselskii) *Let $\mathscr{C}$ be a cone in a Banach space X and $T : \mathscr{C} \to \mathscr{C}$ a compact mapping such that $T(0) = 0$. Assume that there are real numbers $0 < r < R$ and $t_0 > 0$ such that*

(i) $x \neq tTx$ for $0 \leq t \leq 1$ and $x \in \mathscr{C}$, $||x|| = r$, *and*
(ii) There exists a compact mapping $H : \overline{B}_R \times [0, \infty) \to \mathscr{C}$ (where $B_R = \{x \in \mathscr{C} : ||x|| < R\}$) such that

(a) $H(x, 0) = Tx$ for $||x|| = R$,
(b) $H(x, t) \neq x$ for $||x|| = R$ *and* $t \geq 0$
(c) $H(x, t) = x$ *has no solution* $x \in \overline{B}_R$ *for* $t \geq t_0$

Then

$$i_c(T, B_r) = 1, \; i_c(T, B_R) = 0, \; i_c(T, U) = -1,$$

where $U = \{x \in \mathscr{C} : r < ||x|| < R\}$, and i_c denotes the Leray-Schauder index. As a consequence T has a fixed point in U.

When applying this result the main difficulty arises in the verification of condition (b), which is nothing more than an *a priori bound* on the solutions of the system. It is well known that the existence of a priori bounds depends on the growth of the functions f and g as u and v go to infinity. We have seen when treating the variational systems that the nonlinearities were restricted to have polynomial growth. This was a requirement, together with other conditions, in order to get the associated functional defined, as well as a Palais-Smale condition. Recall that all problems considered in these lectures refer to equations in dimension $N \geq 3$. In dimension $N = 2$ the type of nonlinearities allowed is much larger; indeed nonlinearities of exponential type are allowed, and in this context, one uses the Trudinger-Moser estimates for functions in Sobolev spaces defined in subsets of R^2. See for instance [45, 47].

3.1 Estimates Using Hardy-Type Inequalities

Brézis-Turner [21] using an inequality due to Hardy proved a priori bounds for positive solutions of superlinear elliptic equations (the scalar case), namely

$$-\Delta u = f(x, u) \text{ in } \Omega, \quad u = 0 \text{ on } \partial\Omega. \tag{6}$$

In [25], with Clement and Mitidieri, we used the same technique to obtain a priori bounds for solutions of systems

$$\begin{aligned} -\Delta u &= f(x,u,v,\nabla u,\nabla v), \\ -\Delta v &= g(x,u,v,\nabla u,\nabla v) \text{ in } \Omega, \\ u &= v = 0 \text{ on } \partial\Omega \end{aligned} \tag{7}$$

under the following set of conditions:

(f_1) $f : \overline{\Omega} \times \mathbb{R} \times \mathbb{R} \times \mathbb{R}^N \times \mathbb{R}^N \longrightarrow \mathbb{R}$ is continuous,
(f_2) $\liminf_{t\to\infty} \frac{f(x,s,t,\xi,\eta)}{t} > \lambda_1$ uniformly in $(x,s,t,\xi,\eta) \in \Omega \times \mathbb{R} \times \mathbb{R} \times \mathbb{R}^N \times \mathbb{R}^N$,
(f_3) There exist constants $p \geq 1$, $C > 0$ and $\sigma \geq 0$ such that

$$|f(x,s,t,\xi,\eta)| \leq C(|t|^p + |s|^{p\sigma} + 1)$$

for all $(x,s,t,\xi,\eta) \in \Omega \times \mathbb{R} \times \mathbb{R} \times \mathbb{R}^N \times \mathbb{R}^N$,
(g_1) $g : \overline{\Omega} \times \mathbb{R} \times \mathbb{R} \times \mathbb{R}^N \times \mathbb{R}^N \longrightarrow \mathbb{R}$ is continuous
(g_2) $\liminf_{t\to\infty} \frac{g(x,s,t,\xi,\eta)}{t} > \lambda_1$ uniformly in $(x,s,t,\xi,\eta) \in \Omega \times \mathbb{R} \times \mathbb{R} \times \mathbb{R}^N \times \mathbb{R}^N$,
(g_3) There exist constants $q \geq 1$, $C' > 0$ and $\sigma' \geq 0$ such that

$$|g(x,s,t,\xi,\eta)| \leq C(|s|^q + |t|^{q\sigma'} + 1),$$

for all $(x,s,t,\xi,\eta) \in \Omega \times \mathbb{R} \times \mathbb{R} \times \mathbb{R}^N \times \mathbb{R}^N$.

In our work [25] two other hyperbolas appeared instead of the critical hyperbola, due to the limitations coming from the method. This is precisely like in the scalar case in [21] where the exponent $\frac{N+1}{N-1}$ appeared instead of $\frac{N+2}{N-2}$. Observe that the intersection of the two hyperbolas below is the Brézis-Turner exponent $\frac{N+1}{N-1}$:

$$\begin{aligned} \tfrac{1}{p+1} + \tfrac{N-1}{N+1}\tfrac{1}{q+1} &= \tfrac{N-1}{N+1}, \\ \tfrac{N-1}{N+1}\tfrac{1}{p+1} + \tfrac{1}{q+1} &= \tfrac{N-1}{N+1}. \end{aligned} \tag{8}$$

Theorem 3.2 *Let Ω be a smooth bounded domain in $\mathbb{R}^N$, with $N \geq 4$. Assume that the conditions f_1, f_2, f_3, g_2, g_3 hold with p, q being the coordinates of a point below both of the above hyperbolas. Suppose that σ, σ' are given by*

$$\sigma = \frac{L}{\max(L,K)}, \quad \sigma' = \frac{K}{\max(L,K)},$$

where

$$K = \frac{p}{p+1} - \frac{2}{N} > 0, \quad L = \frac{q}{q+1} - \frac{2}{N} > 0.$$

Then the positive solutions of the system (3.2) are bounded in L^∞.

Remark 3.1 If $N = 3$, then $K, L > 0$ imply $p, q > 2$ which is not compatible with (8). So the case $N = 3$ needs a special treatment.

Remark 3.2 A priori estimates for solutions of systems of type (3) using Hardy-Sobolev inequalities in the spirit of [21] has also been considered by Cosner [29]; however, he requires that the growth of nonlinearities has to be (separately) below the Brezis-Turner exponent $\frac{N+1}{N-1}$.

In the proof of the above theorem an essential tool is the proposition below, which is proved using the inequality

$$\left\| \frac{u}{\varphi_1^\tau} \right\|_{L^r} \leq C \|Du\|_{L^q}, \quad \forall \; u \in H_0^1,$$

where $\frac{1}{r} = \frac{1}{q} - \frac{1-\tau}{N}$.

This last inequality is an interpolation between a Sobolev inequality and the usual Hardy inequality [21, 63].

$$\left\| \frac{u}{\varphi_1} \right\|_{L^r} \leq C \|Du\|_{L^q}, \quad \forall \; u \in W_0^{1,q},$$

where $q > 1$ and φ_1 is the eigenfunction associated with the first eigenvalue of $(-\Delta, H_0^1(\Omega))$.

Proposition 3.1 *Let $r_0 \in (1, \infty]$, $r_1 \in [1, \infty)$ and $u \in L^{r_0}(\Omega) \bigcap W_0^{1,r_1}$. Then for all $\tau \in [0, 1]$ we have*

$$\frac{u}{\varphi_1^\tau} \in L^r(\Omega), \text{ where } \frac{1}{r} = \frac{1-\tau}{r_0} + \frac{\tau}{r_1}.$$

Moreover

$$\left\| \frac{u}{\varphi_1^\tau} \right\|_{L^r} \leq C \|u\|_{L^{r_0}}^{1-\tau} \|u\|_{W^{1,r_1}}^{\tau},$$

where the constant C depends only on τ, r_0, and r_1.

Sketch of Proof via Hardy Inequality

The proof given in [25] is rather very technical. So below we just mention the main steps.

Conditions (f_2) and (g_2) (superlinearity) imply that the projections of the positive solutions over the first eigenspace are bounded:

$$\int_\Omega u\varphi_1 dx < const, \quad \int_\Omega v\varphi_1 dx < const.$$

It then follows that

$$\int_\Omega f(.)\varphi_1 dx < const, \quad \int_\Omega g(.)\varphi_1 dx < const.$$

This is then used to estimate

$$\int_\Omega |\Delta u|^{\frac{p+1}{p}} = \int_\Omega |f|^{\frac{p+1}{p}} = \int_\Omega |f|^\alpha \varphi^\alpha |f|^{1-\alpha+\frac{1}{p}} \varphi^{-\alpha},$$

where $0 < \alpha < 1$ is chosen later. In the above relation use Hölder and Hardy, and choose α in order to obtain at the end:

$$\| u \|^{\frac{p+1}{p}}_{W^{2,\frac{p+1}{p}}} \leq C \left(\| u \|^{\frac{p+1}{p}\gamma_1}_{W^{2,\frac{p+1}{p}}} + \| v \|^{\frac{q+1}{q}\gamma_2}_{W^{2,\frac{q+1}{q}}} + 1 \right),$$

$$\| v \|^{\frac{q+1}{q}}_{W^{2,\frac{q+1}{q}}} \leq C \left(\| u \|^{\frac{p+1}{p}\gamma_3}_{W^{2,\frac{p+1}{p}}} + \| v \|^{\frac{q+1}{q}\gamma_4}_{W^{2,\frac{q+1}{q}}} + 1 \right),$$

with $0 < \gamma_i < 1$, $i = 1, \ldots, 4$. This gives

$$\| u \|_{W^{2,\frac{p+1}{p}}} \leq C, \quad \| v \|_{W^{2,\frac{q+1}{q}}} \leq C,$$

and complete the proof by using a bootstrap procedure.

3.2 *Estimates Using Moving Planes*

The procedure by this method parallels our work with Lions and Nussbaum [39] done for the scalar case.

For systems, one has to use Troy's extension [90] of Gidas–Ni–Nirenberg [58] for cooperative systems. See also [33].

The main point is to estimate the gradients of eventual solutions of the system near the boundary, and then use Pohozaev-type (cf. [68, 75]) identities to get bounds in some L^p norms. Finally bootstrap.

By this method we consider the problem

$$-\Delta u = f(v), \quad -\Delta v = g(u) \text{ in } \Omega, \quad u = v = 0 \text{ on } \partial\Omega,$$

under the following hypotheses:

(f_1) $f, g : \mathbb{R}^+ \to \mathbb{R}$, C^1 with $f', g' \geq 0$

(f_2) $\exists \alpha, \beta \in (0, \infty)$ $1 < p, q < \infty$, s.t. :

$$\lim_{s\to\infty} \frac{f(s)}{s^p} = \alpha, \quad \lim_{s\to\infty} \frac{g(s)}{s^q} = \beta.$$

The following result is in our paper with Clement and Mitidieri [24].

Theorem 3.3 *Under above hypotheses, Ω convex, and*

$$\frac{1}{p+1}+\frac{1}{q+1}>1-\frac{2}{N}$$

the positive solutions of the system below are L^∞ bounded:

$$-\Delta u=f(v),\quad -\Delta v=g(u)\text{in }\Omega,\quad u=v=0\,\text{on}\,\partial\Omega.$$

3.3 The Blow-Up Method

The other technique used to obtain a priori bounds for solutions of systems is the *Blow-up Method*, first used by Gidas-Spruck in [59] to treat scalar equations. Since there will be many symmetry assumptions regarding the behavior of the nonlinearities with respect to the unknowns u, v, it will be more convenient, henceforth in this section, to replace them by u_1, u_2. So, let us consider the system in the form:

$$\begin{cases} -\Delta u_1 = f(x,u_1,u_2,\nabla u_1,\nabla u_2) & \text{in } \Omega \\ -\Delta u_2 = g(x,u_1,u_2,\nabla u_1,\nabla u_2) & \text{in } \Omega \\ u_1=u_2=0 & \text{on } \partial\Omega, \end{cases} \tag{9}$$

where we look for solutions u_1, u_2 that are real-valued functions defined on a smooth bounded domain Ω in $\mathbb{R}^N$, $N \geq 3$, and f and g are real-valued functions defined in $\overline{\Omega}\times R\times R\times\mathbb{R}^N\times\mathbb{R}^N$. As far as the regularity of the solutions, we aim for them to be in $C^0(\overline{\Omega})\cap C^2(\Omega)$.

We then write the system as follows, assuming that the leading parts of f and g involve just pure powers of u_1 and u_2, and we have

$$\begin{cases} -\Delta u_1 = a(x)u_1^{\alpha_{11}}+b(x)u_2^{\alpha_{12}}+h_1(x,u_1,u_2,\nabla u_1,\nabla u_2) \\ -\Delta u_2 = c(x)u_1^{\alpha_{21}}+d(x)u_2^{\alpha_{22}}+h_2(x,u_1,u_2,\nabla u_1,\nabla u_2). \end{cases} \tag{10}$$

All along this subsection we assume the following conditions:

(A1) The coefficients $a, b, c, d : \overline{\Omega}\to[0,\infty)$ are continuous functions.
(A2) The exponents $\alpha_{ij}\geq 0 \quad i,j=1,2$.
(A3) There exist positive constants c_1 and c_2 such that

$$|h_1(x,s,t,\xi,\eta)|\leq c_1(1+|s|^{\beta_{11}}+|t|^{\beta_{12}}+|\xi|^{\gamma_{11}}+|\eta|^{\gamma_{12}})$$
$$|h_2(x,s,t,\xi,\eta)|\leq c_2(1+|s|^{\beta_{21}}+|t|^{\beta_{22}}+|\xi|^{\gamma_{21}}+|\eta|^{\gamma_{22}})$$

where

$$0 \le \beta_{ij} < \alpha_{ij} \quad i, j = 1, 2.$$

Later on, we state hypotheses on the exponents γ in order to obtain some estimates.

The Blow-up Method consists in assuming, by contradiction, that there is no a priori bound for the solutions of the system (10). So assume that there exists a sequence $(u_{1,n}, u_{2,n})$ of positive solutions of (10) such that at least one of the sequences $u_{1,n}$ or $u_{2,n}$ tends to infinity in the L^∞-norm. Without loss of generality, we may suppose that

$$\max_{x\in\Omega} ||u_{1,n}||^{\beta_2} \ge \max_{x\in\Omega} ||u_{2,n}||^{\beta_1},$$

where β_1, β_2 are positive constants to be chosen later. Let $x_n \in \Omega$ be a point where $u_{1,n}$ assumes its maximum: $u_{1,n}(x_n) = \max_{x\in\Omega} u_{1,n}(x)$. Then the sequence $\lambda_n = u_{1,n}(x_n)^{-\frac{1}{\beta_1}}$ is such that $\lambda_n \to 0$. The functions

$$v_{i,n}(x) = \lambda_n^{\beta_i} u_{i,n}(\lambda_n x + x_n),$$

satisfy $v_{1,n}(0) = 1, \quad 0 \le v_{i,n} \le 1$ in Ω. One also verifies that the functions $v_{1,n}$ and $v_{2,n}$ satisfy

$$\begin{cases} -\Delta v_{1,n} = a(\cdot)\lambda_n^{\beta_1+2-\beta_1\alpha_{11}}\, v_{1,n}^{\alpha_{11}} + b(\cdot)\lambda_n^{\beta_1+2-\beta_2\alpha_{12}}\, v_{2,n}^{\alpha_{12}} + \widetilde{h}_1(\cdot) \\ -\Delta v_{2,n} = c(\cdot)\lambda_n^{\beta_2+2-\beta_1\alpha_{21}}\, v_{1,n}^{\alpha_{21}} + d(\cdot)\lambda_n^{\beta_2+2-\beta_2\alpha_{22}}\, v_{2,n}^{\alpha_{22}} + \widetilde{h}_2(\cdot), \end{cases} \tag{11}$$

in the domain $\Omega_n = \dfrac{1}{\lambda_n}(\Omega - x_n)$, where the dot stands for $\lambda_n x + x_n$.
The idea of the method is then to pass to the limit as $n \to \infty$ in (11) and obtain a system either in $\mathbb{R}^N$ or in $\mathbb{R}^N_+$, which can be proved that it has only the trivial solution. This would contradict the fact that the limit of $v_{1,n}$ has value 1 at the origin. By compactness the sequence (x_n) or a subsequence of it converges to a point x_0. We observe that the limiting system is defined in $\mathbb{R}^N$ or in $\mathbb{R}^N_+$, accordingly to this limit point (x_0) being a point in Ω or in $\partial\Omega$. In the next proposition we make precise these statements.

Proposition 3.2 *The sequences $(v_{1,n})$ and $(v_{2,n})$ converge in $W^{2,p}_{\text{loc}}$, with $2 \le p < \infty$, to functions $v_1, v_2 \in C^2(G)\bigcap C^0(\overline{G})$, satisfying the limiting system of (11) in $G = \mathbb{R}^N$ or in $G = \mathbb{R}^N_+$, provided all the powers of λ_n in (11) are non-negative. This limiting system is obtained by removing the terms in (11) where the powers of λ_n are strictly positive, the terms where the coefficients vanishes at x_0, and the lower order terms.*

In [72] and [32] special classes of systems were studied. In [41] with Sirakov we considered more general systems and did a complete discussion of the systems (1) weakly coupled and (2) strongly coupled. The terminology, which we do below, is explained by the type of system obtained after the passage to the limit. We next analyze these two classes.

Definition 3.1 System (10) is *weakly coupled* if there are positive numbers β_1, β_2 such that

$$\begin{aligned} \beta_1 + 2 - \beta_1\alpha_{11} = 0 \ , \quad \beta_1 + 2 - \beta_2\alpha_{12} > 0 \\ \beta_2 + 2 - \beta_1\alpha_{21} > 0 \ , \quad \beta_2 + 2 - \beta_2\alpha_{22} = 0 \end{aligned} \tag{12}$$

Definition 3.2 System (10) is *strongly coupled* if there are positive numbers β_1, β_2 such that

$$\begin{aligned} \beta_1 + 2 - \beta_1\alpha_{11} > 0 \ , \quad \beta_1 + 2 - \beta_2\alpha_{12} = 0 \\ \beta_2 + 2 - \beta_1\alpha_{21} = 0 \ , \quad \beta_2 + 2 - \beta_2\alpha_{22} > 0 \end{aligned} \tag{13}$$

Remark 3.3 It follows that if the system (10) is weakly coupled then necessarily we should have

$$\beta_1 = \frac{2}{\alpha_{11} - 1} \text{ and } \beta_2 = \frac{2}{\alpha_{22} - 1}, \tag{14}$$

which requires that $\alpha_{11} > 1, \alpha_{22} > 1$ and

$$\alpha_{12} < \frac{\alpha_{22} - 1}{\alpha_{11} - 1}\alpha_{11} \text{ and } \alpha_{21} < \frac{\alpha_{11} - 1}{\alpha_{22} - 1}\alpha_{22}. \tag{15}$$

Remark 3.4 If the system (10) is strongly coupled, then

$$\beta_1 = \frac{2(\alpha_{12} + 1)}{\alpha_{12}\alpha_{21} - 1} \text{ and } \beta_2 = \frac{2(\alpha_{21} + 1)}{\alpha_{12}\alpha_{21} - 1}, \tag{16}$$

which requires that $\alpha_{12}\alpha_{21} > 1$ and

$$\alpha_{11} < \frac{\alpha_{21} + 1}{\alpha_{12} + 1}\alpha_{12} \text{ and } \alpha_{22} < \frac{\alpha_{12} + 1}{\alpha_{21} + 1}\alpha_{21}. \tag{17}$$

Remark 3.5 In order to take care of the gradients, we assume further the following conditions:

(A4) If (10) is weakly coupled, $\gamma_{ij}, i, j = 1, 2$ satisfy

$$\gamma_{11} < \frac{2\alpha_{11}}{\alpha_{11} + 1}, \quad \gamma_{22} < \frac{2\alpha_{22}}{\alpha_{22} + 1}$$

$$\gamma_{12} < \frac{2\alpha_{11}(\alpha_{22} - 1)}{(\alpha_{11} - 1)(\alpha_{22} + 1)}, \quad \gamma_{21} < \frac{2\alpha_{22}(\alpha_{11} - 1)}{(\alpha_{22} - 1)(\alpha_{11} + 1)};$$

(A5) If (10) is strongly coupled, $\gamma_{ij}, i, j = 1, 2$ satisfy

$$\gamma_{11} < \frac{2\alpha_{12}(\alpha_{21}+1)}{2\alpha_{12}+\alpha_{12}\alpha_{21}+1}, \quad \gamma_{22} < \frac{2\alpha_{21}(\alpha_{12}+1)}{2\alpha_{21}+\alpha_{12}\alpha_{21}+1}$$

$$\gamma_{12} < \frac{2\alpha_{12}(\alpha_{21}+1)}{2\alpha_{21}+\alpha_{12}\alpha_{21}+1}, \quad \gamma_{21} < \frac{2\alpha_{21}(\alpha_{12}+1)}{2\alpha_{12}+\alpha_{12}\alpha_{21}+1}.$$

We observe that the requirements that $\alpha_{11}, \alpha_{22} > 1$ and $\alpha_{12}\alpha_{21} > 1$ are known as *super-linearity* conditions.

Weakly Coupled System After the Blow-up, the limiting system becomes, using a scaling of the solutions v_1, v_2:

$$\begin{aligned} -\Delta w_1 &= w_1^{\alpha_{11}} \\ -\Delta w_2 &= w_2^{\alpha_{22}} \ , \ \text{ in } \mathbb{R}^N, \end{aligned} \tag{18}$$

and

$$\begin{aligned} -\Delta w_1 &= w_1^{\alpha_{11}}, \\ -\Delta w_2 &= w_2^{\alpha_{22}} \ \text{ in } \mathbb{R}^N_+ \\ w_1 = w_2 &= 0 \ \text{ on } \ x_N = 0. \end{aligned} \tag{19}$$

The existence or not of positive solutions for such systems is the object of the so-called Liouville type theorems. They will be discussed in the next section. For the time being we anticipate that

1. the equations in system (18) have only the trivial solution if $0 < \alpha_{11}, \alpha_{22} < \frac{N+2}{N-2}$
2. the equations in system (19) have only the trivial solution if $1 < \alpha_{11}, \alpha_{22} < \frac{N+1}{N-3}$, if the dimension $N > 3$, cfr [34].

So the following result holds. Conditions (A1), (A2), (A3), (A4) are assumed.

Theorem 3.4 *Let (10) be a weakly coupled system with continuous coefficients* a, b, c, d, *exponents* α*'s* ≥ 0, *and such that* $a(x), d(x) \geq c_0 > 0$ *for* $x \in \overline{\Omega}$. *Assume also that* $0 < \alpha_{11}, \alpha_{22} < (N+2)/(N-2)$. *Then there is a constant* $C > 0$ *such that*

$$||u_1||_{L^\infty}, ||u_2||_{L^\infty} \leq C$$

for all positive solutions $u_1, u_2 \in C^2(\Omega) \cap C^0(\overline{\Omega})$ *of system (10).*

Strongly Coupled System As in the case of a weakly coupled system, the limiting systems are

$$-\Delta\omega_1 = \omega_2^{\alpha_{12}}, \quad -\Delta\omega_2 = \omega_1^{\alpha_{21}} \ \text{ in } \mathbb{R}^N \tag{20}$$

and

$$-\Delta\omega_1 = \omega_2^{\alpha_{12}}, \quad -\Delta\omega_2 = \omega_1^{\alpha_{21}} \text{ in } (\mathbb{R}^N)^+ \tag{21}$$

with

$$\omega_1(x', 0) = \omega_2(x', 0) = 0.$$

So, a contradiction comes if the exponents are such that (20) and (21) have only the trivial solution $\omega_1 = \omega_2 \equiv 0$. In summary, the following result holds with conditions (A1), (A2), (A3), (A5) assumed.

Theorem 3.5 *Let (10) be a strongly coupled system with continuous coefficients a, b, c, d, such that $b(x), c(x) \geq c_0 > 0$ for $x \in \overline{\Omega}$. Assume also that the α exponents are non-negative. Assume further that the following conditions hold:*

(L1) *The exponents α_{12} and α_{21} are such that the only non-negative solution of*

$$-\Delta\omega_1 = \omega_2^{\alpha_{12}}, \quad -\Delta\omega_2 = \omega_1^{\alpha_{21}} \text{ in } \mathbb{R}^N$$

is $w_1 = \omega_2 \equiv 0$.

(L2) *The only non-negative solution of*

$$-\Delta\omega_1 = \omega_2^{\alpha_{12}}, \quad -\Delta\omega_2 = \omega_1^{\alpha_{21}} \text{ in } \mathbb{R}_+^N$$

with $\omega_1(x', 0) = \omega_2(x', 0) = 0$ is $\omega_1 = \omega_2 \equiv 0$. Then there is a constant $C > 0$ such that

$$||u_1||_{L^\infty}, ||u_2||_{L^\infty} \leq C$$

for all non-negative solutions (u_1, u_2) of system (10).

Remark 3.6 Which conditions should be required on the exponents α_{12} and α_{21} in such a way that (L1) and (L2) holds? Again these are Liouville type theorems for systems, which will be described in the next section.

4 Liouville Theorems

Next we make comments on types of Liouville theorems that are necessary for completing the proofs of the a priori estimates done by Blow-up in the previous subsection. An extensive discussion of Liouville theorems can be seen in our paper [34]. See also [69, 81]

We start with the scalar case:

$$-\Delta u = u^p \tag{22}$$

For $\mathbb{R}^N$, $N \geq 3$, we have the following result:

Theorem 4.1 *Let u be a non-negative C^2 function defined in the whole of $\mathbb{R}^N$, such that (22) holds in $\mathbb{R}^N$. If $0 < p < (N+2)/(N-2)$, then $u \equiv 0$.*

This result was proved by Gidas-Spruck [60] in the case $1 < p < (N+2)/(N-2)$. A simpler proof using the method of moving parallel planes was given by Chen-Li [23], and it is valid in the whole range of p. An elementary proof valid for $p \in [1, \frac{N}{N-2})$ was given by Souto [87].

Theorem 4.2 *Let $u \in C^2(\mathbb{R}^N_+) \cap C^0(\mathbb{R}^N_+)$ be a non-negative function such that*

$$\begin{cases} -\Delta u = u^p & \text{in } \mathbb{R}^N_+ \\ u(x', 0) = 0 \end{cases} \tag{23}$$

If $1 < p \leq (N+2)/(N-2)$, then $u \equiv 0$.

Remark 4.1 This theorem was proved in [59]. It is remarkable that in the case of the half-space the exponent $(N+2)/(N-2)$ is not the right one for theorems of Liouville type. Indeed, Dancer [30] has proved the following result.

Theorem 4.3 *Let $u \in C^2(\mathbb{R}^N_+) \cap C^0(\mathbb{R}^N_+)$ be a non-negative bounded solution of (23). If $1 < p < (N+1)/(N-3)$ for $N \geq 4$ and $1 < p < \infty$ for $N = 3$, then $u \equiv 0$.*

Remark 4.2 If $p = (N+2)/(N-2)$, $N \geq 3$, then (22) has a two-parameter family of bounded positive solutions:

$$U_{\varepsilon, x_0}(x) = \left[\frac{\varepsilon \sqrt{N(N-2)}}{\varepsilon^2 + |x - x_0|^2} \right]^{\frac{N-2}{2}},$$

which are called *instantons*.

Liouville for Systems Defined in the Whole of $\mathbb{R}^N$

We start considering systems of the form

$$-\Delta u = v^p, \quad -\Delta v = u^q. \tag{24}$$

In analogy with the scalar case, the dividing line here between existence and nonexistence of positive solutions (u, v) defined in the whole of $\mathbb{R}^N$ should be the *critical hyperbola* [24, 62]. Such hyperbola associated with problems of the form (24) is defined by

$$\frac{1}{p+1} + \frac{1}{q+1} = 1 - \frac{2}{N}, \quad p, q > 0. \tag{25}$$

Continuing the analogy with the scalar case, one may *conjecture* that (24) has no bounded positive solutions defined in the whole of $\mathbb{R}^N$ if p, q are below the critical hyperbola, namely

$$\frac{1}{p+1} + \frac{1}{q+1} > 1 - \frac{2}{N}, \quad p, q > 0. \tag{26}$$

To our knowledge, this conjecture has not been settled in full so far. Why such a conjecture? In answering it, let us remind some facts, already contained in the previous sections. The critical hyperbola appeared in the study of existence of positive solutions for superlinear elliptic systems of the form

$$-\Delta u = g(v), \quad -\Delta v = f(u) \tag{27}$$

subject to Dirichlet boundary conditions in a bounded domain Ω of $\mathbb{R}^N$. If $g(v) \sim v^p$ and $f(u) \sim u^q$ as $u, v \to \infty$, then system (24) is said to be sub-critical if p, q satisfy (26). For such systems [in analogy with sub-critical scalar equations, $-\Delta u = f(u)$, $f(u) \sim u^p$ and $1 < p < (N+2)/(N-2)$] one can establish in many cases a priori bounds of positive solutions, prove a Palais-Smale condition and put through an existence theory by a topological or a variational method. This sort of work initiated in [24] and [76] has been continued. We have surveyed some of this work in the previous sections. Recall that the problem in the critical scalar case (that is, $-\Delta u = |u|^{2^*-2}u$ in Ω, $u = 0$ on $\partial\Omega$) has no solution $u \neq 0$ if Ω is a star-shaped bounded domain in $\mathbb{R}^N$, $N \geq 3$. In the case of systems, the critical hyperbola appears in the statement: if Ω is a bounded star-shaped domain in $\mathbb{R}^N$, $N \geq 3$, the Dirichlet problem for the system below has no non-trivial solution:

$$-\Delta u = |v|^{p-1}v, \quad -\Delta v = |u|^{q-1}u$$

if, p, q satisfy (25). This follows from an identity of Pohozaev-type, see Mitidieri [68]; also Pucci-Serrin [75] for general forms of Pohozaev-type identities.

Next we describe several Liouville-type theorem for systems.

Theorem 4.4 *Let $p, q > 0$ satisfying (26). Then system (24) has no non-trivial radial positive solutions of class $C^2(\mathbb{R}^N)$.*

Remark 4.3 This result settles the conjecture in the class of *radial* functions. It was proved in [68] for $p, q > 1$, and for p, q in the full range by Serrin-Zou [78]. The proof explores the fact that eventual positive radial solutions of (24) have a definite decay at ∞; this follows from an interesting observation (cf. Lemma 6.1 in [68]), namely:

If $u \in C^2(\mathbb{R}^N)$ is a positive radial superharmonic function, then

$$ru'(r) + (N-2)u(r) \geq 0, \text{ for all } r > 0.$$

Theorem 4.4 is sharp as far as the critical hyperbola is concerned. Indeed, there is the following existence result of Serrin-Zou [80].

Theorem 4.5 *Suppose that* $p, q > 0$ *and that*

$$\frac{1}{p+1} + \frac{1}{q+1} \leq 1 - \frac{2}{N}. \tag{28}$$

Then there exist infinitely many values $\xi = (\xi_1, \xi_2) \in R^+ \times R^+$ *such that system (24) admits a positive radial solution* (u, v) *with central values* $u(0) = \xi_1, \ v(0) = \xi_2$. *Moreover* $u, v \to 0$ *as* $|x| \to \infty$. *So the solution is in fact a ground state for (24).*

The next result extends, as compared with the previous results, the region under the critical hyperbola where the Liouville theorem holds.

Theorem 4.6

(A) If $p > 0$ *and* $q > 0$ *are such that* $p, q \leq (N+2)/(N-2)$, *but not both equal to* $(N+2)/(N-2)$, *then the only non-negative solution of (24) is* $u = v = 0$.
(B) If $p = q = (N+2)/(N-2)$, *then* u *and* v *are radially symmetric with respect to some point of* $\mathbb{R}^N$.

This theorem is due to de Figueiredo-Felmer [40]. The proof uses the method of Moving Planes. A good basic reference of this method is [13]. The idea in the proof of the above theorem is to use Kelvin transform in the solutions u, v of (24), which a priori has no known (or prescribed) behavior at infinite. By means of Kelvin's u and v are transformed in new unknowns w and z satisfying

$$\begin{aligned} -\Delta w &= \frac{1}{|x|^{N+2-p(N-2)}} z^p(x), \\ -\Delta z &= \frac{1}{|x|^{N+2-q(N-2)}} w^q(x), \end{aligned} \tag{29}$$

which now have a definite decay at ∞, provided (p, q) satisfy the conditions of Theorem 4.6. It is precisely at this point that we cannot take $p > \frac{N+2}{N-2}$, because then one would lose the right type of monotonicity of the coefficients necessary to put the Moving Plane method to work. So having this correct monotonicity of the coefficients the method of moving planes can start. This result has been extended by Felmer [54] to systems with more than two equations.

In Theorems 4.7 and 4.9 below we assume $pq > 1$ and introduce the notation

$$\alpha = \frac{2(p+1)}{pq-1}, \quad \beta = \frac{2(q+1)}{pq-1}.$$

The next result is due to Busca-Manasevich [18] and extends further, as compared with Theorem 4.6, the region of values of p, q where the Liouville theorem for system (24) holds.

Theorem 4.7 *Suppose that* $p, q > 1$ *and*

$$\alpha, \beta \in \left(\frac{N-2}{2},\ N-2\right). \tag{30}$$

Then system (24) has no non-trivial solution of class $C^2(\mathbb{R}^N)$.

If some behavior of u and v at ∞ is known, the Liouville theorem can be established for all (p, q) below the critical hyperbola, as in the next result.

Theorem 4.8 *Let* $p > 0$ *and* $q > 0$ *satisfying (26), then there are no positive solutions of (24) satisfying*

$$u(x) = o(|x|^{-\frac{N}{q+1}}),\quad v(x) = o(|x|^{-\frac{N}{p+1}}),\ \text{as } |x| \to \infty. \tag{31}$$

The above result is due to Serrin-Zou [78],

Remark 4.4 Observe that Theorem 4.8 extends Theorem 4.4, since radial positive solutions have a decay at infinity.

Liouville Theorems for Systems Defined in Half-Spaces

Now we look at the system below and state some results on the nonexistence of non-trivial solutions and also of supersolutions.

$$\begin{cases} -\Delta u = v^p & \text{in } \mathbb{R}^N_+ \\ -\Delta v = u^q & \text{in } \mathbb{R}^N_+ \\ u, v \geq 0 & \text{in } \mathbb{R}^N_+ \\ u, v = 0 & \text{on } \partial\mathbb{R}^N_+ \end{cases} \tag{32}$$

Theorem 4.9 *Let* $p, q > 1$ *satisfying*

$$\max(\alpha, \beta) \geq N - 3. \tag{33}$$

Then the system (32) has only the trivial solution.

Remark 4.5 This result is due to Birindelli-Mitidieri [15].

A Liouville Theorem for a Full System

Now we consider the following system:

$$\begin{aligned} -\Delta u_1 &= u_1^{\alpha_{11}} + u_2^{\alpha_{12}}, \\ -\Delta u_2 &= u_1^{\alpha_{21}} + u_2^{\alpha_{22}} \ \text{in } \mathbb{R}^N. \end{aligned} \tag{34}$$

In order to state the next result we introduce the following notation:

$$\overline{\alpha} = \frac{2(\alpha_{12}+1)}{\alpha_{12}\alpha_{21}-1},\quad \overline{\beta} = \frac{2(\alpha_{21}+1)}{\alpha_{12}\alpha_{21}-1}.$$

Theorem 4.10 *System (34) has only the trivial solution if the following conditions hold:*

$$\alpha_{11}, \alpha_{22} < \frac{N+2}{N-2}, \quad min\{\overline{\alpha}, \overline{\beta}\} > \frac{N-2}{2}. \tag{35}$$

This result is due to de Figueiredo-Sirakov [41], and it relies heavily on results, which are also proved in [41]: the first one is an extension of a result by Dancer [30], proved for the scalar case, and the second one is an extension of a result by Busca-Manasevich in [18]. More details in [34]. See also [71].

Final Remarks on Liouville Theorem for Systems

1. The conjecture on the validity of a Liouville theorem in the whole of $\mathbb{R}^N$ for all p and q below the critical hyperbola seems to be unsettled at this moment. In dimension $N = 3$ the conjecture was proved by Serrin and Zou in [78]. In dimension $N = 4$ the conjecture has been proved recently by Souplet in [86]. See also Theorem 4.8 above, where the conjecture is proved provided one supposes that u or v has at most algebraic growth.

Theorem 4.11 *Let $u, v \in C^2(\mathbb{R}^N_+) \cap C^0(\overline{\mathbb{R}^N_+})$ be non-negative solutions of (6.28) with $u = v = 0$ on $\partial\mathbb{R}^N_+$. If $1 \leq p, q \leq \dfrac{N+2}{N-2}$ then $u = v \equiv 0$.*

2. Liouville-type theorems for systems of p-Laplacians have been studied recently by Mitidieri-Pohozaev [70].
3. Liouville theorems for equations with a weight have been considered in Berestycki, Capuzzo Dolcetta-Nirenberg [12].

References

1. H. Amann, Fixed Point Equation and nonlinear Eigenvalue Problems in Ordered Banach Spaces, SIAM Review 18 (1976), 630–709.
2. A. Ambrosetti and P. H. Rabinowitz, Dual variational methods in critical point theory and applications, J. Fctl. Anal. 14 (1973), 349–381.
3. C. Azizieh and Ph. Clément, A priori estimates and continuation methods for positive solutions of p-Laplace equations, J. Diff. Eq. 179 (2002), 213–245
4. C. Azizieh and Ph. Clément, Existence and a priori estimates for positive solutions of p-Laplace systems, J. Diff. Eq. 184 (2002), 422–442
5. C. Bandle and M.Essen, On positive solutions of Emden equations in cones. Arch. Rat. Mech. Anal. 112 (1990), 319–338
6. T. Bartsch, Infinitely many solutions of symmetric Dirichlet problems, Nonl. Anal. TMA 20 (1993), 1205–1216.
7. T. Bartsch and M. Clapp, Critical point theory for indefinite functionals with symmetries, J. Fctl. Anal. 138 (1996), 107–136.
8. T. Bartsch and D.G. de Figueiredo, Infinitely many solutions of nonlinear elliptic systems, Progress in Nonlinear Differential Equations and their Applications, Vol. 35 (The Herbert Amann Anniversary Volume) (1999), 51–68. (1996), 107–136.

9. T. Bartsch and M. Willem, Infinitely many nonradial solutions of a Euclidean scalar field equation, J. Fctl. Anal. 117 (1993), 447–460.
10. V. Benci and P.H. Rabinowitz, Critical Point Theorems for Indefinite Functionals, Invent. Math. (1979), 241–273.
11. T. B. Benjamin, A Unified Theory of Conjugate Flows, Phil. Trans. Royal Soc. 269 A (1971), 587–643.
12. H. Berestycki, I. Capuzzo-Dolcetta and L. Nirenberg, Superlinear indefinite elliptic problems and nonlinear Liouville theorems, Top. Meth. Nonl. Anal. 4 (1995), 59–78.
13. H. Berestycki and L. Nirenberg, On the method of moving planes and the sliding method, Bull. Soc. Bras. Mat. 22 (1991), 1–22.
14. M.F. Bidault-Veron and P. Grillot, Singularities in elliptic systems with absorption terms.
15. I. Birindelli and E. Mitidieri, Liouville Theorems for Elliptic Inequalities and Applications. Proc. Royal Soc. Edinb. 128A (1998), 1217–1247.
16. L. Boccardo and D.G. de Figueiredo, Some remarks on a system of quasilinear elliptic equations, NODEA Nonlinear Differential Equations Appl. Vol 9 (2002), 309–323.
17. L. Boccardo, J. Fleckinger and F. de Thélin, Elliptic Systems with various growths. Preprint.
18. J. Busca and R. Manasevich, A Liouville type theorem for Lane-Emden systems, Indiana Math. J. 51 (2002), 37–51.
19. J. Busca and B. Sirakov, Symmetry results for semilinear elliptic systems in the whole space, J. Diif. Eq. 163 (2000), 41–56.
20. J. Busca and B. Sirakov, Harnack type estimates for nonlinear elliptic systems and applications. Ann. Inst. H. Poincaré. (2004), (21)5, 543–590
21. H. Brézis and R. E. L. Turner, On a Class of Superlinear Elliptic Problems, Comm. Part. Diff. Eq. 2(1977), 601–614.
22. G. Cerami, Un criterio de esistenza per i punti critici su varietà ilimitate, Istituto Lombardo di Scienze et Lettere 112(1978), 332–336.
23. W. Chen and C. Li, Classification of solutions of some nonlinear elliptic equations Duke Math. J. 63(1991), 615–622.
24. Ph. Clément, D. G. de Figueiredo and E. Mitidieri, Positive Solutions of Semilinear Elliptic Systems, Comm. Part. Diff. Eq. 17(1992), 923–940.
25. Ph. Clement, D. G. de Figueiredo and E. Mitidieri, A priori estimates for positive solutions of semilinear elliptic systems via Hardy-Sobolev inequalities. Pitman Res. Notes in Math.(1996) 73–91.
26. Ph. Clément, R. Manásevich, E. Mitidieri, Positive Solutions For a Quasilinear System via Blow Up, Comm. Part. Diff. Eq. 18(1993) 2071–2106.
27. D.G. Costa and C.A. Magalhães, A Variational Approach to Subquadratic Perturbations of Elliptic Systems, J. Diff. Eq. 111(1994), 103–122.
28. D.G. Costa and C.A. Magalhães, A Unified Approach to a Class of Strongly Indefinite Functionals, J. Diff. Eq. 122(1996), 521–547.
29. C. Cosner, Positive solutions of Semi-linear Elliptic Systems without Variational Structure, Nonlinear Analysis T.M.A., Vol 8 , 12 (1994), 1427–1436.
30. E. N. Dancer, Some notes on the method of moving planes, Bull Austral. Math. Soc. 46(1992), 425–434.
31. D. G. de Figueiredo, Positive Solutions of Semilinear Elliptic Equations, Springer Lecture Notes in Mathematics 957 (1982), 34–87.
32. D. G. de Figueiredo, Semilinear Elliptic Systems, Nonl. Funct. Anal. Appl. Diff. Eq., World Sci. Publishing, River Edge, (1998), 122–152.
33. D. G. de Figueiredo, Monotonicity and symmetry of solutions of elliptic systems in general domains. Nodea 1 (1994), 119–123.
34. D.G. de Figueiredo, Semilinear Elliptic Systems: Existence, Multiplicity, Symmetry of Solutions, Handbook of Differential Equations (Stationary PDE, vol3) Edited by M.Chipot, Elsevier (2008), 01–48
35. D. G. de Figueiredo and Y.H. Ding, Strongly Indefinite Functionals and Multiple Solutions of Elliptic Systems", Transactions of the American Mathematical Society, vol. 355 (2003), 2973–2989.

36. D.G. de Figueiredo and P. L. Felmer, On Superquadratic Elliptic Systems, Trans. Amer. Math. Soc. 343 (1994), 119–123.
37. D.G. de Figueiredo and C. A. Magalhães, On Nonquadratic Hamiltonian Elliptic Systems, Advances in Diff. Eq., 1(1996), 881–898.
38. D.G. de Figueiredo and M. Ramos, On Linear Perturbations of Superquadratic Elliptic Systems, Reaction-Diffusion Systems, Lectures Notes in Pure and Applied Mathematics Series 194 (1998), p.121–130.
39. D. G. de Figueiredo, P.-L.Lions and R. Nussbaum, A priori Estimates and Existence of Positive Solutions of Semilinear Elliptic of Positive Solutions of Semilinear Elliptic Equations, J. Math. Pures et Appl. 61 (1982), 41–63.
40. D.G. de Figueiredo and P. L. Felmer, A Liouville-type Theorem for Systems, Ann. Sc. Norm. Sup. Pisa, XXI (1994), 387–397.
41. D.G. de Figueiredo and B. Sirakov, Liouville Type Theorems, Monotonicity Results and a priori Bounds for positive Solutions of Elliptic Systems, Rel. Pesq. 07/04 IMECC-UNICAMP. (2004)
42. D.G. de Figueiredo and J. Yang, Decay, Symmetry and Existence of positive solutions of semilinear elliptic Systems, Nonl. Anal., TMA, Vol. 33 (1988), 211–234.
43. D.G. de Figueiredo and J. Yang, A priori bounds for positive solutions of a non-variational elliptic system, Comm. on PDE, Vol. 26 (2011), 2305–2324.
44. D.G. de Figueiredo, J.M.B.do O, and B. Ruf, On an inequality by Trudinger and Moser and related elliptic equation, Comm. Pure App. Mathematics, vol 55, (2002), 135–152.
45. D.G. de Figueiredo, J.M.B.do O, and B.Ruf, Critical and Subcritical Elliptic Systems in Dimension two, Indiana Univ. Mayh. Journal Vol. 53, (2004), 1037–1054.
46. D.G. de Figueiredo, J.M.B.do O, and B.Ruf, An Orlicz Approach to Superlinear Elliptic Systems, Journal of Functional Analysis. Vol. 7, (2005), 471–496.
47. D.G. de Figueiredo, J.M.B.do O, and B. Ruf, Non-variational Elliptic Systems in dimension two: a priori bounds and existence of positive solutions. Journal of Fixed Point Theory and its Applications, Vo. 187 (2008), 531–545
48. K. Deimling, Nonlinear Functional Analysis, Springer Verlag (1985).
49. F. deThelin, Premiére valeur propre dun système elliptique non linéaire, C. R. Acad. Sci. Paris 311 (1990), 603–606.
50. E. DiBenedetto, $C^{1+\alpha}$ local regularity of weak solutions of degenerate elliptic equations, Nonlin. Anal., TMA 7(1983), 827–850.
51. E. DiBenedetto, Partial Differential Equations, Birkhäuser (1995).
52. Y.H. Ding, Infinitely many entire solutions of elliptic systems with symmetry, Top. Meth. Nonl. Anal. 9 (1997), 313–323.
53. P.L. Felmer, Periodic Solutions of "Superquadratic" Hamiltonian Systems, J. Diff. Eq. 17(1992), 923–940.
54. P.L. Felmer, Nonexistence and Symmetry theorems for elliptic systems in $\mathbb{R}^N$, Rendiconti del Circ. Mat. Palermo XLIII (1994), 259–284.
55. M. Garcia-Huidobro and C. S. Yarur, Classification of Positive Singular Solutions for a class of semilinear Elliptic Systems.
56. M. Garcia-Huidobro, R. Manasevich, E. Mitidieri and C. S. Yarur, Existence and nonexistence of Positive Singular Solutions for a class of semilinear Elliptic Systems, Arch. Rat. Mech. Anal. (1997), 253–284.
57. B. Gidas, Symmetry properties and isolated singularities of positive solutions of nonlinear elliptic equations, Eds R. Sternberg, A. Kalinowski and J. Papadakis Proc. Conf. Kingston (1979). Lect. Notes on Pure and Appl. Math. 54(1980), 255–273.
58. B. Gidas, W.M. Ni and L. Nirenberg, Symmetry and related properties via maximum principles. Comm. Math. Phys. 68 (1979), 209–243.
59. B. Gidas and J. Spruck, A priori bounds for positive solutions of nonlinear elliptic equations, Comm. Part. Diff. Eq. 6(1981), 883–901.
60. B. Gidas and J. Spruck, Global and Local Behaviour of Positive Solutions of Nonlinear Elliptic Equations, Comm. Pure Appl. Math. 34(1981), 525–598.

61. D. Gilbarg and N. S. Trudinger, Elliptic Partial Differential Equations of Second Order, Springer Verlag 2nd Edition (1983).
62. J. Hulshof and R.C.A.M. van der Vorst, Differential Systems with Strongly Indefinite Variational Structure, J. Fctl. Anal. 114(1993), 32–58.
63. O. Kavian, Inegalité de Hardy-Sobolev et Application, Thèse de Doctorat de 3$^{\grave{e}me}$ cycle, Université de Paris VI (1978).
64. M. A. Krasnosels'kii, Positive Solutions of Operator Equations. P. Noordhoff Groningen (1964).
65. G.G. Laptev, Absence of global positive solutions of systems of semilinear elliptic inequalities in cones. Izv. Math. 64(2000), 1197–1215.
66. J.Q. Liu and S.Li, Some existence theorems on multiple critical points and their applications, Kexue Tongbao 17 (1984)
67. S. Li and M. Willem, Application of local linking to critical point theory, J. Math. Anal. Appl. 189 (1995), 6–32.
68. E. Mitidieri, A Rellich Type Identity and Applications, Comm. Part. Diff. Eq.18(1993), 125–151.
69. E. Mitidieri, Nonexistence of Positive Solutions of Semilinear Elliptic Systems in $\mathbb{R}^N$, Diff. Int. Eq. 9 (1996), 465–479.
70. E. Mitidieri and S.I. Pohozaev, A priori estimates and the absence of solutions of nonlinear partial differential equations and inequalities. Proceedings of the Steklov Institute of Mathematics 2001, 234:1–375.
71. E. Mitidieri, G. Sweers and R. van der Vorst, Nonexistence theorems for systems of quasilinear partial differential equations, Diff. Int. Eq. 8(1995), 1331–1354.
72. M. S. Montenegro, Criticalidade, superlinearidade e sublinearidade para sistemas elíptico semilineares, Tese de Doutoramento, Unicamp (1997).
73. D.C. de Morais Filho and M.A.S. Souto, Systems of p-Laplacian equations involving homogeneous nonlinearities with critical Sobolev exponent degrees,
74. R. Nussbaum, Positive Solutions of Nonlinear Elliptic Boundary Value Problems, J. Math. Anal. Appl. 51(1975), 461–482.
75. P. Pucci and J. Serrin, A general Variational Identity, Indiana Univ. Math. J. 35(1986), 681–703.
76. L. A. Peletier and R. C. A. M. van der Vorst, Existence and Non-existence of Positive Solutions of Non-linear Elliptic Systems and the Biharmonic Equation, Diff. Int. Eq. 5(1992), 747–767.
77. J. Fernandez Bonder and J.D. Rossi, Existence results for the p-Laplacian with nonlinear boundary conditions, J. Math. Anal. Appl. 263(2001), 195–223.
78. J. Serrin and H. Zou, Nonexistence of Positive Solutions of Semilinear Elliptic Systems, Discourses in Mathematics and Its Applications 3, Dept. of Mathematics, Texas A-M University (1994), 55–68.
79. J. Serrin and H. Zou, Non-existence of Positive Solutions of Lane- Emden System, Diff. Int. Eq. 9(1996), 635–653.
80. J. Serrin and H. Zou, Existence of Positive Solutions of Lane-Emden System, to appear in Atti del Sem. Mat. Univ. Modena (1997).
81. J. Serrin and H. Zou, The existence of Positive Entire Solutions of Elliptic Hamiltonian Systems.
82. E.A.B. Silva, Critical point theorems and applications to differential equations, Thesis University Wisconsin-Madison !1988)
83. E.A.B. Silva, Linking theorems and applications to nonlinear elliptic problems at resonance. Nonl. Anal. TMA (1991) 455–477.
84. B. Sirakov, On the existence of solutions of Hamiltonian systems in $\mathbb{R}^N$, Advances in Diff. Eq.
85. M. A. S. Souto, Sobre a existência de soluções positivas para sistemas cooperativos não-lineares, Tese de doutorado 1992, UNICAMP.
86. P.Souplet, The proof of the Lane-Emden conjecture in four space dimensions, Advances in Mathematics 221 (2009), 1409–1427.

87. M. A. S. Souto, A priori estimates and existence of positive solutions of nonlinear cooperative elliptic systems. Diff. Int. Eq. 8(1995), 1245–1258.
88. K. Tanaka, Multiple Positive Solutions for Some Nonlinear Elliptic Systems. Top. Meth. Nonlinear Anal. 10(1997),15–45.
89. P. Tolksdorf, Regularity for a more general class of quasilinear elliptic equations, J. Diff. Eq. 51(1984), 126–150.
90. W.C. Troy, Symmetry properties in systems of semilinear elliptic equations. J. Diff. Eq. 42 (1981) 400–413.
91. J. L. Vasquez, A strong maximum principle for some quasilinear elliptic, equations. Appl. Math. and Optim.12(1984), 191–202.
92. J. Vélin and F. de Thélin, Existence and Non-existence of non-trivial solutions for some nonlinear elliptic systems. Rev. Math. Univ. Compl. Madrid 6(1993), 153–194.
93. C. Yarur, Nonexistence of positive singular solutions for a class of semilinear elliptic systems, Elect. JDE 1996:8 (1996), 1–22

Perfect Simulation and Convex Mixture of Context Trees

Nancy L. Garcia and Sandro Gallo

Abstract Chains with unbounded memory have attracted lot of attention since the 30s and the pioneering work of Onicescu and Mihoc (Bull Sci Math 59(2):174–192, 1935) and Doeblin and Fortet (Bull Soc Math France 65:132–148, 1937). The construction of perfect simulation algorithm for these chains was first presented in the beginning of the century, and the particular case of discontinuous cases was first studied in the 2010s. The present paper presents a particular approach to perfect simulation of possibly discontinuous chains with unbounded memory. The main idea is to use a representation of the kernel through a convex mixture of probabilistic context trees.

1 Introduction

Stochastic chains with unbounded memory extend Markov chains in a natural way. The idea is that the conditional probabilities with respect to the past may depend on an unbounded part of the past, contrarily to Markov chains, which have bound memory. Usually, we are given a probability kernel (the equivalent to transition matrix of Markov chains) and we ask basic questions, such as (i) existence, (ii) uniqueness, and (iii) statistical properties of the invariant measures specified by the kernel.

The above questions were originally addressed for these chains, in the stochastic processes literature, by Onicescu and Mihoc [35] and Doeblin and Fortet [14]. Since then, the literature expanded through several areas, and stochastic chains with unbounded memory appear with a variety of names. "Chains with complete

N. L. Garcia (✉)
Institute of Mathematics, Statistics and Scientific Computing, University of Campinas, Campinas, SP, Brazil
e-mail: nancy@ime.unicamp.br

S. Gallo
Center of Sciences, Federal University of São Carlos, São Carlos, SP, Brazil
e-mail: sandro.gallo@ufscar.br

C. Lavor, F. A. M. Gomes (eds.), *Advances in Mathematics and Applications*,
https://doi.org/10.1007/978-3-319-94015-1_7

connections" [2, 13, 18] originally coined by Onicescu and Mihoc [35], "chains of infinite order" [8, 26], "chains with infinite memory" [15, 33], and "g-measures" [4, 27, 29, 30, 40] in the ergodic theory literature. This reference list is not exhaustive but can be a starting point for the interested reader. They also are, in an indirect manner, present in information theory as "ergodic sources" [20, 36, 39], the theory of stochastic recursive sequences [3], and naturally in statistical physics, since one-dimensional Gibbs states can be seen as stochastic process, if we see $\mathbb{Z}$ as time instead of "space" (see [17]).

In their seminal paper, Doeblin and Fortet [14] wrote the following:

> It seems to us that a hypothesis at the same time natural and fruitful for the study of chains with complete connections would be to suppose that the conditional probabilities that the chain enters a state a given an infinite path $\underline{a}$ depend very few on the remote states of $\underline{a}$, and that in the limit, the transition probabilities do not depend at all on the infinitely ancient experiences. There exist several nonequivalent ways to translate this hypothesis mathematically (...).

It turns out that, to answer questions (i)–(iii), the papers in the literature commonly assume that the conditional probabilities are continuous with respect to the past (continuity here is to be understood in the infinite product topology as we will explain later). This is a particular way to translate the hypothesis of [14]. It also has the advantage that it allows to apply classical methods such as the Ruelle–Perron–Frobenius transfer operator, or the variational principle, both present in dynamical systems as well as statistical physics. However, when we do not assume continuity, even the very basic question of existence becomes harder to solve. In fact, these methods do not work anymore, at least in the form we find them in the literature, and extending them to a larger class of dynamics is an interesting problem in itself. This problem is comparable to that of non-Gibbsianity in statistical physics [25], and the Dobrushin restoration program.

A nice constructive approach to solve these questions is to present explicitly the invariant measure (or a finite sample of it). Algorithms that sample from the invariant measure are called *perfect simulation algorithms*. Not only does the construction of a perfect simulation algorithm naturally prove existence but also, as almost direct consequences, proves uniqueness and other results. Therefore, designing such an algorithm is, besides interesting in its own matter, a way to answer basic questions concerning chains with unbounded memory.

In this work, we will be interested in one particular class of algorithms called *coupling from the past* (CFTP) algorithms. The history of coupling from the past (CFTP) algorithms for Markov chains started with the seminal paper of Propp and Wilson [37]. Their idea was, instead of running the Markov into the future starting from one fixed initial condition, as it is done for MCMC algorithms, starting the chain from the past, from any possible initial state, and running all coupled trajectories up to time 0. If, for some starting point in the past, all the trajectories coalesce at time 0, then the sample at 0 is drawn from the stationary measure.

For chains with unbounded memory, the first CFTP algorithm was constructed by Comets et al. [9], under the assumption of uniform continuity. They implicitly took advantage of the fact that continuous transition kernels (conditional probabilities) may be represented as convex mixture of Markov kernels (of increasing, yet finite, order). This representation is originally due to Berbee [2] and Kalikow [28]. For their algorithm to work, the continuity rate of the conditional probabilities has to vanish rapidly to 0, uniformly on the pasts. Gallo [21] constructed a CFTP for chains for which the continuity rate could vanish very slowly along some pasts (or even not vanish at all, yielding a discontinuity), but for the other pasts, this continuity rate had to suddenly fall to 0 after a certain portion of the past. Such chains are called (unbounded) *variable length memory chains*, and their dependence on the past is encoded by a *probabilistic context tree*. Chains with variable memory were originally introduced in information theory by Rissanen [38] for data compression, and then popularized in the statistical literature by Bühlmann and Wyner [5]. However, [21] seems to be the first paper to consider the model from a strictly probabilistic point of view.

Gallo and Garcia [22] constructed a CFTP algorithm generalizing at the same time the algorithm proposed by Comets et al. [9] and the one of Gallo [21]. The main objective of the present paper is to explain, in a detailed way, the difference between these three algorithms. We will focus on the notion of convex mixture of probabilistic context trees, which is the main feature behind the algorithm presented in [22], extending the method of the paper [9]. We adopt an approach close to a mini-course, in the hope of explaining clearly the differences and similarities among the algorithms.

2 Stochastic Chains with Unbounded Memory

The present paper is concerned with discrete time stochastic chains, that is, sequences of random variables $\ldots, X_{-1}, X_0, X_1, \ldots$ defined on a probability space $(\Omega, \mathscr{A}, \mathbb{P})$, seen as a temporal evolution. We are interested in stationary sequences, that is, for any n, the joint distribution of $(X_i, X_{i+1}, \ldots, X_{i+n})$ does not depend on i. In this temporal evolution, the conditional distribution of X_i given the past $\ldots, X_{i-2}, X_{i-1}$ may depend on the whole past. We assume that X_i's take value in some finite set A (called *alphabet*). We can think of A as the set of possible states of a given physical system, undergoing some random temporal evolution. The simplest interesting case (excluding the case of independent sequences in which the system does not depend on its previous states) for our purposes is that of (k-steps) Markov chains, for which, conditionally on the past, the distribution of X_i only depends on $X_{i-k}, \ldots, X_{i-1}$:

$$\mathbb{P}(X_i = a_i | X_{-\infty}^{i-1} = a_{-\infty}^{i-1}) = \mathbb{P}(X_i = a_i | X_{i-k}^{i-1} = a_{i-k}^{i-1}) \tag{1}$$

for any $i \in \mathbb{Z}$ and any left-infinite sequence $a_{-\infty}^{i} := \ldots a_{i-1}a_{i-1}$ of elements of A. We sometimes read in the literature that Markov chains are those satisfying (1) for $k = 1$, but this is slightly misleading since a k-steps Markov chain can always be seen as a 1-step Markov chain on the extended alphabet A^k (cardinal product of sets), which is still a finite set in our case. So for us, there is only one distinction: Markov chains, satisfying (1) for some finite k, and non-Markov chains, or chains with unbounded memory.

Example 1 The simplest example of stochastic chains with unbounded memory is the undelayed renewal sequences (see [31], for instance). It is defined using a sequence of i.i.d. $\{1, 2, \ldots\}$-valued random variables $B_i, i \in \mathbb{Z}$ as follows:

- $X_0 = 2$,
- For $i \geq 1$, $X_i = 2$ if there exists $k \geq 0$ such that $i = \sum_{j=0}^{k} B_j$, $X_i = 1$ otherwise
- For $i \leq -1$, $X_i = 2$ if there exists $k \geq 1$ such that $i = -\sum_{j=1}^{k} B_{-j}$, $X_i = 1$ otherwise.

In other words, the sequence $B_i, i \in \mathbb{Z}$, specifies marks on $\mathbb{Z}$ and we put 2 on the marks and 1 elsewhere. Depending on the common distribution of the B_i's, this is not a Markov chain of any order. Indeed, for any $k \geq 1$, and $i \neq 0$, if $a_{-\infty}^{-1}$ is such that $a_{-k} = 2$ and $a_j = 1$ for $j = -k+1, \ldots, -1$ and $b_{-\infty}^{-1}$ is such that $b_{-k-1} = 2$ and $b_j = 1$ for $j = -k, \ldots, -1$, then

$$\mathbb{P}(X_i = 2 | X_{-\infty}^{i-1} = a_{-\infty}^{-1}) = \mathbb{P}(B = k+1 | B \geq k+1)$$

while

$$\mathbb{P}(X_i = 2 | X_{-\infty}^{i-1} = b_{-\infty}^{-1}) = \mathbb{P}(B = k+2 | B \geq k+2).$$

These quantities are equal if B has geometric distribution, but not in general. This means that, for any $k \geq 1, \ldots, X_{-1}, X_0, X_1, \ldots$ does not follow the evolution of a k-steps Markov chain.

Fixing $X_0 = 2$ implies that the chain is not stationary; however, it is easy to make it stationary by a random translation of the origin; see, for example, Theorem I.2.20, [39].

Let us conclude this example mentioning that its definition is usually done on the alphabet $A = \{0, 1\}$. However, to keep our notation homogeneous along the paper, we use $A = \{1, 2\}$.

2.1 Transition Kernels and Compatible Chains

In order to formalize the notion of unbounded memory, we need to define the *transition kernel*, sometimes called *family of transition probabilities*. Let us denote by $A^{-\mathbb{N}}$ the set of left-infinite sequences of symbols in A (pasts) and $\underline{a} = \ldots a_{-2}a_{-1}$ an element of $A^{-\mathbb{N}}$. A probability transition kernel on A is a function

$$\begin{aligned} p : A \times A^{-\mathbb{N}} &\to [0, 1] \\ (a, \underline{a}) &\mapsto p(a|\underline{a}) \end{aligned} \tag{2}$$

such that

$$\sum_{a \in A} p(a|\underline{a}) = 1 \,, \quad \forall \underline{a} \in A^{-\mathbb{N}}.$$

If there exists a $k \geq 1$ such that for any $\underline{a}$, $p(a|\underline{a})$ only depends on a_{-k}^{-1}, this definition corresponds to a k-steps Markov transition matrix.

In this paper, we will concentrate on *weakly non-null* kernels, that is,

$$\sum_{a} \inf_{\underline{a}} p(a|\underline{a}) > 0. \tag{3}$$

Given a transition kernel p, starting from a fixed past $\underline{a}$, we can construct a chain $(X_n^{(\underline{a})})_{n\geq 0}$ by applying iteratively p. In other words, the chain on $\mathbb{Z}$ is defined through $X_n^{(\underline{a})} = a_n$ for $n \leq -1$ and for any $n \geq 0$,

$$\mathbb{P}(X_n^{(\underline{a})} = b|X_{n-1}^{(\underline{a})} = x_{-1}, X_{n-2}^{(\underline{a})} = x_{-2}, \ldots) = p(b|\underline{x}). \tag{4}$$

(Observe that we swap the time ordering in the conditioning event.) The resulting chain is not stationary since it started from a *fixed past* $\underline{a}$. For the case of nonperiodic Markov chains in finite alphabet, there is always a stationary process with stationary marginal distribution given by the limit as $n \to \infty$ of (4). This is not always the case for transition kernels that depend on an unbounded part of the past $\underline{a}$, which are precisely the focus of this paper. Unless explicitly mentioned, the transition kernels p of the present paper will always have this property.

Definition 1 A stationary stochastic chain $\mathbf{X} = (X_n)_{n\in\mathbb{Z}}$ on A having law μ is said to be *compatible* with a kernel p if the latter is a regular version of the conditional probabilities of the former, that is

$$\mu(X_0 = a|X_{-\infty}^{-1} = \underline{a}) = p(a|\underline{a}) \tag{5}$$

for every $a \in A$ and μ-almost every $\underline{a}$ in $A^{-\mathbb{N}}$.

Even with a finite alphabet, existence of the stationary measure (or equivalently, chain) compatible with a given kernel is not granted.

Example 2 (Example 1 Revisited) Renewal sequences can be seen as a stochastic process compatible with the following transition kernel: for any $\underline{a}$ such that $a_{-k-1} = 1$ and $a_j = 0$ for $j = -k, \ldots, -1$,

$$p(2|\underline{a}) = 1 - p(1|\underline{a}) = \mathbb{P}(B = k + 1|B \geq k + 1) =: p_k \tag{6}$$

It remains to specify $p_\infty := p(2|\underline{a})$ when $\underline{a} = 0^\infty$, that is, $a_i = 0$ for all i. Depending on the value of p_∞, we have very distinct situations. An extremal situation occurs when $\sum_n \prod_{k=0}^{n-1}(1 - p_k) = \infty$ and $p_\infty > 0$. In this case (see [6]), there exists no stationary process specified by p. Relaxing either condition leads to existence. Taking $p_\infty = 0$ that the Dirac distribution δ_{0^∞}, associating measure 1 to a single configuration, the whole 0 sequence is a stationary measure. On the other hand, assuming $\sum_n \prod_{k=0}^{n-1}(1 - p_k) < \infty$ also implies existence, of a nontrivial measure, which is mutually singular to δ_{0^∞}.

As shown by the example above, some assumptions are necessary in order to ensure existence. Most commonly, the literature focus on continuous kernels.

Definition 2 The kernel p is *continuous* at a point (past) $\underline{a}$, in the product topology, if

$$p(a|\underline{x}a_{-k}^{-1}) \to p(a|\underline{a})$$

as $k \to \infty$ for any $\underline{x}$. This is equivalent to ask that the continuity rate at $\underline{a}$, defined by

$$\beta_k(\underline{a}) := \sup_{a \in A} \sup_{\underline{y},\underline{z}} |p(a|a_{-k}^{-1}\underline{y}) - p(a|a_{-k}^{-1}\underline{z})|,$$

converges to 0 as $k \to \infty$.

So, a kernel p is continuous if, and only if, $\beta_k(\underline{a})$ vanishes as k diverges, for any $\underline{a}$. Since A is finite, continuity everywhere is equivalent to uniform continuity, which writes

$$\beta_k := \sup_{\underline{a}} \beta_k(\underline{a}) \to 0$$

as $k \to \infty$. When the alphabet is finite, as we assume in this paper, continuity everywhere implies existence of the stationary measure compatible with p, by standard machinery (a fixed point argument in the compact set of stationary measures, see [29], for instance). The second step, once existence is granted, is to inquire about uniqueness. The first example of nonuniqueness (phase transition) was given by Bramson and Kalikow [4]. They presented a continuous kernel p uniformly bounded away from zero, i.e., there exists $\epsilon > 0$ such that $\inf_{a,\underline{a}} p(a|\underline{a}) \geq \epsilon$ which specifies more than one stationary process. A sufficient condition for uniqueness is that $\sum_k \beta_k^2 < \infty$, as proved in [27].

Example 3 (Example 1 Revisited) Observe that if we assume both, $\sum_n \prod_{k=0}^{n-1}(1 - p_k) < \infty$ and $p_\infty = 0$, we technically have two stationary measures, but these measures are mutually singular. This is referred as *non-irreducibility*, in the literature of Markov chains, and it is not as interesting as true phase transitions.

In the example presented in [4], all stationary measures are mutually absolutely continuous due to the positivity assumption.

It is easy to find examples of discontinuous kernels for which there exists a unique stationary measure.

Example 4 (Example 1 Revisited) The continuity rate of this chain can be easily computed as

$$\begin{aligned}\beta_k &= \sup_{\underline{a}} \sup_{\underline{y},\underline{z}} |p(2|a_{-k}^{-1}\underline{y}) - p(2|a_{-k}^{-1}\underline{z})| \\ &= \sup_{1_{-k}^{-1}} \sup_{\underline{y},\underline{z}} |p(2|a_{-k}^{-1}\underline{y}) - p(2|a_{-k}^{-1}\underline{z})| \\ &= \sup_{l,m\geq k} |p_l - p_m|.\end{aligned}$$

The second equality follows from the fact that $p(2|a_{-k}^{-1}\underline{y}) = p(2|a_{-k}^{-1}\underline{x})$ whenever $2 \in a_{-k}^{-1}$. (For finite strings $w, v \in A$, $|w| \leq |v|$, we use the (abuse of) notation $w \in v$ (*resp.* $w \notin v$) which means "w is (*resp.* is not) a substring of v").

So, we understand that the renewal chain is continuous if, and only if, p_k is convergent. The case of nonexistence, exhibited above in Example 2, is indeed discontinuous. However, it is easy to see that discontinuity does not imply nonexistence. For instance, consider the case in which $p_{2k} = \epsilon$ and $p_{2k+1} = 1 - \epsilon$, in such a way that p_k does not converge. Existence follows from $\sum_n \prod_{k=0}^{n-1}(1 - p_k) < \infty$. If we further assume $p_\infty > 0$, we have uniqueness (see [6]).

In the next section, we introduce probabilistic context trees, a class of kernels that include the one presented in Example 1. This class of kernels is the base of our method to study discontinuous kernels.

2.2 Probabilistic Context Tree

We say that a subset of $\tau \subset \cup_{i\geq 1} A^{\{-i,\dots,-1\}} \cup A^{-\mathbb{N}}$ is a *context tree* if it satisfies the following property. For any $\underline{a} \in A^{-\mathbb{N}}$, there exists *a unique* element $v \in \tau$ such that $a_{-|v|}^{-1} = v$, with the convention that $|v| = \infty$ if v is a left-infinite sequence, in which case $\underline{a} = v$. This element is called the *context* of $\underline{a}$ in τ and denoted $c_\tau(\underline{a})$.

According to this definition, we can identify the set $\tau = \{c_\tau(\underline{a})\}_{\underline{a}\in A^{-\mathbb{N}}}$ with the set of leaves of a rooted tree where each node has either $|A|$ sons (internal node) or 0 sons (leaf).

We say that a kernel p is a *probabilistic context tree* if

$$p(a|\underline{a}) = p(a|\underline{b}) \text{ whenever } c_\tau(\underline{a}) = c_\tau(\underline{b}).$$

A probabilistic context tree is an ordered pair (τ, p_τ) where τ is a context tree and $p_\tau = \{p_\tau(a|v)\}_{a\in A, v\in\tau}$ is a set of transition probabilities associated to each element

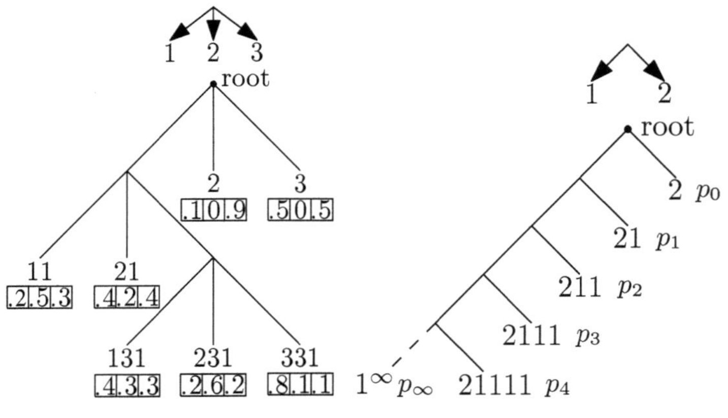

(a) (b)

Fig. 1 Examples of probabilistic context trees

of τ. Thus, the probabilistic context tree (τ, p_τ) *represents* the kernel p if for all $\underline{a} \in A^{-\mathbb{N}}$ and $a \in A$

$$p(a|\underline{a}) = p_\tau(a|c_\tau(\underline{a})).$$

Examples of probabilistic context trees are shown in Fig. 1(a) (for the bounded case) and (b) (for the unbounded case). In the first one, at each leaf (context) of the tree, we associate three boxes representing the transition probabilities to each symbols of A given this context. In the second one, we only need to specify the probability $p_i := p(2|1^i2)$; the transition probabilities to 1 are simply $1 - p_i$. This latter example is the probabilistic context tree of Example 1.

Stochastic chains $\mathbf{X}$ compatible, in the sense of (5), with probabilistic context trees are called *chains with variable length memory*.

In terms of continuity, the context tree assumption amounts to say that pasts $\underline{a}$ with finite context are continuous in a strong sense, since $\beta_k(\underline{a}) = 0$ for any $k \geq |c_\tau(\underline{a})|$. However, nothing is assumed for the remaining pasts, which may be discontinuity points. In this regard, chains with variable length are a very nice laboratory, source of examples and counterexamples, as shown by Example 1.

3 Convex Mixtures of Kernels

In this section, we describe our main tool to construct CFTP algorithms, which is to decompose potentially complicated, and quite general, kernels into a convex mixture of simple kernels. This section aims also at uniformizing the notation and compare/contrast the approaches of Comets et al. [9], Gallo [21], and Gallo and

Garcia [22]. The superscripts "CFF", "G," and "GG" in the notation help to make the parallel.

3.1 Continuous Case: Convex Mixture of Markov Kernels

Berbee [2] and Kalikow [28] proved that a transition probability kernel p is continuous if, and only if, it can be represented as a convex mixture of Markov kernels. That is, there exist probability distributions $\{p_0(a)\}_{a\in A}$ and $\{\lambda_k\}_{k\geq 0}$ and a sequence of Markov kernels $\{p_k\}_{k\geq 1}$ such that, for any $a \in A$ and $\underline{z} \in A^{-\mathbb{N}}$

$$p(a|\underline{z}) = \lambda_0 p_0(a) + \sum_{k\geq 1} \lambda_k p_k(a|z_{-k}^{-1}). \tag{7}$$

This decomposition is not unique. For perfect simulation purposes, there is an optimal (in terms of the λ_k's) decomposition described below [9]. For any $a \in A$ and $a_{-k}^{-1} \in A^k$, consider the functions

$$\alpha_0(a) := \inf_{\underline{z}} p(a|\underline{z}) \ , \quad \alpha_0 := \sum_{a\in A} \alpha_0(a)$$

and the sequence $\{\alpha_k^{\mathrm{CFF}}\}_{k\geq 1}$ defined by

$$\alpha_k^{\mathrm{CFF}} := \inf_{a_{-k}^{-1}\in A^k} \sum_{a\in A} \inf_{\underline{z}} p(a|a_{-k}^{-1}\underline{z}). \tag{8}$$

These are, as they say, "probabilistic threshold for memories limited to k preceding instants." To assume continuity is equivalent to assume that α_k^{CFF} converges to 1 as k diverges, and to assume pointwise continuity at $\underline{a}$ is equivalent to assume that

$$\alpha_k(\underline{a}) := \sum_{a\in A} \inf_{\underline{z}} p(a|a_{-k}^{-1}\underline{z})$$

converges to 1 as k diverges. Under the continuity assumption, we can choose the probability distribution $\{\lambda_k\}_{k\geq 1}$ used in (7) to be $\lambda_0 = \lambda^{\mathrm{CFF}} = \alpha_0$ and $\lambda_k^{\mathrm{CFF}} = \alpha_k^{\mathrm{CFF}} - \alpha_{k-1}^{\mathrm{CFF}}$ for $k \geq 1$.

Let us now explain the meaning of decomposition (7). Define a random variable L^{CFF} taking values on $\mathbb{N}$ with probability law $\{\lambda_k^{\mathrm{CFF}}\}_{k\geq 0}$. To choose the next symbol looking at the whole past $\underline{z}$ using the distribution $\{p(a|\underline{z})\}_{a\in A}$ is equivalent to the following two-step procedure:

(I) Choose L^{CFF},
(II) (i) If $L^{\mathrm{CFF}} = 0$, choose the next symbol according to $\{p_0(a)\}_{a\in A}$,

(ii) If $L^{\mathrm{CFF}} = k > 0$, choose the next symbol looking at z_{-k}^{-1} and using $\{p_k^{\mathrm{CFF}}(a|z_{-k}^{-1})\}_{a\in A}$.

Observe that L^{CFF} is independent of everything (in particular, it does not depend on $\underline{z}$). This two-step procedure justifies the terminology "random Markov processes" introduced by Kalikow [28] regarding continuous chains.

3.2 Dropping the Continuity Assumption

To fix ideas, in the remaining of this section, let us assume the particular case that p is a transition probability kernel on $A = \{1, 2\}$ with a single discontinuity point at $\underline{1} := 1^{-\mathbb{N}}$. Then, $\alpha_k^{\mathrm{CFF}}(\underline{a})$ goes to 1 as k diverges if, and only if, $\underline{a} \neq \underline{1}$. In this case, α_k^{CFF} does not converge to 1 and the decomposition into Markov kernels cannot be done.

3.2.1 The Context Tree Assumption

Gallo [21] assumed that p is represented by the probabilistic context tree (τ, p_τ), with

$$\tau = \underline{1} \cup \bigcup_{i\geq 0} \bigcup_{a_{-\ell^2(i)}^{-1}\in A^{\ell^2(i)}} a_{-\ell^2(i)}^{-1}\, 2\, 1^i, \tag{9}$$

where $\ell^2 : \mathbb{N} \to \mathbb{N}$ is a deterministic function. This context tree is represented in Fig. 2.

Making a parallel with the CFF case, we can decompose such kernel as follows, for any $a \in A$ and $\underline{z} \in A^{-\mathbb{N}}$

$$p(a|\underline{z}) = p_\tau(a|c_\tau(\underline{z})) = \lambda_0 p_0(a) + (1 - \lambda_0)p'_\tau(a|c_\tau(\underline{z}))$$

where

$$p'_\tau(a|c_\tau(\underline{z})) := \frac{p_\tau(a|c_\tau(\underline{z})) - \lambda_0 p_0(a)}{1 - \lambda_0}.$$

Define an $\mathbb{N}$-valued random variable L^G which takes value 0 w.p. λ_0 or $|c_\tau(\underline{z})|$ w.p. $1 - \lambda_0$. The context tree assumption for p means the following. To choose the next symbol looking at the whole past $\underline{z}$ using the distribution $\{p(a|\underline{z})\}_{a\in A}$ is equivalent to the following two-step procedure:

(I) Choose L^G,
(II) (i) If $L^G = 0$, choose the next symbol w.p. $\{p_0(a)\}_{a\in A}$,

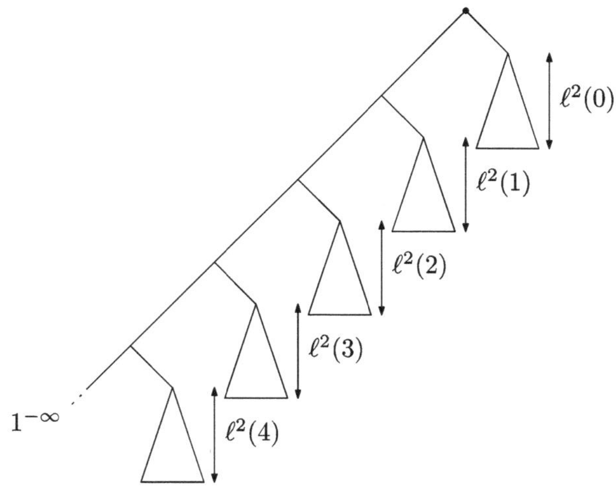

Fig. 2 Graphical representation of context tree τ given by (9)

(ii) If $L^G = |c_\tau(\underline{z})|$, choose the next symbol looking at $c_\tau(\underline{z})$ and using $\{p'_\tau(a|c_\tau(\underline{z})\}_{a\in A}$.

Observe that the random variable L^G is a deterministic function of the past $\underline{z}$ whenever its value is not 0: $L^G = |c_\tau(\underline{z})|$.

3.2.2 Convex Mixture of Probabilistic Context Trees

So far, two extreme cases have been considered: L^G is a deterministic function of the past, and L^{CFF} is a random variable totally independent of the past. It is the objective of the present paper to explain the in-between case proposed by Gallo and Garcia [22] which combines the two preceding approaches. It allows us to consider kernels p which are neither necessarily represented by a probabilistic context tree nor necessarily continuous. This approach is based on the assumption that (continuing with the case where $\underline{1}$ is the unique discontinuity point)

$$\alpha_k^{\text{GG}} := \inf_{i\geq 0} \inf_{a_{-k}^{-1}\in A^k} \sum_{a\in A} \inf_{\underline{z}} p(a|1^i 2a_{-k}^{-1}\underline{z}) \overset{k\to\infty}{\longrightarrow} 1. \tag{10}$$

The α_k^{GG}'s are probabilistic thresholds for memories going until the kth instant preceding the last occurrence of symbol 2 in the past. In this case also, we have that $\sum_{a\in A} \inf_{\underline{z}} p(a|a_{-k}^{-1}\underline{z})$ goes to 1 as k diverges for any $\underline{a} \neq \underline{1}$ and not necessarily for $\underline{1}$. Notice that the probabilistic context tree assumption introduced in Sect. 3.2.1 satisfies $\inf_{a_{-k}^{-1}\in A^k} \sum_{a\in A} \inf_{\underline{z}} p(a|1^i 2a_{-k}^{-1}\underline{z}) = 1$ for $k > \ell^2(i)$, which is slightly different (neither weaker nor stronger) than (10). Under assumption (10), it will be shown in the next section that there exist a probability distribution $\{\lambda_k^{\text{GG}}\}_{k\geq 0}$ and a sequence of probabilistic context trees $\{(\tau_k, p_{\tau_k}^{\text{GG}})\}_{k\geq 0}$ such that

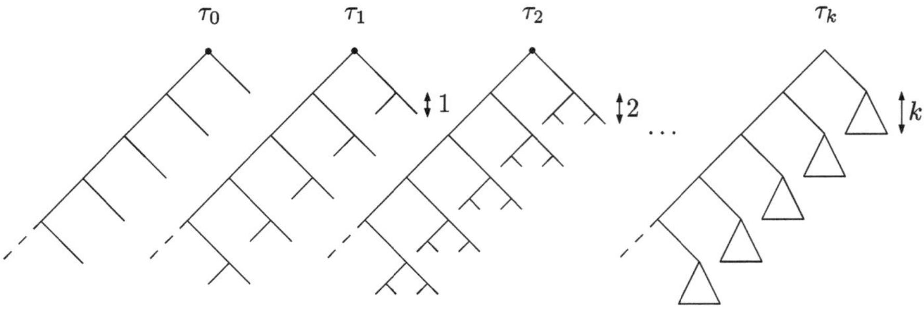

Fig. 3 Graphical representation of the context trees τ_k given by (12)

$$p(a|\underline{z}) = \lambda_0 p_0(a) + \sum_{k\geq 0} \lambda_k^{\mathrm{GG}} p_{\tau_k}^{\mathrm{GG}}(a|c_{\tau_k}(\underline{z})). \tag{11}$$

The kth context tree of decomposition (11) is given by

$$\tau_k := \underline{1} \cup \bigcup_{i\geq 0} \bigcup_{a_{-k}^{-1}\in A^k} a_{-k}^{-1} 2\, 1^i. \tag{12}$$

The sequence of context trees $\{\tau_k\}_{k\geq 0}$ for the present particular case is illustrated in Fig. 3. Define a random variable K^{GG} taking values -1 w.p. λ_0 and k w.p. λ_k^{GG} for $k \geq 0$. One more time, let us translate this decomposition into a two-step procedure:

(I) Choose K^{GG},
(II) (i) If $K^{\mathrm{GG}} = -1$, put $L^{\mathrm{GG}} = 0$ and choose the next symbol w.p. $\{p_0(a)\}_{a\in A}$,
(ii) If $K^{\mathrm{GG}} = k \geq 0$, put $L^{\mathrm{GG}} = |c_{\tau_k}(\underline{z})|$, choose the next symbol looking at $c_{\tau_k}(\underline{z})$ and using $\{p_k^{\mathrm{GG}}(a|c_{\tau_k}(\underline{z})\}_{a\in A}$.

Observe that this time, the random variable L^{GG} depends on the past $\underline{z}$, but through a random mechanism using the distribution $\{\lambda_k^{\mathrm{GG}}\}_{k\geq 0}$.

In the next section, we state a result in a general framework in which the role played above by symbol 2 can be played by any finite string w. In this case, we allow p to have discontinuities at every point $\underline{z} \in A^{-\mathbb{N}}$ which does not have w as subsequence.

4 Perfect Simulation Based on Convex Mixture of Unbounded Probabilistic Context Trees

In this section, we first decompose a locally continuous transition kernel as a convex mixture of unbounded probabilistic context trees, and then use this decomposition to construct a perfect simulation algorithm that stops after a $\mathbb{P}$-a.s. finite number of steps. This decomposition relies on the existence of a reference string w that "identifies" discontinuous pasts. The perfect simulation scheme described here is a particular case of the one studied in [22], in which (using their terminology) the *probabilistic skeleton* has terminal string w.

4.1 The Convex Mixture of Unbounded Probabilistic Context Trees

Given a finite string w, let

$$m^w(\underline{a}) := \inf\{k \geq 0 : a_{-k}^{-k+|w|-1} = w\}, \tag{13}$$

with the convention that $m^w(\underline{a}) = +\infty$ if the set of indexes is empty. This is the size of the smallest suffix of $\underline{a}$ containing w. Using this definition, define the context tree

$$\tau_0^w := \{a_{-m^w(\underline{a})}^{-1}\}_{\underline{a}}.$$

Theorem 1 *Consider a probability kernel p, and assume that there exists a finite reference string w for which*

$$\alpha_k^w := \inf_{v\in\tau_0^w} \inf_{c_{-k}^{-1}\in A^k} \sum_{a\in A} \inf_{\underline{z}} p(a|v\, c_{-k}^{-1}\, \underline{z}) \overset{k\to+\infty}{\longrightarrow} 1. \tag{14}$$

Then, there exist two probability distributions $\{\lambda_k^w\}_{k\geq -1}$ and $\{p_{-1}^w(a)\}_{a\in A}$, and a sequence of probabilistic context trees $\{(\tau_k^w, p_{\tau_k}^w)\}_{k\geq 0}$ such that

$$p(a|\underline{z}) = \lambda_{-1}^w p_{-1}^w(a) + \sum_{k\geq 0} \lambda_k^w p_{\tau_k}^w(a|c_{\tau_k^w}(\underline{z})). \tag{15}$$

Observe that, in Example 1 with $p_{2i} = \epsilon$ and $p_{2i+1} = 1-\epsilon$, we have $\alpha_k^2 = 1$ for any $k \geq 0$; however, α_k^{CFF} alternates between 1 and 2ϵ. On the other hand, it is clear that if α_k^{CFF} converges to 1, this is also the case of α_k^w for any reference string w. Thus, our definition is strictly more general than the original one of [9]. In particular, this means that Theorem 1 extends to convex mixture of finite Markov kernels. Under the new assumptions, we obtain a Kalikow-type decomposition of our kernels as a mixture of unbounded probabilistic context trees. The fact that our decomposition involves unbounded probabilistic context trees instead of Markov kernels is "the price to pay" to allow discontinuities at some points.

Let us mention that this result is a simple instance of a more general decomposition used (although not explicitly mentioned) by Gallo and Garcia [22]. A result in the same vein was also obtained in [34] (see Sect. 6).

This result, although not completely new, was not explicitly proved in the literature, so for completeness we include a proof here. Once we are given the reference string w, Theorem 1 states that there exists a triplet of parameters (which is not unique): two probability distributions $\{\lambda_k^w\}_{k\geq -1}$ and $\{p_{-1}^w(a)\}_{a\in A}$, and a sequence of probabilistic context trees $\{(\tau_k^w, p_k^w)\}_{k\geq 0}$ that controls the decomposition of the kernel. Section 4.2 is dedicated to the definition of such a triplet of parameters.

4.2 Proof of Theorem 1: Construction of a Triplet

The definition of our triplet is based on two partitions of $[0, 1[$ inspired by Comets et al. [9] and that we now explain. These partitions are of particular interest because we will use them in the construction of our perfect simulation algorithm in Sect. 4.3. To avoid overloaded notation, we will omit the superscript w in all the quantities depending on it, when no ambiguity is possible.

4.2.1 Definition of the First Partition of [0,1[

Recall that the reference string w is fixed, and everything is done according to this reference string. In particular, it is related to this string that the tree τ_0^w is constructed (see, for instance, Fig. 3 for $w = 1$ and Fig. 5 for $w = 12$).

Suppose we are given an entire past $\underline{a} \in A^{-\mathbb{N}}$ such that $c_{\tau_0}(\underline{a}) = v$ with $|v| < \infty$. We now explain how the length of the intervals constituting the first partition of $[0, 1[$ represented in Fig. 4 is chosen. Notice first that $\underline{a}$ has to be of the form $\underline{b}v$ for some $\underline{b}$. For any $a \in A$, the interval $I(a)$ has size $\inf_{\underline{z}} p(a|\underline{z})$ independently of everything. The remaining intervals have length

$$|I(a, \underline{b}v, 0)| = \inf_{\underline{z}} p(a|v\underline{z}) - \inf_{\underline{z}} p(a|\underline{z})$$

$$|I(a, \underline{b}v, 1)| = \inf_{\underline{z}} p(a|vb_{-1}\underline{z}) - \inf_{\underline{z}} p(a|v\underline{z})$$

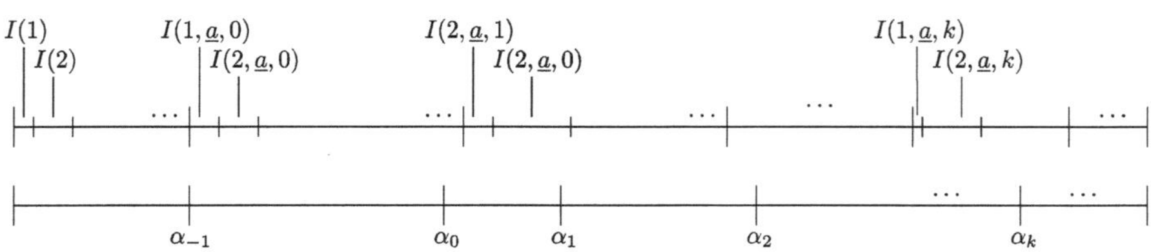

Fig. 4 Illustration of the first partition (upper part) for a given past $\underline{a}$ having finite $c_{\tau_0}(\underline{a})$ and of the second partition (lower part) which does not depend on the past

$$|I(a, \underline{b}v, k)| = \inf_{\underline{z}} p(a|vb_{-k}^{-1}\underline{z}) - \inf_{\underline{z}} p(a|vb_{-k+1}^{-1}\underline{z}) \ , \quad k \geq 2.$$

Due to our assumption that pasts having a w as substring are continuous, we have

$$\inf_{\underline{z}} p(a|vb_{-k}^{-1}\underline{z}) \nearrow p(a|v\underline{b}), \tag{16}$$

for any $a \in A$, $v \in \tau_0$ with $|v| < \infty$, and $\underline{b} \in A^{-\mathbb{N}}$.

As a consequence of (16), we constructed a partition of [0, 1[, which satisfies

$$\left| I(a) \cup \bigcup_{k \geq 0} I(a, \underline{a}, k) \right| = p(a|\underline{z}) \ , \quad \forall a \in A. \tag{17}$$

Another important property of this partition is that we can construct the interval $I(a, \underline{a}, k)$ knowing only the suffix $a_{-(k+c_{\tau_0}(\underline{a}))}^{-1}$.

4.2.2 Definition of the Second Partition of [0,1[

For any $k \geq 1$, let us define

$$\alpha_k := \inf_{v \in \tau_0} \inf_{b_{-k}^{-1}} \sum_a \inf_{\underline{z}} p(a|vb_{-k}^{-1}\underline{z}),$$

as well as

$$\alpha_0 := \inf_{v \in \tau_0} \sum_a \inf_{\underline{z}} p(a|v\underline{z}).$$

Once again, our assumptions imply that $\{\alpha_k\}_{k \geq 0}$ is a [0, 1]-valued nondecreasing sequence which converges to 1 as k diverges. It follows that denoting $\alpha_{-1} := \sum_{a \in A} \alpha(a)$, and using the convention that $\alpha_{-2} = 0$, the sequence of intervals $\{[\alpha_{k-1}, \alpha_k[\}_{k \geq -1}$ constitutes a partition of [0, 1[. See the second partition of Fig. 4.

4.2.3 Definition of the Triplet of Parameters $\left(\{\lambda_k\}_{k \geq -1}, \{p_{-1}(a)\}_{a \in A}, \{(\tau_k, p_{\tau_k})\}_{k \geq 0}\right)$

Let U be a random variable with uniform distribution in [0, 1[. We now introduce one triplet $\left(\{\lambda_k\}_{k \geq -1}, \{p_{-1}(a)\}_{a \in A}, \{(\tau_k, p_{\tau_k})\}_{k \geq 0}\right)$ that will give the decomposition stated in Theorem 1. Define

- For any $k \geq -1$,

$$\lambda_k := \mathbb{P}(U_0 \in [\alpha_{k-1}, \alpha_k[). \tag{18}$$

- For any $a \in A$,

$$p_{-1}(a) := \mathbb{P}(U_0 \in I(a) \big| U_0 \in [0, \alpha_{-1}[) = \alpha(a)/\alpha_{-1}. \tag{19}$$

- For any $k \geq 0$, let

$$\tau_k := \bigcup_{v \in \tau_0 : |v| = \infty} v \cup \bigcup_{v \in \tau_0 : |v| < \infty} \bigcup_{b_{-k}^{-1} \in A^k} b_{-k}^{-1} v \ , \quad k \geq 0. \tag{20}$$

Then, due to the observation made right after Eq. (17), it makes sense to define, for any finite $v \in \tau_k$, the conditional probability

$$p_{\tau_k}(a|v) := \mathbb{P}\left(U_0 \in I(a) \cup \bigcup_{l=0}^{k} I(a, v, l) \Big| U_0 \in [\alpha_{k-1}, \alpha_k[\right). \tag{21}$$

Two examples of sequences of context trees $\{\tau_k^w\}_{k \geq 0}$ on $A = \{1, 2\}$ are given in Figs. 3 and 5, the first one with $w = 2$, and the second one with $w = 12$.

4.2.4 What About $v \in \tau_0$ Having Infinite Size?

The reader who arrived here may be asking why we only considered, so far, the finite contexts of τ_0. Finite contexts correspond to continuous pasts for p. The remaining contexts correspond to pasts having no occurrence of the sequence w, and as such, possibly discontinuous. In fact, if we are only interested in the perfect simulation algorithm, we could very well not define anything for these pasts, since, due to positivity assumption $\inf_{a,\underline{a}} p(a|\underline{a}) > 0$, these pasts have null probability of occurrence. Nevertheless, for the sake of definiteness, let us explain how we could proceed. The effect of discontinuity along $\underline{a}$ is that (16) does not hold, and therefore, our partitions are not well defined. Indeed, we only have a partition up to $\sum_a \lim_k \inf_{\underline{z}} p(a|vb_{-k}^{-1}\underline{z})$ which is smaller than 1. But, this is no problem since the size of the contexts for these pasts is fixed to infinity. We simply define $p_{\tau_k}(a|v) = p(a|v)$ for those v's, and this can easily be done through a partition.

4.2.5 Proof of Theorem 1

The convex mixture representation stated in the theorem is a direct consequence of the definition of the triplet, using a uniformly distributed random variable.

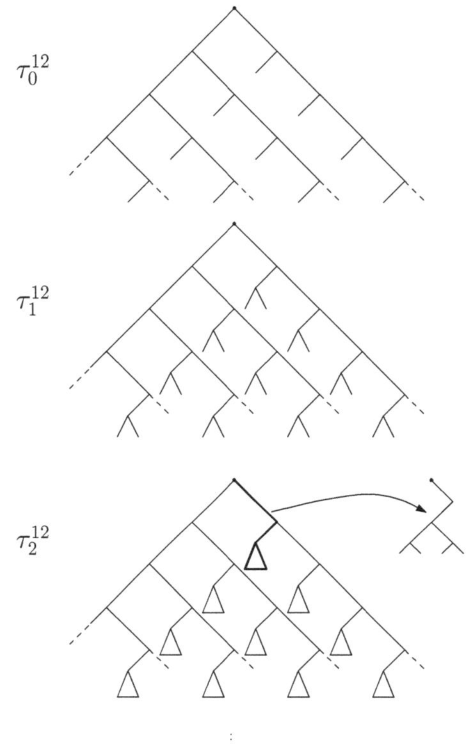

Fig. 5 Graphical representation of the context trees τ_k^w with reference string $w = 1, 2$ in Theorem 1

4.3 *Coupling from the Past*

The perfect simulation algorithm presented here works as a mixture of the algorithms presented in [9, 21].

Suppose, for simplicity, that the given kernel is strictly positive, that is, there exists $\epsilon > 0$ such that

$$\inf_{a,\underline{a}} p(a|\underline{a}) = \epsilon.$$

In addition, consider that p satisfies the condition of Theorem 1 with reference string w. In order to simplify the notation, we will omit the superscript w in most of the quantities that depend on this string.

Let us introduce an i.i.d. sequence $\mathbf{U} = (U_i)_{i\in\mathbb{Z}}$ of random variables uniformly distributed in $[0, 1[$. We denote by $(\Omega, \mathscr{F}, \mathbb{P})$ the corresponding probability space.

We want to get a deterministic measurable function $X : [0, 1[^{\mathbb{Z}} \to A^{\mathbb{Z}}, \mathbf{U} \mapsto X(\mathbf{U})$ such that the law $\mathbb{P}(X(\mathbf{U}) \in \cdot)$ is compatible with p in the sense of (5). The idea is to use the sequence $\mathbf{U}$ together with the partitions of $[0, 1[$ introduced before (and illustrated in Fig. 4) to mimic the two-step procedure described in Sect. 3.2.2.

In particular, for any $n \in \mathbb{Z}$, we put $[X(\mathbf{U})]_n = a$ whenever $U_n \in I(a)$. Suppose that for some time index $n \in \mathbb{Z}$ there exists a string $a_{-k}^{-1} \in A^k$ such that $U_{n-i} \in I(a_{-i})$, $i = 1, \ldots, k$, in this case, we put

$$[X(\mathbf{U})]_{n-k}^{n-1} = a_{-k}^{-1}.$$

We say that this sample has been *spontaneously* constructed. Now, suppose $U_n \in [\alpha_{l-1}, \alpha_l[$ for some $l \geq 0$. This means that we pick up the context tree τ_l in the convex mixture representation of p, and look whether or not there exists a context in τ_l which is suffix of $[X(\mathbf{U})]_{n-k}^{n-1} = a_{-k}^{-1}$. If such context exists, then we put

$$[X(\mathbf{U})]_n = \sum_{a \in A} a \cdot \mathbf{1}\left\{U_n \in \bigcup_{j=0}^{l} I(a, a_{-k}^{-1}, j)\right\}.$$

If there is no such context (we will write $c_{\tau_l}(a_{-k}^{-1}) = \emptyset$), we cannot construct the state $[X(\mathbf{U})]_n$: we need further knowledge of the past.

In the first case, $[X(\mathbf{U})]_{n-k}^{n}$ has been constructed independently of $U_{-\infty}^{n-k-1}$ and $U_{n+1}^{+\infty}$.

Now, suppose we want to construct $[X(\mathbf{U})]_0$. We generate, backward in time, the U_i's until the first time $k \leq 0$ such that we can perform the above construction from time k up to time 0 using only U_k^0. A priori, there is no reason for k to be finite. Theorem 2 gives sufficient conditions for $\mathbb{P}$-almost sure finiteness.

To formalize the above description, let us define for any $u \in [0, 1[$

$$\ell(u) := \sum_{k \geq -1} k \cdot \mathbf{1}\{u \in [\alpha_{k-1}, \alpha_k[\}.$$

By Theorem 1, $\ell(U_i) = -1$ means that we can choose the state of $X(\mathbf{U})_i$ according to distribution $p_{-1}(\cdot)$, and independently of everything else. On the other hand, $\ell(U_i) = l \geq 0$ means that we have to use the context tree (τ_l, p_{τ_l}) in order to construct the state of $X(\mathbf{U})_i$. In particular, we recall that for any $l \geq 0$ the size of the context $c_{\tau_l}(a_m^n)$ is $|c_{\tau_0}(a_m^n)| + l$.

Let us denote by $A^\star$ the set of finite strings of letters of A. One of the inputs for Algorithm 1 (presented below) is the *update function* F. It is a measurable function $F : [0, 1[\times(\emptyset \cup A^\star \cup A^{-\mathbb{N}}) \to A \cup \{?\}$ (the interrogation mark has the sense of indefiniteness) which uses the part of the past we already know, together with the uniform random variable to compute the present state. It is defined as follows, for any $a_m^n \in \emptyset \cup A^\star \cup A^{-\mathbb{N}}$, with $-\infty < n < +\infty$ and $-\infty \leq m \leq n+1$,

$$F(u, a_m^n) := \begin{cases} \sum_{a \in A} a \cdot \mathbf{1}\{u \in I(a)\} & \text{if } \ell(u) = -1 \\ \sum_{a \in A} a \cdot \mathbf{1}\left\{u \in \bigcup_{k=0}^{\ell(u)} I(a, a_m^n, k)\right\} & \text{if } \ell(u) \geq 0 \text{ and } c_{\tau_{\ell(u)}}(a_m^n) \neq \emptyset \\ ? & \text{otherwise} \end{cases} \tag{22}$$

with the convention that $a_{n+1}^{n} = \emptyset$ and for any context tree τ, $c_{\tau}(\emptyset) = \emptyset$. Here, we understand that, if $w \notin a_{m}^{n}$, unless u belongs to $[0, \alpha_{-1}[$, we cannot construct the next symbol, and for this reason we put a question mark. This is also the reason why we do not have to care about pasts with not having w as substring.

When we consider an infinite past $\underline{z} \in A^{-\mathbb{N}}$ with w as substring, we have, by (17),

$$\mathbb{P}(F(U, \underline{z}) = a) = \mathbb{P}\left(U \in I(a) \cup \bigcup_{k \geq 0} I(a, \underline{z}, k)\right) = p(a|\underline{z}). \tag{23}$$

When the update function returns the symbol "?", it means that we do not have enough knowledge of the past to compute the present state.

We define, for any $m \leq n$, the $\mathscr{F}(U_{m}^{n})$-measurable function $\mathscr{L} : [0, 1[^{n-m+1} \rightarrow \{0, 1\}$ which takes value 1 if, and only if, we can construct $[X(\mathbf{U})]_{m}^{n}$ independently of $U_{-\infty}^{m-1}$ and $U_{n+1}^{+\infty}$ using the construction described above. Formally,

$$\{\mathscr{L}(U_{m}^{n}) = 1\} := \bigcup_{a_{m}^{n} \in A^{n-m+1}} \bigcap_{i=m}^{n} \{F(U_{i}, a_{m}^{i-1}) = a_{i}\}.$$

Finally, for any $-\infty < m \leq n \leq +\infty$, we define the *regeneration time* for the window $[m, n]$ as the first time before m such that the construction described above is successful until time n, that is

$$\theta[m, n] := \max\{k \leq m : \mathscr{L}(U_{k}^{n}) = 1\} \tag{24}$$

with the convention that $\theta[m] := \theta[m, m]$.

4.4 The Algorithm

This algorithm takes as "input" two integers $-\infty < m \leq n < +\infty$ and the update function F, and returns as "output" the regeneration time $\theta[m, n]$ and the constructed sample $[X(\mathbf{U})]_{\theta[m,n]}^{n}$. The function F contains all the information we need about the kernel p, and we suppose that it is already implemented in the software used for programing the algorithm.

At each time, the set B contains the sites that remain to be constructed. We initialize with $B = \{m, \ldots, n\}$ and a forward procedure (lines 2–8) attempts to construct $[X(\mathbf{U})]_{m}^{n}$ using $U_{m}, \ldots, U_{n}$. If it succeeds, then the algorithm stops and returns $\theta[m, n] = m$ and the constructed sample. If it fails, B is not empty and a backward procedure ("while loop": lines 10–27) begins. In this loop, each time the algorithm cannot construct the next site of B, it generates a new uniform random variable backward in time. At each new generated random variable, the algorithm attempts to go as far as possible in the construction of the remaining sites of B using

Algorithm 1 Perfect simulation algorithm of the sample $[X(\mathbf{U})]_m^n$

```
Input: m, n, F; Output: θ[m, n], ([X(U)]_{θ[m,n]}, ..., [X(U)]_n)
Sample U_m, ..., U_n uniformly in [0, 1[
i ← m, B = {m, ..., n}, θ[m, n] ← m, [X(U)]_m^n ← ?^{n-m+1}
while F(U_i, [X(U)]_m^{i-1}) ∈ A and B ≠ ∅ do
    [X(U)]_i ← F(U_i, [X(U)]_m^{i-1})
    B ← B \ {i}
    i ← i + 1
end while
i ← m
while B ≠ ∅ do
    i ← i - 1
    B ← B ∪ {i}
    Sample U_i uniformly in [0, 1[
    while U_i ≥ α_{-1} do
        i ← i - 1
        B ← B ∪ {i}
        Sample U_i uniformly in [0, 1[
    end while
    [X(U)]_i ← F(U_i, ∅)
    B ← B \ {i}
    t ← min B
    while F(U_t, [X(U)]_i^{t-1}) ∈ A and B ≠ ∅ do
        [X(U)]_t ← F(U_t, [X(U)]_i^{t-1})
        B ← B \ {t}
        t ← min B
    end while
end while
θ[m, n] ← i
return θ[m, n], ([X(U)]_{θ[m,n]}, ..., [X(U)]_n)
```

the uniform random variables that have been previously generated. Theorem 2 gives sufficient conditions for this procedure to stop after a ¶-a.s. finite number of steps.

This theorem is a consequence of Theorem 4.1 and Proposition 5.1 in [22], since it can be seen, using their terminology, as a particular case in which the *probabilistic skeleton* has terminal string w.

Theorem 2 *Consider a kernel p satisfying the conditions of Theorem 1 for some reference string w and assume moreover that*

$$\inf_{a\in A} \inf_{\underline{a}\in A^{-\mathbb{N}}} p(a|\underline{a}) \geq \epsilon > 0.$$

According to $\{\alpha_k^w\}_{k\geq 0}$, we have the following situations.

(i) If $\sum_{k\geq 1}\prod_{j=0}^{k-1}\alpha_k^w = +\infty$, then $\theta[0]$ is $\mathbb{P}$-a.s. finite.
(ii) If $\prod_k \alpha_k^w > 0$, then $\theta[0]$ has summable tail.
(iii) If $\{1-\alpha_k^w\}_{k\geq 0}$ decays exponentially fast to zero, then $\theta[0]$ has exponential tail.

In particular, in each of these regimes, the CFTP with update function F is feasible and the output of Algorithm 1 is a sample of the unique stationary chain compatible with p.

5 Complete Description of a Simple Example on $A = \{1, 2\}$

In this section, we focus on the two-letter alphabet $A = \{1, 2\}$. Compiling results of the literature on the uniform continuity case, we have the following:

(i) If $(1 - \alpha_k) \to 0$, then there exists at least one stationary measure compatible [28, 29].
(ii) If $\sum_k (1-\alpha_k)^2 < \infty$, then there exists a unique stationary measure compatible [27].
(iii) If $\sum_k \prod_{i=0}^{k-1} \alpha_i = \infty$, then we have coupling from the past to perfectly simulate finite windows of the unique stationary measure. In particular, $\theta[0]$ is finite a.s. [9].
(iv) If $\prod_k \alpha_k > 0$, then we can simulate right-infinite vectors of the stationary measure, and the chain can be seen as a concatenation of independent vectors of random size. In particular, $\theta[0]$ has summable tail (finite expectation) [9].

The object of the present section is to obtain the same classification for our non-necessarily continuous chains, those related to the occurrence or not of a fixed reference string w.

Once we fix a string w, we will need the asymptotic number or binary strings of length n having no w as substring. This is a well-known topic in combinatorics, and we refer, for instance, to [19]. Let

$$|\mathcal{N}_n| = \#\{x_1^n \in \{1, 2\}^n;\ w \notin x_1^n\}.$$

The asymptotic behavior of $|\mathcal{N}_n|$ can be obtained by expanding locally the generating function

$$S(z) = \sum_{n=0}^{\infty} |\mathcal{N}_n| z^n.$$

It is known that (see, for example, Figure I.10, p. 50, [19]),

$$S(z) = \frac{c(z)}{z^{|w|} + (1 - 2z)c(z)}$$

where $c(z)$ is the autocorrelation polynomial given by $c(z) = \sum_{j=0}^{|w|-1} c_j z^j$ with c_j equals to 1 if w coincides with its jth shifted version and 0 otherwise. The function $S(z)$ is a rational function and the asymptotic behavior of its coefficients is given by its expansion. If there is a dominant pole $\rho = |\alpha_1| < |\alpha_2| \leq |\alpha_3| \leq \ldots$ with multiplicity r, then

$$|\mathcal{N}_n| = O(\rho^{-n+r}).$$

Notice that all the above quantity were related to w, so in particular we will adopt the notation $\rho(w)$ from now on.

Theorem 3 *Let $\epsilon := \inf_{b,\underline{a}} p(b|\underline{a})$ and w the reference string for p. Then, we have the following results.*

(i) If $\rho(w) \geq 1 - \epsilon$ and $\alpha_k^w \to 1$ as k diverges, then there exists at least one stationary measure compatible.
(ii) If $\rho(w) \geq 1 - \epsilon$ and $\sum_k (1 - \alpha_k^w)^2 < \infty$, then there exists a unique stationary measure compatible.
(iii) If $\sum_k \prod_{i=0}^{k-1} \alpha_i^w = \infty$, then our coupling from the past perfectly simulates finite windows of the unique stationary measure. In particular, $\theta[0]$ is finite a.s.
(iv) If $\sum_k (1 - \alpha_k^w) < \infty$, then our coupling from the past perfectly simulates right-infinite vectors of the stationary measure, and the chain can be seen as a concatenation of independent vectors of random size. In particular, $\theta[0]$ has summable tail (finite expectation).

Proof (Proof of Theorem 3) The items (iii) and (iv) are direct consequences of Theorem 2, and there is no extra condition on w and ϵ, else than being finite for the former and strictly positive for the latter. Items (i) and (ii) follow from [23] (Corollary 1 and Theorem 2, respectively). Consider the set $\mathcal{D}$ of potentially discontinuous pasts of our kernel p, that is the set of infinite branches of τ_0^w. Let us denote $\mathcal{D}^n = \{a_{-n}^{-1}\}_{\underline{a} \in \mathcal{D}}$ and also the upper exponential growth rate of $\mathcal{D}$ as $\bar{g}r(\mathcal{D}) := \limsup_n |\mathcal{D}^n|^{1/n}$. Gallo and Paccaut gave conditions on the set $\mathcal{D}$ which guaranty existence of a stationary compatible measure [23]. They proved (Corollary 1) that, in the case where $\inf_{a \in A} \inf_{\underline{a} \in A^{-\mathbb{N}}} p(a|\underline{a}) = \epsilon > 0$, a sufficient condition is that $\bar{g}r(\mathcal{D}) < (1 - (|A| - 1)\epsilon)^{-1}$. In our case, we have $\mathcal{D}_n = \mathcal{N}_n$, and since we have $|A| = 2$, it is enough that $\bar{g}r(\mathcal{N}) < (1 - \epsilon)^{-1}$. By the discussion preceding the statement of Theorem 3, it is enough that $\rho(w) \geq 1 - \epsilon$. In order to prove (ii) using Theorem 2 of [23], it remains to check their hypothesis (H4) which reads as follows:

$$\sum_{v \in \tau_0^w} \mu(v) \sum_{k \geq |v|} \left(1 - \inf_{a_{-k}^{-1} \in B_v} \sum_{a \in \{1,2\}} \inf_{\underline{z}} p(a|a_{-k}^{-1}\underline{z})\right)^2 < \infty.$$

But, in our case, we have, for any $v \in \tau_0^w$

$$\inf_{a_{-k}^{-1} \in B_v} \sum_{a \in \{1,2\}} \inf_{\underline{z}} p(a|a_{-k}^{-1}\underline{z}) = \alpha_{k-|v|}^w,$$

so

$$\sum_{k\geq|v|}\left(1-\inf_{a_{-k}^{-1}\in B_v}\sum_{a\in\{1,2\}}\inf_{\underline{z}} p(a|a_{-k}^{-1}\underline{z})\right)^2=\sum_{k\geq 1}(1-\alpha_k^w)^2<\infty.$$

This proves that H4 is satisfied under our condition, and therefore we have uniqueness.

The condition $\rho(w)\geq 1-\epsilon$ is not explicit, so let us give some examples.

1. If $w=1$ or 2, then $\mathcal{N}_n=1$ for any $n\geq 1$ and thus $\rho(w)=1$.
2. If $w=12$ or 21, then we have $\mathcal{N}_n=n+1$ for any $n\geq 1$ and thus $\rho(w)=1$ as well.
3. The case where w is composed of a single symbol repeated k times, it can be proven (see p. 309 [19]) that

$$\frac{1}{2}\left(1+\left(\frac{1}{2}\right)^{k+1}\right)<\rho(w)<\frac{1}{2}\left(1+\left(\frac{3}{5}\right)^{k+1}\right).$$

4. For patterns of length 3 or 4, the values of ρ can be computed explicitly (see Figure IV.13, p. 272, [19]).
   (a) For $w=112,122,221,211$, $\rho(w)=0.61803$;
   (b) For $w=121,212$, $\rho(w)=0.56984$;
   (c) For $w=1112,1122,1222,2221,2211,2111$, $\rho(w)=0.54368$;
   (d) For $w=1121,1221,1211,2212,2112,2122$, $\rho(w)=0.53568$; and
   (e) For $w=1212,2121$, $\rho(w)=0.53101$.
5. When $|w|\geq 5$, $\rho(w)$ is the unique root in $(1/2,6/10)$ of the equation $z^{|w|}+(1-2z)c(z)=0$ (see Proposition IV.4 in [19]).

So, we see that, except for the cases where $w=1,2,12,21$, in which no extra condition is necessary on ε (since for any ε we have $\rho(w)=1>1-\varepsilon$ in these cases), all the other cases imply quite strong conditions on ε. For instance, if $w=112$, we have $\rho(w)=0.61803$, thus we need $0.38197<\varepsilon<0.5$.

6 Recent Bibliography and Some Open Problems

The present paper was about perfect simulation for chains with unbounded memory, through convex mixture of probabilistic context tree kernels for discontinuous kernels. Let us conclude by making a more complete compilation of recent results concerning these topics.

Concerning Example 1 This example served as a laboratory to explain and exemplify each notion/result introduced. It turns out that this example is quite recurrent in the literature, and we refer, for instance, to [1, 6, 7, 10, 16, 32] for a non-exhaustive list of references where this process is studied regarding several aspects.

Concerning Perfect Simulation An interesting framework was introduced in [11], making use of an a priori knowledge about the histories, extracted from the **U** sequence used for the CFTP algorithm. Their interest, although completely related to ours (perfect simulation of discontinuous chains with unbounded memory), is slightly different in the method. A comparison between efficacy of the results is presented in [22] through examples. Recently, [12] considered a particular class of unbounded memory chains, that they called *autoregressive processes with noise*. They considered under which conditions on the parameters uniqueness, phase transition, or successful coupling from the past are obtained. Their phase transition is due to nonpositivity of the kernel; that is, as we already mentioned earlier (see Example 3), it is a non-irreducibility situation.

Concerning Existence The only result of the literature which focuses on the issue of existence for discontinuous kernels is the one of [22]. As explained in the proof of Theorem 3, the idea is to put all the discontinuous points into a skeleton context tree, and to ask that the set of infinite branches of this context tree, crossing height n, does not increase too fast in n. Around the same time, [6] studied the relation between chains with variable length and transformation of the interval in dynamical systems. As examples, they studied in detail the existence/uniqueness of stationary measures for two simple examples, one of which is Example 1.

Concerning Convex Mixture of Context Trees Motivated by statistical inference for stochastic chains, [34] extends the idea of Kalikow's convex mixture of Markov kernels, but in a slightly different approach. Assuming the existence of a stationary measure μ compatible with some possibly discontinuous kernel p, an optimal convex mixture of kernels based on μ is constructed. Intuitively, it is optimal in the sense that it minimizes the look-back sizes implied by the mixture. Doing so, in particular, it proves that a kernel is μ-a.s. continuous with respect to the given measure if, and only if, these look-back sizes are μ-a.s. finite.

Some Open Problems

- Concerning perfect simulation, it is still lacking a necessary condition for existence of CFTP algorithms, when continuity is assumed.
- Another interesting question is how to perfectly simulate without the non-nullness condition (3). Under the continuity assumption, [11] gave sufficient conditions. On the other hand, [24] constructed an algorithm able to perfectly simulate for discontinuous kernel not satisfying (3), but it does not give any general sufficient conditions for convergence of the algorithm.
- Concerning existence, it would be interesting to construct a kernel with $\inf_a \inf_{\underline{a} p(a|\underline{a})>0}$ for which there exists no stationary compatible measure.

Acknowledgements SG is supported by an FAPESP fellowship (grant 2009/09809-1), and NG is supported by grants CNPq 302598/2014-6 and FAPESP 2014/26419-0.

References

1. Miguel Abadi, Liliam Cardeño, and Sandro Gallo. Potential well spectrum and hitting time in renewal processes. *Journal of Statistical Physics*, 159(5):1087–1106, 2015.
2. Henry Berbee. Chains with infinite connections: uniqueness and Markov representation. *Probab. Theory Related Fields*, 76(2):243–253, 1987.
3. A. A. Borovkov. *Ergodicity and stability of stochastic processes*. Wiley Series in Probability and Statistics: Probability and Statistics. John Wiley & Sons Ltd., Chichester, 1998. Translated from the 1994 Russian original by V. Yurinsky [V. V. Yurinskiĭ].
4. Maury Bramson and Steven Kalikow. Nonuniqueness in g-functions. *Israel J. Math.*, 84(1-2):153–160, 1993.
5. Peter Bühlmann and Abraham J. Wyner. Variable length Markov chains. *Ann. Statist.*, 27(2):480–513, 1999.
6. Peggy Cénac, Brigitte Chauvin, Frédéric Paccaut, and Nicolas Pouyanne. Context trees, variable length Markov chains and dynamical sources. In *Séminaire de Probabilités XLIV*, pages 1–39. Springer, 2012.
7. Peggy Cénac, Brigitte Chauvin, Frédéric Paccaut, and Nicolas Pouyanne. Uncommon suffix tries. *Random Structures Algorithms*, 46(1):117–141, 2015.
8. P. Collet, D. Duarte, and A. Galves. Bootstrap central limit theorem for chains of infinite order via Markov approximations. *Markov Process. Related Fields*, 11(3):443–464, 2005.
9. Francis Comets, Roberto Fernández, and Pablo A. Ferrari. Processes with long memory: regenerative construction and perfect simulation. *Ann. Appl. Probab.*, 12(3):921–943, 2002.
10. Walter A.F. De Carvalho, Sandro Gallo, and Nancy L. Garcia. Continuity properties of a factor of Markov chains. *Journal of Applied Probability*, 53(1):216–230, 2016.
11. Emilio De Santis and Mauro Piccioni. Backward coalescence times for perfect simulation of chains with infinite memory. *J. Appl. Probab.*, 49(2):319–337, 2012.
12. Emilio De Santis and Mauro Piccioni. One-dimensional infinite memory imitation models with noise. *J. Stat. Phys.*, 161(2):346–364, 2015.
13. JCA Dias and S Friedli. Uniqueness vs. non-uniqueness for complete connections with modified majority rules. *Probability Theory and Related Fields*, 164(3-4):893–929, 2016.
14. Wolfgang Doeblin and Robert Fortet. Sur des chaînes à liaisons complètes. *Bull. Soc. Math. France*, 65:132–148, 1937.
15. Paul Doukhan and Olivier Wintenberger. Weakly dependent chains with infinite memory. *Stochastic Processes and their Applications*, 118(11):1997–2013, 2008.
16. Roberto Fernández, Sandro Gallo, and Grégory Maillard. Regular C-measures are not always Gibbsian. *Electron. Commun. Probab.*, 16:732–740, 2011.
17. Roberto Fernández and Grégory Maillard. *Chains and specifications*. Eurandom, 2004.
18. Roberto Fernández and Grégory Maillard. Chains with complete connections: general theory, uniqueness, loss of memory and mixing properties. *J. Stat. Phys.*, 118(3-4):555–588, 2005.
19. Philippe Flajolet and Robert Sedgewick. *Analytic combinatorics*. Cambridge University press, 2009.
20. Robert G Gallager. *Information theory and reliable communication*, volume 2. John Wiley, 1968.
21. Sandro Gallo. Chains with unbounded variable length memory: perfect simulation and a visible regeneration scheme. *Adv. in Appl. Probab.*, 43(3):735–759, 2011.
22. Sandro Gallo and Nancy L. Garcia. Perfect simulation for locally continuous chains of infinite order. *Stochastic Process. Appl.*, 123(11):3877–3902, 2013.

23. Sandro Gallo and Frédéric Paccaut. On non-regular g-measures. *Nonlinearity*, 26(3):763–776, 2013.
24. Aurélien Garivier. Perfect simulation of processes with long memory: a "coupling into and from the past" algorithm. *Random Structures Algorithms*, 46(2):300–319, 2015.
25. Hans-Otto Georgii. *Gibbs measures and phase transitions*, volume 9 of *de Gruyter Studies in Mathematics*. Walter de Gruyter & Co., Berlin, second edition, 2011.
26. T. E. Harris. On chains of infinite order. *Pacific J. Math.*, 5:707–724, 1955.
27. Anders Johansson and Anders Öberg. Square summability of variations of g-functions and uniqueness of g-measures. *Math. Res. Lett.*, 10(5-6):587–601, 2003.
28. Steve Kalikow. Random Markov processes and uniform martingales. *Israel J. Math.*, 71(1):33–54, 1990.
29. Michael Keane. Strongly mixing g-measures. *Invent. Math.*, 16:309–324, 1972.
30. François Ledrappier. Principe variationnel et systemes dynamiques symboliques. *Probability Theory and Related Fields*, 30(3):185–202, 1974.
31. Torgny Lindvall. *Lectures on the coupling method.* Wiley Series in Probability and Mathematical Statistics: Probability and Mathematical Statistics. John Wiley & Sons Inc., New York, 1992. A Wiley-Interscience Publication.
32. Gusztáv Morvai and Benjamin Weiss. On universal estimates for binary renewal processes. *Ann. Appl. Probab.*, 18(5):1970–1992, 2008.
33. P Ney and E Nummelin. Regeneration for chains with infinite memory. *Probability theory and related fields*, 96(4):503–520, 1993.
34. Roberto Imbuzeiro Oliveira. Stochastic processes with random contexts: a characterization and adaptive estimators for the transition probabilities. *IEEE Trans. Inform. Theory*, 61(12):6910–6925, 2015.
35. Octave Onicescu and Gheorghe Mihoc. Sur les chaînes de variables statistiques. *Bull. Sci. Math*, 59(2):174–192, 1935.
36. Donald Samuel Ornstein and Benjamin Weiss. Entropy and data compression schemes. *IEEE Trans. Inform. Theory*, 39(1):78–83, 1993.
37. James Gary Propp and David Bruce Wilson. Exact sampling with coupled Markov chains and applications to statistical mechanics. In *Proceedings of the Seventh International Conference on Random Structures and Algorithms (Atlanta, GA, 1995)*, volume 9, pages 223–252, 1996.
38. Jorma Rissanen. A universal data compression system. *IEEE Trans. Inform. Theory*, 29(5):656–664, 1983.
39. Paul C. Shields. *The ergodic theory of discrete sample paths*, volume 13 of *Graduate Studies in Mathematics*. American Mathematical Society, Providence, RI, 1996.
40. Peter Walters. Ruelle's operator theorem and g-measures. *Transactions of the American Mathematical Society*, 214:375–387, 1975.

Inference in (M)GARCH Models in the Presence of Additive Outliers: Specification, Estimation, and Prediction

Luiz Koodi Hotta and Carlos Trucíos

Abstract The (M)GARCH models are probably the most widely used to estimate and predict volatility. Estimation and prediction of volatility are very important in many financial applications. One important issue in the application of (M)GARCH models is the frequent presence of outliers in financial time series and their effects in all stages of model application. We present some issues involved in making inference in (M)GARCH models in the presence of additive outliers. Specifically, we present the effects of outliers on specification, estimation of models, and their volatility and volatility prediction. We also present some robust methods to estimate the model and to predict volatility. We emphasize the presentation of robust methods for volatility forecast density.

1 Introduction

The estimation and prediction of asset return's volatility is important in numerous financial applications, such as pricing of financial derivatives, risk assessment, and portfolio management, see, for instance, [15, 28, 35]. Since the introduction of the autoregressive conditionally heteroskedastic (ARCH) model by Engle in 1982 [32] and the generalized ARCH (GARCH) model by Bollerslev in 1986 [10], these models and their variants have become a reference in modeling univariate asset return volatility. Because many, and probably most, finance applications involve more than one asset and the returns of the assets are not independent, the GARCH models were soon extended to a multivariate framework. The first GARCH type model for the conditional covariance matrices was proposed by Bollerslev et al. [13], the VEC-GARCH model. As in the univariate case, many variants soon appeared in

L. K. Hotta (✉)
Institute of Mathematics, Statistics and Scientific Computing, University of Campinas, Campinas, SP, Brazil
e-mail: hotta@ime.unicamp.br

C. Trucíos
São Paulo School of Economics, Getúlio Vargas Foundation, São Paulo, SP, Brazil

C. Lavor, F. A. M. Gomes (eds.), *Advances in Mathematics and Applications*,
https://doi.org/10.1007/978-3-319-94015-1_8

the literature, generally called multivariate GARCH (MGARCH) models. Because multivariate modeling involves extra issues, such as estimability due to the huge number of parameters, positiveness of the volatility matrix, and dependence, many suggestions appeared to simplify or modify the VEC-GARCH model to deal with these problems, see, for instance, [3, 9, 38, 76] for good reviews of MGARCH models.

The success of applying these models, however, can be badly undermined by well-known stylized fact found in financial data, the presence of outliers [8, 77, 86], especially the additive outliers. Many papers have studied their negative effects in all stages of applying the model, from selection or specification to estimation, prediction, and application. In order to minimize their effects, the main approaches are to detect the outliers and consider them in the modeling stage or to adopt robust procedures. In this chapter we present a general review of their effects and robust approaches, both for univariate and multivariate models. Because, as pointed out earlier, a large number of models exist, we discuss mostly the univariate GARCH(1,1) and a version of dynamical conditional correlation (DCC) model proposed by Engle [34], the corrected DCC (cDCC) model [2]. For the sake of brevity and because the problem of measuring the effect of outliers and robust procedures for estimation are better known in the literature, our review will emphasize robust forecasting in the cDCC model. The remainder of the chapter is organized as follows. In Sect. 2 we present the uncontaminated GARCH model and the model contaminated by additive outliers, while in Sect. 3 we present the uncontaminated and contaminated cDCC models. We also discuss briefly the estimation of the volatility and of the model and also the prediction. Section 4 presents the effects of outliers on the specification, estimation, and volatility prediction, and the literature on influential observation techniques applied to GARCH models. Tests to detect outliers are presented in Sect. 5. Section 6 presents robust methods to estimate the models and their volatility and forecast densities. Finally, Sect. 7 presents the final remarks.

2 GARCH Model

This section presents the multivariate and univariate GARCH models including uncontaminated and contaminated models. In the following, the observed return is in fact the observed return filtered by the conditional mean. Besides, when we state conditional mean or (co)variance, we refer to conditional information given by the past observation of the (observed) return series.

2.1 Uncontaminated GARCH Models

The uncontaminated GARCH model was proposed in [10] as a generalization of ARCH model proposed in [32]. The model is commonly used to represent the dynamic dependence in the second-order moments of return in economic and financial time series. The GARCH(1,1) model is defined as

$$r_t = \sigma_t \varepsilon_t, \tag{1a}$$

$$\sigma_t^2 = \omega + \alpha r_{t-1}^2 + \beta \sigma_{t-1}^2, \tag{1b}$$

for $t = 1, \ldots, T$, where r_t is the return observed at time t, σ_t the Volatility, and ε_t is an independent identically distributed process with zero mean and variance one. The parameters are assumed to satisfy the conditions $\omega > 0$, $\alpha, \beta \geq 0$, and $\alpha + \beta < 1$, which are sufficient for stationarity and positiveness of σ_t^2. The initial values of (σ_0^2, r_0) come from the unconditional bivariate distribution.

2.2 Contaminated GARCH Model

The GARCH model contaminated by additive outliers was defined by Hotta and Tsay [52]. The contaminated GARCH(1,1) model is defined as

$$r_t = z_t + sign(z_t) w_t I(t \in A), \tag{2a}$$

$$z_t = \sigma_t \varepsilon_t, \tag{2b}$$

$$\sigma_t^2 = \omega + \alpha z_{t-1}^2 + \beta \sigma_{t-1}^2, \tag{2c}$$

for $t = 1, \ldots, T$, where r_t and all the other terms and parameters are defined as in Sect. 2.1, including the parameter restrictions, w_t represents the size of the outlier at time t, $I(\cdot)$ is the indicator function, and A is the set of contaminated observations.

2.3 Parameter and Volatility Estimation

Parameter estimation is usually done by applying a Gaussian quasi-maximum likelihood (QML) estimator which is based on maximizing the logarithm of the Gaussian likelihood function. Conditional on σ_1^2 and r_1, the Gaussian log-likelihood is given by

$$l(\theta; r) \propto -\left(\sum_{t=2}^{T} \frac{r_t^2}{\sigma_t^2} + \sum_{t=2}^{T} \log(\sigma_t^2) \right), \tag{3}$$

where θ is the vector of unknown parameters, and r_t and σ_t are defined as previously. Hence, the estimated volatility is given by

$$\hat{\sigma}_t^2 = \hat{\omega} + \hat{\alpha} r_{t-1}^2 + \hat{\beta}\hat{\sigma}_{t-1}^2. \tag{4}$$

Note that, no corrective measures to protect again possible outliers are applied. In fact, many authors have shown that this estimator is highly affected by outliers, see, for instance, [18, 68, 82]. Some robust model estimators are presented later in Sect. 6.1, while robust filters to estimate volatility are presented in Sect. 6.2.

2.4 Forecast Densities

One of the main objectives in modeling financial returns and volatilities is to produce forecast. Point forecasting is the focus of many books and guidelines. However, we prefer to emphasize forecast densities instead just point forecast because forecast densities can be even more important than point forecasts. One way to obtain forecast densities[1] in volatility in a frequentist approach is through bootstrap procedures. There are several bootstrap procedures in univariate volatility models but we focus in the procedure of [69] because it has good finite sample properties. Readers interested in other bootstrap procedures can see, for instance, [24, 67, 74].

Pascual, Romo, and Ruiz (PRR) [69] propose a procedure to obtain forecast densities for returns and volatilities in GARCH models. Their bootstrap procedure is based on the QML estimator and the standard volatility equation. As usual in residual-based bootstrap procedures, after fitting the model and obtaining the centered standardized residual, new bootstrap series that will be used in the forecast are obtained. To allow construction of forecast densities for the one-step-ahead volatility, the parameters in each bootstrap series are re-estimated and the h-steps-ahead forecast is obtained. Using this procedure, no assumption about the error distribution is necessary, and additionally, the re-estimation of the parameters makes it possible to handle parameter uncertainty. In this section we will not present the bootstrap algorithm mentioned because in the subsequent sections we will describe a way to make this procedure more robust.

It is important to mention that this procedure has shown good finite sample properties in uncontaminated GARCH models. Implementations and extensions to other univariate volatility models can be found in [45, 54, 81].

[1] In this chapter we will not present Bayesian methodologies. However, the reader interested can see [51, 58, 89] for some references.

3 MGARCH Models

GARCH model were quickly extended to a multivariate framework, see [9, 38, 76] for good reviews of MGARCH models. Two of the most popular MGARCH models, which have become benchmarks in multivariate volatility modeling, are the DCC model of [34] and the cDCC version of [2]. Denote by $r_t = (r_{1,t}, \ldots, r_{p,t})$ the p-dimensional vector of returns observed at time t.

3.1 Uncontaminated cDCC Model

The DCC model was proposed by Engle [34] and the cDCC version by Aielli [2]. The cDCC model, unlike the constant conditional correlation model proposed by Bollerslev [12], considers that the correlation structure evolves over time, relaxing the assumption of a constant correlation structure, which is too restrictive in most of the financial applications. The cDCC model is defined by Aielli [2] as

$$r_t = H_t^{1/2}\varepsilon_t, \tag{5a}$$

$$H_t = D_t R_t D_t, \tag{5b}$$

where H_t is the conditional covariance matrix, $R_t = diag(Q_t)^{-1/2} Q_t diag(Q_t)^{-1/2}$ is the conditional correlation matrix, D_t is a diagonal matrix containing the univariate GARCH(1,1) variances, and ε_t is an independent and identically distributed p-dimensional process with mean zero and identity covariance matrix. For the (1, 1) order, the matrix Q_t is defined as

$$Q_t = (1 - a - b)S + a\, diag(Q_{t-1})^{1/2} v_{t-1} v_{t-1}' diag(Q_{t-1})^{1/2} + b\, Q_{t-1}, \tag{6}$$

where $diag(Q_t)^{1/2} = diag(\sqrt{q_{11,t}}, \ldots, \sqrt{q_{pp,t}})$, $v_t = D_t^{-1} r_t$ and S is the unconditional correlation matrix of $diag(Q_t)^{1/2} v_t$. The parameters are assumed to satisfy $a, b \geq 0$ and the stationary conditions $a + b < 1$.

3.2 Contaminated cDCC Model

The contaminated version of the cDCC model was defined by Boudt et al. [15]. It is defined as

$$r_t = Z_t + A_t I(t \in B) \tag{7a}$$

$$Z_t = H_t^{1/2}\varepsilon_t \tag{7b}$$

$$H_t = D_t R_t D_t, \tag{7c}$$

where A_t is the p-dimensional vector of contaminations and B denotes the set of contaminated observations. For identifiability purposes, each component of $A_t = (A_{1t}, \ldots, A_{pt})'$ is defined following [52] as

$$A_{it} = sign(z_{it})w_{it}I(t \in B_i), \tag{8}$$

where w_{it} is the size of the outlier of the i-*th* series at time t and B_i is the set of contaminated observations on the i-th series.

3.3 Parameters and Model Estimation

The parameters, volatilities, and correlations are usually estimated through the QML estimator of [2]. The procedure of [2] is a three-step estimator based on the maximization of the Gaussian quasi-likelihood, which conditional on D_1, R_1, and r_1 is given by

$$l(\theta, \phi, S) \propto -\frac{1}{2}\sum_{t=2}^{T}\left[2\log(det(D_t)) + \log(det(R_t)) + v_t' R_t^{-1} v_t\right], \tag{9}$$

where $v_t = D_t^{-1} r_t$. In the first step, the univariate GARCH models are estimated separately by QML. Secondly, the parameters a and b are estimated also by QML. Finally, S is estimated, using the estimated parameters, obtained as $\hat{S} = \frac{\sum_{t=1}^{T} \hat{Q}_t^{1/2} \hat{v}_t \hat{v}_t' \hat{Q}_t^{1/2}}{T}$. As usual, the volatility and correlation estimates are obtained replacing the parameters by their estimates. This procedure is not robust to outliers and the non-robustness has been analyzed in [15] and [84]. For more information about the estimation of DCC models using composite likelihood, linear and non-linear Shrinkage procedures see [48, 70] and [36].

3.4 Forecast Densities

Fresoli and Ruiz [41] extend the bootstrap procedure of [69] to the multivariate case. This algorithm provides forecast densities for returns, volatilities, and also for conditional correlations in the cDCC model. This algorithm follows the same idea of the univariate algorithm, but incorporates an appropriate bootstrap procedure in the correlation equation. This procedure, as well as the univariate version, has good finite sample properties when applied to uncontaminated series. However, when applied to contaminated series, the performance of the algorithms is poor. For instance, [81, 82, 84] show that these bootstrap procedures are highly affected by additive outliers.

Outliers are not unusual in time series and their presence can produce devastating effects on the bootstrap procedures presented previously. For this reason, several authors have proposed robust procedures to mitigate the effect of outliers, mainly additive outliers, with consistent results. These procedures are described in the next section.

We will not present in detail the algorithm of Fresoli and Ruiz because in the subsequent sections we will present a robustification of this procedure.

4 Effects of Outliers

This section presents a brief review of the literature on the effects of additive outliers. The effects on specifications are presented in Sect. 4.1, the effects on estimation in Sect. 4.2, and on volatility estimation and prediction in Sect. 4.3. We observed in the introduction that estimation and prediction of volatility are important in several financial applications. However, although it is an important issue to understand the effect of outliers for financial applications of interest, this is not pursued in this chapter.

4.1 Effects on Specification

The presence of outliers in time series can lead to two types of errors: wrongly suggesting conditional heteroskedasticity and failing to detect conditional heteroskedasticity. This was first pointed out by Van Dijk et al. [85], who analyzed the properties of the Lagrange multiplier (LM) test for ARCH models in the presence of additive outliers. They found out that, when the conditional mean has an autoregressive component, the LM test rejects the true null hypothesis of no conditional homoskedasticity too often. Similar conclusions have been reached by other authors theoretically, using simulation, or by analyzing real-time series, see, for instance, [1, 7, 16, 31, 39, 40, 46, 61, 63, 66, 78]. Most of the cited papers in this section propose a robust test to detect conditional heteroskedasticity, see also [39, 47]. Additionally, in a leverage effect context, [19] show that the presence of outliers can affect the identification of the asymmetric response of volatility and could detect spurious asymmetries, asymmetries of the wrong sign or could also hide the true leverage effect.

4.2 Effects on Estimation

The effect of outliers on estimation is described in many works, when using simulation or when analyzing real-time series. References [75] and [66] find a large effect when using simulation to analyze the effect of a single additive outlier on the maximum likelihood (ML) estimator of the parameters of the GARCH(1,1) model. Similar results are found by Verhoeven and McAleer [88], Li and Kao [61], Carnero et al. [16], Welsch and Zhou [90], Muler and Yohai [68], Ardelean [5] and Carnero et al. [18] using different estimators, i.e., a single additive outlier can have a strong effect in the model estimation. As expected the effect is larger when the outlier is not near the end of the series. Reference [16] presents asymptotic results for different estimators of ARCH and GARCH(1,1) models. Finite sample properties are addressed through simulation. The effect persists even when estimated by the QML method using heavy-tailed distributions instead of the Gaussian distribution.

Reference [14] uses simulation to show the large effect of additive outliers on Gaussian QML estimators of the bivariate BEKK [33] model and present a robust M-estimator.

References [43] and [87] show that even outliers of moderate magnitude can have a large effect on the estimation of multivariate GARCH models. They use simulation to study the effect of additive outliers on the diagonal BEKK (D-BEKK), CCC, and DCC models. In all the simulation they use bivariate models and their focus is on estimating the correlation. They consider single and multiple isolated additive outliers and patches of additive outliers and find a larger effect on the CCC and DCC models. References [15] and [80] also find a strong effect on the cDCC model. Reference [15] presents a robust estimator which is presented in Sect. 6.1.5, while [80] also presents a simulation comparing the effect of additive outliers on several robust and non-robust estimators. References [73] and [37] discuss the effect of outliers in asymmetric GARCH-type models.

4.3 *Effects on Volatility Estimation and Prediction*

An additive outlier affects the volatility prediction directly through the autoregressive coefficient in Eq. (1b) when the outlier occurs in the last observation, or through the effect on past volatility when the outlier appears before this. However, an indirect effect can also exist through the effect of the estimation of volatilities used in the prediction and the parameters of the model. Understanding this indirect effect is not simple because the outlier can affect not only the estimation of α, β, and ω, but also the estimation of the volatility of the last observations, which are used in GARCH(1,1) prediction. In general, one expects to have an increase in the estimation of the unconditional variance $\omega/(1-\alpha-\beta)$ and the persistence $(\alpha+\beta)$, but without a clear indication of the overall indirect effect. There are many works dealing with the effect of additive outliers on the estimation and prediction of the volatility, see, for instance, [17, 18, 20, 21, 37, 39, 42, 43, 73, 80, 81, 88]. Most of the papers contain proposal for a robust estimation and prediction methods. These are mostly based on a robust estimation of the parameter models and a robust filter to estimate the volatility. All these papers deal with univariate GARCH models. In a multivariate context, the papers of [14, 44, 84, 87] can be mentioned.

5 Detection of Outliers

There is a huge literature on tests to detect additive outliers for univariate GARCH models. Because of this, we do not present details about the tests, but rather present the main features of some tests, classifying them by the method used. The presentation follows [53].

Besides the tests that are listed in the following sections, there are many other tests. For instance, [75] suggest using the difference between the QML and the proposed robust two-stage S-estimates to detect outliers or leverage points, [72] use excess of kurtosis, [4] check whether the observed return is covered by one-step-ahead interval forecast, [60] use the standardized residuals, [5] present a test based on the cumulative sum of the squared observations, [26] propose a weighted forward approach, and [59] present a iterative procedure for four type of outliers. References [73] and [37] propose tests for asymmetric GARCH-type models.

On the other hand, there are few works addressing the testing for outliers for MGARCH models. A review of some of these tests is presented in Sect. 5.4.

5.1 *Lagrange Multiplier and Likelihood Ratio Tests*

Reference [52] presents a Lagrange multiplier test to detect additive outliers. As in most of the tests proposed to detect outliers, this test is initially developed for a fixed observation. When the position and the number of outliers are not know, the authors suggest using the maximum of the test statistics over the entire period and applying the test iteratively, as in [25]. This type of approach is usually used in almost all tests when the position and the number of outliers are not known. They also suggest using simulation to find critical values.

Reference [79] extends the Lagrange multiplier tests to include outliers in every observation after time τ, which they call level shift outliers while [30] present a likelihood ratio test to detect additive and volatility outliers. The authors present the test for an ARMA-GARCH model, but it can be used for the simple GARCH model with additive outliers. When the positions and number of outliers are not known they suggest the usual approach as given previously. They also suggest critical values based on simulation.

5.2 *Test Based on ARMA Representation*

Reference [39] uses the fact that the square of the series generated by a GARCH process follows an ARIMA process to propose a method to detect additive outliers, following the procedure suggested by Chen and Liu [25] for ARMA models. When the position of the outlier is not known they compute the maximum of the test statistics over the entire period and compare it with a threshold value C. The authors suggest using $C = 4$, while [22] for the same test, following [88], use $C = 10$ for sample size equal to 912. The detection test is used recursively as in [25].

5.3 *Test Based on Wavelets*

Reference [42] proposes a test based on the coefficients of the discrete wavelet transform of the residuals on wavelets to detect outliers in volatility models. The procedure is based on the coefficients from the discrete wavelet transform of the residuals. The statistics based on detail coefficient have power to detect isolated and patches of additive outliers. The detection test is used recursively as in [25]. Reference [23] also uses wavelet-based algorithm to detect outliers when modeling US stock market volatility.

5.4 *Test for MGARCH Models*

Reference [87] extends the test based on wavelets to multivariate GARCH models. The authors translate the multivariate problem to a univariate setting by applying the random projection method. The performance of the tests is evaluated using simulation for the D-BEKK, CCC, and DCC models with Gaussian and Student-t error with 7 d.f. with isolated and patches of outliers. A shorter version of this paper can be found in [44].

5.5 *Influential Observation*

Although influential observations are not always outliers and not all outliers are influential, influential analysis using different perturbation schemes can be used to detect potential outliers. In this sense, we present some papers which deal with influential analysis in (M)GARCH models. Influence diagnostics for GARCH models have been studied by Liu [62] for models with elliptical errors (but without statistical analysis), by Zhang [92] and Zhang and King [93] for models with Gaussian errors, by Zevallos and Hotta [91] for models with Gaussian or Student-t errors, and by Hotta and Zevallos [53] as a particular case of conditional heteroskedastic time series models with Gaussian, Student-t, or generalized exponential distribution errors.

The papers deal mainly with three perturbations schemes: innovative, additive, and data perturbations. The last two perturbations are related to additive outliers. In the additive model's perturbation scheme, the perturbations are proportional to the conditional standard error and for the data perturbation scheme the perturbation is not.

Reference [93] presents the expressions for the slope and curvature statistics for the three perturbations schemes using Gaussian errors, while [91] compute them using Student-t errors and [62] uses elliptical errors. References [93] and [91] propose using simulation to estimate the distribution of the test statistics to find threshold values.

There is only one paper dealing with influential analysis in MGARCH models. Reference [29] presents influential analysis by introducing perturbations to the conditional variances and covariances in two bivariate GARCH models.

6 Robustness

As mentioned in the previous section, outliers may have a strong effect on model parameter estimation as well as on volatility and correlation estimation. Outliers also have a strong impact on the forecast procedures, distorting predictions and giving a misleading picture of what can be expected in the future, leading to incorrect decisions. Because there is a large literature we do not cite or present all the proposed estimators. For instance, we do not discuss bounded influence estimator of [61], the closed form estimator of [6], or the robust procedures of [26, 49, 50, 71] and [73]; the last one for asymmetric GARCH models. In this section, we discuss some alternative approaches meant to be robust to the presence of outliers, showing good finite sample properties.

6.1 Parameter Estimation

There are many estimators available in the literature to estimate GARCH models that are meant to be robust to the presence of additive outliers. We introduce briefly some of the most popular ones to estimate model (2), without taking explicitly the presence of outliers. Robust estimators are proposed to obtain parameter estimates that are not affected by atypical observations and also to mitigate the effect of additive outliers on the volatility and correlation estimates. In general, robustness depends on the choice of an objective function as well as the choice of threshold parameters of the objective function. There are many robust estimators derived from the class of robust M-estimators, see, for instance, [55] for an early simulation comparison of some M-estimators. In fact, some of the estimators presented in the following belong to this class of estimators.

6.1.1 QMLt Estimator

In order to mitigate the influence of atypical observations [11] proposes a QML estimator based on the maximization of the Student-t log-likelihood function.

Because the Student-t distribution has heavier tails than the Gaussian distribution it gives less weight to larger innovations. The Student-t log-likelihood, conditional on σ_1^2 and r_1, is given by

$$l(\theta; r) \propto -\sum_{t=1}^{T} \left(\log \left(t_\nu \left(\frac{r_t}{\sqrt{\sigma_t^2}} \right) \right) - \frac{1}{2} \log \left(\sigma_t^2 \right) \right), \tag{10}$$

where θ is the vector of unknown parameters (including the degrees of freedom ν), r is the vector of observed returns, σ_t^2 is the conditional variance defined in Eq. (2c), and $t_\nu(\cdot)$ is the density of the Student-t distribution with ν degrees of freedom scaled to have unit variance. Then, the QMLt parameter estimates are defined as

$$\hat{\theta} = \underset{\theta}{\operatorname{argmax}} \left(l(\theta; r) \right). \tag{11}$$

6.1.2 BM Estimator

Reference [68] proposes an M-estimator robust to outliers based on a robust filter for the volatility. Conditional on σ_1^2 and r_1, the objective function is defined as:

$$M(\theta; r) = \frac{1}{T-1} \sum_{t=2}^{T} \rho(\log(r_t^2) - \log(\sigma_t^2)), \tag{12}$$

where θ is the vector of unknown parameters and $\rho(x) = m\left(-\log\left(f\left(e^{\frac{x}{2}}\right)e^{\frac{x}{2}}\right)\right)$, with $m(\cdot)$ being a bounded nondecreasing function and $f(\cdot)$ a centered density function. Then, the BM (Bounded M) estimator is defined as

$$\hat{\theta}_{\text{BM}} = \begin{cases} \hat{\theta}_1 = \operatorname{argmin}_\theta M_T, & M_T(\hat{\theta}_1; r) \leq M_{Tk}^*(\hat{\theta}_2; r) \\ \hat{\theta}_2 = \operatorname{argmin}_\theta M_{Tk}^*, & M_T(\hat{\theta}_1; r) > M_{Tk}^*(\hat{\theta}_2; r), \end{cases} \tag{13}$$

where $M_T(\cdot, \cdot)$ and $M_{Tk}^*(\cdot, \cdot)$ are both defined as in Eq. (12). The difference between them is that $M_T(\cdot, \cdot)$ uses σ_t^2 as defined in Eq. (2c) and $M_{Tk}^*(\cdot, \cdot)$ defines σ_t^2 as

$$\sigma_t^2 = \omega + \alpha r_c \left(\frac{r_{t-1}^2}{\sigma_{t-1}^2} \right) \sigma_{t-1}^2 + \beta \sigma_{t-1}^2, \tag{14}$$

with $r_c(\cdot)$ being a robust filter used in the volatility equation to mitigate the influence of atypical observations and is given by

$$r_c(x) = \begin{cases} x, & \text{if } x \leq c \\ c, & \text{if } x > c, \end{cases} \tag{15}$$

with the threshold constant c being a convenient tradeoff between efficiency and robustness. For more references about the choice of the threshold c and the nondecreasing function $m(\cdot)$ see [68].

6.1.3 BQMLt Estimator

The BQMLt estimator, proposed by Carnero et al. [18], follows the same idea of the BM estimator but uses as objective function the Student-t log-likelihood function. The BQMLt estimator is defined as

$$\hat{\theta}_{\text{BQMLt}} = \begin{cases} \hat{\theta}_1, & l(\hat{\theta}_1, r) \leq l^R(\hat{\theta}_2, r) \\ \hat{\theta}_2, & l(\hat{\theta}_1, r) > l^R(\hat{\theta}_2, r), \end{cases} \tag{16}$$

where $l(\theta_1, r)$ and $l^R(\theta_2, r)$ are both Student-t log-likelihood functions. However, $l(\theta_1, r)$ uses the volatility equation as in Eq. (2c) and $l^R(\theta_2, r)$ uses the robust volatility equation defined in Eq. (14). Differently from [68], that uses a robust filter replacing large values by a threshold value, the BQMLt estimator uses the robust filter that replaces large values by their unconditional expectation. The $r_c(\cdot)$ filter is given as

$$r_c(x) = \begin{cases} x, & \text{if } x \leq c \\ 1, & \text{if } x > c, \end{cases} \tag{17}$$

with $c = 9$ for a convenient tradeoff between efficiency and robustness.

6.1.4 MT Estimator

The MT estimator proposed by [65] is an M-estimator for conditional location and scale models. The robust estimating function is obtained from the Gaussian pseudo maximum likelihood score function by a downweighting procedure that limits the potential damaging effects of data points that generate a too large sensitivity of Gaussian pseudo maximum likelihood. The M-estimator, denoted as $\hat{\theta}^{(\text{MT})}$, is defined as the solution of the following estimating equation:

$$T^{-1} \sum_{t=3}^{T} A(\theta) \left(s(r_{t-2}^t; \theta) - \tau_\theta \right) \omega(r_{t-2}^t; \theta) = 0, \tag{18}$$

with

$$A^t(\theta) A(\theta) = \left[T^{-1} \sum_{t=3}^{T} \left(s(r_{t-2}^t; \theta) - \tau_\theta \right) \left(s(r_{t-2}^t; \theta) - \tau_\theta \right)^t \omega^2(r_{t-2}^t; \theta) \right]^{-1}, \tag{19}$$

and $\omega(r_{t-2}^t;\theta) = min(1, c\|A(\theta)\left(s(r_{t-2}^t;\theta) - \tau_\theta\right)\|^{-1})$, $r_{t-2}^t = (r_{t-2}, r_{t-1}, r_t)$, $s(r_{t-2}^t;\theta)$ being the score function and τ_θ a correction factor. The norm $\|\cdot\|$ is the L_2−norm and c is a constant that controls the degree of robustness; [65] use $c = 11$. Furthermore, τ_θ is computed using the two-step procedure described in the appendix of [64]. Finally, the volatility is estimated using Eq. (2c) as usual.

6.1.5 BVT Estimator

Reference [15] proposes a robust estimator based on a robust variance target estimator and on a robust filter of the volatility. This estimator is a modification of the BM estimator of [68]. The first modification is in the estimation of the marginal variance, which is estimated as

$$\hat{\sigma}_z^{2(BVT)} = 1.318\frac{\sum_{t=1}^{T}\left(r_t - \hat{\mu}\right)^2 J_t}{\sum_{t=1}^{T} J_t}, \tag{20}$$

where $J_t = I\left(\frac{(r_t-\hat{\mu})^2}{(1.4826\times MAD_{t,K}(r_t))^2} \le \chi_1^2(95\%)\right)$ with $I(\cdot)$ being the indicator function, $\hat{\mu} = \frac{\sum_{t=1}^{T} r_t I_t}{\sum_{t=1}^{T} I_t}$ with $I_t = I\left(\frac{(r_t-Med_{t,K}(r_t))^2}{(1.4826\times MAD_{t,K}(r_t))^2} \le \chi_1^2(95\%)\right)$ and the statistics $MAD_{t,K}(\cdot)$ and $Med_{t,K}(\cdot)$ being the median absolute deviation and the median estimated in window of size K around r_t, respectively.

The second modification is in the volatility equation, where a constant c_γ is included in the equation to guarantee that the conditional expectation in the absence of outliers is still the conditional variance. Then, the volatility equation is given as

$$\sigma_{t-1}^2 = \omega + \alpha c_\gamma r_c\left(\frac{r_{t-1}^2}{\sigma_{t-1}^2}\right)\sigma_{t-1}^2 + \beta\sigma_{t-1}^2, \tag{21}$$

where $c_\gamma = \frac{E[u]}{E[min(u,k_{\gamma,1})]}$ with u being a chi-squared random variable with one degree of freedom and $k_{\gamma,1}$ is the γ quantile of the same distribution.

The robust filter $r_c(\cdot)$ in the BVT estimator is defined as in Eq. (15). However, alternative filters can also be used, see, for instance, [82] who use the BVT estimator with $r_c(\cdot)$ defined as in Eq. (17).

Reference [82] reports that, in general, the QML estimator has the best performance in the absence of outliers reporting the smallest bias and RMSE. However, this same estimator has the worst performance when outliers are present in the series. Estimators such as QMLt, BM, BQMLt, and MT report an improvement in comparison with the classic QML estimator. Nevertheless, the BVT estimator presents the best performance for series contaminated by isolated or two consecutive outliers.

6.2 Volatility Estimation

In Sect. 6.1 we saw different estimators which are meant to be robust to additive outliers. Different estimators use different volatility equations, which can be divided into two groups: standard volatility equations and robust volatility equations. In the first group no procedure is used to mitigate the propagation of the effect of outliers on the estimation of the volatility, whereas the other group incorporates robust filters in the volatility equation to mitigate the outlier effects. Thus, replacing the parameters by the respective estimated values, the estimated volatility can be obtained, for instance, through

$$\hat{\sigma}_t^2 = \hat{\omega} + \hat{\alpha} r_{t-1}^2 + \hat{\beta}\hat{\sigma}_{t-1}^2, \tag{22a}$$

$$\hat{\sigma}_t^2 = \hat{\omega} + \hat{\alpha} r_c \left(\frac{r_{t-1}^2}{\hat{\sigma}_{t-1}^2} \right) \hat{\sigma}_{t-1}^2 + \hat{\beta}\hat{\sigma}_{t-1}^2, \tag{22b}$$

$$\hat{\sigma}_t^2 = \hat{\omega} + \hat{\alpha} c_\gamma r_c \left(\frac{r_{t-1}^2}{\hat{\sigma}_{t-1}^2} \right) \hat{\sigma}_{t-1}^2 + \hat{\beta}\hat{\sigma}_{t-1}^2, \tag{22c}$$

where Eq. (22a) is the standard volatility and Eqs. (22b) and (22c) are robust volatility equations. The robust volatility Eq. (22b) is used in [68] and [18], whereas the robust volatility Eq. (22c) is used in [15]. This procedure has been extended to a leverage effect context by [60].

Some robust procedures are just robust for the parameter estimates and not for the volatility equation. In these cases, the in-sample volatilities are strongly affected by outliers. In particular it affects the volatility estimate at $t = T$, whose effect is carried out in the prediction. In general, as reported in [82] and [84], estimating the volatility using the robust filter Eq. (17) results in the best performance.

6.3 Correlation Estimation

Reference [15] proposes a robust estimator of the cDCC models in two stages. In the first stage, the volatilities of each univariate series are estimated through the procedure described in Sect. 6.1.5. The second stage estimates the correlation matrix using the residuals obtained in the first stage. The unconditional correlation matrix S is calculated in a local window around v_t as

$$RC = \frac{c_{0.95,p}}{\sum_{t=1}^{T} L_t} \sum_{t=1}^{T} v_t v_t' L_t, \tag{23}$$

with $L_t = I(v_t' SC_t^{-1} v_t < \chi_p^2(0.95))$, $SC_t = 2\,sin\left(\frac{\pi}{6} Sp_t\right)$ and $c_{\gamma,p} = \frac{E[u]}{E[min(u,k_{\gamma,p})]}$, where u has a chi-squared distribution with p degrees of freedom and $k_{\gamma,p}$ is the γ quantile of the same distribution. Thus, $\hat{S}$ is given by

$$\hat{S} = diag(RC_{11}^{-1/2}, \ldots, RC_{pp}^{-1/2}) \times RC \times diag(RC_{11}^{-1/2}, \ldots, RC_{pp}^{-1/2}). \tag{24}$$

To estimate the conditional correlation and the parameters a and b in the cDCC specification, [15] use a robust specification given as

$$\begin{aligned} Q_t = {} & (1 - a - b)S + a \times c_{\delta,p} \times r_c(d_{t-1}) \\ & diag(Q_{t-1})^{1/2} v_{t-1} v_{t-1}' diag(Q_{t-1})^{1/2} + bQ_{t-1}, \end{aligned} \tag{25}$$

where $d_{t-1} = v_{t-1}' R_{t-1}^{-1} v_{t-1}$ (squared Mahalanobis distance). The robust filter $r_c(\cdot)$ in Eq. (25) is defined as

$$r_c(x) = \begin{cases} x, & \text{if } x \le c \\ c/x, & \text{if } x > c. \end{cases} \tag{26}$$

Alternatively, following the ideas of [18] and [82], $r_c(\cdot)$ can also be defined as

$$r_c(x) = \begin{cases} x, & \text{if } x \le c \\ E(X)/x, & \text{if } x > c. \end{cases} \tag{27}$$

Then, the robust M-estimator is obtained as

$$\hat{\phi} = \underset{\phi}{\text{argmax}} \left(-\frac{1}{T} \sum_{t=1}^{T} \left[\log\left(det\left(R_t\right)\right) + \sigma_{p,4} \rho\left(d_t\right) \right] \right), \tag{28}$$

where $\rho(x) = -x + \sigma_{p,4} \rho_{t_{p,4}}(\exp(x))$, $\rho_{t_{p,4}}(u) = (p+4)\log\left(1 + \frac{u}{2}\right)$ and $\sigma_{p,4} = \frac{p}{E\left[\rho_{t_{p,\nu}}'(u)u\right]}$ with u being as described previously. The choice of δ and c is based on a convenient tradeoff between efficiency and robustness. Reference [15] uses $\delta = 0.975$ and $c = \chi_p^2(\delta)$. The estimated conditional correlation matrix is obtained by Eq. (25) replacing the parameters by their robust estimates, so that $\hat{R}_t = diag(\hat{Q}_t)^{-1/2} \hat{Q}_t diag(\hat{Q}_t)^{-1/2}$.

Some other approaches to deal with the estimation of the conditional covariance matrix in a robust framework have been proposed by Boudt and Croux [14], Croux et al. [27], Iqbal [57] and Trucíos et al. [83].

6.4 Forecasting

Forecast densities for returns and volatilities are an useful tool in financial econometrics, because as a by-product it is possible to have forecast intervals, which can be used to measure the uncertainty of return or volatilities. This is helpful to obtain risk measures, such as the value-at-risk. For instance, [56] investigate the performance of some robust estimators in the prediction of value-at-risk. In this section, we will focus only on bootstrap procedures to compute forecast densities.

Forecast densities using bootstrap procedures have been shown to be highly affected by additive outlier, see, for instance, [82] and [81]. We present two alternative robust bootstrap procedures to obtain forecast densities and comment on their finite sample properties.

6.4.1 Mancini and Trojani Algorithm

The robust bootstrap procedure proposed by Mancini and Trojani [65] is based on the robust estimator described in Sect. 6.1.4 and also on a robust procedure to estimate the tails of the innovation distribution. Unlike other residual-based bootstrap procedures, this algorithm does not need re-estimation of the parameters in each bootstrap replication. The algorithm is computationally simple, since re-estimation of the parameters is not needed. The main task of the procedure relies on the estimation of the parameters and tails of the innovations. The algorithm can be summarized in the following steps:

- **Step 1:** Estimate the parameters θ by the estimator described in Sect. 6.1.4, $\hat{\theta} = (\hat{\omega}, \hat{\alpha}, \hat{\beta})$, and obtain the standardized residuals $\hat{\varepsilon}_t = \frac{r_t}{\hat{\sigma}_t}$, $t = 1, \ldots, T$.
- **Step 2:** Estimate the parameters of the generalized Pareto distribution, GPD$(\hat{a}_l, \hat{b}_l)$ and GPD$(\hat{a}_u, \hat{b}_u)$, using the 10% smallest/largest standardized, residuals respectively.
- **Step 3:** For $h = 1, \ldots, H$ obtain the bootstrap forecast densities of $r_{T+h|T}$ and $\sigma_{T+h|T}$ repeating B times the recursion

$$\begin{aligned} \hat{\sigma}^{*2}_{T+h|T} &= \hat{\omega} + \hat{\alpha}\hat{r}^{*2}_{T+h-1|T} + \hat{\beta}\hat{\sigma}^{*2}_{T+h-1|T}, \\ \hat{r}^{*}_{T+h|T} &= \varepsilon^{*}_{\mathrm{MT},T+h}\hat{\sigma}^{*}_{T+h|T}, \end{aligned} \tag{29}$$

where $\hat{r}^{*2}_{T|T} = r^2_T$, $\hat{\sigma}^{*2}_{T|T} = \hat{\sigma}^2_T$ and $\varepsilon^{*}_{\mathrm{MT},T+h}$ is a random draw defined in Eq. (30).

The bootstrap residuals used in the algorithm are obtained as

$$\varepsilon^{*}_{\mathrm{MT},t} = \begin{cases} \varepsilon^{*}_t, & \text{if } l \leq \varepsilon^{*}_t \leq u, \\ u + x_{u,t}, & \text{if } \varepsilon^{*}_t > u, \\ l - x_{l,t}, & \text{if } \varepsilon^{*}_t < l, \end{cases} \tag{30}$$

where l and u are the 10th and 90th percentile of the standardized residuals and x_l and x_u are random draws from GPD($\hat{a}_l, \hat{b}_l$) and GPD($\hat{a}_u, \hat{b}_u$).

This procedure was originally proposed as a robust way to estimate the VaR. Moreover, as described in the previous steps, it is also possible to obtain forecast densities for returns and volatilities. However, it is important to note that in this algorithm there is no source of variability in the one-step-ahead forecast because in this algorithm the estimated model is the same for all bootstrap recursions, i.e., the forecast density is a degenerate distribution.

6.4.2 Trucíos, Hotta, and Ruiz Algorithm

The bootstrap procedure proposed by Trucíos et al. [82] is a robustification of the procedure of [69]. It is based on a robust estimator of the parameters and on a robust filter to estimate the volatility in the entire bootstrap procedure. The algorithm can be summarized in the following steps:

- **Step 1:** Estimate the parameters θ by the estimator described in Sect. 6.1.5, $\hat{\theta} = (\hat{\omega}, \hat{\alpha}, \hat{\beta})$, and obtain the corresponding standardized residuals $\hat{\varepsilon}_t = \frac{r_t}{\hat{\sigma}_t}$, $t = 1, \ldots, T$ where $\hat{\sigma}_t$ is obtained by Eq. (22c). Denote by $\hat{F}_\varepsilon$ the empirical distribution of the centered standardized residuals.
- **Step 2:** Generate a bootstrap series r^*. For $t = 1, \ldots, T$,

$$
\begin{aligned}
r_t^* &= \sigma_t^* \varepsilon_t^*, \\
\sigma_{t+1}^{*2} &= \hat{\omega} + \hat{\alpha}\sigma_t^{*2} c_\gamma r_c\left(\frac{r_t^{*2}}{\sigma_t^{*2}}\right) + \hat{\beta}\sigma_t^{*2},
\end{aligned}
\tag{31}
$$

  where ε_t^* are bootstrap extractions from $\hat{F}_\varepsilon$, $\sigma_1^{*2} = \hat{\sigma}_1^2$ and the filter $r_c(\cdot)$ is defined in Eq. (34). Estimate the parameters, $\hat{\theta}^*$ using the same procedure used in Step 1.
- **Step 3:** Obtain h-steps-ahead forecast for returns and volatilities as

$$
\begin{aligned}
\hat{\sigma}_{T+h|T}^{*2} &= \hat{\omega}^* + \hat{\alpha}^*\hat{\sigma}_{T+h-1|T}^{*2} c_\gamma r_c\left(\frac{r_{T+h-1|T}^{*2}}{\hat{\sigma}_{T+h-1|T}^{*2}}\right) + \hat{\beta}^*\hat{\sigma}_{T+h-1|T}^{*2}, \\
\hat{r}_{T+h|T}^* &= \varepsilon_{T+h}^* \hat{\sigma}_{T+h|T}^*,
\end{aligned}
\tag{32}
$$

  for $h = 1, \ldots, H$, and where $\hat{r}_{T|T}^* = r_T$, ε_{T+h}^* are bootstrap extractions from $\hat{F}_\varepsilon$ and $\hat{\sigma}_{T|T}^{*2}$ is obtained using the recursion

$$
\hat{\sigma}_{t|T}^{*2} = \hat{\omega}^* + \hat{\alpha}^*\hat{\sigma}_{t-1|T}^{*2} c_\gamma r_c\left(\frac{r_{t-1}^2}{\hat{\sigma}_{t-1|T}^{*2}}\right) + \hat{\beta}^*\hat{\sigma}_{t-1|T}^{*2},
\tag{33}
$$

for $t = 2, \ldots, T$, $\hat{\sigma}^{*2}_{1|T} = \hat{\sigma}^2_1$ and $r_c(\cdot)$ defined as

$$r_c(x) = \begin{cases} x, & \text{if } x \leq c \\ \varepsilon^{*2}_t, & \text{if } x > c, \end{cases} \tag{34}$$

with $c = 9$ for a tradeoff between robustness and efficiency in the context of Gaussian errors.

- **Step 4:** Repeat steps 2 and 3 B times to obtain B bootstrap replicates $(\hat{r}^{*(1)}_{T+h|T}, \ldots, \hat{r}^{*(B)}_{T+h|T})$ and $(\hat{s}^{*(1)}_{T+h|T}, \ldots, \hat{s}^{*(B)}_{T+h|T})$ for r_{T+h} and s_{T+h}, respectively.

This bootstrap procedure has good finite sample properties in both contaminated and uncontaminated series. This algorithm is computationally more expensive than the MT procedure since re-estimation in each replication is needed. However, the estimator used in this algorithm is faster than the one used in the MT algorithm. Thus, the time to process this algorithm is not comparatively much longer than the processing time of the MT algorithm.

6.4.3 Trucíos, Hotta, and Ruiz Algorithm: Multivariate Version

In this section, we introduce the multivariate version of the algorithm described in the previous subsection. This procedure proposed by Trucíos et al. [84] extends the bootstrap algorithm of [82] in a multivariate way and the bootstrap procedure of [41] in a robust way. The main idea follows [82], and the algorithm is based on a robust estimator for the parameters, volatilities, and correlations and on robust filters for volatilities and correlations. The algorithm is constructed for the dynamic conditional correlation models but the ideas behind it can be used in other multivariate GARCH models. The robust algorithm can be summarized as

- **Step 1:** Estimate the model parameters ($\hat{\psi}$) by the procedure described in Sect. 6.3 and obtain $\hat{\varepsilon}_t = \hat{H}_t^{-1/2} r_t$. Denote the corresponding empirical distribution function by $\hat{F}_{\hat{\varepsilon}}$.
- **Step 2:** Using $\hat{\psi}$ and $\varepsilon^* \sim \hat{F}_{\hat{\varepsilon}}$, generate multivariate bootstrap series r^*_t
- **Step 3:** Fit the cDCC model on r^*_t and obtain $\hat{\psi}^*$.
- **Step 4:** Compute h-steps-ahead bootstrap forecast for returns, volatilities, and correlations by recursion using the bootstrap estimated parameters $\hat{\psi}^*$ and the original multivariate series r_t,
- **Step 5:** Repeat steps 2–4, B times, and compute $(\hat{r}^{*1}_{T+h|T}, \ldots, \hat{r}^{*B}_{T+h|T})$, $(\hat{D}^{*1}_{T+h|T}, \ldots, \hat{D}^{*B}_{T+h|T})$ and $(\hat{R}^{*1}_{T+h|T}, \ldots, \hat{R}^{*B}_{T+h|T})$ where $h = 1, \ldots, H$.

The main difference between this and the algorithm of [41] is that the model parameters are estimated using a robust procedure instead of the classic approach of [2]. The volatility and correlation are estimated using robust filters instead of the standard filters. Basically, standard equations are replaced by

$$\sigma^{*2}_{i,t} = \hat{\omega}_i + \hat{\alpha}_i r^{*2}_{i,t-1} c_\gamma r_c \left(\frac{r^{*2}_{i,t-1}}{\sigma^{*2}_{i,t-1}} \right) + \hat{\beta}_i \sigma^{*2}_{i,t-1}, \tag{35}$$

and

$$\hat{Q}_t^* = (1 - \hat{a}^* - \hat{b}^*)\hat{S}^* + \hat{a}^*$$
$$\left[diag(\hat{Q}_{t-1}^*)^{\frac{1}{2}} \varepsilon_{t-1}^{\prime *} \varepsilon_{t-1}^* diag(\hat{Q}_{t-1}^*)^{\frac{1}{2}}\right] + \hat{b}^* Q_{t-1}^*, \quad (36)$$

respectively, see [15] and [84] for details.

The algorithms of [69] and [65] overestimate the coverage for returns when outliers are present near to the end of the sample period. However, the robust alternative of [82] presents estimated coverage closer to the nominal value than the other algorithms. The advantage of the procedure of [82] is only marginal for uncontaminated series, and all procedures have a good performance in the absence of outliers. For volatilities, the presence of outliers affects directly the construction of the forecast densities. In some cases the failure coverage is almost 100% when the algorithms of [69] and [65] are used. As reported by Trucíos et al. [82], the distortion of the forecast in the presence of outliers is strong and the results are disastrous when outliers appear near the end of the sample period.

7 Conclusion and Final Remarks

The (M)GARCH models are probably the most used to estimate and predict volatility. Estimation and prediction of volatility are very important in many financial applications. One important issue in the application of (M)GARCH models is the frequent presence of outliers in financial time series and their effects in all stages of model application. Because of these drawbacks, there is a huge literature analyzing the effect of outliers and procedures to mitigate this effect. We present some issues involved in the inference in (M)GARCH models in the presence of additive outliers, which are the most important type of outliers found in financial time series. Because there is already an extensive literature on the subject we had to focus on only some issues. We decided to emphasize the presentation of the literature on the effect of the outliers and on robust inference, but mostly on the prediction intervals. This decision is based mostly on the importance of prediction intervals and the fact this problem has been investigated less. Although many papers have been published dealing with outliers, there is still a lot to be done, mainly on MGARCH models and procedures, depending on the final application.

Acknowledgements The first author acknowledges financial support from São Paulo Research Foundation (FAPESP), grants 2013/00506-1 and 2013/22930-0. The second author is also grateful for financial support from FAPESP, grants 2012/09596-0 and 2016/18599-4. Both authors acknowledge the support of the Centre of Applied Research on Econometrics, Finance and Statistics (CAREFS).

References

1. Aggarwal, R., Inclan, C., Leal, R.: Volatility in emerging stock markets. J. Financ. Quant. Anal. **34.1**, 33–55 (1999)
2. Aielli, G.P.: Dynamic conditional correlation: on properties and estimation. J. Bus. Econ. Stat. **31.3**, 282–299 (2013)
3. Almeida, D., Hotta, L. K., Ruiz, E.: MGARCH models: Tradeoff between feasibility and flexibility. Int. J. Forecast. **34.1**, 45–63 (2018)
4. Ané, T., Loredana, U.R., Gambet, J.B., Bouverot, J.: Robust outlier detection for Asia-Pacific stock index returns. J. Int. Finan Markets, Inst. Money **18.4**, 326–343 (2018)
5. Ardelean, V.: Detecting outliers in time series. No. 05/2012. Friedrich-Alexander-Universität Erlangen-Nürnberg, Institut für Wirtschaftspolitik und Quantitative Wirtschaftsforschung (IWQW) (2012)
6. Bahamonde, N., Veiga, H.: A robust closed-form estimator for the GARCH (1, 1) model. J. Stat. Comput. Simul. **86.8**, 1605–1619 (2016)
7. Balke, N. S., Fomby, T. B.: Large shocks, small shocks, and economic fluctuations: Outliers in macroeconomic time series. J. Appl. Econom. **9.2**, 181–200 (1994)
8. Ballester, C., Furió, D.: Effects of renewables on the stylized facts of electricity prices. Renew. Sustain. Energy Rev. **52.1**, 1596–1609 (2015)
9. Bauwens, L., Laurent, S., Rombouts, J.V.K.: Multivariate GARCH models: a survey. J. Appl. Econom. **21.1**, 79–109 (2006)
10. Bollesrlev, T.: Generalized autoregressive conditional heteroskedasticity. J. Econom. **31.3**, 307–327 (1986)
11. Bollesrlev, T.: A conditional heteroskedastic time series model for speculative prices and rates of return. Rev. Econ. Stat. **7.1**, 297–305 (1987)
12. Bollerslev, T.: Modelling the coherence in short-run nominal exchange rates: a multivariate generalized ARCH model. Rev. Econ. Stat. **72.3**, 498–505 (1990)
13. Bollerslev, T., Engle, R. F., Wooldridge, J. M.: A capital asset pricing model with time-varying covariances. J. Political Econ. **96.1**, 116–131 (1988)
14. Boudt, K., Croux, C.: Robust M-estimation of multivariate GARCH models. Comput. Stat. Data Anal. **54.11**, 2459–2469 (2010)
15. Boudt, K., Danielsson, J., Laurent, S.: Robust forecasting of dynamic conditional correlation GARCH models. Int. J. Forecast. **29.2**, 244–257 (2013)
16. Carnero, M., Peña, D., Ruiz, E.: Effects of outliers on the identification and estimation of GARCH models. J. Time Ser. Anal. **28.4**, 471–497 (2007)
17. Carnero, M. A., Peña, D., Ruiz, E.: Estimating and forecasting GARCH volatility in the presence of outliers. Working Papers of the Instituto Valenciano de Investigaciones Económicas, Universidad de La Rioja, Spain (2008)
18. Carnero, M. A., Peña, D., Ruiz, E.: Estimating GARCH volatility in the presence of outliers. Econ. Lett. **114.1**, 86–90 (2012)
19. Carnero, M. A, Perez, A., Ruiz, E.: Identification of asymmetric conditional heteroscedasticity in the presence of outliers. SERIEs. **7.1**, 179–201 (2016)
20. Catalán, B, Trívez, F. J.: Forecasting volatility in GARCH models with additive outliers. Quant. Financ. **7.6**, 591–596 (2007)
21. Charles, A.: Forecasting volatility with outliers in GARCH models. J. Forecast. **27.7**, 551–565 (2008)
22. Charles, A., Darné, O.: Outliers and GARCH models in financial data. Econ. Lett. **86.3**, 347–352 (2005)
23. Chatzikonstanti, V.: Breaks and outliers when modelling the volatility of the US stock market. Appl. Econ **49.46**, 4704–4717 (2017)
24. Chen, B., Gel, Y. R., Balakrishna, N., Abraham, B.: Computationally efficient bootstrap prediction intervals for returns and volatilities in ARCH and GARCH processes. J. Forecast. **30.1**, 51–71 (2011)

25. Chen, C., Liu, L.: Joint estimation of model parameters and outlier effects. J. Am. Stat. Assoc. **88.421**, 284–29 (1993)
26. Crosato, L., Grossi, L.: Correcting outliers in GARCH models: a weighted forward approach. Stat. Pap. https://doi.org/10.1007/s00362-017-0903-y (in press)
27. Croux, Ch., Gelper, S., Mahieu, K.: Robust exponential smoothing of multivariate time series. Comput. Stat. Data Anal. **54.12**, 2999–3006 (2010)
28. Danielsson, J., James, K. R., Valenzuela, M., Zer, I.: Model risk of risk models. J. Financ. Stab. **23.1**, 79–91 (2016)
29. Dark, J., Zhang, X., Qu, N.: Influence diagnostics for multivariate GARCH processes. J. Time Ser. Anal. **31.4**, 278–291 (2010)
30. Doornik, J. A., Ooms, M.: Outlier detection in GARCH models. No. 05-092/4. Amsterdam: Tinbergen Institute (2005)
31. Duchesne, P.: On robust testing for conditional heteroscedasticity in time series models. Comput. Stat. Data Anal. **46.2**, 227–256 (2004)
32. Engle, R. F.: Autoregressive conditional heteroscedasticity with estimates of the variance of United Kingdom inflation. Econometrica **50.4**, 987–1007 (1982)
33. Engle, R. F., Kroner, K. F.: Multivariate simultaneous generalized ARCH. Econom. Theory **11.1**, 122–150 (1995)
34. Engle, R.: Dynamic conditional correlation: A simple class of multivariate generalized autoregressive conditional heteroskedasticity models. J. Bus. Econ. Stat. **20.3**, 339–350 (2002)
35. Engle, R.: *Anticipating Correlations: A New Paradigm for Risk Management.* Princeton University Press. New Jersey (2009)
36. Engle, R. F., Ledoit, O., Wolf, M.: Large dynamic covariance matrices. J. Bus. Econ. Stat. https://doi.org/10.1080/07350015.2017.1345683 (in press)
37. Fernández, M. A. C, Espartero, A. P.: Outliers and misleading leverage effect in asymmetric GARCH-type models. Working Papers Serie AD2018–01. Instituto Valenciano de Investigaciones Económicas, SA (2018)
38. Franq, Ch., Zokoian, J.: GARCH Models: Structure, Statistical Inference and Financial Applications. Ed John Wiley & Sons, (2011)
39. Franses, P. H., Ghijsels, H.: Additive outliers, GARCH and forecasting volatility. Int. J. Forecast. **15.1**, 1–9 (1999)
40. Franses, P. H., Van Dijk, D., Lucas, A.: Short patches of outliers, ARCH and volatility modelling. Appl. Financial Econ. **14.4**, 221–231 (2004)
41. Fresoli, D. E., Ruiz, R.: The uncertainty of conditional returns, volatilities and correlations in DCC models. Comput. Stat. Data Anal. **100.1**, 170–185 (2016)
42. Grané, A., Veiga, H.: Wavelet-based detection of outliers in financial time series. Comput. Stat. Data Anal. **54.11**, 2580–2593 (2010)
43. Grané, A., Veiga, H.: Outliers, GARCH-type models and risk measures: A comparison of several approaches. J. Empir. Financ. **26.1**, 26–40 (2014)
44. Grané, A., Veiga, H., Martín-Barragáan, B.: Additive Level Outliers in Multivariate GARCH Models. In V. Melas, S. Mignani, P. Monari, and L. Salmaso (Eds.), Topics in Statistical Simulation, Volume 114 of Springer Proceedings in Mathematics & Statistics, pp. 247–255. Springer (2014)
45. Grigoletto, M., Lisi, F.: Practical implications of higher moments in risk management. Stat. Methods Appl. **20.4**, 487–506 (2011)
46. Grossi, L., Laurini, F.: Analysis of economic time series: Effects of extremal observations on testing heteroscedastic components. Appl. Stoch. Model. Bus. Ind. **20.2**, 115–130 (2004)
47. Grossi, L., Laurini, F.: A robust forward weighted Lagrange multiplier test for conditional heteroscedasticity. Comput. Stat. Data Anal. **53.6**, 2251–2263 (2009)
48. Hafner, C. M., Reznikova, O.: On the estimation of dynamic conditional correlation models. Comput. Stat. Data Anal. **56.11**, 3533–3545 (2012)
49. Hill, J. B.: Robust estimation and inference for heavy tailed GARCH. Bernoulli **21.3**, 1629–1669 (2015)
50. Hill, J. B., Prokhorov, A.: GEL estimation for heavy-tailed GARCH models with robust empirical likelihood inference. J. Econom. **190.1**, 18–45 (2016)

51. Hoogerheide, L., van Dijk, H. K.: Bayesian forecasting of value at risk and expected shortfall using adaptive importance sampling. Int. J. Forecast.**26.2**, 231–247 (2010)
52. Hotta, L. K.; Tsay, R. S.: Outliers in GARCH processes. In: Bell, W., Hollan, S., McElroy, T. (eds.) Economic Time Series: Modeling and Seasonality, pp. 337–358. CRC Press, Boca Raton (2012)
53. Hotta, L. K.; Zevallos, M.: Test of outliers and influential observations in GARCH models: A review. Estadística **65.184**, 99–119 (2013)
54. Huang, T. H., Wang, Y. H.: The volatility and density prediction performance of alternative GARCH models. J. Forecast. **31.2**, 157–171 (2012)
55. Iqbal, F., Mukherjee, K.: M-estimators for some GARCH-type models; Computation and application. Stat. Comput. **20.4**, 435–445 (2010)
56. Iqbal, F., Mukherjee, K.: A study of Value-at-Risk based on M-estimators of the conditional heteroscedastic models. J. Forecast. **31.5**, 377–390 (2012)
57. Iqbal, F.: Robust Estimation for the Orthogonal GARCH Model. The Manchester School. **81.6**, 904–924 (2013)
58. Jacquier, E., Olson, N. G., Rossi, P. E.: Bayesian analysis of stochastic volatility models. J. Bus. Econ. Stat. **12.4**, 371–380 (1994)
59. Kamranfar, H. Chinipardaz, R., Mansouri, B.: Detecting outliers in GARCH (p, q) models. Commun. Stat. Simul. Comput. **46.10**, 7844–7854 (2017)
60. Laurent, S., Lecourt, Ch., Palm, F. C.: Testing for jumps in conditionally Gaussian ARMA-GARCH models, a robust approach. Comput. Stat. Data Anal. **100.1**, 383–400 (2016)
61. Li, J., Kao, C.: A bounded influence estimation and outlier detection for ARCH/GARCH models with an application to foreign exchange rates. Manuscript. Finance and Insurance group, Northeastern University (2002)
62. Liu, S.: On diagnostics in conditionally heteroskedastic time series models under elliptical distributions. J. Appl. Probab. **41.1**, 393–405 (2004)
63. Lumsdaine, R. L., Ng, S.: Testing for ARCH in the presence of a possibly misspecified conditional mean. J. Econom. **93.2**, 257–279 (1999)
64. Mancini, L., Ronchetti, E., Trojani, F.: Optimal conditionally unbiased bounded-influence inference in dynamic location and scale models. J. Am. Stat. Assoc. **100.470**, 628–641 (2005)
65. Mancini, L., Trojani, F.: Robust value at risk prediction. J. Financ. Econom. **9.2**, 281–313 (2011)
66. Mendes, B. V. D. M.: Assessing the bias of maximum likelihood estimates of contaminated GARCH models. J. Stat. Comput. Simul. **67.4**, 359–376 (2000)
67. Miguel, J. A., Olave, P.: Bootstrapping forecast intervals in ARCH models. Test **8.2**, 345–364 (1999)
68. Muler, N., Yohai, V. J.: Robust estimates for GARCH models. J. Stat. Plan. Inference **138.10**, 2918–2940 (2008)
69. Pascual, L., Romo, J., Ruiz, E.: Bootstrap prediction for returns and volatilities in GARCH models. Comput. Stat. Data Anal. **50.9**, 2293–2312 (2006)
70. Pakel, C., Shephard, N., Sheppard, K., Engle, R. F.: Fitting vast dimensional time-varying covariance models. NYU Working Paper No. FIN-08-009. Available at SSRN: https://ssrn.com/abstract=1354497 (2014)
71. Park, B. J.: An outlier robust GARCH model and forecasting volatility of exchange rate returns. J. Forecast. **21.5**, 381–393 (2002)
72. Rakesh, B., Guirguis, H.: Extreme observations and non-normality in ARCH and GARCH. Int. Rev. Econ. Financ. **16.3**, 332–346 (2007)
73. Raziq, A., Iqbal, F., Talpur, G. H.: Effects of additive outliers on asymmetric GARCH models. Pak J. Statist. **33.1**, 63–74 (2017)
74. Reeves, J. J.: Bootstrap prediction intervals for ARCH models. Int. J. Forecast. **21.2**, 237–248 (2005)
75. Sakata, S., White, H.: High breakdown point conditional dispersion estimation with application to S& P 500 daily returns volatility. Econometrica **66.3**, 529–567 (1998)

76. Silvennoinen, A., Teräsvirta, T.: Multivariate GARCH models in Handbook of Financial Time Series. Ed. Springer, pp. 201–229 (2009)
77. Smith, J. Q., Santos, A. A. F.: Second-order filter distribution approximations for financial time series with extreme outliers. J. Bus. Econ. Stat. **24.3**, 329–337 (2006)
78. Tolvi, J.: The effects of outliers on two nonlinearity tests. Commun. Stat. Simul. Comput. **29.3**, 897–918 (2000)
79. Trívez, F. J., Catalán, B.: Detecting level shifts in ARMA-GARCH (1, 1) Models. J. Appl. Stat. **36.6**, 679–697 (2009)
80. Trucíos, C.: Bootstrap forecast densities in univariate and multivariate volatility models. Ph.D Thesis, University of Campinas (2016)
81. Trucíos, C., Hotta, L. K.: Bootstrap prediction in univariate volatility models with leverage effect. Math. Comput. Simul. **120**, 91–103 (2016)
82. Trucíos, C., Hotta, L. K., Ruiz, E.: Robust bootstrap forecast densities for GARCH returns and volatilities. J. Stat. Comput. Simul. **87.16**, 3152–3174 (2017)
83. Trucíos, C., Hotta, L. K., Pereira, P. L. V.: On the robustness of the principal volatility components. CEQEF Working Paper Series 47 available at SSRN: https://ssrn.com/abstract=3143870 (2018)
84. Trucíos, C., Hotta, L. K., Ruiz, E.: Robust Bootstrap Densities for Dynamic Conditional Correlations: Implications for Portfolio Selection and Value-at-Risk. J. Stat. Comput. Simul. **88.10**, 1976–2000 (2018)
85. Van Dijk, D., Franses, P. H., Lucas, A.: Testing for ARCH in the presence of additive outliers. J. Appl. Econom. **14.5**, 539–562 (1999)
86. Van Hui, Y., Jiang, J.: Robust modelling of DTARCH models. Econom. J. **8.2**, 143–158 (2005)
87. Veiga, H., Martín-Barragán, B., Grané, A.: Outliers in Multivariate GARCH Models: Effects and Detection. UC3M Working Paper Statistics and Econometrics Series **14.5** (2014)
88. Verhoeven, P., McAleer, M.: Modelling outliers and extreme observations for ARMA-GARCH processes. Working Paper, University of Western Australia (2000)
89. Vrontos, I. D., Dellaportas, P., Politis, D. N.: Full Bayesian inference for GARCH and EGARCH models. J. Bus. Econ. Stat. **18.2**, 187–198 (2000)
90. Welsch, R. E., Zhou, X.: Application of robust statistics to asset allocation models. Revstat Stat. J. **5.1**, 97–114 (2007)
91. Zevallos, M., Hotta, L. K.: Influential observations in GARCH models. J. Stat. Comput. Simul. **82.11**, 1571–1589 (2012)
92. Zhang, X.: Assessment of local influence in GARCH processes. J. Time Ser. Anal. **25.2**, 301–313 (2004)
93. Zhang, X., King, M. L.: Influence diagnostics in generalized autoregressive conditional heteroscedasticity processes. J. Bus. Econ. Stat. **23.1**, 118–129 (2005)

Notes on Newton's Method After 1960

José Mario Martínez

Abstract Some Newtonian ideas will be reported with respect to research areas that emerged in numerical Mathematics after, approximately, 1960. For the problems of solving nonlinear equations and unconstrained optimization, Quasi-Newton methods, which stayed in the mainstream of numerical optimization for more than 30 years, will be motivated and discussed. The topic of complexity in unconstrained optimization will be introduced and some fundamental results will be rigorously proved. Newtonian algorithmic schemes in Linear Programming, which emerged after 1984 and presently represent competitive alternatives for large-scale problems, will be commented. Finally, surprising negative results concerning the capacity of Newton's method to detect approximate solutions of constrained optimization problems will be reported.

1 Introduction

Every undergraduate student of Mathematics, Physics, or Engineering learns that Newton's method is a powerful tool for solving equations and that this method converges very fast if the initial approximation is reasonably close to the solution. Moreover, in most situations, fast convergence occurs even if the initial approximation is poor. Fast convergence means, in general, "quadratic" convergence, a property that guarantees that the number of correct digits at some iteration approximately doubles the corresponding number at the previous one. Later, students learn that there are many methods in numerical mathematics that are called "Newtonian" or, simply, "Newton" for solving different practical problems. Popular knowledge about these methods guarantee that they all are "good," in the sense that they are very fast close to the solutions and that enjoy other theoretical and practical excellent properties. More recently, it became accepted the idea that, not only "every Newton

J. M. Martínez (✉)
Institute of Mathematics, Statistics and Scientific Computing, University of Campinas, Campinas, SP, Brazil
e-mail: martinez@ime.unicamp.br

C. Lavor, F. A. M. Gomes (eds.), *Advances in Mathematics and Applications*,
https://doi.org/10.1007/978-3-319-94015-1_9

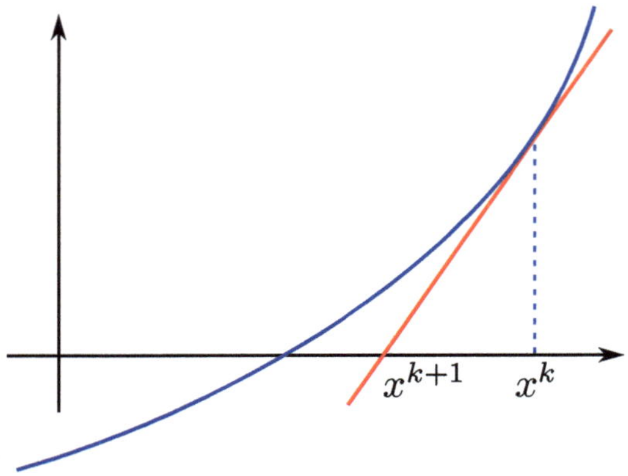

Fig. 1 Newton's iteration for solving a scalar equation

is good" but, also, "every good is Newton," because several methods that were well known as being effective and having nice convergence properties were shown to be, in one sense or another, versions of Newton's method [22].

Newton's idea is the following: Given a complicated problem and some approximation to its solution, one builds a simpler and solvable problem and we postulate that its solution is a better approximation to the solution of the original problem than the previously computed approximation. The simpler problem is built using information available at the current approximation.

The Newtonian paradigm can be applied to many problems, even nonmathematical ones. The most simple case consists of finding a solution of a scalar equation $g(x) = 0$. If x^k is an approximate solution, the presumably better approximation x^{k+1} is obtained by solving (when possible) the linear equation $g(x^k) + g'(x^k)(x - x^k) = 0$. See Fig. 1.

In the same way, we define the Newtonian iteration when the problem is to solve a nonlinear system of equations $g(x) = 0$, where $g : \mathbb{R}^n \to \mathbb{R}^n$. In this case, the Jacobian $g'(x^k)$ is an $n \times n$ matrix and finding the new iterate involves the solution of an $n \times n$ linear system of equations.

One of the most popular applications of solving nonlinear systems comes from the unconstrained minimization of scalar functions. If $f : \mathbb{R}^n \to \mathbb{R}$, $g = \nabla f$ is its gradient, and $H = g' = \nabla^2 f$ is its Hessian, the iterate defined by the solution of $g(x^k) + g'(x^k)(x - x^k) = 0$ defines a stationary point (perhaps a minimizer) of the quadratic approximation $f(x^k) + \langle g(x^k), x - x^k \rangle + \frac{1}{2}(x - x^k)^T H(x^k)(x - x^k)$.

Thus, the simple problem that corresponds to the minimization of an n-dimensional scalar function f consists of minimizing a quadratic function, a problem that, in turn, is roughly equivalent to solving a linear system of equations.

We use to say that the simple problem associated with every iteration of a Newtonian method is a Model of the original problem. In unconstrained optimization, Newton deals with quadratic models, although different interpretations are possible, as we will see later. Before 1960, it was believed that minimizing functions with more than 10 variables employing Newton's method was very hard because of complications solving "big" linear systems and the computation of second derivatives. These complications motivated the upraise of the quasi-Newton age, as we will see in Sect. 2.

2 Quasi-Newton Age

The quasi-Newton age arose around 1960 associated to the unconstrained minimization problem:

$$\text{Minimize } f(x), \tag{1}$$

where $f : \mathbb{R}^n \rightarrow \mathbb{R}$. The Steepest Descent method (or Cauchy's method, or Gradient method [10]) for solving (1) proceeds, at each iteration, computing the gradient $g(x^k)$ and performing a "line search" along that direction with the aim of obtaining a better approximate solution:

$$x^{k+1} = x^k - t_k g(x^k), \tag{2}$$

where $t_k > 0$ is such that, at least, $f(x^{k+1}) < f(x^k)$. If $g(x^k)$ does not vanish, this condition is always verified if t_k is small enough because the direction $-g(x^k)$ is a "descent direction." Many alternatives exist for deciding the most convenient value of t_k. Cauchy's method is easy to implement and relatively cheap since performing one iteration only needs computation of function values and a gradient (no Hessians), while linear algebra calculations associated with (2) are trivial. Moreover, memory requirements for the implementation of (2) are minimal.

However, the sequences generated by the Cauchy method usually converge to stationary points of (1) (points where $g(x) = 0$) very slowly. This is because Cauchy's method reflects a "greedy" way of taking decisions. According to the steepest descent point of view, the decision maker stays at x^k, verifies the characteristics of its problem in a very small neighborhood of the present approximation, and takes a decision based only on such "myopic" observation. Of course, problems do not behave far from the actual approximation in the same way as they do close to it. For this reason, the number of iterations needed to achieve good solutions could be unacceptably large.

On the other hand, Newton's method seems to work in a very different way. Instead of taking a quick decision based on local considerations, Newton "stops to think" about the choice of a good model that, perhaps, should reflect problem features in a smarter way. This model will correspond to the minimization of the quadratic that coincides with the objective function f up to its second derivatives (not only the first ones). The consequence is that, in general, the goal of obtaining very good approximate solutions in a small number of iterations is achieved but, on the other hand, the computational effort to perform an iteration is considerably bigger than the one required by Cauchy.

The dream of the pioneers of the quasi-Newton age [13, 18] was to devise algorithms for solving (1) that, at the first iterations, behave as Cauchy and, at the end, behave as Newton. The rationale behind this idea is that, at the beginning, when we are probably far from the solution, there is no reason to lose a lot of time building or solving a good model, whereas, close to the solution (at the end), the Newtonian model reflects very accurately the original problem and, therefore, Newtonian iterations produce very fast approximations. (The local convergence properties of Newton's methods for solving nonlinear systems were known several decades ago.) On the other hand, quasi-Newton iterations should involve considerably less computational effort than Newton steps.

By (2), the gradient method with line searches takes the form

$$x^{k+1} = x^k - t_k H_k g(x^k) \tag{3}$$

with $H_k = I$ (the Identity matrix) for all $k \in \mathbb{N}$. Moreover, a line-search version of Newton's method also has the form (3) with $H_k = \nabla^2 f(x^k)^{-1}$. Should it be possible to devise a method of the form (3) in which $H_0 = I$ and $H_k \approx \nabla^2 f(x^k)^{-1}$ for k large with moderate computational cost per iteration? Methods with such purpose were ultimately called "quasi-Newton methods" and their development and analysis dominated mainstream research in computational optimization for more than three decades.

By the Mean Value Theorem, we have that:

$$\left[\int_0^1 \nabla^2 f(x^k + ts^k)dt\right]s^k = y^k, \tag{4}$$

where

$$s^k = x^{k+1} - x^k \text{ and } y^k = g(x^{k+1}) - g(x^k).$$

Then, since the matrix $[\int_0^1 \nabla^2 f(x^k + ts^k)dt]$ is an average of the Hessians of f in the segment $[x^k, x^{k+1}]$, it turns out that the Hessian $\nabla^2 f(x^{k+1})$ approximately satisfies the "secant equation"

$$Bs^k = y^k. \tag{5}$$

The secant system has n equations and n^2 unknowns (the entries of B). The number of unknowns can be reduced to $(n + 1)n/2$ considering that Hessians are symmetric matrices and (5) defines an affine subspace in the space of matrices. If B_k is an approximation to $\nabla^2 f(x^k)$, it is natural to define B_{k+1}, the approximation to $\nabla^2 f(x^{k+1})$, as some kind of projection of B_k on the affine subspace defined by the secant equation. For example, the BFGS method, which is the most popular quasi-Newton method for unconstrained minimization, is defined by:

$$x^{k+1} = x^k - t_k B_k^{-1} g(x^k)$$

and

$$B_{k+1} = B_k + \frac{y^k (y^k)^T}{(y^k)^T s^k} - \frac{B_k s^k (s^k)^T B_k}{(s^k)^T B_k s^k}. \tag{6}$$

The interpretation of (6) as a (variable with respect to x^k) projection on the set of solutions of (5) may be found in the classical book by Dennis and Schnabel [15].

The BFGS method may be defined without explicitly mentioning the inverse of any matrix, since the inverse of B_{k+1} in (6) can be computed in terms of the inverse of B_k by means of a judicious application of the Sherman–Morrison formula [19].

The line-search parameter t_k is used to guarantee sufficient descent of $f(x^{k+1})$ with respect to $f(x^k)$. Algorithms for choosing t_k differ in degrees of sophistication and cause different numerical behaviors of the methods so far implemented.

Quasi-Newton methods were generalized to solving arbitrary nonlinear systems of equations by Broyden [5] and many followers. Generalizations include taking advantage of specific structures (for example, nonlinear least squares problems), using sparsity patterns of Hessians or Jacobians, direct updates of factorizations [25], nonlinear systems coming from constrained optimization, and many others.

Roughly speaking, as one can expect quadratic local convergence from Newton's method, superlinear convergence is usually observed, and many times proved, in quasi-Newton algorithms. The pioneers' project of devising methods that smoothly evolve from Cauchy behavior to Newtonian behavior was only partially successful. Most practitioners believe that, when it is affordable to use Newton in unconstrained optimization or nonlinear systems, the Newton alternative is more efficient than quasi-Newton ones. The motivation for quasi-Newton methods decreased with the development of algorithmic differentiation [21], sparse matrix techniques [16], and the use of iterative methods for solving the Newtonian linear equation [14]. However, quasi-Newton ideas emerge frequently in modern optimization in combination with new techniques for multiobjective problems, equilibrium problems, constrained and nonsmooth optimization, and many others.

3 Linear Programming

Linear Programming is the problem of minimizing a linear function subject to linear inequalities and equalities. Every Linear Programming problem can be reduced to the Standard Form:

$$\text{Minimize } c^T x \text{ subject to } Ax = b \text{ and } x \geq 0. \tag{7}$$

A point $x \in \mathbb{R}^n$ is a solution of (7) if and only if it satisfies the KKT conditions:

$$c + A^T y - z = 0, x_j z_j = 0 \text{ for all } j = 1, \ldots, n, \tag{8}$$

for some $y \in \mathbb{R}^m$ and $z \geq 0$, together with the feasibility conditions

$$Ax = b, x \geq 0. \tag{9}$$

Moreover, if the Linear Programming problem has a solution, then one of its solutions is a vertex of the polytope defined by (9). See [28] and many other textbooks.

The latter property motivates the best known method for solving Linear Programming problems: The Simplex Method, invented by George Dantzig in 1949 [12], proceeds visiting vertices of the polytope (9) always reducing the objective function value. Since the number of vertices is finite, the Simplex Method finds a solution of (7), when such a solution exists, in a finite number of steps.

The Simplex Method was the standard procedure for solving Linear Programming problems until 1984 and, perhaps, still is. However, at least from the theoretical point of view, this method has a drawback: In the worst case, it may need to visit all the vertices of a polytope for finding a solution and, since the number of vertices grows exponentially with the number of variables, the computer time needed to solve a large problem may be, in the worst case, unaffordable. This drawback motivated, in 1979, the introduction of a new method by Khachiyan [24] who showed that a solution with arbitrary chosen precision can be found in polynomial time. However, Khachiyan's method was shown very soon to be ineffective in practical computations.

In 1984, Karmarkar [23] introduced a new method for Linear Programming, enjoying similar convergence properties as Khachiyan's method, for which he claimed that, especially for large problems, the performance was orders of magnitude better than the performance of the Simplex algorithm. His results and claims attracted the attention of the whole optimization community. Karmarkar's method, whose practical performance could not be reproduced by independent experiments, introduced new ideas, as projective transformations, approximation by means of interior points, and potential functions, that seemed to be in the kernel of polynomiality proofs and practical performance. Later, it was verified that the only new idea that was crucial both for proofs and for practical behavior was the interiority of the sequence of iterates generated by the method. See [20].

Independently of the eventual discard of the original Karmarkar's method, his work had the merit of motivating a lot of fruitful research that showed that challenging alternatives to the Simplex method may exist. Ultimately, the challenge of the so-called Interior Point methods motivated an enormous improvement in Simplex implementations.

Modern descriptions of Interior Point methods are closely related to the Newton paradigm. In fact, from (8) and (9) we may extract the nonlinear system of equations:

$$c + A^T y - z = 0, Ax = b, x_j z_j = 0 \text{ for all } j = 1, \ldots, n. \tag{10}$$

If (x, y, z) is a solution of (10) such that $x \geq 0$ and $z \geq 0$, we have that x is a solution of (7).

But, (10) is a nonlinear system of equations with $2n+m$ equations and unknowns, then using Newton's method is an interesting alternative for its solution. On the other hand, we are interested only in solutions such that $x \geq 0, z \geq 0$, which justifies the decision of starting with $x^0 > 0, z^0 > 0$ and to maintain $x^k > 0, z^k > 0$ throughout the calculations. It is not recommendable to admit $x_j^k = 0$ or $z_j^k = 0$ because, in this case, Newton's method would maintain $x_j^{k+1} = 0$ (or $z_j^{k+1} = 0$) for all k. Therefore, the pure Newtonian iterations must be modified in order to prefer the positivity (interiority) of x^k and z^k. This is usually done by means of the introduction of (close to 1) damping parameters. Namely, if (x^k, y^k, z^k) (with $x^k > 0$ and $z^k > 0$) is the current iteration and (d_x, d_y, d_z) is the increment computed by one iteration of Newton's method for solving the nonlinear system (10), we will compute:

$$(x^{k+1}, y^{k+1}, z^{k+1}) = (x^k + \theta_k d_x, y^k + \theta_k d_y, z^k + \theta_k d_z),$$

in such a way that the new iterate remains interior and the difference with respect to the pure Newton iterate is cautiously small.

The procedure described above is called Primal-Dual Affine-Scaling method. In Newtonian terms, this is a damped Newton method that preserves interiority. This method behaves well except when some variable x_j (or z_j) becomes close to zero when it must be positive at the solution. In order to understand the best succeeded procedures for improving the robustness of the primal-dual affine-scaling method, let us assume first that (x^k, y^k, z^k) is such that

$$c + A^T y^k - z^k = 0, Ax^k = b, x^k > 0, \text{ and } z^k > 0.$$

Clearly, (x^k, y^k, z^k) is a solution of the nonlinear system

$$c + A^T y - z = 0, Ax = b, x_j z_j = x_j^k z_j^k, j = 1, \ldots, n.$$

This means that we already know a solution of the system

$$c + A^T y - z = 0, Ax = b, x_j z_j = t x_j^k z_j^k, j = 1, \ldots, n \tag{11}$$

with $x > 0, z > 0$, for $t = 1$, whereas we wish a solution for $t = 0$. The Primal-Dual Affine-Scaling (Newton) step is an aggressive attempt of achieving the solution for $t = 0$. If this attempt is considered to be unsuccessful (for some more or less theoretical justified criterion), the natural procedure is to try a less ambitious value of $t > 0$. For approximating the solution of (11) for the new value of t, starting from an iterate (x^k, y^k, z^k), a Newton-like iteration is also employed that may use the same matrix factorization as the one employed for finding the Primal-Dual Affine-Scaling step. Variations of this idea define the best succeeded modern Interior Point methods for Linear Programming. It is remarkable that a problem traditionally solved by means of a combinatorial procedure as Simplex, later challenged by the nonstandard ideas of Khachiyan and Karmarkar, eventually found in the Newton paradigm one of the most promising solution tools for many difficult, especially large-scale, situations.

4 Convergence and Complexity in Unconstrained Optimization

Numerical methods for solving general continuous optimization problems are iterative. Since finding global minimizers without employing the specific structure of the problems is very difficult, we generally rely on methods that guarantee convergence to points that satisfy necessary optimality conditions (hopefully, local minimizers). Classical convergence theories analyze the sequences generated by optimization methods and prove that the sequence of gradients tend to zero or that the gradient vanishes at limit points. These "global" theories say nothing about the speed of convergence. Many times they are complemented with "capture theorems" that say that, when an iterate is close enough to a local minimizer with good properties, convergence to the local minimizer takes place with satisfactory convergence rate.

Only recently, it has been considered to be relevant to compute bounds for the computer effort that is necessary to achieve a predetermined precision ε. For example, if one assumes that "precision ε" means that the norm of the gradient is smaller than ε, the question is about the number of iterations and function-gradient evaluations that are necessary to achieve such precision, as a function of ε, the functional value at the initial point, characteristics of the problem, and parameters of the method.

Being a bit more formal than in the previous sections, we will assume here that $f : \mathbb{R}^n \to \mathbb{R}$ has continuous first derivatives $g(x) = \nabla f(x)$ and that a Lipschitz inequality for the gradient holds. As a consequence, by Elementary Calculus, there exists $\gamma > 0$ such that

$$f(x+s) \leq f(x) + g(x)^T s + \gamma \|s\|^2. \tag{12}$$

This assumption is "slightly" weaker than saying that f has bounded second derivatives on $\mathbb{R}^n$. We are going to analyze the worst-case complexity of a version of Cauchy's method, with the aim of relating this analysis with analogous analyses concerning Newton's method.

There is a reason for considering that Cauchy's method is also a Newton-like method: In Newton, for minimizing functions, we use to say that the objective function is approximated, locally, by a quadratic model. Analogously, in Cauchy we may think that we approximate the objective function, locally, by a linear model.

Equivalently, in Newton we approximate the gradient by a linear model whereas in Cauchy we approximate the gradient by a constant vector, namely, the gradient at the current iterate.

The difficulty in this point of view about Cauchy is that, in general, linear functions do not admit minimizers. Therefore, the "Newtonian subproblem" cannot be solved. We will fix this inconvenience observing that, although the linear model $g(x^k)^T(x - x^k)$ does not have a minimizer, the "regularized" version of this model: $g(x^k)^T(x - x^k) + \frac{\rho}{2}\|x - x^k\|^2$ has a unique solution independently of the value of the regularizing parameter $\rho > 0$. Moreover, if we choose $\|\cdot\|$ as the Euclidian norm $\|\cdot\|_2$, the minimizer of $g(x^k)^T(x - x^k) + \frac{\rho}{2}\|x - x^k\|^2$ is given by $x = x^k - \frac{1}{\rho}g(x^k)$. This idea is formalized in the following algorithm.

Algorithm 4.1
Let $x^0 \in \mathbb{R}^n$ and $\alpha > 0$ be given. Initialize $k \leftarrow 0$.

Step 1 Set $\rho \leftarrow 1/2$.
Step 2 Solve the subproblem

$$\text{Minimize } g(x^k)^T s + \rho\|s\|^2,$$

obtaining the solution s^{trial}. (Note that $s^{trial} = -\frac{1}{2\rho}g(x^k)$ if $\|\cdot\| = \|\cdot\|_2$.)
Step 3 (Test the sufficient descent condition)

If

$$f(x^k + s^{trial}) \leq f(x^k) - \alpha\|s^{trial}\|^2, \tag{13}$$

set $s^k = s^{trial}$, $x^{k+1} = x^k + s^k$, $k \leftarrow k + 1$, and go to Step 1.

Otherwise, set $\rho \leftarrow 2\rho$ and go to Step 2.

When $\|\cdot\|$ is the Euclidian norm, Algorithm 4.1 is Cauchy's method with the most simple line-search procedure (backtracking dividing the trial step by 2) and the clothes of regularization.

Lemma 4.1 *If $\rho \geq \gamma + \alpha$, the sufficient descent condition (13) is fulfilled.*

Proof By (12), the hypothesis of this lemma, and Step 2 of the algorithm,

$$\begin{aligned} f(x^k + s) &\leq f(x^k) + g(x^k)^T s + \gamma\|s\|^2 \\ &= f(x^k) + g(x^k)^T s + (\gamma + \alpha)\|s\|^2 - \alpha\|s\|^2 \\ &\leq f(x^k) + g(x^k)^T s + \rho\|s\|^2 - \alpha\|s\|^2 \leq f(x^k) - \alpha\|s\|^2. \end{aligned}$$

This completes the proof. □

By Lemma 4.1, the first term in the sequence $\{1/2, 1, 2, 4, 8\ldots\}$ bigger than $\gamma + \alpha$ necessarily defines a value of ρ for which (13) holds. As a consequence, the following corollary holds.

Corollary 4.1 *At each iteration of Algorithm 4.1, after a maximum of* $1+\log_2(\gamma+\alpha)$ *tests (backtrackings, functional evaluations) we necessarily obtain the descent condition and the final* ρ *for which (13) holds and satisfies:*

$$\rho < 2(\gamma+\alpha).$$

For the sake of simplicity, assume now that $\|\cdot\| = \|\cdot\|_2$. Then, $s^{trial} = -\frac{1}{2\rho} g(x^k)$ and, so, by Corollary 4.1, $\|s^k\| \geq \frac{1}{4(\gamma+\alpha)} \|g(x^k)\|$. Then, by the sufficient descent condition,

$$f(x^{k+1}) \leq f(x^k) - \frac{\alpha}{16(\gamma+\alpha)^2} \|g(x^k)\|^2.$$

Then, if $\|g(x^k)\| \geq \varepsilon$,

$$f(x^{k+1}) \leq f(x^k) - \frac{\alpha}{16(\gamma+\alpha)^2} \varepsilon^2. \tag{14}$$

Assume that $f_{target} < f(x^0)$ is arbitrary. Then, (14) implies that the number of iterations at which $\|g(x^k)\| \geq \varepsilon$ and $f(x^k) > f_{target}$ is bounded by:

$$[f(x^0) - f_{target}] \frac{16(\gamma+\alpha)^2}{\alpha} \varepsilon^{-2}. \tag{15}$$

Thus, by Corollary 4.1, the number of evaluations is bounded by:

$$[f(x^0) - f_{target}][1 + \log_2(\gamma+\alpha)] \frac{16(\gamma+\alpha)^2}{\alpha} \varepsilon^{-2}. \tag{16}$$

Both expressions (15) and (16) have the form $c\varepsilon^{-2}$, where c is a constant that only depends on characteristics of the problem (γ), parameters of the algorithm (α), the initial point x^0, and, of course, the target with respect to which we desired to estimate the computational effort. The dependence on the precision required is represented by ε^{-2}. For this reason, we generally say that the complexity of the algorithm is $O(\varepsilon^{-2})$.

"Gradient-related" methods for unconstrained optimization are characterized by the generation of directions d^k that are related to $g(x^k)$ by means of angle and relative-size conditions as:

$$g(x^k)^T d^k \leq -\theta \|g(x^k)\|_2 \|d^k\|_2 \text{ and } \|d^k\| \geq \beta \|g(x^k)\|, \tag{17}$$

where $\theta \in (0, 1)$ and $\beta > 0$ are algorithmic parameters. These conditions are sufficient to show that gradient-related methods have complexity $O(\varepsilon^{-2})$.

Quasi-Newton methods (as BFGS) also enjoy the worst-case complexity $O(\varepsilon^{-2})$ when conveniently safeguarded in order to satisfy gradient-related conditions. The standard BFGS method (without safeguards) does not satisfy such property. In fact, there exist counterexamples that show that all the limit points generated by this popular method may be such that the norm of the gradient does not vanish at all [11, 26]. Of course, this prevents the possibility of satisfactory complexity results.

Newton's method with line searches may also generate sequences with associated gradients that are bounded away from zero [27]. This cannot happen when Newton's method is coupled with a "trust-region" strategy, which guarantees that every limit point is stationary. In spite of this, it has been shown that even Newton's method with the robust trust-region strategy has worst-case complexity not better than $O(\varepsilon^{-2})$ [6].

It is disappointing that, with Newton's method plus traditional globalization procedures, a complexity better than $O(\varepsilon^{-2})$ cannot be obtained. Fortunately, the reason is that the "traditional globalization procedures" *are not* the natural globalization procedures that should be used for Newton. Mimicking the complexity proof given for Cauchy, we will show that a better complexity result may be obtained for Newton, if one replaces quadratic regularization with cubic regularization and quadratic sufficient descent with cubic convergence descent with respect to $\|s^{trial}\|$. Analogous results with variations with respect to the sufficient descent criterion were given in [3, 7, 29].

In order to define Algorithm 4.2, assume that the Hessian $\nabla^2 f(x)$ exists for all $x \in \mathbb{R}^n$.

Algorithm 4.2
Let $x^0 \in \mathbb{R}^n$ and $\alpha > 0$ be given. Initialize $k \leftarrow 0$.

Step 1 Set $\rho \leftarrow 0$.
Step 2 Solve the subproblem

$$\text{Minimize } g(x^k)^T s + \frac{1}{2} s^T \nabla^2 f(x^k) s + \rho \|s\|^3,$$

obtaining the solution s^{trial}. If the subproblem has no solution (which may occur only if $\rho = 0$), reset $\rho \leftarrow 1$ and repeat Step 2.

Step 3 (Test the sufficient descent condition)

If

$$f(x^k + s^{trial}) \le f(x^k) - \alpha \|s^{trial}\|^3, \tag{18}$$

set $s^k = s^{trial}$, $x^{k+1} = x^k + s^k$, $k \leftarrow k + 1$, and go to Step 1.

Otherwise, set $\rho \leftarrow 2\rho$ and go to Step 2.

The complexity proof for Algorithm 4.2 needs to assume that the Hessian $\nabla^2 f$ is Lipschitz continuous for all $x \in \mathbb{R}^n$. This implies, as in the case of (12), that there exists $\gamma_2 > 0$ such that, for all $x, s \in \mathbb{R}^n$,

$$f(x + s) \le f(x) + g(x)^T s + \frac{1}{2} s^T \nabla^2 f(x) s + \gamma_2 \|s\|^3 \tag{19}$$

and

$$\|g(x+s)\| \le \|g(x) + \nabla^2 f(x)s\| + \gamma_2\|s\|^2. \tag{20}$$

Lemma 4.2 below is entirely analogous to Lemma 4.1.

Lemma 4.2 *If $\rho \ge \gamma_2 + \alpha$, the sufficient descent condition (18) is fulfilled.*

Proof By (19), the hypothesis of this lemma, and Step 2 of the algorithm,

$$\begin{aligned} f(x^k + s) &\le f(x^k) + g(x^k)^T s + \frac{1}{2} s^T \nabla^2 f(x^k) s + \gamma_2 \|s\|^3 \\ &= f(x^k) + g(x^k)^T s + \frac{1}{2} s^T \nabla^2 f(x^k) s + (\gamma_2 + \alpha)\|s\|^3 - \alpha\|s\|^3 \\ &\le f(x^k) + g(x^k)^T s + \frac{1}{2} s^T \nabla^2 f(x^k) s + \rho\|s\|^3 - \alpha\|s\|^3 \\ &\le f(x^k) - \alpha\|s\|^3. \end{aligned}$$

This completes the proof. □

Moreover, as in Corollary 4.1, we have:

Corollary 4.2 *At each iteration of Algorithm 4.2, after a maximum of $1+\log_2(\gamma_2+\alpha)$ tests we necessarily obtain the descent condition (18) with*

$$\rho < 2(\gamma_2 + \alpha).$$

By Corollary 4.2, after computer time that only depends on γ_2 (characteristic of the problem) and α (characteristic of the algorithm), we obtain a decrease at least $\alpha\|s^{trial}\|^3$. Recall that in Algorithm 4.1 the corresponding decrease was $\alpha\|s^{trial}\|^2$. In Algorithm 4.1, our proof finished showing that $\|s^k\|$ was bigger than a multiple of $\|g(x^k)\|$. Here, we will show that $\|s^k\|$ is bigger than a multiple of $\|g(x^k+s^k)\|^{\frac{1}{2}}$. In other words, we will prove that $\|g(x^k+s^k)\|$ is smaller than a multiple of $\|s^k\|^2$. In fact, by (20),

$$\|g(x^k + s^k)\| \le \|g(x^k) + \nabla^2 f(x^k)s^k\| + \gamma_2\|s^k\|^2.$$

So, assuming, for simplicity, that $\|\cdot\| = \|\cdot\|_2$, using that $\nabla(\|s\|^3) = 3s\|s\|$, and the fact that gradient of the objective function of the subproblem must vanish at s^k, we have that:

$$\begin{aligned} \|g(x^k + s^k)\| &\le \left\|g(x^k) + \nabla^2 f(x^k)s^k + 3\rho s^k\|s^k\|\right\| + 3\rho\|s^k\|^2 + \gamma_2\|s^k\|^2 \\ &= (3\rho + \gamma_2)\|s^k\|^2. \end{aligned}$$

Then, since the final ρ accepted at (18) is smaller than $2(\gamma + \alpha)$,

$$\|g(x^k + s^k)\| \le [6(\gamma_2 + \alpha) + \gamma_2]\|s^k\|^2.$$

Thus,

$$\|s^k\| \ge \frac{\|g(x^{k+1})\|^{\frac{1}{2}}}{\sqrt{[6(\gamma_2 + \alpha) + \gamma_2]}}.$$

By (18), this implies that, at each iteration of Algorithm 4.2,

$$f(x^{k+1}) \le f(x^k) - \alpha \frac{\|g(x^{k+1})\|^{3/2}}{[6(\gamma_2 + \alpha) + \gamma_2]^{3/2}}.$$

Therefore, the number of iterations for which $\|g(x^{k+1})\| \ge \varepsilon$ and $f(x^{k+1}) \ge f_{target}$ is bounded above by $c[f(x^0) - f_{target}]\varepsilon^{-3/2}$, where c is a constant that only depends on γ_2 and α. As in the case of Algorithm 4.1, by Corollary 4.2, this implies that the worst-case complexity of the Newtonian Algorithm 4.2 is $O(\varepsilon^{-3/2})$.

This rather simple result, as several analogous ones [3, 7, 17, 29], confirms the intuition that some version of Newton's method should have better worst-case complexity than every gradient-related method.

A straightforward generalization of Algorithm 4.2 consists of replacing the subproblem with the minimization of the q-th Taylor polynomial plus a regularization of the form $\rho\|s\|^{q+1}$ and replacing (18) with

$$f(x^k + s^{trial}) \le f(x^k) - \alpha\|s^{trial}\|^{q+1}.$$

Assuming Lipschitz conditions on the derivatives of order q and following, mutatis mutandi, the proof for the case $q = 2$, we obtain an algorithm with complexity $O(\varepsilon^{(q+1)/q})$. Slight variations of this algorithm have been given in [3].

5 Newton in Constrained Optimization

The smooth constrained optimization problem consists of minimizing a smooth function $f(x)$ subject to $h(x) = 0$ and $g(x) \le 0$, where $h : \mathbb{R}^n \to \mathbb{R}^m$ and $g : \mathbb{R}^n \to \mathbb{R}^p$ are, also, sufficiently smooth. For simplicity, this section will be restricted to the case in which there are not inequality constraints, although all the arguments apply straightforwardly to the case $p > 0$. Then, the problem considered here is

$$\text{Minimize } f(x) \text{ subject to } h(x) = 0. \tag{21}$$

In unconstrained optimization, it is quite natural to require, as stopping criterion for computer algorithms, the condition $\|\nabla f(x^k)\| \leq \varepsilon$ because $\nabla f(x) = 0$ is a necessary condition for every local minimizer. However, in constrained optimization we have the additional requirement on the feasibility of the approximate solution and, moreover, a computable necessary condition based on the gradients of f and the constraints does not exist. In fact, the Lagrange conditions (called KKT in the presence of inequalities) establish that

$$\nabla f(x) + \sum_{i=1}^{m} \lambda_i \nabla h_i(x) = 0 \tag{22}$$

should hold for suitable multipliers $\lambda \in \mathbb{R}^m$, but these conditions are guaranteed to hold at a minimizer only if such point satisfies a "constraint qualification." For example, the problem of minimizing x subject to $x^2 = 0$ has an obvious global minimizer at $x = 0$, but (22) does not hold.

This inconvenience raises the question about the practical convergence test that should be used in numerical algorithms designed to solve (21). Some authors employ stopping criteria based on "scaled KKT conditions." Instead of requiring that

$$\left\| \nabla f(x) + \sum_{i=1}^{m} \lambda_i^k \nabla h_i(x) \right\| \leq \varepsilon \tag{23}$$

they stop their algorithms when

$$\frac{1}{\max\{1, \|\lambda^k\|_\infty\}} \left\| \nabla f(x) + \sum_{i=1}^{m} \lambda_i^k \nabla h_i(x) \right\| \leq \varepsilon, \tag{24}$$

a weaker condition than (23) that may hold close to a minimizer at which constraint qualifications are not fulfilled [8, 9]. However, (24) may hold in simple problems at points that are arbitrarily far from the solution. For example, consider the problem of minimizing x^2 subject to $0x = 0$ and take $x^k = 10^{20}$. Clearly, (24) holds with $\varepsilon = 10^{-10}$ and $\lambda^k = 2 \times 10^{30}$. Thus, the criterion (23) may be useful to save computer work when convergence of an algorithm is in fact occurring to a correct minimizer but may also lead to incorrect decisions when the iterate is far from a solution.

Fortunately, an interesting result concerning the approximate fulfillment of (22) at local minimizers exists. Although local minimizers may not satisfy (22), they do satisfy the approximate version of this system of equations. By this we mean that, if x^* is a local minimizer, given $\varepsilon > 0$ arbitrary small, there exist $x \in \mathbb{R}^n$ and $\lambda \in \mathbb{R}^m$ such that $\|x - x^*\| \leq \varepsilon$, $\|h(x)\| \leq \varepsilon$, and $\|\nabla f(x) + \sum_{i=1}^{m} \lambda_i \nabla h_i(x)\| \leq \varepsilon$. See [1, 4] and other papers that study Sequential Optimality Conditions.

As a consequence, the following is a well-justified stopping criteria for algorithms that aim to solve (21):

$$\|h(x^k)\| \le \varepsilon, \left\| \nabla f(x^k) + \sum_{i=1}^{m} \lambda_i^k \nabla h_i(x^k) \right\| \le \varepsilon. \tag{25}$$

The natural question that arises is: Given a particular algorithm for solving (21) that converges to a minimizer x^*, is it possible to prove that, for all $\varepsilon > 0$, an iterate x^k, associated with suitable multipliers λ^k, exists? If the answer is positive, the algorithm will eventually stop satisfying (25). It has been proved that this is the case of penalty and Augmented Lagrangian algorithms [4]. Surprisingly, it can be shown in very simple examples that, when Newton's method is applied to the nonlinear system that includes $h(x) = 0$ and (22), the resulting sequence x^k may converge to a minimizer of (21) but the KKT-residual $\|\nabla f(x^k) + \sum_{i=1}^{m} \tilde{\lambda}_i^k \nabla h_i(x^k)\|$ is bounded away from zero independently of the value of $\tilde{\lambda}^k$. This means that, even converging to the solution, Newton's method would never detect that such convergence occurs [2]. Therefore, no complexity results associated with the condition (25) is possible for Newton's method. The example given in [2] consists of minimizing x_1 subject to $\|x\|_2^2 = 0$, with $n \ge 2$. Starting with $\lambda^1 > 0$, the sequence generated by Newton's method converges to the solution $x = 0$ but the norm of the KKT-residual is bounded away from zero (bigger than $(\sqrt{5}-1)/4$ if $n = 2$) for most initial choices of x^0. The simplicity of this example is amazing and suggests that this "failure" of Newton's method might occur frequently in practical problems in which optimality cannot be easily detected by other means.

Acknowledgements I am indebted to Lúcio Tunes dos Santos and Luiz Rafael Santos for revising the first version of this manuscript.

This work was supported by Cepid-Cemeai-Fapesp, PRONEX-CNPq/FAPERJ E-26/111.449/2010-APQ1, FAPESP (grants 2010/10133-0, 2013/03447-6, 2013/05475-7, and 2013/07375-0), and CNPq.

References

1. R. Andreani, G. Haeser, and J. M. Martínez, On sequential optimality conditions for smooth constrained optimization, *Optimization* **60**, pp. 627–641 (2011).
2. R. Andreani, J. M. Martínez, and L. T. Santos, Newton's method may fail to recognize proximity to optimal points in constrained optimization, To appear in *Mathematical Programming*.
3. E. G. Birgin, J. L. Gardenghi, J. M. Martínez, S. A. Santos, and Ph. L. Toint, Worst-case evaluation complexity for unconstrained nonlinear optimization using high-order regularized models, to appear in *Mathematical Programming*.
4. E. G. Birgin and J. M. Martínez, *Practical Augmented Lagrangian Methods for Constrained Optimization*, SIAM Publications, Series: Fundamentals of Algorithms, Philadelphia, 2014.
5. C. G. Broyden, A class of methods for solving nonlinear simultaneous equations, *Mathematics of Computation* **19**, pp. 577–593 (1965).

6. C. Cartis, N. I. M. Gould, and Ph. L. Toint, On the complexity of steepest descent, Newton's and regularized Newton's methods for nonconvex unconstrained optimization, *SIAM Journal on Optimization* **20**, pp. 2833–2852 (2010).
7. C. Cartis, N. I. M. Gould, and Ph. L. Toint, Adaptive cubic regularization methods for unconstrained optimization, *Mathematical Programming* **127**, pp. 245–295 (2011).
8. C. Cartis, N. I. M. Gould, and Ph. L. Toint, On the complexity of finding first-order critical points in constrained nonlinear optimization, *Mathematical Programming* **144**, pp. 93–106 (2014).
9. C. Cartis, N. I. M. Gould, and Ph. L. Toint, Corrigendum: On the complexity of finding first-order critical points in constrained nonlinear optimization, Preprint RAL-P-2014-013 (2014), Rutherford Appleton Laboratory, Chilton, England.
10. A. Cauchy, Méthode générale pour la résolution des systèmes d'équations simultaneés, *Computes Rendus de l'Académie des Sciences* **27**, pp. 536–538 (1847).
11. Y.-H. Dai, Convergence properties for the BFGS algorithm, *SIAM Journal on Optimization* **13**, pp. 693–701 (2002).
12. G. B. Dantzig, *Linear Programming and Extensions*, Princeton University Press, Princeton, N. J., 1963.
13. W. C. Davidon, Variable metric method for minimization, AEC Research and Development Report, Argonne National Laboratory ANL-5990, 1959.
14. R. S. Dembo, S. C. Eisenstat, and T. Steihaug, Inexact Newton Methods, *SIAM Journal on Numerical Analysis* **19**, pp. 400–408 (1982).
15. J. E. Dennis, Jr. and R. B. Schnabel, *Numerical Methods for Unconstrained Optimization and Nonlinear Equations*, volume 16 of *Classics in Applied Mathematics*, SIAM, Philadelphia, 1996.
16. I. S. Duff, A. M. Erisman, and J. K. Reid, *Direct Methods for Sparse Matrices*, Oxford University Press, New York and Oxford, 1986.
17. J. P. Dussault, Simple unified convergence proofs for the trust-region and a new ARC variant, Technical Report, University of Sherbrooke, Sherbrooke, Canada, 2015.
18. R. Fletcher and M. J. D. Powell, A rapidly convergent descent method for minimization, *Computer Journal* **6**, pp. 163–168 (1963).
19. G. H. Golub and Ch. F. Van Loan, *Matrix computations*, Johns Hopkins University Press, Baltimore, 2012.
20. C. C. Gonzaga, Path-Following Methods for Linear Programming, *SIAM Review* **34**, pp. 167–224 (1992).
21. A. Griewank, *Automatic Differentiation*, Princeton Companion to Applied Mathematics, Nicolas Higham Ed., Princeton University Press, 2014.
22. A. F. Izmailov and M. V. Solodov, *Newton-Type Methods for Optimization and Variational Problems*, Springer Series in Operations Research and Financial Engineering, New York, 2014.
23. N. Karmarkar, A new polynomial-time algorithm for linear programming, *Combinatorics* **4**, pp. 373–395 (1984).
24. L. G. Khachiyan, A polynomial algorithm in linear programming, *Doklady Akad. Nauk. USSR* **244**, pp. 1093–1096 (1979).
25. J. M. Martínez, A family of quasi-Newton methods with direct secant updates of matrix factorizations, *SIAM Journal on Numerical Analysis* **27**, pp. 1034–1049 (1990).
26. W. F. Mascarenhas, The BFGS method with exact line searches fails for non-convex objective functions, *Mathematical Programming* **99**, pp. 49–61 (2004).
27. W. F. Mascarenhas, On the divergence of line search methods, *Computational and Applied Mathematics* **26**, pp. 129–169 (2007).
28. J. Nocedal and S. J. Wright, *Numerical Optimization*, Springer Series in Operations Research and Financial Engineering, New York, Second Edition, 2006.
29. Y. Nesterov and B. T. Polyak, Cubic regularization method and its global performance, *Mathematical Programming* **108**, pp. 177–205 (2006).

Minimal Surfaces and Their Gauss Maps

Francesco Mercuri and Luquesio P. M. Jorge

Abstract In this paper we will discuss some classical results in minimal surfaces theory, related to the Gauss map of such surfaces. In the last section we will comment on some work in progress and some open problems related to one of these results.

1 Introduction

It is generically accepted that the theory of minimal surfaces starts with the work of the Italian mathematician J. N. Lagrange who, in 1760, posed the following problem:

> consider a bounded open set $\Omega \subseteq \mathbb{R}^2$ with smooth boundary $\partial\Omega$ and a smooth function $\phi : \partial\Omega \longrightarrow \mathbb{R}$. Find a smooth function $f : \Omega \longrightarrow \mathbb{R}$ such that $f_{|\partial\Omega} = \phi$ and the graph of f has area smaller or equal to the area of the graph of any other smooth function $g : \Omega \longrightarrow \mathbb{R}$ such that $g_{|\partial\Omega} = \phi$.

Lagrange approach to the problem is the, by now, basic approach of the calculus of variations. Suppose that f is a solution of the problem. Consider a function $\eta : \Omega \longrightarrow \mathbb{R}$ such that $\eta_{|\partial\Omega} = 0$. Then the function $f_t = f + t\eta$ agrees with ϕ on $\partial\Omega$ and the area of its graph is

$$A_\eta(t) = \int_{\Omega} \left[1 + \left(\frac{\partial f_t}{\partial x} \right)^2 + \left(\frac{\partial f_t}{\partial y} \right)^2 \right]^{\frac{1}{2}} \mathrm{d}x\mathrm{d}y.$$

F. Mercuri (✉)
Institute of Mathematics, Statistics and Scientific Computing, University of Campinas, Campinas, SP, Brazil
e-mail: mercuri@ime.unicamp.br

L. P. M. Jorge
Center of Sciences, Federal University of Ceará, Fortaleza, CE, Brazil
e-mail: luquesio.jorge@ufc.br

C. Lavor, F. A. M. Gomes (eds.), *Advances in Mathematics and Applications*,
https://doi.org/10.1007/978-3-319-94015-1_10

Since f is supposed to be a solution of the problem, the function $A_\eta(t)$ has a minimum at $t = 0$, hence $A'_\eta(0) = 0$. A simple calculation gives

$$A'_\eta(0) = -\int_{\overline{\Omega}} \operatorname{div}\left(\frac{\nabla f}{\sqrt{1+\|\nabla f\|^2}}\right) \eta \mathrm{d}x\mathrm{d}y = 0.$$

Since $A'_\eta(0) = 0$, $\forall\, \eta$ with $\eta|_{\partial\Omega} = 0$, it follows that solutions of the problem are solutions of the equation

$$\operatorname{div}\left(\frac{\nabla f}{\sqrt{1+\|\nabla f\|^2}}\right) = 0. \tag{1}$$

Equation (1) is called the *minimal surface equation* or the *Euler Lagrange equation of the problem.*

Remark 1.1 It turns out, from the regularity theory for elliptic partial differential equations, that a solution of (1) is a real analytic function (see also Lemma 2.7).

Some years later J. B. Meusnier was looking for a good concept of curvature of a regular surface M in $\mathbb{R}^3$. He considered a point $p \in M$ and a unit normal vector $N \in [T_pM]^\perp$. For a unit tangent vector $v \in T_pM$ he considered the plane determined by N and v and passing through p. Cutting M with such a plane he obtain a plane curve and he denoted by $k_p(v)$ the (oriented) curvature of this curve. It turns out that the function $k_p(v)$ has a unique minimum value, $k_m(p)$, and a unique maximum value, $k_M(p)$. These two numbers are called the *principal curvatures* of M at p and were introduced earlier by Euler.

Given the principal curvatures we can define

- the *Gaussian curvature* of the surface, $K(p) = k_m(p)k_M(p)$,
- the *mean curvature*, $H(p) = \frac{1}{2}(k_m(p) + k_M(p))$.

Meusnier showed that a function f is a solution of (1), if and only if its graph has vanishing mean curvature. This leads to the following definition.

Definition 1.2 A regular surface in $\mathbb{R}^3$ is called a *minimal surface* if its mean curvature vanishes identically.

Remark 1.3 Following Gauss, a concept should be considered only if it is "pregnant with theorems." He certainly had many important results involving the Gaussian curvature, but not so many involving the mean curvature. So he never seriously considered the latter concept.

We will take a slightly more general approach. Consider a surface M, i.e., a two-dimensional differentiable manifold, that, for simplicity, we will assume connected and *oriented by a positive atlas* of smooth charts $\{(\Omega_\alpha, \psi_\alpha)\}$.[1] Let $f : M \longrightarrow \mathbb{R}^3$ be an immersion. Then f induces a Riemannian metric on M,

$$\langle X, Y\rangle_p := \langle \mathrm{d}f(p)(X), \mathrm{d}f(p)(Y)\rangle, \quad X, Y \in T_pM,$$

which make f an isometric immersion. We can define the *Gauss map*

$$\underline{\mathbf{n}} : M \longrightarrow S^2 = \{x \in \mathbb{R}^3 : \|x\| = 1\},$$

where $\underline{\mathbf{n}}(x) \in [\mathrm{d}f(x)(T_xM)]^\perp \subseteq \mathbb{R}^3$ is the unit vector such that $(\psi_u, \psi_v, \underline{\mathbf{n}}(x))$ is a positively oriented basis of $T_{f(x)}\mathbb{R}^3$. If (Ω, ψ) is a positive chart, then

$$\underline{\mathbf{n}}(\psi(u, v)) = \frac{\psi_u \wedge \psi_v}{\|\psi_u \wedge \psi_v\|}$$

so $\underline{\mathbf{n}}$ is a well-defined smooth map.

Remark 1.4 If f is an immersion, then for all $x \in M$ there exists a neighborhood U of x such that $f(U)$ is a regular surface in $\mathbb{R}^3$. So, for local considerations, we can identify M with $f(M)$.

Consider the differential of the Gauss map

$$\mathrm{d}\underline{\mathbf{n}}(x) : T_xM \longrightarrow T_{\underline{\mathbf{n}}(x)}S^2 = T_xM$$

and the operator $A_x = -\mathrm{d}\underline{\mathbf{n}}(x)$. The operator A_x is called the *shape operator* at x or the *second fundamental form*. It turns out that A_x is a symmetric operator whose eigenvalues are exactly the principal curvatures.

Lagrange did not give any example of solutions of Eq. (1) (except for the trivial ones, i.e., the affine functions). There where many efforts to produce examples and characterize minimal surfaces with special properties. We will recall now some results in this direction.

Theorem 1.5 (Meusnier) *If f is a solution of Eq. (1) whose level curves are straight lines segments, then*

$$f(x, y) = A \arctan \frac{y - y_0}{x - x_0} + B, \quad x_0, y_0, A, B \in \mathbb{R}$$

i.e., the graph is an open part of a helicoid (figure a below).

[1] That is, an atlas such that the change of coordinates has positive Jacobian.

Theorem 1.6 (Sherk) *If f is a solution of Eq. (1) of the form $f(x, y) = g(x) + h(y)$, then*

$$f(x, y) = a^{-1} \log \left(\frac{\cos ax}{\cos ay} \right), \quad a \in \mathbb{R}$$

(figure b below).

Theorem 1.7 (Euler) *A minimal surface of revolution is an open part of the catenoid*

$$(\cos v \sinh u, \sin v \cosh u, u),$$

up to rigid motions, or of a plane (figure c below).

Theorem 1.8 (Catalan) *A ruled minimal surface is an open part of a helicoid or of a plane.*

We refer to [2] for proofs and further information.

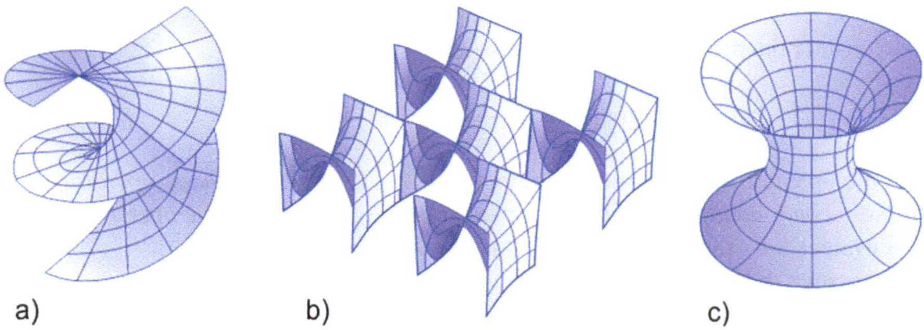

Around 1866 A. Enneper and K. Weierstrass gave a special parametrization for minimal surface, today known as the Weierstrass representation formula, which turns out to be a basic tool for producing examples of minimal surfaces. We will discuss this parametrization in Sect. 3.

We point out that the above results are, essentially, locally in nature. Probably the first global result is due to Bernstein who, around 1915, proved the following

Theorem 1.9 (Bernstein) *Let $f : \mathbb{R}^2 \longrightarrow \mathbb{R}$ be a solution of the minimal surface equation. Then f is an affine function, i.e., its graph is an affine plane.*

Remark 1.10 Bernstein's theorem should be compared with Liouville's theorem which states that a bounded harmonic function $f : \mathbb{R}^2 \longrightarrow \mathbb{R}$ is constant. However, in the first one, there are no conditions on the behavior of the function at infinity.

We will discuss in the next sections a couple of results that generalize Bernstein's theorem (see Remarks 2.35 and 4.2).

Remark 1.11 A natural question is if a similar theorem holds for functions $f : \mathbb{R}^n \longrightarrow \mathbb{R}$ which verify Eq. (1). Surprising enough the answer is positive if $n \leq 7$, but there are counterexamples for $n \geq 8$.

A basic question in the theory is the existence of a minimal surface whose boundary is a given simple closed curve Γ. This question has a long and rich history dating from the experiments of the Belgian physicist J. Plateau in 1847. He showed that dipping a wire on a soap solution we obtain a soup film which is a minimal surface and it is stable for small perturbations, i.e., is a *local minimum* for the area functional. So we have an "experimental proof" of the existence of minimal solution spanning a given boundary. But a "mathematical proof" proved to be a much more difficult task. It was only in 1930 that we have a first general answer to the question when, independently, Douglas (see [6]) and Radó (see [18]) proved the existence of a "minimal map" from a disk to $\mathbb{R}^3$, mapping the boundary of the disk onto a given rectifiable Jordan curve. A proof that the map is in fact an immersion, i.e., a minimal surface, appeared around 1960. There are still various problems under investigation, for example, the problem of uniqueness. If we consider the set of Jordan curves in $\mathbb{R}^3$, with a suitable topology, the subset for which the solution of the Plateau problem is unique is dense. Strangely enough the complement of this subset is also dense, and in fact there exist curves that bound an uncountable number of solutions. Naturally we are talking of *geometrically distinct* solutions, i.e., solutions up to a reparametrization. Also the existence of solutions with more complicate topology is an interesting field of investigation. We will not treat these questions here and refer to [4, 12] and the references therein for an introduction to these problems.

2 Stability

Let M be a surface, that, for simplicity we will assume connected and oriented, and let $f : M \longrightarrow \mathbb{R}^3$ be an immersion. A *domain* $D \subseteq M$ will be a connected, *relatively compact* open set such that the boundary is a finite union of disjoint piecewise smooth curves.

Definition 2.1 Let $D \subseteq M$ de a domain. A *(proper) variation* of f, supported on D, is a smooth function $F : (-\epsilon, \epsilon) \times M \longrightarrow \mathbb{R}^3$ such that:

(1) $F(0, x) = f(x)$,
(2) the restriction of F to $\{t_0\} \times M$ is an immersion,
(3) $F(t, x) = x$ if $x \notin D$.

When clear from the context we will simply say that F is a variation of f.

Given a variation F, the *variational vector field* is the vector field

$$V_F(x) := \mathrm{d}F(0, x)\left(\frac{\partial}{\partial t}\right).$$

Clearly V is a vector field along f[2] vanishing outside D. Set

$$F_t : M \longrightarrow \mathbb{R}^3, \quad F_t(x) = F(t, x).$$

Since F_t is an immersion, we can consider in M the induced metric and we will denote by $A_F(t)$ the area of D with respect to the induced metric. Then

Lemma 2.2 (First Variational Formula)

$$\frac{\mathrm{d}A_F(t)}{\mathrm{d}t}(0) = -2\int_M \langle H\underline{\mathbf{n}}, V_F\rangle \mathrm{d}M,$$

where H is the mean curvature, and V_F is the variational vector field.

In particular if D has minimal area for all variations, $f_{|D}$ is a *minimal surface*. But in general a minimal surface is just a critical point of the area functional, not necessarily a minimum, not even a relative minimum.[3] In order to decide if a minimal surface is a relative minimum of the area functional we have to look at the second derivative of the area functional. To compute this derivative we need some preliminaries.

Let M be a Riemannian surface. If $u : M \longrightarrow \mathbb{R}$ is a smooth map, the *Laplacian* of u, Δu, is defined as

$$\Delta u = \operatorname{div} \nabla u,$$

where the gradient ∇u and the divergence are taken in relation to the metric of M.

It is useful, sometimes, to work in special local coordinates. Let $U \subseteq \mathbb{R}^2$ be an open set with coordinates (u, v).

Definition 2.3 A local parametrization $\psi : U \longrightarrow M$ of a Riemannian surface M is *isothermal* if

$$\|\psi_u\| = \|\psi_v\| = \lambda, \quad \langle \psi_u, \psi_v\rangle = 0,$$

where $\lambda : U \longrightarrow \mathbb{R}$ is a positive smooth function and subscripts denoted derivatives.

The coordinates (u, v) will be called *isothermal coordinates* (or *isothermal parameters*).

The following is well known

Theorem 2.4 *Let M be a Riemannian surface. Then $\forall\, p \in M$ there are isothermal coordinates in a neighborhood of p.*

Remark 2.5 In the case of minimal surfaces a simpler proof can be found in [16].

---

[2]That is, a map $V : M \longrightarrow T\mathbb{R}^3$ such that $V(x) \in T_{f(x)}\mathbb{R}^3$.

[3]That is, a minimum with respect to nearby surfaces bounding the same curve.

Remark 2.6 If M is a connected oriented Riemannian surface, we can choose a *positive* atlas of isothermal coordinates. Once we do this, the changes of coordinates are conformal, hence holomorphic (where defined). A differentiable surface with such an atlas is called a *Riemann surface*.

In terms of isothermal coordinates the Laplacian is given by

$$\Delta = \lambda^{-2}\left(\frac{\partial^2}{\partial^2 u} + \frac{\partial^2}{\partial^2 v}\right),$$

and the Gaussian curvature is given by

$$K = -\Delta \log \lambda.$$

The following lemma is easy to prove

Lemma 2.7 *If* $f : M \longrightarrow \mathbb{R}^3$ *is an isometric immersion,* $f(u, v) = (f_1(u, v), f_2(u, v) f_3(u, v))$, *then*

$$\Delta f = (\Delta f_1, \Delta f_2, \Delta f_3) = 2H\underline{\mathbf{n}},$$

hence f *is minimal if and only if the coordinate functions are harmonic. In particular the coordinate functions of a minimal immersion are real analytic functions, and so are the Gaussian and mean curvature functions.*

Let $D \subseteq M$ be a domain. We will assume, for simplicity, that M is oriented and let $\underline{\mathbf{n}} : D \longrightarrow S^2$ be the Gauss map associated with the (fixed) orientation. We will denote by $H = H(D)$ the space of continuous functions on M, vanishing outside D, whose gradient exists almost everywhere and its norm is square integrable.

Remark 2.8 The space H has a natural norm given by

$$\|u\|^2 = \int_D u^2 \mathrm{d}M + \int_D \|\nabla u\|^2 \mathrm{d}M.$$

The subspace of smooth functions is dense in H, so, in many cases, we can assume that a function in H is smooth.

If $u \in H$ we consider the *normal variation*

$$F_u(t, x) = f(x) + tu(x)\underline{\mathbf{n}}(x),$$

whose variational vector field is

$$V = V_F = u\underline{\mathbf{n}}.$$

Theorem 2.9 (Second Variational Formula) *The second derivative of the area functional in the V direction is*

$$I(V, V) = \int_{D} u(-\Delta u + 2Ku)\mathrm{d}M.$$

We refer to [12] for a proof.

The quadratic form I is called the *index form* for D. If the index form is not positive semi-definite, the domain is *not* a relative minimum of the area functional. It turns out that, for local minimality questions, it is not restrictive to consider only normal variations. In particular if the index form of D is positive definite, then D is a local minimum of the area functional.

Definition 2.10 Let $f : M \longrightarrow \mathbb{R}^3$ be a minimal immersion and let $D \subseteq M$ be a domain. We will say that D is *stable* (resp. *strongly stable*) if the index form of D is positive semi-definite (res. positive definite).

An important concept related to stability is the following:

Definition 2.11 A normal field $J = u\underline{\mathbf{n}}$ is called a *Jacobi field* if

$$-\Delta u + 2uK = 0. \tag{2}$$

Definition 2.12 Let $f : M \longrightarrow \mathbb{R}^3$ be an immersion and let $D \subseteq M$ be a domain.

1. ∂D is said to be a *conjugate boundary* if there exists a Jacobi field $u\underline{\mathbf{n}}$ with $u \in H \setminus \{0\}$.
2. The *multiplicity* or *nullity* of a conjugate boundary, denoted by $\nu(D)$, is the dimension of the space of Jacoby fields $u\underline{\mathbf{n}},\ u \in H$.
3. ∂D is a *first conjugate boundary* if it is a conjugate boundary and for all domains $D' \subseteq D$, $\partial D'$ is a conjugate boundary if and only if $D = D'$.

Remark 2.13 In the theory of geodesics we have analogous concepts. If M is a Riemannian manifold and $\gamma : [0, a] \longrightarrow M$ is a geodesic, a vector field J along γ is a Jacobi field if $\ddot{J} + R(\dot{\gamma}, J)\dot{\gamma} = 0$ where $\ddot{J}$ is the second covariant derivative of J along γ and R is the curvature tensor. A point $t_0 \in [0, a]$ is conjugate to 0 if there exists a non-trivial Jacoby field vanishing at 0 and t_0. Then the geodesic γ is a local minimum for the energy functional acting on curves joining $\gamma(0)$ and $\gamma(a)$ if there are no conjugate values in $[0, a)$. The corresponding assertion is true in our context, i.e., if for all domains $D' \subsetneq D$ $\partial D'$ is not a conjugate boundary, then D is stable.

We will give now a short proof of the following fact:

Theorem 2.14 *If $D \subseteq M$ is a domain and ∂D is a first conjugate boundary, then* $\nu(D) = 1$.

Proof Set $\mathbb{J} = \{u \in H : -\Delta u + 2uK = 0\}$. Since ∂D is conjugate, $\mathbb{J} \neq \{0\}$. Fix a point $p \in D$. Define the linear map

$$L : \mathbb{J} \longrightarrow \mathbb{R}, \quad L(u) = u(p).$$

Consider $u \in \mathbb{J} \setminus \{0\}$. It is known that either $u(p) \neq 0$ or u change sign in D. In the latter case consider a connected component D' of the complement of the zero set of u. This is a domain properly contained in D, and u vanishes on $\partial D'$. Hence $\partial D'$ is a conjugate boundary, contradicting the fact that ∂D is a first conjugate boundary. Hence $u(p) \neq 0$ and so L is injective and surjective, hence an isomorphism. It follows that $\dim \mathbb{J} = 1 = \nu(D)$. □

Definition 2.15 Let $\mathbb{E}$ be a real vector space and let $I : \mathbb{E} \longrightarrow \mathbb{R}$ be a quadratic form. The *index* of I is the superior of the dimensions of subspaces of $\mathbb{E}$ on which I is negative definite.

Definition 2.16 If $f : M \longrightarrow \mathbb{R}^3$ is a minimal immersion and $D \subseteq M$ is a domain, we define the *index* of D, $i(D)$, as the index of the index form of D.

In the theory of geodesics, the celebrated theorem of Morse states that if M is a Riemannian manifold and $\gamma : [0, a] \subseteq \mathbb{R} \longrightarrow M$ is a geodesic, the index of γ, i.e., the index of the second derivative of the energy functional, is the number of instants $t \in [0, a)$ conjugate to 0, counted with multiplicity. This result has been generalized for minimal surfaces by Smale. He considered a domain D in a minimal surface M and *a flow of contractions*, i.e., a family of diffeomorphisms $\phi_t : M \longrightarrow M,\ t \geq 0$, such that

- $\phi_0 = \mathbb{1}_M$,
- $\phi_t(D) \subset \phi_s(D)$ if $t > s$,
- $\lim_{t \longrightarrow \infty} A(\phi_t(D)) = 0$.

Theorem 2.17 (Morse-Smale) *Let D and ϕ_t be as above and set $D_t = \phi_t(D)$. Then*

$$i(D) = \sum_{t>0} \nu(D_t).$$

For $\lambda \in \mathbb{R}$ we consider the space

$$\Sigma_\lambda = \{u \in H : \Delta u + \lambda u = 0\}.$$

If $\dim \Sigma_\lambda = n_\lambda > 0$ then λ is an eigenvalue of the operator $-\Delta$. It follows from the spectral theory of such an operator, that the eigenvalues of $-\Delta$ form a countable set of positive numbers and we can order them in such a way that

$$0 < \lambda_1 \leq \lambda_2 \leq \cdots \leq \lambda_n \leq \cdots .$$

Remark 2.18 When dealing with more than one domain we will write $\lambda_i(D)$ in order to avoid confusions.

We state now some well-known basic properties of the eigenvalues of $-\Delta$.

Theorem 2.19

(1) *If* $u \in \Sigma_\lambda$ *the* u *is analytic.*
(2) *If* $u \in H \setminus \{0\}$,

$$\lambda_1 \leq \frac{\int_M \|\nabla u\|^2 \mathrm{d}M}{\int_M u^2 \mathrm{d}M}.$$

and equality holds if and only if $u \in \Sigma_{\lambda_1}$.
(3) *If* $u \in \Sigma_{\lambda_1}$ *then* $u(x) \neq 0,\ \forall\, x \in D$.
(4) *If* $u \in \Sigma_{\lambda_i},\ i > 1$, *then* u *changes sign in* D.
(5) *If* $D' \subseteq D$ *then* $\lambda_1(D') \geq \lambda_1(D)$, *and equality holds if and only if* $D' = D$.

Example 2.20 Consider the unit sphere $S^2 \subseteq \mathbb{R}^3$. Let $u : S^2 \longrightarrow \mathbb{R},\ u(x, y, z) = z$ (a spherical harmonic of the first kind) and consider the restriction $\tilde{u}$ of u to the half sphere $S^2_+ = \{(x, y.z) \in S^2 : z > 0\}$. It is easily seen that $\Delta\tilde{u} + 2\tilde{u} = 0$. The function $\tilde{u}$ vanishes on ∂S^2_+ and is positive in the interior of S^2_+, hence, by items (3) and (4) of Theorem 2.19, $\lambda(S^2_+) = 2$. Also, for any proper subdomain $D \subset S^2_+$, $\lambda_1(D) > 2$, by item (5) of the same theorem.

The following fact, of interest in itself, will be useful later.

Theorem 2.21 *The spherical caps of* S^2 *minimize the first eigenvalue of* $-\Delta$ *among all domains with the same area.*

Proof See [17] □

Let $f : M \longrightarrow \mathbb{R}^3$ be a minimal immersion. Since the Gauss curvature is analytic (see Lemma 2.7) either the Gauss curvature vanishes identically, in which case $f(M)$ is an open subspace of an affine plane, or the zeros of K are isolated, hence finite in number on every domain in M.[4] We will assume that K is not identically zero, and will set

$$M_0 = M \setminus \{x \in M : K(x) = 0\}.$$

If $x \in M_0$, $\mathrm{d}\underline{\mathbf{n}}(x)$ is an isomorphism and we can define a new metric in M_0 setting

$$\langle X, Y\rangle_0 = \langle \mathrm{d}\underline{\mathbf{n}}(X), \mathrm{d}\underline{\mathbf{n}}(Y)\rangle, \quad \forall\, X, Y \in TM_0.$$

[4] Since a domain is relatively compact according to our definition.

We will denote by $\mathrm{d}M_0$ the volume form with respect to this metric and by Δ_0 the Laplace operator of this metric. Then, as it is easily seen,

$$\Delta = -K\Delta_0, \tag{3}$$

$$\mathrm{d}M_0 = -K\mathrm{d}M. \tag{4}$$

Let $D \subseteq M$ be a domain and $u \in H(D)$. We will denote by $\tilde{u}$ the restriction of u to $M_0 \cap \overline{D}$.

Lemma 2.22 *Let $X = u\underline{\mathbf{n}}$. Then*

(1) *The index form is given by*

$$I(X, X) = -\int_{\overline{D}\cap M_0} (\tilde{u}\Delta_0\tilde{u} + 2\tilde{u}^2)\mathrm{d}M_0,$$

(2) *X is a Jacobi field if and only if*

$$\Delta_0\tilde{u} + 2\tilde{u} = 0.$$

Proof The first assertion follows from $\mathrm{d}M_0 = -K\mathrm{d}M$ and the fact that $M \setminus M_0$ has measure zero. The second one follows from $\Delta = -K\Delta_0$ and continuity.

Next we will prove a first result relating stability and eigenvalues of the Laplacian.

Theorem 2.23 (Schwarz) *Let $f : M \longrightarrow \mathbb{R}^3$ be a minimal immersion and let $D \subseteq M$ be a domain. Assume that $\underline{\mathbf{n}}(D)$ is a domain in S^2 with first eigenvalue of $-\Delta_0$ smaller than 2. Then D is not stable.*

Proof Since the zero curvature points in $\overline{D}$ are finite in number, $\underline{\mathbf{n}} : \overline{D} \longrightarrow \underline{\mathbf{n}}(\overline{D})$ is a branched covering map and $\underline{\mathbf{n}}(\partial D) \subseteq \partial\underline{\mathbf{n}}(\overline{D})$. Let v be a function in the first eigenspace of $\underline{\mathbf{n}}(D)$. Consider $u = v \circ \underline{\mathbf{n}}$. Then u is a function in $H(D)$ and

$$\Delta_0 u + \lambda_1 u = 0.$$

Consider the vector field along D, $X = u\underline{\mathbf{n}}$ and denote, as before, by $\tilde{u}$ the restriction to $D \cap M_0$. Then, by Lemma 2.22,

$$I(X, X) = -\int_{\overline{D}\cap M_0} [\tilde{u}\Delta_0\tilde{u} + 2\tilde{u}^2]\mathrm{d}M_0 = (\lambda_1 - 2)\int_{\overline{D}\cap M_0} \tilde{u}^2\mathrm{d}M_0 \quad < 0.$$

The operator Δ_0 is, essentially, the Laplacian on the sphere S^2 and we will use the same symbol for the two operators. It follows from Theorem 2.23, that a domain whose image by the Gauss map contains properly an hemisphere is not stable.

The next result relates stability with the image of the Gauss map of a domain.

Theorem 2.24 (Barbosa-do Carmo) *If $f : M \longrightarrow \mathbb{R}^3$ is a minimal immersion and $D \subseteq M$ is a domain such that $A(\underline{\mathbf{n}}(D)) < 2\pi$, then D is strongly stable.*

Proof (Sketch) The proof is by contradiction. Suppose D is not stable. Then, by Theorem 2.17, there is a domain $D' \subseteq D$ such that $\partial D'$ is a first conjugate boundary. The heart of the proof is the existence of a function $v \in H(\underline{\mathbf{n}}(D'))$ such that

$$\int_{\underline{\mathbf{n}}(D')} \|\nabla v\|^2 \leq 2 \int_{\underline{\mathbf{n}}(D')} v^2. \tag{5}$$

Once we have such a function we proceed as follows. From Theorem 2.19 (2), we have $\lambda_1(\underline{\mathbf{n}}(D')) \leq 2$. Consider a spherical cup C such that $A(C) = A(\underline{\mathbf{n}}(D')) < 2\pi$. Then $\lambda_1(C) > 2$ (see Example 2.20) and, by Theorem 2.21,

$$\lambda_1(\underline{\mathbf{n}}(D')) \geq \lambda_1(C) > 2,$$

a contradiction.

We will sketch the construction of the function v.

Since $\partial D'$ is a first conjugate boundary, we have a Jacobi field $u\underline{\mathbf{n}}$, $u \in H(D')$, u positive in D'. If $p \in \overline{D'}$, the Gauss map in a suitable neighborhood of p looks like $\underline{\mathbf{n}}(z) = z^n$, with respect to a complex local coordinate z (see also Sect. 5). We set $\nu(p) = n$. Observe that $\nu(p) = 1$ if $K(p) \neq 0$. Given $q \in \underline{\mathbf{n}}(\overline{D'})$ the set $\underline{\mathbf{n}}^{-1}(q) \cap \overline{D'}$ is finite and we define

$$v(q) = \sum_{p \in \underline{\mathbf{n}}^{-1}(q) \cap \overline{D'}} \nu(p) u(p).$$

Then v vanishes on $\overline{\partial \underline{\mathbf{n}}(D')}$ and is positive in the interior. The proof that v verify Eq. (5) is not simple and we refer to [1] for it.

Essentially the same proof shows the following "companion" if Theorem 2.23.

Theorem 2.25 *Let $D \subseteq M$ be a domain such that $\lambda_1(\underline{\mathbf{n}}(D)) > 2$. Then D is strongly stable.*

Example 2.26 A good example to keep in mind is the catenoid. Consider the domain of the catenoid between the panes $z = -\epsilon,\ z = N,\ 0 \leq \epsilon \leq N$. Then the image of this domain is a spherical ring shaped domain bounded by two parallels. If $\epsilon = 0$ then the area of the spherical image is less than 2π, so the domain is stable. If $\epsilon > 0$, then, by a limiting argument, if N is sufficiently large, the first eigenvalue is smaller than 2. Hence the domain is not stable. In particular the estimates in the above results are sharp.

Next we will consider stability from a global point of view.

Definition 2.27 Let $f : M \longrightarrow \mathbb{R}^3$ be a minimal immersion. We will say that M is *stable* if every domain in M is stable.

Before describing the next result, which characterizes complete stable minimal surfaces, we will recall a basic fact in Riemann surfaces theory, the *uniformization theorem*.

Theorem 2.28 (Uniformization Theorem) *Let M be a Riemann surface. Then there is a conformal covering map $\pi : \tilde{M}_c \longrightarrow M$, where $\tilde{M}_c$ is either the complex plane $\mathbb{C}$, the unit disk $\mathbb{D} = \{z \in \mathbb{C}; |z| < 1\}$ or the sphere S^2.*

Definition 2.29 The space $\tilde{M}_c$ in the theorem above is called the conformal universal covering of M.

Remark 2.30 The disk and the plane are obviously diffeomorphic but they are not conformally equivalent since there are no non-constant conformal maps from the plane to the disk, by Liouville's theorem.

Remark 2.31 If $f : M \longrightarrow \mathbb{R}^3$ is an isometric immersion, with M compact and without boundary, then there exists a point in M where the Gaussian curvature is positive (for example, a point $p \in M$ such that $f(p)$ has maximal distance from the origin). If f is a minimal immersion, its Gaussian curvature is non-positive and so M cannot be compact. Since $\pi : \tilde{M}_c \longrightarrow M$ is conformal and the coordinate functions of f are harmonic, so are the ones of $f \circ \pi$, hence $f \circ \pi$ is a minimal immersion and $\tilde{M}_c$ cannot be compact. In particular its conformal universal covering of a minimal surface is either the complex plane $\mathbb{C}$ or the unit disc $\mathbb{D}$.

The following result is due to do Carmo and Peng (see [5]) and, independently, to Fisher Colbrie and Shoen (see [8]).

Theorem 2.32 *Let $f : M \longrightarrow \mathbb{R}^3$ be a complete minimal immersion. Then, if M is stable, $f(M)$ is a plane.*

Proof (Sketch) We will start with the following:

Lemma 2.33 *Let $\pi : \tilde{M} \longrightarrow M$ be a conformal covering map. Then $f \circ \pi : \tilde{M} \longrightarrow \mathbb{R}^3$ is a complete stable minimal surface.*

Proof By Remark 2.31 $f \circ \pi$ is minimal. Also it is well known that $\tilde{M}$, with the covering metric, is complete. It remains to show that it is stable. Suppose that $\tilde{D}$ is a domain which is not stable. Then there exists a domain $\tilde{D}' \subseteq \tilde{D}$ such that $\partial \tilde{D}'$ is a first conjugate boundary. Hence we have a function $\tilde{u} \in H(\tilde{D}')$, positive in $\tilde{D}'$, such that $\Delta \tilde{u} - 2K\tilde{u} = 0$. Consider $q \in \pi(\tilde{D}') := D'$. Since $\tilde{D}'$ is relatively compact, $\pi^{-1}(q) \cap \tilde{D}'$ is a finite set of points say $\{p_1, \ldots, p_k\}$. Set

$$u(q) = \sum_1^k \tilde{u}(p_i).$$

Then $u \in H(D')$ and is positive in D'. As in Theorem 2.24, it is possible to show that

$$\int_{\overline{D'}} \|\nabla u\|^2 \mathrm{d}M \leq 2\int_{\overline{D'}} -Ku^2 \mathrm{d}M$$

and the same argument as in Theorem 2.24 shows that D' is not stable, a contradiction.

By the lemma, we can assume that M is simply connected. Then, by the uniformization theorem, M is conformally equivalent to either the complex plane $\mathbb{C}$ or the unit disk $\mathbb{D}$. Assume the latter and let $\mathrm{d}s^2 = \lambda^2|\mathrm{d}z|^2$ be the induced metric.

We will proceed by contradiction supposing that the Gaussian curvature is not identically zero.

Set $\phi = \lambda^{-1}$. Then we have

$$K = -\phi^2\Delta_0\phi^-1, \;\; \nabla = \phi\nabla_0, \;\; \Delta = \phi^2\Delta_0, \;\; \mathrm{d}M = \phi^2\mathrm{d}A, \tag{6}$$

where $\nabla_0, \;\Delta_0, \mathrm{d}A$ are the gradient, the Laplacian, and the area form with respect to the flat metric.

Since M is stable, we have, for every piecewise smooth compactly supported function $u : M \longrightarrow \mathbb{R}$,

$$\int_M (u\Delta u - 2u^2K)\mathrm{d}M \le 0. \tag{7}$$

Using (6) we have that (7) can be written as

$$\int_D (u\Delta_0 u + u^2\Delta_0 \log\lambda^2)\mathrm{d}A \le 0. \tag{8}$$

Replacing u by ϕu in Eq. (6) we have

$$\int_D (\phi u\Delta_0(\phi u) + u^2\phi^2\Delta_0 \log\phi^{-2})\mathrm{d}A \le 0. \tag{9}$$

Since $u\phi$ has compact support, integration by parts give

$$\int_D \phi u\Delta_0\phi u\mathrm{d}A = -\int_D (u^2|\nabla_0\phi|^2 + \phi^2|\nabla_0 u|^2 + 2\langle\nabla_0 u, \nabla_0\phi\rangle_0)\mathrm{d}A, \tag{10}$$

$$\int_D u^2\phi^2\Delta_0 \log\phi^{-2}\mathrm{d}A = 4\int_D (\phi u\langle\nabla_0 u, \nabla_0\phi\rangle_0 + u^2|\nabla_0\phi|^2)\mathrm{d}A. \tag{11}$$

where $|\cdot|$ and $\langle\cdot,\cdot\rangle_0$ are the norm and scalar product of the flat metric. Adding (10) and (11), we get

$$3\int_D |\nabla_0\phi|^2\mathrm{d}A \le \int_D \phi^2|\nabla_0 u|^2\mathrm{d}A - 2\int_D \phi u\langle\nabla_0 u, \nabla_0\phi\rangle_0\mathrm{d}A. \tag{12}$$

Using the inequality

$$|\phi u \langle \nabla_0 u, \nabla_0 \phi \rangle_0| \leq \epsilon |\nabla_0 \phi|^2 u^2 + \epsilon^{-1} |\nabla_0 u|^2 \phi^2, \quad \forall\ \epsilon > 0,$$

we obtain

Lemma 2.34 *There exists a positive constant b such that*

$$\int_M |\nabla \phi|^2 u^2 \mathrm{d}M \leq b \int_M \phi^2 |\nabla u|^2 \mathrm{d}M.$$

Let B_r be the geodesic ball of radius r, and $\theta \in (0, 1)$. Let $u : M \longrightarrow \mathbb{R}$ be a function which vanishes outside B_r, it is 1 in $B_{\theta r}$ and is linear in $B_r \setminus B_{\theta r}$. From Lemma 2.34 we have

$$\int_{B_r} |\nabla \phi|^2 \mathrm{d}M \leq \frac{b}{(1-\theta)^2 r^2} \int_M \phi^2 \mathrm{d}M = \frac{b}{(1-\theta)^2 r^2} \int_{B_r} \mathrm{d}A = \frac{\pi b}{(1-\theta)^2 r^2}.$$

Letting $r \longrightarrow \infty$, we get $|\nabla \phi| = 0$, hence $\phi = constant$ and the metric $\mathrm{d}s^2$ is not complete, a contradiction.

With similar techniques we can treat the case in which M is conformally equivalent to $\mathbb{C}$.

Remark 2.35 Let $f : \mathbb{R}^2 \longrightarrow \mathbb{R}$ be a solution of the minimal surfaces equations. Then the graph of f is a complete minimal surface. Moreover, given a domain in the graph of f, the image of this domain by $\underline{\mathbf{n}}$ is strictly contained in a hemisphere and has area smaller than 2π. Hence, by Theorem 2.25, the domain is stable. Hence, as a corollary of Theorem 2.32, we have that f is affine, i.e., the Bernstein's Theorem 1.9.

3 The Weierstrass Representation Formula

The Weierstrass representation formula is a basic tool in the study of minimal surfaces in $\mathbb{R}^3$ because, on one hand, it is a "machine" to produce examples of minimal surfaces and, on the other hand, it allows to use the powerful theory of holomorphic functions to treat theoretical problems. We will start with the local version.

Consider $\mathbb{C} \cong \mathbb{R}^2$ with the complex coordinate $z = u + iv$, $i = \sqrt{-1}$ and the differential operators

$$\frac{\partial}{\partial z} = \frac{1}{2}\left(\frac{\partial}{\partial u} - i\frac{\partial}{\partial v}\right), \quad \frac{\partial}{\partial \bar{z}} = \frac{1}{2}\left(\frac{\partial}{\partial u} + i\frac{\partial}{\partial v}\right)$$

In particular, if $U \subseteq \mathbb{C}$ is an open set, a function $g : U \longrightarrow \mathbb{C}$ is *holomorphic* if and only if

$$\frac{\partial g}{\partial \overline{z}} = 0.$$

The local version of the Weierstrass representation formula can be stated as follows.

Theorem 3.1 (Weierstrass Representation) *Let $\Omega \subseteq \mathbb{C}$ be an open set and let $f : \Omega \longrightarrow \mathbb{R}^3$ be a* conformal *minimal immersion. Consider the "complex tangent vector"*

$$\frac{\partial f}{\partial z} = \sum \phi_i e_i,$$

where e_i is the standard basis of $\mathbb{R}^3$ and the ϕ_i's are complex valued functions. Then

(1) $\sum |\phi_i|^2 \neq 0$,
(2) $\sum \phi_i^2 = 0$,
(3) $\dfrac{\partial \phi_i}{\partial \overline{z}} = 0$.

Conversely, given functions ϕ_i verifying the condition above, if Ω is simply connected, the function $f : \Omega \longrightarrow \mathbb{R}^3$, given by

$$f_i(z) = 2\mathfrak{Real} \int_{\gamma} \phi_i \mathrm{d}z, \quad i = 1, 2, 3,$$

is a well-defined conformal minimal immersion. Here $\mathfrak{Real}$ stands for the real part and the integral is taken along a curve γ in Ω joining a fixed point z_0 to z.

Remark 3.2 In Theorem 3.1, the first condition tells us that f is an immersion, the second one that f is conformal and the last one that f is minimal.

The second condition in Theorem 3.1 says that, essentially, one of the three function depends only on the other two. Set

$$\omega = \phi_1 - i\phi_2, \quad g = \frac{\phi_2}{\omega}.$$

Then ω is a holomorphic function and g a meromorphic one. Moreover, given ω and g, we can recover the ϕ_i's:

$$2\phi_1 = (1 - g^2)\omega, \;\; 2\phi_2 = (1 + g^2)\omega, \;\; \phi_3 = g\omega. \tag{13}$$

Remark 3.3 To make sense to the above procedure we have to ask that g has a pole of order m at z if and only if ω has a zero order $2m$ at z.

Definition 3.4 The pair (g, ω) is called the Weierstrass data of f.

The geometry of the immersion can be described in terms of Weierstrass data. For example, the metric is given by

$$ds^2 = |\omega|^2(1 + |g|^2)^2|dz|^2,$$

and the Gaussian curvature by

$$k = -\left(\frac{2|g'|^2}{|\omega|(1 + |g|^2)^2}\right)^2.$$

The function g has a very interesting geometric interpretation. Let $\sigma : S^2 \longrightarrow \mathbb{C} \cup \{\infty\}$ be the stereographic projection from the north pole. The following is easy to prove:

Lemma 3.5 $g = \sigma \circ \underline{\mathbf{n}}$.

Let M be a Riemann surface, with an atlas of isothermal coordinates, and let $f : M \longrightarrow \mathbb{R}^3$ be a conformal minimal immersion. The function g is a well-defined meromorphic function from M to $\mathbb{C}$, by Lemma 3.5. It turns out that the locally defined holomorphic 1-forms ωdz coincide in the intersection of the domains and so they define a global holomorphic 1-form that we still denote by ω. Now, if we suppose that the zeros of ω are related to the poles of g as in Remark 3.3 and that the forms below in (14) have no real periods,[5] we can recover f from the Weierstrass data by integration:

$$f(z) = \mathfrak{Real} \int_{z_0}^{z} ((1 - g^2)\omega,\ (1 + g^2)\omega,\ 2g\omega), \tag{14}$$

where the integral is along a smooth curve joining a fixed point z_0 to z.

Remark 3.6 In the construction of examples using the formula above, the hard point is, in general, the proof that the forms in question have no real periods.

Example 3.7 Consider $M = \mathbb{C},\ g(z) = -ie^z,\ \omega(z) = e^{-z}dz$. Then g has no poles and ω has no zeros. Since the domain is simply connected, the forms in (14) have no periods and we have

$$f(u, v) = (\cos(v)\sinh(u), \sin(v)\sinh(u), v), \quad z = u + iv$$

i.e., a helicoid.

[5] A *period* of a 1-form is the value of the integral of the form along a closed curve.

Example 3.8 Let $M = D$ be the unit disk, $g(z) = z$, $\omega = 4\mathrm{d}z(1 - z^4)^{-1}$. Again g has no poles, ω has no zeros, and the domain is simply connected. After integration, and some calculations, we have

$$f(u, v) = \left(u, v, \log \frac{\cos v}{\sin u}\right),$$

i.e., the Scherk's surface.

Example 3.9 Consider $M = \mathbb{C}$, $g(z) = -e^x$, $\omega = -e^{-z}\mathrm{d}z$. Again we are in a "good situation" and integration gives

$$f(u, v) = (\cos v \sinh u, \sin v \cosh u, u) - (0, 0, 1),$$

i.e., a catenoid (up to a translation). Such a parametrization wraps the plane around the (geometric) catenoid infinitely many times.

An alternative way to obtain a catenoid is the following: take $M = \mathbb{C} \setminus \{0\}$, $g(z) = z$, $\omega(z) = z^{-2}\mathrm{d}z$. Since M is not simply connected we have to check that the forms have no real periods. Since $\pi_1(M) \cong \mathbb{Z}$, and the forms are closed, it is sufficient to consider the integrals of those forms on the unit circle $\gamma(t) = (\cos t, \sin t)$, $t \in [0, 2\pi]$. A simple calculation shows that the first two forms have no periods and the third one has purely imaginary periods. After integration we get

$$f(u, v) = \left(-u\left(1 + \frac{1}{u^2 + v^2}\right) + 1, -v\left(1 + \frac{1}{u^2 + v^2}\right),\ \log(u^2 + v^2)\right),$$

which is a parametrization of the catenoid, up to a translation.

Remark 3.10 The Weierstrass representation formula holds, mutate mutandis, for minimal surfaces in $\mathbb{R}^n$, $n \geq 3$. The important point is that the domain is two-dimensional.

Remark 3.11 The Weierstrass representation formula asked, for a long time, for a generalization to the case of minimal surfaces in more general spaces. The first two conditions in Theorem 3.1 have an obvious extension, while the third one is replaced by an integral differential equation involving the Riemannian connection of the ambient manifold. This equation is, in general, very difficult to solve explicitly, but, depending on the ambient manifold, arguments ad hoc can be used to produce explicit solutions hence examples and general results. This is still an active field of investigation.

4 On the Image of the Gauss Map: The General Case

As we have seen in Sect. 3 the Gauss map of a minimal surface is a meromorphic map. A classical problem in minimal surface theory is to know which results from the classical complex function theory remain true for the Gauss map. For example, there is a Picard type theorem for the Gauss map of a complete minimal surface? Questions like this were asked since the middle of the last century and still puzzle researchers in the field. L. Nirember conjectured, around 1950, that the image of the Gauss map of a complete non-flat minimal surfaces in $\mathbb{R}^3$ is dense. A positive answer to this conjecture was given by Osserman (see [15] and [14]). We will sketch now Osserman's proof.

Theorem 4.1 (Osserman) *Let $f : M \longrightarrow \mathbb{R}^3$ be a complete, non-flat, minimal surface. Then the image of the Gauss map is dense.*

Proof (Sketch) The map $f \circ \pi : \tilde{M}_c \longrightarrow \mathbb{R}^3$ is a complete minimal immersion, if we consider in $\tilde{M}_c$ the covering metric. By Remark 2.31 $\tilde{M}_c$ is either the plane or the disk. The Gauss maps of $f \circ \pi$ and of f have the same image so we may consider the Gauss map as a meromorphic map defined on $\tilde{M}_c$. If $\tilde{M}_c = \mathbb{C}$ the theorem follows from the classical Liouville's (or Picard) theorem. So we suppose that $\tilde{M}_c$ is the unit disk. Let us suppose that the Gauss map misses a neighborhood of a point, that, without loss of generality, we can assume to be $e_3 = (0, 0, 1)$. Then, for the Weierstrass data we have:

- there exists a constant $A < \infty$ such that $|g(z)| < A \; \forall \; z \in \tilde{M}_c$,
- $\omega(z) \neq 0, \; \forall \; z \in \tilde{M}_c$, since g has no poles (see Remark 3.3).

Consider the map $F : \tilde{M}_c \longrightarrow \mathbb{C}$ defined by

$$F(z) = \int_0^z \omega(\xi) d\xi.$$

Since $\tilde{M}_c$ is simply connected and ω is holomorphic, F is well defined, $F(0) = 0$, and F is locally invertible since $F'(z) = \omega(z) \neq 0$. Let H be a local inverse defined in a neighborhood of 0. Observe that H cannot be defined in the all of $\mathbb{C}$ otherwise it would be a bounded entire holomorphic function, hence constant. So

$$R = \sup\{r \in \mathbb{R} : H \text{ is defined for all } z \text{ with } |z| < r\} < \infty.$$

In particular there is a $v \in \mathbb{C}$ with $|v| = R$ such that H cannot be defined in a neighborhood of v. Consider the curve $\sigma(t) = tv, \; t \in [0, 1)$ and let $\gamma(t) = H(\sigma(t))$. Then is not difficult to prove that

- γ is a divergent curve, i.e., for every compact set $K \subseteq \tilde{M}_c$ there exist t_0 with $\gamma(t_0) \notin K$,
- the length of γ is $R < \infty$.

But these two facts contradict the completeness of the metric and the theorem follows.

Remark 4.2 We observe that, in particular, Theorem 4.1 generalizes Bernstein theorem, since an entire graph is complete and its Gauss map covers at most an hemisphere. Hence, if it is minimal, has to be flat.

Remark 4.3 Really Osserman showed a slight more general result: the complement of the image of the Gauss map of a complete non-flat minimal surface has zero logarithmic capacity.

Subsequently Xavier in [20] improves considerably Osserman result showing that the Gauss map of a complete, non-flat, minimal surface omits at most 6 points. Finally Fujimoto proved in [7] that the Gauss map of such a surface omits at most 4 points. Fujimoto result is sharp since the Gauss map of the Sherk surface misses exactly 4 points.

5 On the Image of the Gauss Map: The Finite Total Curvature Case

In [14] Osserman studies the size of the complement of the image of the Gauss map for the class of complete minimal surfaces of finite total curvature, i.e., minimal surface for which

$$\int_M k dv > -\infty.$$

For this class of minimal immersions he proved the following basic properties:

Theorem 5.1 *Let $f : M \longrightarrow \mathbb{R}^3$ be a complete non-flat minimal surface of finite total curvature. Then*

(1) *M is conformally equivalent to a compact surface $\overline{M}$ minus a finite set of points $E = \{w_1, \dots, w_k\}$.*
(2) *The Gauss map extends to a branched covering map $\underline{\mathbf{n}} : \overline{M} \longrightarrow S^2$.*

Using Theorem 5.1 and the Weierstrass representation formula, he proved the following:

Theorem 5.2 *Let $f : M \longrightarrow \mathbb{R}^3$ be a complete, non-flat, minimal surface of finite total curvature. Then the Gauss map omits at most three points. Moreover, if $\chi(\overline{M}) = 2$ the Gauss map omits at most two points.*

Although not clear from Osserman's proof, the general case and the finite total curvature case are very different in nature. While the general case is a problem in value distribution theory for holomorphic functions in the disk, the finite total curvature case is of topological nature. We will try to explain the last assertion. For this we will introduce a more general class of surfaces.

Definition 5.3 Let M be a complete Riemannian surface and let $f : M \longrightarrow \mathbb{R}^3$ be an isometric immersion. We will say that f is of *finite geometric type* if

(1) M is diffeomorphic to a compact surface $\overline{M}$ minus a finite set of points $E = \{w_1, \ldots, w_k\}$,
(2) the Gauss map of f extends to a branched covering map $\underline{\mathbf{n}} : \overline{M} \longrightarrow S^2$,

Remark 5.4 Let $f : M \longrightarrow \mathbb{R}^3$ be an immersion of finite geometric type. The fact that the Gauss map extends to a branched covering of $\overline{M}$ over S^2 means that this extension is a covering map outside a finite number of points, the *branch points* of the map. The set of such branch points is finite and includes the points of zero Gaussian curvature and, possibly, some of the ends. For such a branch point v, a small punctured neighborhood is mapped onto its image as a covering map of order $\nu(v)$. The number $\beta(v) = \nu(v) - 1$ is called the *branching number* at v. Observe that if v is not a branch point then $\beta(v) = 0$. In particular *M is non-flat and the Gaussian curvature of M vanishes only at a finite set of points.*

The following fact is well known in covering space theory:

Theorem 5.5

$$-2dg(\underline{\mathbf{n}}) = \chi(\overline{M}) + \sum_{w \in \overline{M}} \beta(w) \qquad \textit{(Riemann-Hurevitz relation)}, \tag{15}$$

where $dg(\underline{\mathbf{n}})$ is the degree of the Gauss map and $\chi(\overline{M})$ is the Euler characteristic of $\overline{M}$.

When clear from the context we will also say that M is a *surface of finite geometric type*.

Remark 5.6 Clearly complete, non-flat, minimal surfaces of finite total curvature are immersions of finite geometric type. The latter is a quite wider class. For example, it is stable for local small deformations near points of non-zero Gaussian curvature while minimal surfaces are not.

Remark 5.7 The concept of surface of finite geometric type can be extended to the case of hypersurfaces of $\mathbb{R}^{n+1}$. In this case condition (3) is replaced by the condition that the set of zeros of the Gauss-Kronecker curvature, i.e., the zeros of the determinant of the second fundamental form, does not disconnect M. The basic idea in introducing such hypersurfaces is that while the complex analysis methods for minimal surfaces, i.e., the Weierstrass representation formula, do not extend to the higher dimensional case, some of the topological methods do extend.

Definition 5.8 The points of E or, sometimes, punctured neighborhoods of such points, are called the *ends* of M.

The behavior of an immersion of finite geometric type near the ends is described in [10]. We will recall some basic facts. Let $w \in E$ be an end. The tangent space of $\overline{M}$ at w, as a linear subspace of $\mathbb{R}^3$, is well defined, namely $T_w\overline{M} = [\underline{\mathbf{n}}(w)]^\perp$. It can be shown that, for a sufficiently small neighborhood of w, the immersion, composed with the projection over $T_w\overline{M}$, is a covering map over the complement of a disk, of finite degree $I(w)$.

Definition 5.9 The number $I(w)$ is called the *geometric index* of w.

Remark 5.10 Geometrically, $I(w)$ counts the number of times that f warps a punctured neighborhood of w around the direction $\underline{\mathbf{n}}(w)$. If $I(w) = 1$ there exists a suitable punctured neighborhood W of w such that the projection of $f(W)$ over $[\underline{\mathbf{n}}(w)]^\perp$ is $1 - 1$. Hence $f(W)$ is a graph over the complement of a big ball $B \subseteq [\underline{\mathbf{n}}(w)]^\perp$. In particular $f_{|W}$ is a homeomorphism onto its image, i.e., an embedding. Conversely, if $f_{|W}$ is an embedding, $I(w) = 1$.

We will compute the Euler characteristic of $\overline{M}$ by counting the indexes of a suitable vector field on $\overline{M}$. Let ξ be a fixed unit vector in $\mathbb{R}^3$ such that ξ is a regular value of the Gauss map and $\xi \neq \pm\underline{\mathbf{n}}(w_i)$, $w_i \in E$. Consider the tangent vector field $\eta(x) = P_x(\xi)$ where P_x is the projection onto T_xM. η is the gradient of the height function $h_\xi(x) = \langle f(x), \xi\rangle$, which is a Morse function since ξ is a regular value of $\underline{\mathbf{n}}$. We observe that η extends to a tangent vector field on $\overline{M}$, setting, for $w \in E$, $\eta(w) = \xi - \langle\underline{\mathbf{n}}(w), \xi\rangle\underline{\mathbf{n}}(w)$. Then the singularities of η are the points in $\underline{\mathbf{n}}^{-1}(\pm\xi)$ and, possibly, the ends.

In [3] is shown that the Gauss curvature of a surface of finite geometric type is non-positive. In particular the index of η at a point in $\underline{\mathbf{n}}^{-1}(\pm\xi)$ is -1. For the ends we have the following:

Lemma 5.11 *The index of* η *at an end* w *is* $1 + I(w)$.

Proof We will sketch the proof for the case in which the end is embedded (see Remark 5.10) and refer to [3] for the general case.

The orthogonal projection of η over the complement of $B \subseteq [\underline{\mathbf{n}}(w)]^\perp$ is almost constant. Hence the index on the sphere bounding B is zero. Hence we can extend the projection to a non-vanishing vector field on B. We can take the stereographic projection of $[\underline{\mathbf{n}}(w)]^\perp$ over the sphere S^2. The image of the projection of η, $\tilde{\eta}$ gives a vector field on the sphere with just one singularity at the south pole, hence the index of this singularity is 2. Since the composition of f with the projection into $[\underline{\mathbf{n}}(w)]^\perp$ and the stereographic projection is an orientation preserving diffeomorphism of a small punctured neighborhood of w onto a small neighborhood of the south pole, the conclusion follows.

Adding up the indexes of η we get

Theorem 5.12

$$\chi(\overline{M}) = 2dg(\underline{\mathbf{n}}) + \sum_{w \in E}[I(w) + 1] \qquad (\textit{Total curvature formula}). \tag{16}$$

Remark 5.13 The total curvature formula was first proved by Osserman in [14], as an inequality, using the Weierstrass representation formula. So Osserman proof works only for minimal surfaces. The equality was proved in [10] and in [3] using only the topological properties of a surface of finite geometric type.

At this point a suitable combination of the Riemann-Hurevitz relation and the total curvature formula gives

Theorem 5.14 *The Gauss map of a surface of finite geometric type misses at most 3 points.*

Proof Let $\{\xi_1, \ldots, \xi_l\}$ be the set of points omitted by the Gauss map. Set

$$A_i = \{w \in E : \underline{\mathbf{n}}(w) = \xi_i\}, \quad B = \{w \in E : \underline{\mathbf{n}}(w) \neq \xi_i \ \forall\, i\},$$
$$C = \{q \in M : \nu(q) > 1\}.$$

Let $n = -dg(\underline{\mathbf{n}})$. Then Eq. (15) becomes

$$\chi(\overline{M}) = 2n + \sum_{i=1}^{l} \sum_{p \in A_i} (1 - \nu(p)) + \sum_{p \in B} (1 - \nu(p)) + \sum_{p \in C} (1 - \nu(p)). \tag{17}$$

Observe that

$$\sum_{p \in A_i} \nu(p) = n, \quad \sum_{i=1}^{l} |A_i| + |B| = |E|.$$

Then we have

$$\chi(\overline{M}) = (2 - l)n + |E| - \sum_{p \in B} \nu(p) + \sum_{p \in C} (1 - \nu(p)). \tag{18}$$

Comparing Eq. (18) with Eq. (16), we obtain

$$0 < \sum_{w \in (\cup A_i) \cup B} I(w) = (4 - l)n - \left[\sum_{p \in B} \nu(p) - \sum_{p \in C} (1 - \nu(p)) \right]. \tag{19}$$

Therefore $l < 4$.

There are no examples of surfaces of finite geometric type, in particular complete minimal surface with finite total curvature, whose Gauss map misses three points. There have been various tentatives to prove the following conjecture that we will call the *Osserman conjecture*:

Conjecture 5.15 The Gauss map of a complete minimal surface with finite total curvature omits at most two points.

We will discuss now some results in the direction of giving a positive answer to Osserman conjecture.

Proposition 5.16 *Let $f : M \longrightarrow \mathbb{R}^3$ be a surfaces of finite geometric type. If the Gauss map omits 3 points then $\chi(\overline{M}) \le 0$. Moreover, if $\chi(\overline{M}) = 0$ we have:*

(1) *$l = |E|$, i.e., all ends are omitted,*
(2) *$B = \emptyset = C$,*
(3) *$\sum I(p_i) = |E|$, i.e., all ends are embedded.*

Proof Just combine (19) with the total curvature formula.

A unit vector $\xi \in S^2$ is a regular value of the Gauss map if its inverse image $\underline{\mathbf{n}}^{-1}(\xi)$ does not contain flat points. In particular $\nu(p) = 1 \; \forall \; p \in \underline{\mathbf{n}}^{-1}(\xi)$. In order to extend this concept to an end $w \in E$, we have to take into account first that the curvature goes to zero approaching w and second that the end may not be embedded. The latter fact is measured by the geometric index $I(w)$. These considerations lead to the following definition:

Definition 5.17 We will say that an end $w \in E$ is *non-degenerate* if $\nu(w) \le 1 + I(w)$.

Examples 5.18 Let $f : M \longrightarrow \mathbb{R}^3$ be a minimal surface, w an end with $\underline{\mathbf{n}}(w) = e_3$. Suppose that the end is embedded. Then the end may be parameterized as the graph of a function F, defined on the complement of a disk in the $\{e_1, e_2\}$ plane. If $z = x_1 + ix_2$ is the complex coordinate in this plane, F is of the form

$$F(z) = a \log |z| + b + \langle z_0, z\rangle |z|^{-2} + O(|z|^{-2}),$$

where z_0 is a given vector in the plane (see [19]). If $a \neq 0$ the end is of *catenoid type*. If $a = 0, \; z_0 \neq 0$ we have a *simple flat end*. In both cases the end is non-degenerate.

There are also many examples of non-degenerate ends which are not embedded. For example, for the (unique) end w of the Enneper surface, we have $I(w) = 3, \; \nu(w) = 1$, hence the end is non-degenerate, but not embedded.

Theorem 5.19 *If $f : M \longrightarrow \mathbb{R}^3$ is a surface of finite geometric type and all end are non-degenerate, then the Gauss map omits at most two points.*

Proof Suppose that the Gauss map omits the (distinct) values $\xi_i, \; i = 1, 2, 3$. Assume first that $\xi_1 = -\xi_2$. Computing $\chi(\overline{M})$ (≤ 0 by Corollary 5.16) and counting the singularities of η, we obtain

$$0 \ge \chi(\overline{M}) = \sum_{w \in A_1 \cup A_2} [I(w) + 1 - \nu(w)] + \sum_{w \in A_3 \cup B} [I(w) + 1],$$

which together with the condition $\nu(w) \leq 1 + I(w)$ imply $A_3 \cup B = \emptyset$, a contradiction. Assume now that no two of the ξ_i's are parallel. Then we have

$$0 \geq \chi(\overline{M}) = \sum_{w \in A_1} [I(w) + 1 - \nu(w)] - n + \sum_{w \in A_2 \cup A_2 \cup B} [I(w) + 1],$$

which again lead to the contradiction $A_3 \cup B = \emptyset$.

For minimal surfaces Y. Fang proved, in [9], the following:

Theorem 5.20 *If $f : M \longrightarrow \mathbb{R}^3$ is a complete minimal surfaces with finite total curvature and*

$$\int_M k \geq -20\pi,$$

then the Gauss map misses at most two points.

Remark 5.21 Those results should take care of the hard cases, since, intuitively, the more complicated the topology/geometry is, the "more surjective" the Gauss map should be. But it turns out that this is not the case!

6 Work in Progress and Some Problems

The main idea behind the proof of Theorem 5.14 is to consider the gradient of the function h_ξ which is the projection of the surface onto the ξ-axes. In the last few years we have tried a "dual approach," i.e., considering projection of the surface onto a plane. The general philosophy is that the singularities of such maps are strongly related to the topology of M. There are classical results due to Levine, Whitney, and others that relate the topology of a *compact* surface to the singularities of maps of these surfaces into a plane (see [11, 21] between others). We were able to extend some of these results to the case of surfaces of finite geometric type, but still we will need something finer to give a positive answer to Osserman conjecture.

We are also studying a different approach: find a locally invertible conformal map $\pi : \mathbb{C} \longrightarrow M$. If such a map exists, the composition with the Gauss map will provide a holomorphic function $\phi : \mathbb{C} \longrightarrow S^2$ that, by Picard theorem, misses at most two points. The existence of such a function may be established looking at solutions of a Beltrami type equation

$$\frac{\partial}{\partial \bar{z}} W = \mu \frac{\partial}{\partial z} W,$$

where μ is an expression in the coefficients of the metric. Curvature estimates at the ends should imply that $\sup(|\mu(z)|) < 1$, a fact that guarantees the existence of solutions.

A natural question on this line is the characterization of minimal surfaces of finite total curvature whose Gauss maps miss exactly two points. In [13] Miyaoka and Sato constructed examples of minimal spheres (or tori) punctured tree times whose Gauss maps miss two points that are not antipodal. What can we say if the two missed points are antipodal?

We can also ask the same question for surfaces as above whose Gauss maps miss just one point.

The characterizations above may be intended in terms of the type and number of the ends, the value of the total curvature, and the genus g of the surface. For example, we can ask if there are minimal surfaces of finite total curvature, with one end of Enneper type, i.e., $I(w) = 3$, two ends of catenoid type, i.e., $I(w) = 1$ and total curvature $-4\pi(g+1)$.

References

1. J. L. Barbosa, M. P. do Carmo: *On the size of a stable minimal surface in* $\mathbb{R}^3$, Amer. J. of Math. vol 98 920 515–528.
2. J. L. Barbosa, G. Colares: *Minimal Surfaces in* $\mathbb{R}^3$, Lecture Notes in Mathematics 1195, Springer-Verlag (1986)
3. J. L. Barbosa, R. Fukuoka and F. Mercuri: *Immersions of finite geometric type in Euclidean spaces.* Annals of Global Analysis and Geometry vol. 22, 301–315 (2002)
4. R. Curant: *Dirichlet principle, conformal mappings and minimal surfaces* Interscience, N.Y. (1950)
5. M. P. do Carmo, C. K. Peng: *Stable complete minimal surfaces in* $\mathbb{R}^3$ are planes. Bull. Amer. Math. Soc. Vol. 1, 903–906 (1979).
6. J. Douglas: *Solution of the Plateau problem* Transactions of the AMS, vol 33, 263–321 (1931).
7. Y. Fang: *On the Gauss map of complete minimal surfaces with finite total curvature.* Indiana University Mathematical Journal, vol. 42 (4), 1389–1411 (1993).
8. D. Fischer Colbrie, R, Shoen: *The structure of stable minimal surfaces in 3-manifolds of non-negative scalar curvature.* Communications in Pure and Applied Mathematics, Vol. 33, 199–211 (1980).
9. H. Fujimoto : *On the number of exceptional values of the Gauss map of minimal surfaces.* Michigan Mathematics Journal, vol. 5, 235–247 (1988).
10. L. P. Jorge and W. H. Meeks III: *The topology of complete minimal surfaces of finite total Gaussian curvature.* Topology, Vol. 22 No 2, 203–221 (1983).
11. H. I. Levine: *Mappings of Manifolds into the Plane.* American Journal of Mathematics, vol. 88, No. 2, 357–365 (1966).
12. H. B. Lawson: *Lectures on minimal submanifolds* Mathematical Lectures Series 9, Publish or Perish. Berkeley (1980).
13. R. Miyaoka and K. Sato: *On complete minimal surfaces whose Gauss map misses two points.* Arch. Math., Vol. 63, 565–576 (1994).
14. R. Osserman: *Proof of a conjecture of Nirenberg.* Communication on Pure and Applied Mathematics, vol. 12, 229–232 (1959).
15. R. Osserman: *Global properties of minimal surfaces in* $\mathbf{E}^3$ *and* $\mathbf{E}^n$. Annals of Mathematics, Vol. 80, 340–364 (1964).
16. R. Osserman: *A survey on minimal surfaces* Van Nostrand-Rienhold, N.Y. (1969).
17. J. Peetre: *A generalization of Courant's nodal domain theorem.* Math. Scand. (5), 15–20
18. T. Radó: *On Plateau's problem.* Annals of Math, vol. 31, 457–469 (1930).

19. R. Schoen: *Uniqueness, symmetry and embeddedness of minimal surfaces*, J. Differential Geom. 18, (1983), 791–809.
20. F. Xavier: *The Gauss map of a complete non-flat minimal surface cannot omit 7 points of the sphere.* Annals of Mathematics (2), vol. 113, 211–214 (1981).
21. H. Whitney: *On the singularities of mappings of Euclidean spaces I. Mappings of the plane into the plane* Annals of Mathematics, vol. 62, 374–410 (1955).

Galois Theories: A Survey

Antonio Paques

Abstract We present an overview of the various Galois theories that appeared in the literature since Évariste Galois until to the present day, accompanied with a bit of the inherent history.

1 A Brief Introduction

The celebrated work of Évariste Galois (1811–1832) constitutes a true landmark in the development of mathematics.

On the one hand, it gave a definitive answer to one of the main problems of that time, namely: to decide under what conditions an algebraic equation, regardless of its degree, is solvable by radicals, that is, has roots that can be described in terms of radicals involving its own coefficients.

On the other hand, the ideas contained in that work contributed meaningly to the arising of a modern algebraic language, thanks to the contributions of Richard Dedekind (1831–1916), Leopold Kronecker (1823–1891), and Emil Artin (1898–1962) among other mathematicians, and theories such as the classical field theory (in particular, the finite field theory that allowed the arising of an error-correcting codes theory), the group theory, the linear algebra, the commutative algebra, the algebraic geometry, the algebraic theory of numbers, the arithmetic of fields and, in particular, new Galois theories. It is this last item that we will deal on in this manuscript.

A. Paques (✉)
Institute of Mathematics, Federal University of Rio Grande do Sul, Porto Alegre, RS, Brazil
e-mail: paques@mat.ufrgs.br

C. Lavor, F. A. M. Gomes (eds.), *Advances in Mathematics and Applications*,
https://doi.org/10.1007/978-3-319-94015-1_11

2 A Bit of the Starting History

Galois's work, in spite of its great importance, came to be published only in 1846, by Joseph Liouville (1809–1882), in the Journal of Mathématiques Pures et Appliqueés. However, its 60-page text was still difficult to understand and did not arouse greater interest in the scientific community at the time. Despite the contribution of Enrico Betti (1823–1892) in the sense of to make this text clearer, detailing certain still obscure passages, and completing some demonstrations, the ideas and discoveries of Galois were only known and appreciated in all their amplitude with the publication in 1870 by Camille Jordan (1838–1922) of his "*Traité des substitutions et des équations algébriques*." From there, the theory of Galois gains notoriety and inspires the arising of similar theories in more general by contexts. We mention, by way of illustration, some of the works published in the period 1880–1950, due respectively to

1. Charles Émile Picard (1856–1941) and Ernest Vessiot (1865–1952): a differential Galois theory for homogeneous linear differential equations (see [58]),
2. Wolfgang Krull (1899–1971): a Galois theory for field extensions of infinite dimension [61],
3. Henri Cartan (1904–2008) and Nathan Jacobson (1910–1999): a Galois theory for division rings [20, 52],
4. Jean Dieudonné (1906–1992), Gerhard Hochschild (1915–2010), Goro Azumaya (1920–2010), and Tadasi Nakayama (1912–1964): a Galois theory for simple rings [4, 33, 51, 71, 72].

3 A Comment

Roughly speaking, in his work Galois dealt with the following basic objects:

1. a polynomial $f(X)$ with coefficients, in general, in a subfield K of the complex number field $\mathbb{C}$,
2. the field of the roots (or the splitting field) of $f(X)$, that is, the smallest subfield L of $\mathbb{C}$ including K and all the complex roots of such a polynomial, later called a *Galois extension of* K, and
3. the group $\mathrm{Aut}_K L$ of all K-automorphisms of L, indeed the group of the permutations of the roots of $f(X)$ leaving invariant the coefficients of $f(X)$, later called the *Galois group of* $f(X)$.

Essentially, Galois dealt with a group acting on a field by automorphisms and investigated the correlation between the set of all subfields of such a given field, including the subfield of the invariants under such a given action, and the set of all subgroups of the given group.

Taking into account such remarks, one can even say that the foundations of the theory developed by Galois consist in fact of two main theorems:

- a **definition theorem** that presents a list of equivalent statements that characterize the notion of a Galois extension, and
- a **correspondence theorem** that states a bijection between the subfields of a Galois field extension L of a given field K, including K, and the set of all subgroups of the group $\mathrm{Aut}_K L$, later called **Fundamental Theorem of Galois Theory**.

Our aim is to address the many contributions to generalize such theorems to other contexts that appeared in the literature along of these last seventy years.

4 Definition Theorem

4.1 On Finite Field Extensions

The formalism to enounce this theorem for finite field extensions starts with Artin [3] proving the equivalence of the following statements for any finite field extension L of a field K:

(a) The order of the group $\mathrm{Aut}_K(L)$ is equal to the dimension of L over K.
(b) K is the field of the elements x of L such that $\sigma(x) = x$ for all $\sigma \in \mathrm{Aut}_K(L)$.
(c) L is a splitting field of a separable polynomial with coefficients in K.
(d) L is a normal and separable extension of K.

Such statements are also equivalent to several other ones, in special to the following:

(e) There exists a one-to-one correspondence between the subfields of L including K and the subgroups of $\mathrm{Aut}_K(L)$.

Any finite field extension L of a field K satisfying one of the equivalent statements above enumerated is called a *Galois extension* of K.

Another fact that deserves to be pointed out is the following: any field extension $L \supseteq K$ together with any group G of K-automorphisms of L determine a new algebraic structure, namely the skew group algebra $L\star G$, which is defined as being the L-vector space with basis $\{u_g \mid g \in G\}$, endowed with a multiplication induced by the rule

$$(xu_g)(yu_h) = xg(y)u_{gh}, \quad \text{for all } x, y \in L \text{ and } g, h \in G,$$

as well as a canonical algebra homomorphism φ from $L\star G$ to the algebra $\mathrm{End}_K(L)$ of all K-endomorphisms of L, defined by

$$\varphi(xu_g) : y \mapsto xg(y), \quad \text{for all } x, y \in L \text{ and } g \in G.$$

This map φ is indeed injective, by a result due to R. Dedekind, known in the literature as Dedekind's lemma, which ensures, in particular, that the elements of G as ring endomorphisms of L are free over L. In the case that L is finite dimensional over K and $G = \mathrm{Aut}_K(L)$ the statements above listed are equivalent to

(f) φ is an algebra isomorphism,

which indeed ensures that the product in $L\star G$ corresponds to the product of matrices of order equal to the square of the order of G with entries in L.

This last statement is in particular very special because it has opened the door that allowed to extend the notion of Galois extension to the context of commutative rings.

4.2 On Commutative Ring Extensions

Maurice Auslander (1926–1994) and Oscar Goldman (1925–1986) were the first to introduce in the literature the notion of Galois extension for commutative rings, with the publication in 1960 of their celebrated paper *"The Brauer group of a commutative ring"* [7], which established the foundations of a general theory of separable algebras over a commutative ring.

Following them, given a commutative ring S, a subring R of S and a finite subgroup G of the group $\mathrm{Aut}_R(S)$ of all ring automorphisms of S leaving R elementwise fixed, S is called a *Galois extension of R relative to G* if S is a finitely generated projective R-module (i.e, as R-module S is a direct summand of a free R-module of finite rank), and the map $\varphi : S \star G \to \mathrm{End}_R(S)$ similarly defined as above is a ring isomorphism. In particular, if S and R are fields the corresponding definition coincides with the one given above in the previous subsection. Hence, such definition is indeed the correct generalization of the notion of a Galois extension to the context of commutative rings.

The Galois theory for commutative rings is due to Stephen U. Chase, David K. Harrison, and Alex Rosenberg (1926–2007), published later in 1965 in their celebrated paper *"Galois theory and Galois cohomology of commutative rings"* [23]. In this paper they present a new list of statements characterizing the notion of Galois extension for commutative rings, which we will enumerate below. To do this we need before some complementary information.

On Separability

First of all some words about separability. In the field case this concept concerns to the simplicity of roots of a polynomial, that is, an extension field L of a field K is said to be separable over K if every element of L is a root of multiplicity 1 of its minimal polynomial over K.

In a more general approach the following statements are equivalent:

(a) L is a separable finite field extension of K.
(b) L is a projective $L \otimes_K L$-module via the multiplication map $\mu : L \otimes_K L \to L$.
(c) μ is an (L, L)-bimodule homomorphism that splits.
(d) L is a direct summand of $L \otimes_K L$ as an $L \otimes_K L$-module.

Such characterizations of separability in the field case have induced the following definition in the most general context of algebras (not necessarily commutative) over commutative rings: an algebra A over a commutative ring R is said to be *separable over* R (or R*-separable*) if the multiplication map $\mu : A \otimes_R A \to A$ is an (A, A)-bimodule homomorphism that splits. A more detailed approach about separability, from fields to algebras, can be seen in [75] (see also [74]).

Notice that in the definition of separability for algebras it is not required any kind of finiteness. However, if A and R are fields, then the finite dimension of A over R is forcibly recovered (see, for instance, [62, Proposition III.3.2]). Moreover, the elements of an R-separable algebra in general are not roots of any polynomial with coefficients in R, for instance, the rational field $\mathbb{Q}$ is a $\mathbb{Z}$-separable algebra but none element of $\mathbb{Q} \setminus \mathbb{Z}$ is integral over $\mathbb{Z}$.

On Strong Distinctness

Two algebra homomorphisms $\sigma, \tau : A \to B$ are said to be *strongly distinct* if for every nonzero idempotent $e \in B$ there exists $x \in A$ such that $\sigma(x)e \neq \tau(x)e$. Evidently, these notions "distinct morphisms" and "strongly distinct morphisms" coincide if, for instance, B is either a field, a domain, a local ring or more generally a connected ring. Furthermore, strong distinctness is equivalent to Dedekind's lemma in the context of separable algebras (see [47, Proposition 2.1]).

On $S \star G$-Modules

The notation $S\star G$, where $S \supseteq R$ is a commutative ring extension and $G \subseteq \mathrm{Aut}_R(S)$ is a subgroup, is analogous to the one introduced in the former subsection, that is, $S \star G$ denotes the free S-module with basis $\{u_g \mid g \in G\}$ endowed with an structure of a noncommutative R-algebra via the multiplication induced by the rule: $(xu_g)(yu_h) = xg(y)u_{gh}$, for all $x, y \in S$ and $g, h \in G$.

Every left $S \star G$-module M is in particular an S-module on which the group G acts. We will denote by M^G the R-submodule of the elements of M invariant by the action of G, that is, $M^G = \{m \in M \mid u_g \cdot m = m, \text{ for all } g \in G\}$. In particular S is a left $S \star G$-module via the action $xu_g \cdot y = xg(y)$, for all $x, y \in S$ and $g \in G$, and in this case $S^G = \{y \in S \mid g(y) = y\}$.

The definition theorem to characterize a Galois extension as introduced in [23] by Chase, Harrison and Rosenberg (CHR, for short) is the following:

Theorem 4.1 ([23, Theorem 1.3]) *Let $S \supseteq R$ be a commutative ring extension and G a finite group of ring automorphisms of S. Then, the following statements are equivalent:*

(a) *$S^G = R$, S is separable over R and the elements of G are all pairwise strongly distinct.*
(b) *There exist elements x_i, y_i in S, $1 \leq i \leq n$, such that $\sum_{1 \leq i \leq n} x_i g(y_i) = \delta_{1,g} 1_S$, for all $g \in G$.*
(c) *S is a finitely generated projective R-module and the map $\varphi : S \star G \to \mathrm{End}_R(S)$, given by $\varphi(xu_g)(y) = xg(y)$, is an isomorphism of S-modules and R-algebras.*
(d) *For any left $S \star G$-module M the map $\mu : S \otimes_R M^G \to M$, given by $\mu(x \otimes m) = xm$ is an isomorphism of S-modules.*
(e) *The map $\psi : S \otimes_R S \to \prod_{g \in G} S$, given by $\psi(x \otimes y) = (xg(y))_{g \in G}$, is an isomorphism of S-algebras.*
(f) *$S^G = R$ and for each maximal ideal $\mathfrak{m}$ of S and each $1 \neq g \in G$ there exists $y \in S$ such that $g(y) - y \notin \mathfrak{m}$.*

If a triple (S, R, G) is as in Theorem 4.1, S is called a *Galois extension of R with Galois group G* if one of the equivalent statements listed above is satisfied.

Theorem 4.1 is undoubtedly the closest to the classical definition theorem for fields, one can even say that it is a natural generalization of the previous one. In the context of commutative connected rings, that is, rings whose unique idempotents are 0 and 1, like fields, it is clear how much this generalization is absolutely natural, for in such cases the notions of distinct automorphisms and strongly distinct coincide. However, it is not an absolute generalization for commutative rings because the assumed restriction on the idempotents.

A more general approach to commutative rings is due to Orlando Eugênio Villamayor (1923–1998) and Daniel Zelinsky (1922–2015).

Villamayor and Zelinsky (VZ, for short) dealt with rings with no restriction on their idempotents and developed a theory, so-called *weak Galois theory*, between the years 65–69, published in two articles entitled respectively "*Galois theory with finitely many idempotents*" [90] and "*Galois theory with infinitely many idempotents*" [91]. In their theory a ring extension $S \supseteq R$ is called *weak Galois* if the following three conditions are satisfied:

(a) S is a separable R-algebra,
(b) S is a finitely generated projective R-module, and
(c) $S^{\mathrm{Aut}_R(S)} = R$.

Notice that any Galois extension in the sense of CHR satisfies the above three conditions, that is, it is a Galois extension in the sense of VZ. Also, the definition of weak Galois extension makes no mention of any fixed subgroup of the group $\mathrm{Aut}_R(S)$. Actually, a weak Galois extension turns out to be Galois with respect to several distinct finite subgroups of $\mathrm{Aut}_R(S)$.

4.3 On Ring Extensions

We will divide this subsection into two parts.

4.3.1 Still on Group Actions

In addition to the contributions mentioned in Sect. 2, particularly due to Cartan, Jacobson, Dieudonné, Hochschild, Azumaya, and Nakayama, in specific ring contexts, many other mathematicians have also contributed to the research related to a Galois theory for rings in general. The list of them is long, but particularly we would like to highlight the contributions of DeMeyer [30, 31], Kreimer [59], Kanzaki [55, 56], Ferrero [45, 46], Dress [40], Cohen [26], Cohen, Fischman and Montgomery [27], Montgomery [66], Montgomery and Passman [68], Passman [80], and Kharchenko [57] among many others.

The equivalent statements that characterize the notion of a Galois extension in the general ring case are analogous to those listed in Theorem 4.1. With the necessary adequations to this new context, the definition theorem is enounced as follows:

Theorem 4.2 *Let $S \supseteq R$ be a ring extension and G a finite group of ring automorphisms of S such that $S^G = R$. Then, the following statements are equivalent:*

(a) The map $\psi : S \otimes_R S \to \prod_{g\in G} S$, given by $\psi(x \otimes y) = (xg(y))_{g\in G}$, is an isomorphism of left S-módulos.
(b) There exist elements $x_i, y_i, \in S$, $1 \le i \le n$, such that $\sum_{1\le i\le n} x_i g(y_i) = \delta_{1,g} 1_S$ for all $g \in G$.
(c) S is a finitely generated projective right R-module and the map $\varphi : S \star G \to End(S_R)$, given by $\varphi(xu_g)(y) = xg(y)$, is an isomorphism of left S-modules and right R-modules.
(d) For any left $S\star G$-module M the map $\mu : S\otimes_R M^G \to M$, given by $\mu(x\otimes m) = xm$ is an isomorphism of left S-modules.
(e) The map $\delta : S \otimes_R S \to S \star G$, given by $\delta(x \otimes y) = \sum_{g\in G} xg(y)u_g$, is an epimorphism of $S \star G$-bimodules.
(f) $StS = S \star G$, where $t = \sum_{g\in G} u_g$.
(g) S is a generator for the category of all left $S \star G$-modules. □

With respect to the above statement (f) we observe that S can be seen as an (S, R)-subbimodule of $S \star G$ via $x \hookrightarrow xt$, hence the notation StS makes sense. Actually, StS is an ideal of $S \star G$. The ring S is called a *Galois extension of R with Galois group G* if one of the above statements holds.

It is also interesting to notice that such a definition of Galois extension keeps a very close relation with a suitable Morita context. By Morita context we mean a sixtuple $(A, B, U, V, \gamma, \delta)$ where A and B are rings, U is a (A, B)-bimodule, V is a (B, A)-bimodule, $\gamma : U \otimes_B V \to A$ is an A-bimodule map and $\delta : V \otimes_A U \to B$ is a B-bimodule map, and the following two associative conditions hold:

$$u \cdot \delta(v \otimes u') = \gamma(u \otimes v) \cdot u' \quad \text{e} \quad \delta(v \otimes u) \cdot v' = v \cdot \gamma(u \otimes v'),$$

for all $u, u' \in U$ e $v, v' \in V$. If γ and δ are isomorphisms we say that this context is *strict* and, in this case, the categories ${}_AMod$ and ${}_BMod$ are equivalent via the mutually inverse equivalences $V \otimes_A - : {}_AMod \to {}_BMod$ and $U \otimes_B - : {}_BMod \to {}_AMod$. When it happens we also say that the rings A and B are *Morita equivalent*. If ${}_AU$ is faithfully projective, that is, ${}_AU$ is faithful, projective, and finitely generated, then it is enough the surjectivity of γ and δ in order to have the strictness of the above context [82, Theorems 4.1.4 and 4.1.17].

Now take a triple (S, R, G) as in Theorem 4.2 and notice that S is an R-bimodule (resp., a $S \star G$-bimodule) via the multiplication of S (resp., via the following left and right actions: $xu_g \cdot y = xg(y) = \varphi(xu_g)(y)$ and $y \cdot xu_g = g^{-1}(xy)$). Also, it is straightforward to see that if δ and t are as above defined and $\gamma : S \otimes_{S \star G} S \to R$ is the R-bimodule map given by $\gamma(x \otimes y) = t \cdot (xy)$, then the sixtuple $(R, S \star G, S, S, \gamma, \delta)$ is a Morita context. Furthermore, if S is a Galois extension of R, then the additional statements listed below are also equivalent:

(h) γ is surjective.
(i) $t \cdot S = R$.
(j) S is a generator for the category of all right R-modules.
(l) The Morita context $(R, S \star G, S, S, \gamma, \delta)$ is strict.

It follows from the above that the notion of Galois extension is equivalent to the strictness of the corresponding Morita context, whenever the map γ is surjective. This map γ is also called the trace map because it induces by restriction the map $t_{S/R} : S \to R$ given by $t_{S/R}(x) = t \cdot x = \sum_{g \in G} g(x)$. In the commutative case the map γ is surjective and therefore we have Theorem 4.1 expanded by six additional statements. In particular the notion of Galois extension and the strictness of the above described Morita context are equivalent in the commutative case.

4.3.2 On Hopf Actions

The theories dealt with in the previous subsections can be considered generalizations of the Galois theory for fields in the context of group actions on rings or algebras in general.

In this subsection we will deal with another generalization, the one that extends the classical Galois theory to the context of Hopf algebra actions (shortly, Hopf actions) on algebras. Such a generalization is quite natural considering that, given a fields extension $L \supseteq K$ and a subgroup G of $\mathrm{Aut}_K(L)$, the action of G on L determines univocally an action of the group algebra KG on the K-algebra L via $\lambda g \cdot x := \lambda g(x)$, for all $\lambda \in K$, $g \in G$ e $x \in L$. Group algebras are perhaps the simplest examples of Hopf algebras.

In all what follows K will denote a commutative ring. A Hopf algebra is a K-algebra H provided with two K-algebra homomorphisms $\Delta : H \to H \otimes_K H$ (called *comultiplication*) and $\varepsilon : H \to K$ (called *counit*), and a K-algebra anti-homomorphism $S : H \to H$ (called *antipode*) satisfying the following properties:

1. $(\Delta \otimes I_H) \circ \Delta = (I_H \otimes \Delta) \circ \Delta$,
2. $(\varepsilon \otimes I_H) \circ \Delta \equiv I_H \equiv (I_H \otimes \varepsilon) \circ \Delta$,
3. $S * I_H = 1_H \varepsilon = I_H * S$,

where I_H denotes the identity map of H and, for all $f, g \in \mathrm{End}_K(H)$, $f * g = \mu \circ (f \otimes g) \circ \Delta$, with $\mu : H \otimes_K H \to H$ denoting the multiplication of H.

For simplicity we will use the Heyneman-Sweedler notation for Δ, that is, $\Delta(h) = h_1 \otimes h_2$ (summation understood), for all $h \in H$.

In order to illustrate the above definition we will consider the following three classical examples:

Exemples 4.3

(1) For any group G, the group algebra KG is a free K-module with basis G provided with a multiplication induced by the rule $(ag)(bh) = abgh$, for all $a, b \in K$ e $g, h \in G$, whose identity element is $1_K 1_G$. Furthermore, KG is a Hopf algebra with comultiplication, counit, and antipode given, respectively, by the maps

$$\Delta(g) = g \otimes g, \quad \varepsilon(g) = 1_K \quad \text{e} \quad S(g) = g^{-1},$$

for all $g \in G$.

(2) For any finite group G, the dual KG^* of the group algebra KG is a free K-module with basis $\{p_g \mid g \in G\}$ given by the rule $p_g(h) = \delta_{g,h}$ for all $g, h \in G$, and has a Hopf algebra structure with multiplication induced by the rule $p_g * p_h = \delta_{g,h} p_g$, whose identity element is the counit ε_{KG} of KG. Notice that $\varepsilon_{KG} = \sum_{g \in G} p_g$. The comultiplication, counit, and antipode of KG^* are, respectively, given by the maps

$$\Delta_{KG^*}(p_g) = \sum_{h \in G} p_h \otimes p_{h^{-1}g}, \quad \varepsilon_{KG^*}(p_g) = p_g(1_G) = \delta_{1_G, g}$$

and

$$S_{KG^*}(p_g) = p_{g^{-1}},$$

for all $g \in G$.

(3) The enveloping $U(\mathfrak{g})$ of a Lie algebra $\mathfrak{g}$ is also a Hopf algebra with comultiplication, counit, and antipode, respectively, defined as follows:

$$\Delta(x) = x \otimes 1 + 1 \otimes x, \quad \varepsilon(x) = 0 \quad \text{and} \quad S(x) = -x,$$

for all $x \in \mathfrak{g}$, with multiplicative extension to $U(\mathfrak{g})$.

The notion of Galois-Hopf extension has its roots in the CHR Galois theory, whose ideas were initially extended by S.U. Chase and M.E. Sweedler to the context of coactions of Hopf algebras (shortly Hopf coactions) on algebras in [24]. The general definition is due to Kreimer and Takeuchi and appears in [60]. To present it, we need some preparation.

Let A be a K-algebra and H a Hopf algebra. We say that A is a (left) *H-module algebra* if there exists a linear map $\cdot : H \otimes_K A \to A$ (called a left *action* of H on A) such that

1. A is a left H-module via the action $h \otimes a \mapsto h \cdot a$,
2. $h \cdot (ab) = (h_1 \cdot a)(h_2 \cdot b)$,
3. $h \cdot 1_A = \varepsilon(h)1_A$,

for all $a, b \in A$ e $h \in H$. It can be easily seen that the set

$$A^H = \{a \in A \mid h \cdot a = \varepsilon_H(h)a, \text{ for all } h \in H\},$$

of the elements of A *invariant* by $\cdot$ is a subalgebra of A.

We say that A is a (right) *H-comodule* if there exists a linear map $\rho : A \to A \otimes_K H$ (called a right *coaction* of H on A) such that

1. $(I_A \otimes \Delta) \circ \rho = (\delta \otimes I_A) \circ \rho$,
2. $(I_A \otimes \varepsilon) \circ \rho = \otimes 1_K$,

where I_A denotes the identity map of A.

We say that A is a (right) *H-comodule algebra* if

1. A is a right H-comodule via the coaction $\rho : a \mapsto a_0 \otimes a_1$,
2. ρ is an algebra homomorphism.

The set

$$A^{coH} = \{a \in A \mid \rho(a) = a \otimes 1_H\},$$

of the elements of A *coinvariant* by ρ is also a subalgebra of A.

If H is a finitely generated projective K-module, then its dual $H^* = Hom_K(H, K)$ is also a Hopf algebra with

- multiplication given by $(f * f')(h) = f(h_1)f'(h_2)$,
- unit given by $1_{H^*} = \varepsilon_H$,
- comultiplication given by $\Delta_{H^*}(f) = f_1 \otimes f_2 \iff f(gh) = f_1(g)f_2(h)$
- counit given by $\varepsilon_{H^*}(f) = f(1_H)$,

for all $g, h \in H$ e $f, f' \in H^*$.

For the sequel we will assume that H is a Hopf algebra that as a K-module is finitely generated and projective.

In this context the notions of H-module algebra and H-comodule algebra are, respectively, dual, that is, an algebra A is a left H-module algebra if and only if it is a right H^*-comodule algebra. In this case the coaction $\rho : A \to A \otimes_K H^*$, of H^*

on A, is given by

$$\rho(a) = a_0 \otimes a_1 \quad \Longleftrightarrow \quad h \cdot a = a_1(h)a_0,$$

for all $a \in A$ e $h \in H$. Moreover, $A^{coH^*} = A^H$.

Example 4.4 Let G be a finite group of K-automorphisms of a K-algebra A. Then A is a left KG-module algebra via the action given by $h \cdot a = h(a) = \sum_{g \in G} p_g(h)g(a)$, for all $a \in A$ and $h \in G$, if and only if A is a right KG^*-comodule algebra via the coaction given by $\rho(a) = \sum_{g \in G} g(a) \otimes p_g$, for all $a \in A$. Furthermore, $a \in A^{coKG}$ if and only if $\sum_{g \in G} g(a) \otimes p_g = \rho(a) = a \otimes \varepsilon_{KG} = \sum_{g \in G} a \otimes p_g$, if and only if $g(a) = a$, for all $g \in G$, if and only if $a \in A^G = A^{KG}$.

Now we are in conditions to introduce the definition of a Hopf-Galois extension.

Let A be a right H-comodule algebra and $\rho : A \to A \otimes_K H$ the coaction of H on A. We say that A is a right H-*Galois* extension of A^{coH} if the map

$$\beta : A \otimes_{A^{coH}} A \to A \otimes_K H, \quad a \otimes b \mapsto (a \otimes 1_H)\rho(b)$$

is bijective. Analogously, if A is a left H-module algebra then A is a right H^*-comodule algebra, $A^{coH^*} = A^H$ and A is a right H^*-Galois extension of A^H if the corresponding map $\beta : A \otimes_{A^H} A \to A \otimes_K H^*$ is bijective. Notice that the map β is a homomorphism of left A-modules. The definition of a left H-Galois extension is similar.

Example 4.5 Consider A and G as in Example 4.4. Observe that $A \otimes_K KG^*$ and $\prod_{g \in G} A$ are isomorphic K-algebras via the map $\sum_{g \in G} a_g \otimes p_g \mapsto (a_g)_{g \in G}$. Em particular, $\beta(a \otimes b) = \sum_{g \in G} ag(b) \otimes p_g \mapsto (ag(b))_{g \in G}$, hence A is a KG^*-Galois extension of A^{KG} if and only if A is a Galois extension of A^G, with Galois group G (see Theorem 4.2(a)).

There are several equivalent definitions of a Hopf-Galois extension, similar to those from CHR theory (see Theorem 4.1) coming from the contributions of Kreimer and Takeuchi in [60], Ulbrich in [89], Doi and Takeuchi in [38], Cohen, Fischman and Montgomery in [27], and Ouyang in [73]. In order to enumerate such definitions we need to introduce the notions of integral and smash product.

Un element t of a Hopf algebra H is called a right (resp., left) *integral* in H if $th = \varepsilon(h)t$ (resp., $ht = \varepsilon(h)t$), for all $h \in H$. The set of the right (rep., left) integrals in H is a submodule of H and is denoted by $\int_H^r$ (resp., $\int_H^l$). For instance, if $H = KG$ (resp., $H = KG^*$, with G being finite), then então $\int_H^r = \int_H^l = Kt$ with $t = \sum_{g \in G} g$ (resp., $t = p_{1_G}$).

Given a left H-module algebra A, the *smash product* of A by G is the noncommutative K-algebra, denoted by $A\#G$, which as a K-module coincides with the tensor product $A \otimes_K H$ and has multiplication induced by the rule

$$(a\#h)(b\#l) = a(h_1 \cdot b)\#h_2 l,$$

for all $a, b \in A$ and $h, l \in H$, whose unit is $1_A\#1_H$. For instance, if $H = KG$, then $A\#H$ and $A \star G$ are isomorphic as K-algebras.

The algebras A and H can be seen as subalgebras of $A\#H$, via the respective immersions $a \hookrightarrow a\#1_H$ and $h \hookrightarrow 1_A\#h$, for all $a \in A$ and $h \in H$. Also, if M is a left $A\#H$-module, then M also is a left A-module, a left H-module and

$$M^H = \{m \in M \mid h \cdot m = \varepsilon(h)m, \text{ for all } h \in H\}$$

is a left A^H-submodule of M.

Theorem 4.6 *Let A be a left H-module algebra and assume that H as a K-module is finitely generated and projective. Then the following statements are equivalent:*

(a) A is a H^*-Galois extension of A^H.
(b) There exists elements $x_1, \ldots, x_m, y_1, \ldots, y_m \in A$ and $T \in \int_{H^*}^r$ such that $\sum_{1 \leq i \leq m} x_i(h \cdot y_i) = T(h)1_A$, for all $h \in H$.
(c) A is a right finitely generated and projective A^H-module and the map $\varphi : A\#H \to \mathrm{End}(A_{A^H})$, given by $\varphi(a \otimes h)(x) = a(h \cdot x)$, for all $a, x \in A$ and $h \in H$, is an isomorphism of algebras.
(d) For any left $A\#H$-module M, the map $\mu : A \otimes_{A^H} M^H \to M$, given by $\mu(a \otimes m) = a \cdot m$, for all $a \in A$ and $m \in M^H$, is an isomorphism of left $A\#H$-modules.
(f) If $0 \neq t \in \int_H^l$, then the map $[,] : A \otimes_{A^H} A \to A\#H$, given by $[a, b] = atb$, is surjective.
(g) A is a generator in the category of all left $A\#H$-modules. □

Notice that if one takes $H = KG$ in Theorem 4.6, with G being finite, then Theorem 4.2 in the K-algebras context is recovered.

5 Correspondence Theorem

5.1 On Group Actions

5.1.1 In the Classical Galois Theory for Field Extensions

As it is well known, in this context the theorem of correspondence is the following:

Theorem 5.1 (Fundamental Theorem of Galois Theory) *Let L be a Galois extension field of a field K and $G = \mathrm{Aut}_K(L)$. Then there exists a one-to-one correspondence between the subgroups of G and the subfields of L including K.*

Such a correspondence associates to each subgroup H of G the subfield of L given by

$$L^H = \{x \in L \mid h(x) = x, \textit{ for all } \ h \in H\}$$

and to each subfield F of L including K the subgroup of G given by

$$H_F = \{g \in G \mid g(x) = x, \textit{ for all } x \in F\} = \mathrm{Aut}_F(L).$$

□

As seen in Sect. 4.2, any Galois extension field L of a field K is a K-separable algebra and any of its subfields containing K is a K-separable subalgebra of L. Hence, under such an approach one can say that indeed Theorem 5.1 assures the existence of a bijection between subgroups of G and subalgebras of L that are separable over K. Also, distinct restrictions of distinct elements of G to any subalgebra of L are strongly distinct. This is the approach in the CHR Galois theory.

5.1.2 In the CHR Galois Theory for Commutative Ring Extensions

Let R, S, and G be as in Theorem 4.1. For each subalgebra T of S and each subgroup H of G we denote

$$H_T = \{g \in G \mid g(t) = t, \text{ for all } t \in T\} \text{ and } S^H = \{s \in S \mid h(s) = s, \text{ for all } h \in H\}.$$

Clearly H_T is a subgroup of G as well as S^H is a subalgebra of S that contains R.

A subalgebra T of S is called *G-strong* if distinct restrictions of any distinct two elements of G to T are strongly distinct as maps from T to S.

In the sequel we have the correspondence theorem called the fundamental theorem of the Galois theory due to Chase, Harrison and Rosenberg (shortly, CHR Galois theory). In fact, this theorem is a natural consequence of a more general theorem of correspondence due to Grothendieck for group actions on sets, which we will see in the next subsection.

Theorem 5.2 (Fundamental Theorem of CHR Galois Theory) *Let S be a Galois extension of R with Galois group G.*

Then there exists a one-to-one correspondence between the subgroups of G and the subalgebras of S including R that are G-strong and separable over R.

Such a correspondence associates to each subgroup H of G the subalgebra S^H *of S and to each G-strong and R-separable subalgebra T of S including R the subgroup* H_T *of G.* □

5.1.3 In the Grothendieck's Approach for Group Actions on Sets

The theory of Galois due to Grothendieck, in its totality, is contextualized in the schema language [50] (also see [2]). A version of this theory in the specific context of fields was presented by Dress in [39] and by Borceux and Janelidze in [15, Chap. 2]. Basically this theory presents a new interpretation of the classical Galois theory for field extensions in terms of K-algebras that are L-split (hence, finite) and G-sets, where L is a Galois extension of K with $G = \mathrm{Aut}_K(L)$.

Our purpose in this subsection is to present a generalization of Theorem 5.2 following the Grothendieck's approach for group actions on sets. As in the previous subsection rings and algebras are assumed to be commutative.

Let $S \supseteq R$ be a ring extension and A an R-algebra. We say that A is *weakly S-split* if there is $n > 0$ such that $S \otimes_R A$ is isomorphic to S^n as S-algebras. The following example gives us a good illustration of this concept.

Example 5.3 It is well known that if a field L is a finite and separable extension of a field K, then there exists an element α in L such that $L = K[\alpha] \simeq \frac{K[X]}{(m_{\alpha,K}(X))}$ where $m_{\alpha,K}(X)$ denotes the minimal polynomial of α over K. Besides this, if N is a splitting field of $m_{\alpha,K}(X)$ including L, then there exist pairwise distinct elements $\alpha_1 = \alpha, \alpha_2, \ldots, \alpha_n \in N$ such that $m_{\alpha,K}(X) = \prod_{1\le i\le n}(X - \alpha_i)$, which implies the following sequence of algebra isomorphisms

$$N \otimes_K L \simeq N \otimes_K \frac{K[X]}{(m_{\alpha,K}(X))} \simeq \frac{N[X]}{\prod_{1\le i\le n}(X-\alpha_i)} \simeq \prod_{1\le i\le n} \frac{N[X]}{(X-\alpha_i)} \simeq N^n.$$

Hence, L is a weakly N-split algebra.

If $S \supseteq R$ is a ring extension, then any weakly S-split R-algebra A gives rise to the finite set $X(A)$ of all maps from A to S given by the following composition of maps:

$$\varphi_i = \pi_i \circ \varphi \circ \iota : A \xrightarrow{\iota} S \otimes A \xrightarrow{\varphi} S^n \xrightarrow{\pi_i} S$$

where $\iota : a \mapsto 1_R \otimes a$ is the canonical immersion of A into $S \otimes A$ (notice that $1_R = 1_S$), φ is the algebra isomorphism ensured by the assumption on A, and π_i is the canonical projection from S^n onto its ith-summand. By construction, $\varphi(s \otimes a) = (s\varphi_i(a))_{1\le i\le n}$, for all $s \in S$ and $a \in A$.

Given a group G and a nonempty set X, we say that X is a *G-set* if G acts on X by permutations of its elements, that is, if there exists a group homomorphism from G into the group $\mathfrak{S}_X$ of all permutations of X. For instance, G itself or more generally the set G/H of all left cosets of any subgroup H in G is a G-set via the action given by $g' \cdot gH = g'gH$, for all $g, g' \in G$.

If $S \supseteq R$ is a ring extension with the additional condition that S is faithful, projective, and finitely generated as R-module (shortly, faithfully projective), then given any weakly S-split R-algebra A and any group G of R-automorphisms of S, it is straightforward to check that the set $X(A)$, as above constructed, is a G-set if and only if φ_i and $g\varphi_j$ are strongly distinct for all $1 \leq i, j \leq n$ and $g \in G$.

Furthermore, S is a G-set in an obvious way and if X is a given G-set then the set Map(X, S) of all maps from X to S is also a G-set via the action $g \cdot f = gfg^{-1}$ for all $f \in$ Map(X, S) and $g \in G$. Such a set is also an S-algebra with the usual pointwise operations of addition and multiplication, and, in particular, the set $A(X)$ of all invariants in Map(X, S) by the action of G is an R-subalgebra. Besides this, if X is finite and S is a Galois extension of R with Galois group G in the sense of CHR, then $A(X)$ is weakly S-split via the map $\varphi : S \otimes_R A(X) \to S^{\#X}$ given by $\varphi(s \otimes f) = (sf(x))_{x \in X}$ [47, Lemma 3.3].

An R-algebra A is called *S-split*, for any ring extension $S \supseteq R$, if A is weakly S-split and $X(A)$ is a G-set.

Let us denote by $\mathscr{S}(R)$ the category whose objects are S-split R-algebras and whose morphisms are algebra homomorphisms, and by $\mathscr{F}(G)$ the category whose objects are finite G-sets and whose morphisms are maps between G-sets that commute with the action of G.

Theorem 5.4 (Galois-Grothendieck Correspondence Theorem) *Let $S \supseteq R$ be a Galois extension of R with Galois group G in the sense of CHR. The map*

$$\mathbf{A} : \mathscr{F}(G) \to \mathscr{S}(R), \quad X \mapsto A(X),$$

is a (contravariant) functor that induces an (anti) equivalence between such categories, with inverse given by the functor

$$\mathbf{X} : \mathscr{S}(R) \to \mathscr{F}(G), \quad A \mapsto X(A) = \{\varphi_i \mid 1 \leq i \leq n\},$$

where the φ_i's are the maps from A to S such that $S \otimes_R A \overset{\varphi}{\to} S^n$, $s \otimes a \mapsto (s\varphi_i(a))_{1 \leq i \leq n}$, is an isomorphism of S-algebras.

Theorem 5.2 is indeed a consequence of Theorem 5.4. In order to see this it is enough to check firstly that if $S \supseteq R$ is a Galois extension of R with Galois group G in the sense of CHR and H is any subgroup of G then $A(\frac{G}{H})$ and S^H are isomorphic as R-algebras via the map $\theta : A\left(\frac{G}{H}\right) \to S^H$ given by $\theta(f) = f(H)$, for all $f \in A\left(\frac{G}{H}\right)$. Secondly, under the same assumption on S, R, and G, and using the previous result, to check that any subalgebra T of S is G-strong and R-separable if and only if $T = S^{H_T} \left(= A\left(\frac{G}{H_T}\right)\right)$.

5.1.4 In the VZ Galois Theory for Commutative Ring Extensions

As seen in the Sect. 5.1.2, the CHR Galois theory establishes, for a given Galois extension $S \supseteq R$ with Galois group G, a bijection between all the subgroups of G and (not all) R-separable subalgebras of S.

Differently, the VZ Galois theory developed in [90] establishes a bijection between all R-separable subalgebras of S and suitable (not all) subgroups (called "fat") of $\mathrm{Aut}_R(S)$.

The VZ Galois theory in [91] applies to any commutative ring, without any restriction on the idempotents. The strategy used by Villamayor and Zelinsky in [91] to develop such a theory consisted in using boolean localization to get Galois extensions for which the theory developed in [90] could be applied. However, the notion of Galois extension, as considered in [90], was not good enough for such a strategy to work well (see the second example in [91, Sect. 4]). Hence, it was necessary to use a weaker notion. The good notion of Galois extension compatible with the use of Boolean localization, adopted in [91] for the desired end, is the following: a ring extension $S \supseteq R$ is called a *weak Galois extension* of R if

1. S is R-separable,
2. S is projective and finitely generated as an R-module, and
3. there exists a finite set W of automorphisms of S such that $S^W = R$.

The Galois correspondence in [91] establishes a bijection between all R-separable subalgebras of S and the subgroups (not all) of G satisfying an appropriate "closing condition."

Inspired by the ideas and results of Grothendieck, Villamayor, and Zelinsky, Magid developed in [65] a completely general Galois-Grothendieck theory for commutative rings that generalizes both the theories due respectively to CHR and VZ.

5.1.5 In the Noncommutative Ring Context

As far as we know, in the noncommutative ring context there is not a Galois correspondence theorem like as in the Galois theory due to Chase, Harrison, and Rosenberg, except perhaps in some specific situation like, the one of semiprime algebras.

A correspondence theorem in the setting of semiprime algebras is due to Vladislav Kharchenko who, in order to obtain his result, introduced in [57] the notion of X-inner automorphisms (the "X" corresponding to the first letter of Kharchenko's name in Russian language) by using the extension of an automorphism of a ring to the corresponding Martindale quotient ring.

To recall Kharchenko's Galois correspondence, let R be a prime algebra over a fixed field K and $\mathcal{Q}$ the symmetric quotient algebra of R. An automorphism of R is called *X-inner* if it is inner as an automorphism of $\mathcal{Q}$, otherwise it is called *X-outer*.

A group G of automorphisms of R is called *X-outer* if the unique X-inner automorphism in G is the identity map of R.

Given a group G of automorphisms of R, as usual R^G denotes the subalgebra of the invariants in R by the action of G. If in particular G is a finite X-outer group of automorphisms of R, then the group of all automorphisms of R fixing

R^G elementwise coincides with G and in this case R is called a *Galois extension of* R^G.

A subring R' of R is called *rationally complete* or (according to Montgomery and Passman in [68]) *ideal-cancellable* if for any nonzero ideal I of R' and $x \in R$, the inclusion relation $Ix \subseteq R'$ implies $x \in R'$.

Theorem 5.5 (Kharchenko's Galois Correspondence Theorem) *Let R be a prime ring and G be a finite group of X-outer automorphisms of R. Then the map $H \mapsto R^H$ gives a one-to-one correspondence between the subgroups of G and the rationally complete subrings of R including R^G.* □

5.2 On Hopf Actions

There is a natural correspondence between the actions of a group G on a K-algebra R and the actions of the corresponding group algebra KG over R. As seen in part 2 of Sect. 4.3.2, KG is a Hopf algebra. Furthermore, KG is cocommutative (i.e., its comultiplication is invariant by the canonical flip map), and, if K is a field, KG is also pointed (i.e., its left or right comodules are all one dimensional).

All such above facts stimulated investigations on Hopf actions in order to extend to this context the Kharchenko's ideas and results. Specifically, to extend the notion of X-outer automorphisms to the context of finite-dimensional pointed Hopf algebras, not necessarily cocommutative, in order to get a generalization of Kharchenko's Galois Correspondence Theorem for group actions. Contributions came from several Hopf-algebraists including, in particular, A. Milinski, S. Westreich, A. Masuoka, T. Yanai, V. Ferreira, L. Murakami, and the author. Their results concern to the Galois theory for Hopf actions on prime algebras and appeared in several papers, specially in [69, 92, 93] and [44].

Contributions from Milinski, Westreich, Masuoka, and Yanai converged to a generalized correspondence theorem (see [94, Theorem 2.13]) which Ferreira, Murakami, and the author applied to prove a one-to-one correspondence theorem for homogeneous and faithful Hopf algebras actions on free algebras [44]. This last mentioned theorem will be presented in the sequel.

Recalling notations, F^I denotes the subalgebra of the invariants in F by the action of I, for all free subalgebra F of R and all right coideal subalgebra I of H. Following [44, Corollary 3.2], if R is a free algebra and H is a Hopf algebra acting homogeneously on R then R^I is a free subalgebra of R, for all right coideal subalgebra I of H.

Theorem 5.6 ([44, Theorem 1.2]) *Let X be a nonempty and nonsingular set and $R = K\langle X\rangle$ the free algebra on X over a field K. Let H be a finite-dimensional pointed Hopf algebra acting homogeneously and faithfully on R.Then the maps*

$$\phi : F \mapsto F^H, \mathcal{F} \to \mathcal{I} \quad \textit{and} \quad \psi : I \mapsto R^I, \mathcal{I} \to \mathcal{F}$$

give a one-to-one correspondence between

- $\mathcal{F}$: *the set of all free subalgebras of R including* R^H
- $\mathcal{I}$: *the set of all right coideal subalgebras of H including* $K.1_H$ □

For more about correspondence theorems in the setting of Hopf actions, we recommend the interesting survey by Montgomery [67].

6 Partial Actions

6.1 Partial Group Actions

Partial actions of groups on algebras is a very young theory which had its origin in the paper by Exel [41] related to the study of operators algebras. His main purpose in that paper was to develop a method that would allow him to describe the structure of C^*-algebras under actions of the circle group. For more detailed information on this subject, we recommend his recently published book "*Partial Dynamical Systems, Fell Bundles and Applications*" [43].

The first approach of partial group actions on algebras, in a purely algebraic context, appears later in a paper by Dokuchaev and Exel [36].

A *partial action* of a group G on a (not necessarily unital) algebra S over a commutative ring K is a collection α of ideals S_g ($g \in G$) of S and isomorphisms of (not necessarily unital) K-algebras $\alpha_g : S_{g^{-1}} \to S_g$ such that:

(i) $S_1 = S$ and α_1 is the identity map of S,
(ii) $\alpha_g \alpha_h \leq \alpha_{gh}$, for all $g, h \in G$.

Condition (ii) of the above definition means that α_{gh} is an extension of $\alpha_g \alpha_h$, that is, the domain $S_{(gh)^{-1}}$ of the second map contains the domain $\alpha_{h^{-1}}(S_h \cap S_{g^{-1}})$ of the first one and both the maps coincide on this last set. If $S_g = S$ and $\alpha_g \alpha_h = \alpha_{gh}$ for all $g, h \in G$, then α is called *global*.

Partial group actions can be easily obtained by restrictions from global ones in the following way: take a global action β of G on a given K-algebra S' and an ideal S of S', and put $S_g = S \cap \beta_g(S)$ and $\alpha_g = \beta_g|_{S_{g^{-1}}}$, for all $g \in G$. It is straightforward to check that the collection $\alpha = (S_g, \alpha_g)_{g \in G}$ is a partial action of G on S.

It is natural to ask whether partial actions are all of the above type. In the topological context the answer is affirmative and due to Abadie [1, Theorem 1.1]. Nevertheless, in the purely algebraic context some restrictive assumption is required, namely, the partial given action α must be *unital*, that is, the corresponding ideals S_g, $g \in G$, must be unital. This result is due to Dokuchaev and Exel and appeared in [36, Theorem 4.5].

In all what follows we will deal only with partial actions $\alpha = (S_g, \alpha_g)_{g \in G}$ of G on S where each ideal S_g is unital with identity element denoted 1_g. It is clearly seen that each 1_g is a central idempotent in S, for all $g \in G$. Notice that in such a situation each partial isomorphisms α_g turns out to become an endomorphism of S given by $x \mapsto \alpha_g(x1_{g^{-1}})$ for all $x \in S$.

A Galois theory for unital partial group actions was developed by Dokuchaev, Ferrero, and the author in [37], extending the CHR Galois theory to this new setting.

In order to present the corresponding definition and correspondence theorems we need some preparation. Actually, such a preparation consists of simple adaptations to the partial situation of the necessary concepts that appear in the global case.

The partial version of the global skew group ring is given by the following direct sum of ideals

$$S \star_\alpha G = \oplus_{g \in G} S_g \delta_g,$$

(where the δ_g's are simply placeholders), endowed with the usual addition and the multiplication induced by the rule

$$(x\delta_g)(y\delta_h) = a\alpha_g(y1_{g^{-1}})\delta_{gh}, \quad \text{for all } g, h \in G, \quad x \in S_g \text{ and } y \in S_h.$$

Two elements $g, h \in G$ are called *strongly distinct* with respect to the partial action α of G on S (shortly, *α-strongly distinct*) if for every nonzero idempotent $e \in S_g \cap S_h$ there exists $x \in S$ such that $\alpha_g(x1_{g^{-1}})e \neq \alpha_h(x1_{h^{-1}})e$. A subalgebra T of S is called *α-strong* if the restrictions to T (via α) of any two distinct elements of G are α-strongly distinct.

Given a subalgebra T of S, let

$$H_T = \{g \in G \mid \alpha_g(x1_{g^{-1}}) = x1_g \text{ for all } x \in T\}.$$

In general H_T is not a subgroup of G, we refer to [37, Sect. 6] for examples.

For any subgroup H of G the partial action α of G on S induces by restriction a partial action $\alpha_H = (S_h, \alpha_h)_{h \in H}$ of H on S. We denote by S^{α_H} the subalgebra of the invariants in S by the action of α_H, that is,

$$S^{\alpha_H} = \{x \in S \mid \alpha_h(x1_{h^{-1}}) = x1_h \text{ for all } h \in H\}.$$

If $H = G$, we denote such a subalgebra by S^α.

The algebra S is called an *α-partial Galois extension* of $R = S^\alpha$ if G is finite and there exist elements $x_i, y_i \in S$, $1 \leq i \leq n$, such that $\sum_{1 \leq i \leq n} x_i \alpha_g(y1_{g^{-1}}) = \delta_{1,g}$ for all $g \in G$.

As in the global case the map $\varphi : S \star_\alpha G \to \mathrm{End}(S_R)$, given by $\varphi(x_d\delta_g)(y) = x\alpha_g(y1_{g^{-1}})$, is a homomorphism of left S-modules and right R-modules.

Every left $S \star_\alpha G$-module M is a left S-module and

$$M^G = \{x \in M \mid 1_g\delta_g \cdot x = 1_g x, \ \text{for all } g \in G\}$$

is a left R-module. Notice that S is a left $S \star_\alpha G$-module via φ and in this case S^G coincides with S^α.

Theorem 6.1 *Let $\alpha = (S_g, \alpha_g)_{g\in G}$ be a partial action of a finite group G on a K-algebra S and $S^\alpha = R$. Then the following statements are equivalent:*

(a) S is an α-partial Galois extension of R.
(b) S is finitely generated and projective as right R-module and φ is an isomorphism.
(c) For any left $S\star_\alpha G$-module M the map $\mu : S\otimes_R M^G \to M$, given by $\mu(x\otimes m) = xm$ is an isomorphism of left S-modules.
(d) The map $\psi : S\otimes_R S \to \prod_{g\in G} S_g$, given by $\psi(x\otimes y) = (x\alpha_g(y1_{g^{-1}}))_{g\in G}$, is an isomorphism of left S-módulos.
(e) The map $\delta : S\otimes_R S \to S\star G$, given by $\delta(x\otimes y) = \sum_{g\in G} x\alpha_g(y1_{g^{-1}})\delta_g$, is an epimorphism of $S\star_\alpha G$-bimodules.
(f) $StS = S\star_\alpha G$, where $t = \sum_{g\in G} 1_g\delta_g$.
(g) S is a generator for the category of all left $S\star_\alpha G$-modules. □

A Morita context connecting R and $S\star_\alpha G$ similar to the one constructed in the global case also exists in the partial case and the equivalence of statements analogous to the global ones (h)–(j) also holds, provided that S is an α-partial Galois extension of R.

A correspondence theorem is given only in the commutative ring context.

Theorem 6.2 *Let S be an α-partial Galois extension of $R = S^\alpha$. Then the maps*

$$H \mapsto S^H \quad \textit{and} \quad T \mapsto H_T$$

give a one-to-one correspondence between the subgroups H of G and the separable subalgebras T of S including R, which are α-strong and such that H_T is a group. □

6.2 Partial Hopf Actions

One can say that the papers [36] and [37], on which the previous subsection is based, constitute the starting point of the motivation for investigations on partial actions by Hopf algebras, among other structures such as semigroup algebras, groupoid algebras, weak Hopf algebras, Hopf algebras of multipliers, and also Hopf categories. In this subsection we will deal only with partial Hopf actions.

The notion of partial Hopf action was inspired by that of partial group action. Indeed, any unital partial action α of a group G on an algebra A over a commutative ring K gives rise to the K-linear map

$$\alpha : KG\otimes_K A \to A, \quad \text{denoted by } \alpha(u_g\otimes a) = u_g \triangleright a = \alpha_g(1_{g^{-1}})$$

satisfying the following conditions:

1. $u_1 \triangleright a = a$
2. $(u_g \triangleright (ab) = (u_g \triangleright a)(u_g \triangleright b)$
3. $u_g \triangleright (u_h \triangleright a) = (u_g \triangleright 1_A)(u_{gh} \triangleright a) = (u_{gh} \triangleright a)(u_g \triangleright 1_A)$,

for all $g, h \in G$ and $a, b \in A$.

Conversely, any K-linear map $* : KG \otimes_K A \to A$, denoted by $*(u_g \otimes a) = u_g * a$, and satisfying conditions similar to those above ones, determines a unital partial action $\alpha = (A_g, \alpha_g)_{g \in G}$ of G on A, where $A_g = u_g * A$, with identity $1_g = u_g * 1_A$, and $\alpha_g : A_{g^{-1}} \to A_g$ given by $\alpha_g(x 1_{g^{-1}}) = u_g * x$.

A (left) *partial action* (or a *partial Hopf action*, for short) of a Hopf algebra H on a K-algebra A is a K-linear map $\triangleright : H \otimes_K A \to A$, denoted by $\triangleright(h \otimes x) = h \triangleright x$, that satisfies the following conditions:

1. $1_h \triangleright a = a$
2. $h \triangleright (ab) = (h_1 \triangleright a)(h_2 \triangleright b)$
3. $h \triangleright (k \triangleright a) = (h_1 \triangleright 1_A)(h_2 k \triangleright a)$,

for all $h, k \in H$ and $a, b \in A$. The pair $(\triangleright, A)$ (or simply A) is called a (left) *partial H-module algebra*. It is straightforward to check that $\triangleright$ is global if and only if $h \triangleright 1_A = \varepsilon(h) 1_A$.

If, in addition, $\triangleright$ also satisfies

(4) $h \triangleright (k \triangleright a) = (h_1 k \triangleright a)(h_2 \triangleright 1_A)$,

it is called *symmetric*.

As seen above there is a one-to-one correspondence between unital partial group actions of a group G on a K-algebra A and (left) symmetric partial Hopf actions of KG on A.

Right partial Hopf actions are defined symmetrically.

Examples of partial H-module algebra can be obtained from a global one by restriction in the following way: take an H-module algebra B via a (left) global action $b \mapsto h \cdot b$, for all $b \in B$ and $h \in H$, and A a unital ideal of B with identity element 1_A. Then $A = 1_A B$ and it becomes a partial H-module algebra via the action $h \triangleright a = 1_A(h \cdot a)$. Actually any partial H-module algebra is of this type by [6, Theorem 1].

The subalgebra of the *invariants* in A by $\triangleright$ is given by

$$A^{\underline{H}} = \{a \in A \mid h \triangleright (ab) = a(h \triangleright b)\,, \text{ for all } h \in H \text{ and } b \in A\}.$$

It is straightforward to check that

$$a \in A^{\underline{H}} \Leftrightarrow h \triangleright a = a(h \triangleright 1_A)$$

and if, in addition, $\triangleright$ is symmetric the following also holds

$$a \in A^{\underline{H}} \Leftrightarrow h \triangleright a = (h \triangleright 1_A)a, \text{ for all } h \in H.$$

In particular, in this case, $H \triangleright 1_A$ is contained in the centralizer of $A^{\underline{H}}$ in A.

As in the global case, for the definition of a partial Hopf-Galois extension we need to introduce the notion of (right) partial H-comodule algebra.

By a (right) *partial coaction* of H on an algebra A we mean a K-linear map $\rho : A \to A \otimes H$, denoted $\rho(a) = a_0 \otimes a_1$ (Sweedler notation), such that

1. $\rho(ab) = \rho(a)\rho(b)$
2. $(I_A \otimes \varepsilon)\rho(a) = a$
3. $(\rho \otimes I_H)\rho(a) = (\rho(1_A) \otimes 1_H)((I_A \otimes \Delta)\rho(a))$

for all $a, b \in A$. The pair (A, ρ) (or simply A) is called a (right) *partial H-comodule algebra*. Notice that A is a right (global) H-comodule algebra if and only if $\rho(1_A) = 1_A \otimes 1_H$.

Examples of partial Hopf coactions can also be obtained from global ones. Indeed, if (B, ρ) is a right global H-comodule algebra and A is a right ideal of B with identity 1_A, then the map $\bar{\rho} : A \to A \otimes H$ given by $\bar{\rho}(a) = (1_A \otimes 1_H)\rho(a)$ induces a right partial coaction of H on A. And, any right partial Hopf coaction is of this type by [6, Theorem 4].

The subalgebra of the *coinvariants* of A under ρ is given by

$$A^{\underline{coH}} = \{a \in A \mid \rho(ab) = a\rho(b), \text{ for all } b \in A\}.$$

It is easy to check that

$$a \in A^{\underline{coH}} \Leftrightarrow \rho(a) = a\rho(1_A) \Leftrightarrow \rho(ba) = \rho(b)a, \text{ for all } b \in A.$$

In the sequel we will assume that H is projective and finitely generated as K-module. Under such an assumption any left partial action $\triangleright$ of H on A induces a right partial coaction $\rho_{\triangleright}$ of H^* on A such that

$$h \triangleright a = a_1(h)a_0, \text{ whenever } \rho_{\triangleright}(a) = a_0 \otimes a_1, \text{ for all } a \in A \text{ and } h \in H.$$

Moreover, if H coacts (partially) on A via ρ on the right side, then H^* acts (partially) on A via $\triangleright_\rho$ on the left side by

$$f \triangleright_\rho a = f(a_1)a_0, \text{ whenever } \rho(a) = a_0 \otimes a_1, \text{ for all } a \in A \text{ and } f \in H^*.$$

And, in this case, $A^{\underline{H^*}} = A^{\underline{coH}}$.

A (right) partial H-comodule algebra (A, ρ) is called a *partial H-Galois extension of* $A^{\underline{coH}}$ (or equivalently a (left) partial H^*-Galois extension of $A^{\underline{H^*}}$) if the map

$$can : A \otimes_{A^{\underline{coH}}} A \to A\underline{\otimes}H \text{ induced by } a \otimes b \mapsto (a \otimes 1_H)\rho(b)$$

is bijective, where $A\underline{\otimes}H = (A \otimes_K H)\rho(1_A)$.

Such a definition was firstly given by Caenepeel and Janssen in [19], with some other equivalent definitions in a categorical language. Some other equivalent definitions due to Alves and Batista also appeared in [5]. A complete list of equivalent statements that characterize a partial Hopf-Galois extension, in a way similar to those presented in the previous subsection, is still under construction and will appear in a forthcoming paper by F. Castro, D. Freitas, G. Quadros, and the author.

7 Final Comments

In order not to overstretch this manuscript we omitted many examples, some essential references indicative of the advances of the current research on Galois theory, and even sketches of the proofs of the results presented. In an attempt to remedy a little bit this situation we will indicate below some additional references.

1. For the classical Galois theory for field extensions we refer, for instance, to [70, 81] and [83].
2. For the reader interested in learning more about separability, we refer to [70] (in the field case) and [7, 32] and [62] (in the general ring case).
3. A detailed proof of Teorema 5.2, as presented by Chase, Harrison, and Rosenberg, can be seen in [23] and also in [32] and [74].
4. Examples of Galois extensions for group actions on commutative rings can be found in [74] and [75].
5. The Galois-Grothendieck theory, as presented in this manuscript, was totally inspired in [39] and [47] and extend the corresponding results of [39] to the commutative ring setting.
6. Partial results about Galois correspondence of CHR-type, in the setting of not necessarily comutative rings can be found, for instance, in [85, 86] and [87].
8. There are in the literature some nice survey-type papers. We refer to [67] and [94] for Hopf actions, [76] for partial Hopf actions and [14, 34], and [35] for partial actions in general.
9. In the last ten years new advances related to the study of new Galois theories have been made in both the contexts of partial and global actions. As references we recommend, for instance, [8–13, 16–19, 21, 22, 25, 28, 29, 42, 48, 49, 53, 54, 63, 64, 77–79, 84] and [88].

References

1. F. Abadie, *Enveloping actions and Takay duality for partial actions*, J. Funct. Analysis 197 (2003), 14–67.
2. M. Artin, *Grothendieck Topologies*, Notes on a Seminar, Harvard University, 1962.

3. E. Artin, *Galois Theory*, Notre Dame Mathematical Lectures 2, First edition in 1942, Dover Books on Mathematics, Sixth edition in 1971.
4. G. Azumaya, *Galois theory for uni-serial rings*, J. Math. Soc. Japan 1 (1949), 130–137.
5. M. M. S. Alves, E. Batista, *Partial Hopf actions, partial invariants and a Morita context*, Algebra Disc. Math. no. 3, (2009), 1–19.
6. M. M. S. Alves, E. Batista, *Enveloping Actions for Partial Hopf Actions*, Comm. Algebra 38 (2010),2872–2902.
7. M. Auslander and O. Goldman, *The Brauer group of a commutative ring*, Trans. AMS. 97 (1960), 367–409.
8. J. Ávila Guzmán, J. Lazzarin, *A Morita context related to finite groups acting partially on a ring*, Algebra Disc. Math. 3. (2009), 49–61.
9. D. Azevedo, E. Campos, G. Fonseca and G. Martini, *Partial (Co)Actions of Multiplier Hopf Algebras: Morita and Galois Theories*, arXiv: 1709.08051, (2017). 32 pages.
10. D. Bagio, W. Cortes, M. Ferrero and A. Paques, *Actions of inverse semigroups on algebras*, Comm. Algebra 35 (2007), 3865–3874.
11. D. Bagio, D. Flores and A. Paques, *Partial actions of ordered groupoids on rings*, J. Algebra Appl. 9 (2010), (3), 501–517.
12. D. Bagio, J. Lazzarin and A. Paques, *Crossed products by twisted partial actions: separability, semisimplicity and Frobenius properties*, Comm. Algebra 38 (2010), 496–508.
13. D. Bagio and A. Paques, *Partial groupoid actions: globalization, Morita theory and Galois theory*, Comm. Algebra 40 (2012), 3658–3678.
14. E. Batista, *Partial actions: What they are and why we care*, Bull. Belg. Math. Soc. -Simon Stevin, 24, (2017), (1), 35–71 (arXiv:1604.06393v1).
15. F. Borceux and G. Janelidze, *Galois Theories*, Cambridge Univ. Press, 2001.
16. S. Caenepeel, *Galois corings from the descent theory point of view*, Fields Inst. Comm. 43 (2004), 163–186.
17. S. Caenepeel and E. De Groot, *Corings applied to partial Galois theory*, Proceedings of the International Conference on Mathematics and Applications, ICMA 2004, Kuwait Univ. (2005), 117–134.
18. S. Caenepeel and E. De Groot, *Galois theory for weak Hopf algebras*, Rev. Roumaine Math. Pures Appl. 52 (2007), 51–76.
19. S. Caenepeel and K. Janssen, *Partial (co)actions of Hopf algebras and partial Hopf-Galois theory*, Comm. Algebra 36 (2008), 2923–2946.
20. H. Cartan, *Théorie de Galois pour les corps non-commutatifs*, Ann. Scien. de l'École Norm. Sup. 65 (1948), 60–77
21. F. Castro, A. Paques, G. Quadros and A. Sant'Ana, *Partial actions of weak Hopf algebras: Smash product, globalization and Morita theory*, J. Pure Appl. Algebra, 219, (2015), 5511–5538.
22. F. Castro and G. Quadros, *Galois Theory for Partial Comodule Coalgebras*, Preprint.
23. S. U. Chase, D. K. Harrison and A. Rosenberg, *Galois theory and Galois cohomology of commutative rings*, Memoirs AMS 52 (1965), 1–19.
24. S. U. Chase and M. E. Sweedler, *Hopf Algebras and Galois Theory*, LNM 97, Springer Verlag, Berlin, 1969.
25. C. Cibils and A. Solotar, *Galois coverings, Morita equivalence and smash extensions of categories over a field*, Doc. Math., 11, (2006), 143–159.
26. M. Cohen, *A Morita context related to finite automorphism groups of rings*, Pacific J. Math. 98 (1) (1982).
27. M. Cohen, D. Fischman and S. Montgomery, *Hopf-Galois extensions, smash products and Morita equivalence*, J. Algebra 133 (1990), 351–372.
28. W. Cortes and E. Marcos, *Descriptions of partial actions*, Preprint, arXiv:1708.01330, (2017).
29. W. Cortes and T. R. Tamusiunas, *A characterisation for a groupoid Galois extension using partial isomorphisms*, Bull. Aust. Math. Soc. (2017), 1–10.
30. F. DeMeyer, *Galois Theory in Rings and Algebras*, PhD thesis, Univ. of Oregon, 1965, 59 pp.

31. F. DeMeyer, *Some notes on the general Galois theory of rings*, Osaka J. Math. 2 (1965), 117–127.
32. F. DeMeyer and E. Ingraham, *Separable Algebras Over Commutative Rings*, LNM 181, Springer Verlag, Berlin, 1971.
33. J. Dieudonné, *La théorie de Galois des anneaux simples et semi-simples*, Comment. Math. Helv. 21 (1948), 154–184.
34. M. Dokuchaev, *Partial actions: A survey*. Contemporary Math. 537 (2011), 173–184.
35. M. Dokuchaev, *Recent developments around partial actions*, arXiv:1801.09105v1 (2018), 60 pages.
36. M. Dokuchaev and R. Exel, *Associativity of crossed products by partial actions, enveloping actions and partial representations*, Trans. AMS 357 (2005) 1931–1952.
37. M. Dokuchaev, M. Ferrero and A. Paques, *Partial actions and Galois theory*, J. Pure Appl. Algebra 208 (2007), 77–87.
38. Y. Doi and M. Takeuchi, *Hopf-Galois extensions of algebras, the Miyashita-Ulbrich action, and Azumaya algebras*, J. Algebra 121 (1989), 448–516.
39. A. W. Dress, *One more shortcut to Galois theory*, Adv. Math. 110 (1995), 129–140.
40. A. W. Dress, *Basic non-commutative Galois Theory*, Adv. Math. 110 (1995), 141–173.
41. R. Exel, *Twisted partial actions: a classification of regular $C^\star$-algebraic bundles*, Proc. London Math. Soc. 74 (3) (1997) 417–443.
42. R. Exel, *Partial actions of groups and actions of semigroups*, Proc. AMS 126 (1998) 3481–3494.
43. R. Exel, *Partial Dynamical Systems, Fell Bundles and Applications*, Mathematical Surveys and Monographs, American Mathematical Society, vol. 224, 321 pages, 2017.
44. V. Ferreira, L. Murakami and A. Paques, *A Hopf-Galois correspondence for free algebras*, J. Algebra 276 (2004), 407–416.
45. M. Ferrero, *Teoria de Galois para Anillos Graduados*, PhD Thesis, Univ. de Buenos Aires, 1970.
46. M. Ferrero, *On the Galois theory over non-comutative rings*, Osaka J. Math. 7 (1970), 81–88.
47. M. Ferrero and A. Paques, *Galois theory of commutative rings revisited*, Beiträge zur Algebra und Geometrie 38 (1997), 299–410.
48. G. L. Fonseca, *Galois, Dedekind e Grothendieck*, MSc dissertation, UFRGS, Brazil, 2015.
49. D. Freitas and A. Paques, *On partial Galois Azumaya extensions*, Algebra Disc. Math., 11, (2011), 64–77.
50. A. Grothendieck, *Revêtements étales et groupe fondamental*, SGA 1, exposé V, LNM 224, Springer Verlag, (1971).
51. G. Hochschild, *Double vetor spaces over division rings*, Amer. J. Math. 71 (1949), 443–460.
52. N. Jacobson, *The fundamental theorem of Galois theory for quasi-fields*, Annals Math. 41 (1940), 1–7.
53. X.-L. Jiang and G. Szeto, *Galois extensions induced by a central idempotent in a partial Galois extension*, Intern. J. Algebra, 8, (11), (2014), 505–510.
54. X.-L. Jiang and G. Szeto,*The group idempotents in a partial Galois extension*, Gulf J. Math. 3 (2015), 42–47.
55. T. Kanzaki, *On commutator rings and Galois theory of separable algebras*, Osaka J. Math. 1 (1964), 103–115.
56. T. Kanzaki, *On Galois extension of rings*, Nagoya Math. J. 27 (1966), 43–49.
57. V. Kharchenko, *Galois theory of semiprime rings*, Algebra i Logica 16 (1977), 313–363 (English translation 1978, 208–258).
58. E. R. Kolchin, *Picard-Vessiot theory of partial differential fields*, Proc. AMS 3 (1952), 596–603.
59. H. F. Kreimer, *A Galois theory of non-commutative rings*, Trans. AMS 127 (1967), 29–41.
60. H.F. Kreimer and M. Takeuchi, *Hopf algebras and Galois extensions of an algebra*, Indiana Univ. Math. J. 30 (1981), 675–692.

61. W. Krull, *Galoissche Theorie der unendlichen algebraschen Erweiterungen*, Math. Ann. 100 (1928), 687–698.
62. M-A. Knus and M. Ojanguren, *Théorie de la Descente et Algèbres d'Azumaya*, LNM 389, Springer Verlag, Berlin, 1974.
63. J.-M. Kuo and G. Szeto, *The structure of a partial Galois extension. II*, J. Algebra Appl. 15, No. 4, Article ID 1650061, 12 p. (2016).
64. J.-M. Kuo and G. Szeto, *Partial group actions and partial Galois extensions*, Monatsh. Math (to appear).
65. A. R. Magid, *The Separable Galois Theory of Commutative Rings*, Marcel Dekker, 1974.
66. S. Montgomery, *Fixed Rings of Finite Automorphism Groups of Associative Rings*, LNM 818, Springer Verlag, 1980.
67. S. Montgomery, *Hopf Galois theory: A survey*, Geo. Top. Monographs 16 (2009), 367–400.
68. S. Montgomery and D. S. Passman *Outer Galois theory of prime rings*, Rocky Mountain J. Math. 14 (1984), 305–318.
69. A. Masuoka and T. Yanai, *Hopf module duality applied X-outer Galois theory*, J. Algebra 265 (2003), 229–246.
70. P. J. McCarthy, *Algebraic Extensions of Fields*, Dover Pub., Inc., New York, 1991.
71. T. Nakayama, *Galois theory for general rings with minimum condition*, J. Math. Soc. Japan 1 (1949), 203–216.
72. T. Nakayama, *Galois theory of simple rings*, Trans. AMS 73 (1952), 276–292.
73. M. Ouyang, *Azumaya extensions and Galois correspondence*, Algebra Colloquium 7 (2000), 43–57.
74. A. Paques, *Teoria de Galois Sobre Anillos Conmutativos*, Univ. Los Andes, Mérida, Venezuela, 1999.
75. A. Paques, *Teorias de Galois*, Mini-course Notes (in portuguese), XXII Brazilian Algebra Meeting, UFBA, Brazil, 2012.
76. A. Paques, *Galois, Morita and the trace map: a survey*, São Paulo J. Math. Sc. 10 (2) (2016), 372–383.
77. A. Paques, V. Rodrigues and A. Sant'Ana, *Galois correspondences for partial Galois Azumaya extensions*, J. Algebra Appl., 10, (5), (2011), 835–847.
78. A. Paques and A. Sant'Ana, *When is a crossed product by a twisted partial action is Azumaya?*, Comm. Algebra, 38, (2010), 1093–1103.
79. A. Paques and T. R. Tamusiunas, *A Galois-Grothendieck-type correspondence for groupoid actions*, Algebra Disc. Math. 17 (2014), 80–97.
80. D. Passman *Group Rings, Crossed Products and Galois Theory*, CBMS 64, AMS, Providence, RI, 2000.
81. J. Rotman, *Galois Theory*, Springer-Verlag, New York, 1990.
82. L. Rowen, *Ring theory*, Academic Press, Inc., Boston, MA, (1991)
83. I. Stewart, *Galois Theory*, Chapman and Hall, London, 1987.
84. G. Szeto and L. Xue, *The Galois algebra with Galois group which is the automorphism group*, J. Algebra 293 (2005), 312–318.
85. G. Szeto and L. Xue, *On Galois algebras satisfying the fundamental theorem*, Comm. Algebra 35 (2007), 3979–3985.
86. G. Szeto and L. Xue, *On Galois extensions satisfying the fundamental theorem*, Int. Math. Forum 36 (2007), 1773–1777.
87. G. Szeto and L. Xue, *On Galois extensions with a one-to-one Galois map*, Int. J. Algebra 5 (2011), 801–807.
88. T. R. Tamusiunas, *Teorias de Galois para ação de grupóides*, Tese de Doutorado, UFRGS, Brasil, 2012.
89. K. H. Ulbrich, *Galois erweiterungen von nicht-kommutativen ringen*, Comm. Algebra 10 (1982), 655–672.
90. O. E. Villamayor and D. Zelinsky, *Galois theory with finitely many idempotents*, Nagoya math. j. 27 (1966), 721–731.

91. O. E. Villamayor and D. Zelinsky, *Galois theory with infinitely many idempotents*, Nagoya Math. J. 35 (1969), 83–98.
92. S. Westreich and T. Yanai, *More about a Galois-type correspondence theory*, J. Algebra 246 (2001), 629–640.
93. T. Yanai, *Correspondences theory of Kharchenko and X-outer actions of pointed Hopf algebras*, Comm. Algebra 25 (1997), 1713–1740.
94. T. Yanai, *Galois correspondence theorem for Hopf algebra actions*, Contemp. Math. 376 (2005), 393–411.

On the Geometry and Topology of the Commutator of Unit Quaternions

Alcibiades Rigas and Dan A. Agüero Cerna

Abstract This is a narrative rather than a survey of the research relevant to the subject of the title, done basically at IMECC, Unicamp, from 1982 to 2005. Many people, including students, as well as the first author, contributed. The second author compiled the material in his master's dissertation (Cerna, On the geometry and topology of the commutator of unit quaternions. Master's Dissertation, IMECC, Unicamp, 2016) and he is not responsible for imprecisions, for forgetting to give credit where due, etc. We consider known a little basic homotopy, like the exact sequence of a fibration, a few basic homotopy groups of spheres, and the Bott periodicity for $Spin$, U, and Sp, and also basic definitions of Riemannian geometry.

1 Introduction and Our Motivation

In the early 1970s, Jeff Cheeger and Detleff Gromoll [7] published their famous "soul theorem":

Every complete open Riemannian manifold M with nonnegative sectional curvature ($K \geq 0$) is diffeomorphic to the total space of a vector bundle over a totally geodesic compact submanifold S, the soul, embedded as the zero section: $\mathbb{R}^n \dashrightarrow M \longrightarrow S$.

Consequently, the totally geodesic soul has $K \geq 0$ as well. Obviously, there was a search for examples and as far as I know, the immediate question "do all vector bundles over spheres (regular, not exotic) admit complete Riemannian metrics with $K \geq 0$," is still not answered completely [15, 22, 42]. There were some partial answers back then and one of them [36] showed that if you add a large enough trivial bundle to any vector bundle over any sphere, then the Whitney sum admits a complete Riemannian metric with $K \geq 0$. Some of the attention shifted to vector bundles away from the stable range. Let us take a quick look.

A. Rigas (✉) · D. A. A. Cerna
Institute of Mathematics, Statistics and Scientific Computing, University of Campinas, Campinas, SP, Brazil
e-mail: rigas@ime.unicamp.br

C. Lavor, F. A. M. Gomes (eds.), *Advances in Mathematics and Applications*,
https://doi.org/10.1007/978-3-319-94015-1_12

Over S^1 we have the trivial cylinder $S^1 \times \mathbb{R}$ and the infinite Möbious band $S^1 \times_{\mathbb{Z}_2} \mathbb{R}$, both of which can be nonnegatively curved.

The $SO(n)$-principal bundles over S^2 are easily obtained from the lens spaces $S^3/\mathbb{Z}_n$ through the free quotient by S^1 employing the commutativity of the complex numbers. That you get all principal ones with group S^1 follows from the homotopy ladder of the pullback diagram, which implies that π_1(Total space) classifies. So, since $\pi_1(S^3/\mathbb{Z}_n) = \mathbb{Z}_n$, the boundary operator $\partial : \pi_2(S^2) \longrightarrow \pi_1(S^1)$, which in this case is essentially the classifying map, has $\partial(1) = n$, so we get them all. All these lens spaces have $K > 0$ and the total space of the associated vector bundle $S^3/\mathbb{Z}_n \times_{S^1} \mathbb{R}^2$ inherits $K \geq 0$ by O'Neill's Riemannian submersion theorem. So, all vector bundles over S^2 admit $K \geq 0$.

All vector bundles over S^3 are trivial due to the basic fact that $\pi_2 G = 0$ for all compact semisimple Lie groups G, so they admit the product metric with $K \geq 0$.

The first nontrivial case is bundles over S^4. Let us restrict to principals, with simply connected group if the fiber has three or more dimensions, for brevity. Note that there is an added motivation here from Theoretical Physics, since S^4 is the one point compactification of space–time and it seems like a reasonable assumption that potentials die out at infinity. So, doing physics or geometry over the 4-sphere is meaningful.

Case 1. The principal S^1 or $SO(2)$ bundles over S^4 are classified by $\pi_4\mathbb{C}P^\infty \cong \pi^3 S^1 = 0$, and there is only the trivial one.

Case 2. $S^3 \dashrightarrow P \longrightarrow S^4$ are classified by $\pi_4 QP^\infty \cong \pi_3 S^3 \cong \mathbb{Z}$: much to look for here but let us take one more step first.

Case 3. $S^3 \times S^3 \dashrightarrow T \longrightarrow S^4$, classified by $\pi_3(S^3 \times S^3) \cong \mathbb{Z} \times \mathbb{Z}$. Models for these bundles were usually described (except for $T = Sp(2)$) by joining two copies of $D^4 \times (S^3 \times S^3)$ along their common boundary using a map $\partial D^4 = S^3 \longrightarrow S^3 \times S^3$ defined by the two integers (m, n) like $q \longmapsto (q^m, q^n)$.

Note that Case 2 is contained in Case 3 and the Hopf fibration $S^3 \dashrightarrow S^7 \longrightarrow S^4$ relates Case 3 to the principal bundles $S^3 \dashrightarrow P \longrightarrow S^7$. Following the same classification ritual as above, there are $\pi_6(S^3)$-many isomorphism classes of these principal bundles. It was known since 1950, through the work of Serre [27], that this group is $\mathbb{Z}_{12}$ and that it is generated by the homotopy class of the **commutator of quaternions**: $S^3 \times S^3 \ni (p, q) \longmapsto pqp^{-1}q^{-1} \in S^3$, this map factors through $S^3 \wedge S^3 = S^6 \longrightarrow S^3$ and its homotopy class in $\pi_6 S^3$ generates this group of principal bundles. We will see further on using elementary means how this works. For the time being, we register that in 1983 [37] algebraic models for all elements in Case 3 (and consequently Case 2) were constructed as sub-bundles of $Sp(n) \dashrightarrow Sp(n+1) \longrightarrow S^{4n+3}$. To include the bundles over S^7 in the picture, we pull back the principal $S^3 \dashrightarrow P_n \longrightarrow S^4$ by the Hopf map $h : S^7 \longrightarrow S^4$ and the basic diagram is

$$
\begin{array}{ccccccccc}
 & & S^3 & & S^3 & & S^3 & & \\
 & & \downarrow & & \downarrow & & \downarrow & & \\
Sp(1)^n & \dashrightarrow & \widetilde{P}_n & \longrightarrow & P_n & \longrightarrow & S^7 & & \\
 & & \downarrow & & \downarrow & & \downarrow -h & & \\
S^3 & \dashrightarrow & S^7 & \longrightarrow & S^4 & \longrightarrow & S^4 & \longrightarrow & BS^3 \\
 & & & h & & f_n & & j &
\end{array}
$$

where f_n is a map of degree n and j is the inclusion of $S^4 = QP^1 \subset \lim_{n\to\infty} QP^n = BS^3$, the classifying map for $\widetilde{P}_n$ is $j \circ f_n \circ h$. Here, $Sp(1)^n$ means the group $Sp(1) \cong S^3$ acting by the diagonal inclusion in the group of quaternionic matrices $Sp(n)$, i.e., as $q \mapsto diag(q, q, \ldots, q)$. The accounting of the $\widetilde{P}_n$'s, as related to the homotopy class of a power of the commutator of quaternions, was wrong originally and was later corrected by Barros [2].

Let us take a look at the details of the simplest cases.

For $n = 1$, $f_1 = id_{S^4}$, we have $P_1 = Sp(2)$. This is due to the definition of the group $Sp(n) = \{A \in Q(n \times n) \mid AA^* = I\}$ and this is equivalent to $A^*A = I$. In the case of $Sp(2) \ni \begin{pmatrix} a & c \\ b & d \end{pmatrix}$, we get $a\overline{a} + b\overline{b} = c\overline{c} + d\overline{d} = a\overline{a} + c\overline{c} = 1$ and $a\overline{b} + c\overline{d} = 0$, $\overline{a}b + \overline{c}d = 0$. The free action by the $\begin{pmatrix} 1 & 0 \\ 0 & \overline{q} \end{pmatrix}$, $q \in Sp(1)$, subgroup from the right defines a principal $Sp(1) \equiv S^3$ bundle over the first column, i.e., $Sp(1) \dashrightarrow Sp(2) \longrightarrow S^7$. The S^3 action by the diagonal $\begin{pmatrix} p & 0 \\ 0 & p \end{pmatrix}\begin{pmatrix} a & c \\ b & d \end{pmatrix}$ as a subgroup commutes with the right action by $\overline{q}$ and so it descends to a (free) action on the orbit space (first column) S^7. The quotient is $QP^1 \equiv S^4$ and the projection is classically written as the Hopf map or Hopf fibration, with formula $h\begin{pmatrix} a \\ b \end{pmatrix} = \begin{pmatrix} |a|^2 - |b|^2 \\ 2a\overline{b} \end{pmatrix} \in S^4$, the unit sphere in $\mathbb{R} \oplus \lim ImQ$. The relations from the definition of $Sp(2)$ above show that $Sp(2) = h^*(-h)$, pullback, where the horizontal map covering h is the projection to the second column $\begin{pmatrix} a & c \\ b & d \end{pmatrix} \longmapsto \begin{pmatrix} c \\ d \end{pmatrix} \in S^7$. There is an algorithmic construction in [36] that extends this diagram

$$
\begin{array}{ccccc}
 & & S^3 & & S^3 \\
 & & \downarrow & & \downarrow \\
S^3 & \dashrightarrow & Sp(2) & \longrightarrow & S^7 \\
 & & \downarrow & & \downarrow -h \\
S^3 & \dashrightarrow & S^7 & \longrightarrow & S^4
\end{array}
$$

to the big diagram above for every $n \in \mathbb{Z}$. For example, $\widetilde{P}_3$ is the ten-dimensional submanifold of $Sp(3) \ni \begin{pmatrix} a & -b\,|b|^2 & x \\ b & b\bar{a}b & y \\ 0 & a\sqrt{1+|b|^2} & z \end{pmatrix}$. Note that S^3 acts by the diagonal $\begin{pmatrix} q & & \\ & q & \\ & & q \end{pmatrix}$ from the left and S^3 acts by quaternionic multiplication from the right on the last column. These two commuting actions determine a free $S^3 \times S^3 = Spin(4)$ action with quotient S^4. This is the meaning of the large diagram above. One keeps adding zeros to the first column and the rest of the columns, except the last one, are made out of expressions like $a\bar{b}a$ or $b\bar{a}b$ or $a\,|a|^2$, etc., equivariant with respect to the diagonal action by S^3 from the left. There is an algorithm that constructs $\widetilde{P}_{n+1}$ from $\widetilde{P}_{n.}$ Some of these bundles are trivial, isomorphic to $S^7 \times S^3$, which ones we learn from Barros' accounting. The formulas for this equivalence depend on a homotopy, essentially the commutator of quaternions to the 12th power homotopic to a constant, which is true from J.-P. Serres' work. From this one gets some exotic 7-spheres as free quotients $S^3 \dashrightarrow S^7 \times S^3 \longrightarrow \Sigma^7$. This fact is not original, it follows from Wall's work [41] that $S^7 \times S^3$ is diffeomorphic to $\Sigma^7 \times S^3$ for every homotopy sphere $\Sigma^7 \in \mathbb{Z}_{28}$ the group Γ_7 of manifolds homeomorphic to the Euclidean 7-sphere. This fact immediately implies that there are free quotients as stated above, for every Σ^7. The novelty is that for some exotic 7-spheres the action would be explicitly described by a formula once we know **how** $(pqp^{-1}q^{-1})^{12}$ is homotopic to a constant as a map : $S^6 \longrightarrow S^3$. Curiously, this seems to be still unknown.

2 Duran's Idea

In the late 90's, Carlos Duran gave new life to these problems by reshaping the presentation of the quaternionic commutator using Differential Geometry [12]. We take a quick look at his idea and use the formulas to present elementary proofs of some facts that were known for decades employing relatively heavy machinery. Afterwards, we will get back to the consequences of Duran's work in the description of exotic phenomena by simple formulas.

We begin with $Sp(1) \dashrightarrow Sp(2) \longrightarrow S^7$ first column projection as described above.

First, let us take a quick look at the trivialization of this bundle using the quaternionic algebra. Let $U = \left\{\binom{a}{b} \in S^7 \mid a \neq 0\right\}$ and $V = \left\{\binom{a}{b} \in S^7 \mid b \neq 0\right\}$. A partial section of the bundle over U can be of the form $\begin{pmatrix} a & x \\ b & \bar{a} \end{pmatrix} \in Sp(2)$. Using $A^*A = AA^* = I$, we have $\bar{a}x + \bar{b}\bar{a} = 0$ we get $x = -a\bar{b}\bar{a}\,|a|^{-2}$. So, the bundle is $\begin{pmatrix} a & -(a\overline{b}\overline{a}\,|a|^{-2})\cdot p \\ b & \overline{a}\cdot p \end{pmatrix}$ over U, $p \in Sp(1) = S^3$. Similarly, over V one can trivialize as $\begin{pmatrix} a & \overline{b}\cdot q \\ b & -(b\overline{a}\overline{b}\,|b|^{-2})\cdot q \end{pmatrix}$, $q \in Sp(1)$. Over $U \cap V = \{ab \neq 0\}$, setting equal the (2,1) coordinates we get $q = -bab\overline{a}\,|a|^{-2}\,|b|^{-2}$ which is essentially the commutator of the two unit quaternions $b|b|^{-1}$ and $a|a|^{-1}$. This says that a certain transition function is the commutator of quaternions. This, by itself, is practically nothing, since transition functions do not classify principal bundles. It was proved, however, by nonelementary means, first by Borel and Serre [4] and then by James [26], that in this case the homotopy class of this commutator in $\pi_6 S^3$, generates this group which is isomorphic to $\mathbb{Z}_{12}$.

Given any principal bundle $G \dashrightarrow P \longrightarrow M$, where G is a numerable topological group, M is a numerable topological space, and P is the pullback bundle of the universal $G \dashrightarrow EG \longrightarrow BG$ by some $f : M \longrightarrow BG$. Homotopic maps pull back isomorphic bundles and EG is always contractible. So, if M is a sphere, $M = S^n$, the isomorphism classes of $G \dashrightarrow P \longrightarrow S^n$ are in 1-1 correspondence with $\pi_n BG$ and this in turn is isomorphic to $\pi_{n-1} G$. In our case $G = S^3$ and $M = S^7$, so $\partial(1) \in \pi_6 S^3$ classifies the principal $S^3 \dashrightarrow P \longrightarrow S^7$. According to J.-P. Serre, there are 12 such isomorphism classes and we say that P generates this group if $\partial(1)$ is a generator of $\mathbb{Z}_{12} = \pi_6 S^3$. Note now that $\pi_6 Sp(2) = 0$ since it is already stable. So, $\partial(1)$ generates $\pi_6 S^3$, as follows from the exact homotopy sequence of $S^3 \dashrightarrow Sp(2) \longrightarrow S^7$, and we can choose it to be $= 1$. A geometric approach to constructing a map $S^6 \longrightarrow S^3$ whose class is $\partial(1)$, where 1 is the class of id_{S^7}, was Duran's idea. A geometer sees a sphere as made out of geodesics, each of length π, joining the north pole N to the south pole S. These are parametrized by their unit tangent vectors at N, say $X \in T_N S^7$. Take $N = 1$, $S = -1$ in $S^7 \subset \mathbb{C}a$. So, $X \in S^6$ can be taken to be a unit vector in lim $Im\mathbb{C}a$. (The Cayley algebra, $\mathbb{C}a$, and its imaginary subspace will be described next). If we can lift the geodesic $\gamma_X(t)$ continuously to curves $\Gamma_X(t)$ from, say, $I \in Sp(2)$ all the way to the fiber over -1, which is $(projection)^{-1}(-1) = S^3(-1) = \left\{ \begin{pmatrix} -1 & 0 \\ 0 & q \end{pmatrix}, q \in S^3 \right\}$, then the homotopy class of $S^6 \ni X \mapsto \Gamma_X(1) \in S^3$ will be $\partial(1) = \partial[id_{S^7}] \in \pi_6 S^3$. This is, basically, a consequence of the definition of the boundary map of a fibration applied to our bundle, for $\partial : \pi_7 S^7 \longrightarrow \pi_6 S^3$. A tool in the Riemannian geometry of fibrations is a Riemannian submersion. In our case, the projection to the first column of $Sp(2)$, call it pr, preserves the metric on a smoothly defined horizontal complement H_A of the vertical space V_A, which in turn is the tangent space of the fiber at every point (matrix) $A \in Sp(2)$. The vertical distribution is born integrable: the fibers are submanifolds of $Sp(2)$. The horizontal distribution is, in general, not integrable, unless the bundle is trivial. The metric on the total space makes $H_A \perp V_A$, so $T_A Sp(2) = H_A \oplus V_A$, and the Riemannian metric on the base (here S^7) is such that $\|pr_* X\|_{S^7} = |X|_{Sp(2)}$ for any horizontal vector X. Of course, $pr_* V = 0$ for any vertical V.

The useful property of a Riemannian submersion is that one can lift curves from the base to the total space, so that the tangent vectors of the lifted curve are horizontal at every point. We call such a lift a horizontal lift. Moreover, a simple differential geometry argument shows that a geodesic that is horizontal at one point is everywhere horizontal. So, one can lift Euclidean geodesics in S^7 to horizontal geodesics in $Sp(2)$, provided that we have a submersion metric on $Sp(2)$ that projects to the Euclidean metric on S^7. It helps calculations if the Euclidean metric is of sectional curvature $= 1$, then we have a good formula for the geodesics. Duran noticed that the usual bi-invariant metric on $Sp(2)$ does not project to a Euclidean metric on S^7 (it projects to an ellipsoidal one). This had been noticed before and I, for one, had given up on this. As we will see, "this" relates to understanding some of the geometry of a generator of the homotopy 7-spheres, the Gromoll–Meyer sphere [24].

Let us get back to Duran. His idea was to change the metric on $Sp(2)$ so that we get a Riemannian submersion to $(S^7, K = 1)$ as we would like. It is worth reproducing this idea here.

The tangent space $T_I Sp(2) = \left\{X = \begin{pmatrix} x & -\overline{w} \\ w & s \end{pmatrix}, x, s \in \lim Im\mathbb{Q}, w \in \mathbb{Q}\right\}$. The action of $Sp(1) = S^3$ is from the right as a subgroup: $\begin{pmatrix} a & c \\ b & d \end{pmatrix} \bullet \begin{pmatrix} 1 & 0 \\ 0 & \overline{q} \end{pmatrix} = \begin{pmatrix} a & c\overline{q} \\ b & d\overline{q} \end{pmatrix}$, with $pr \begin{pmatrix} a & c \\ b & d \end{pmatrix} = \begin{pmatrix} a \\ b \end{pmatrix} \in S^7$.

The square length of X relative to the bi-invariant metric on $Sp(2)$ is $|x|^2 + 2\,|w|^2 + |z|^2$ and the Euclidean square length of $(pr)_* X$ is $|x|^2 + |w|^2$. We want to change the metric on $Sp(2)$ to project to the $K = 1$ metric on S^7.

Note that the action of $Sp(2)$ on itself, by left multiplication commutes with the $\bullet$-action of S^3, so it induces an action on the base S^7, and so does the action of the subgroup $S^3 \ni \begin{pmatrix} \overline{p} & 0 \\ 0 & 1 \end{pmatrix}$ from the right. The first remark, together with the fact that $Sp(2) \cong Spin(5)$ acts on the Euclidean S^7 by isometries, suggests that a left invariant metric on $Sp(2)$ should work. A Kaluza–Klein metric (K-K) is constructed on a bundle, roughly, following the steps below:

Assign a metric to each fiber. Here, we have fiber $= S^3$ and we give all of them the Euclidean $K = 1$ metric.

Define a smooth connection, i.e., a complementary space H_A to the tangent space of the fiber V_A at each point $A \in Sp(2)$. Here, we take H_A to be the orthogonal space to each fiber relative to the bi-invariant metric of $Sp(2)$.

Declare H orthogonal to V at each point and put a metric (smoothly) on each H. Here, we take the metric on H_A to be the one that makes the isomorphism $(pr)_* : H_A \longrightarrow T_{prA} S^7$ an isometry.

In our case (metric on V_A independent of $pr(A)$), the fibers are totally geodesic and pr is a Riemannian submersion on $(S^7, K = 1)$.

Trivializing $TSp(2)$ through left translations $(L_B, B \in Sp(2))$ we get $T_BSp(2) = V_B + H_B$ that goes by applying left translation by $B^{-1} = B^*$ to $V_I \oplus H_I = \left\{\begin{pmatrix} 0 & 0 \\ 0 & s \end{pmatrix}, s \in \lim Im\mathbb{Q}\right\} \oplus \left\{\begin{pmatrix} x & -\overline{w} \\ w & 0 \end{pmatrix}, x \in \lim Im\mathbb{Q}, w \in \mathbb{Q}\right\}$, with $L_{B^*}(V_B) = V_I$ and $L_{B^*}(H_B) = H_I$. This is an easy consequence of the definition of the K-K metric on $Sp(2)$ above. In short, $\forall B \in Sp(2)$, L_B acts by bundle maps preserving $V \oplus H$, so the connection is preserved together with its metric as well as the fiber metric and induces an isometric action on $(S^7, K = 1)$.

Remember that we are looking for a formula for the boundary map of $S^3 \dashrightarrow Sp(2) \longrightarrow S^7$, that is expected from lifting horizontally to $Sp(2)$ Euclidean geodesics of S^7 from 1 to -1. As we saw, this amounts to following the horizontal geodesics $\gamma_X(t)$ in $Sp(2)$ with its K-K metric up to the fiber over $-1 \in S^7$. Here, $X \in H_I$, above. Take a unitary X. The end points of all these geodesics compose a map from the unit 6-sphere $S^6 \ni X \mapsto \gamma_X(\pi) \in \begin{pmatrix} -1 & 0 \\ 0 & S^3 \end{pmatrix} \equiv S^3$. This Duran does in two steps: first he deals with the case $\begin{pmatrix} x & -r \\ r & 0 \end{pmatrix} \in H_I$, $r \in \mathbb{R}$ and then replaces r with any $w \in \mathbb{Q}$. The calculations are elementary and the final formula is

$$\gamma_X(t) = \begin{pmatrix} \cos(t) + \sin(t)x & -\sin(t)e^{tx}\overline{w} \\ \sin(t)w & \frac{w}{|w|}(\cos(t) - \sin(t)x)e^{tx}\frac{\overline{w}}{|w|} \end{pmatrix}$$

so $\gamma_X(\pi) = \begin{pmatrix} -1 & 0 \\ 0 & -\frac{w}{|w|}e^{\pi x}\frac{\overline{w}}{|w|} \end{pmatrix}$, and $S^6 \ni X = \begin{pmatrix} x \\ w \end{pmatrix} \mapsto \frac{w}{|w|}e^{\pi x}\frac{\overline{w}}{|w|} \in S^3$. This is the geometric version of the classifying map of the bundle and also the commutator of quaternions. At first glance, it looks like we are dividing by zero at the points where $w = 0$, x a unitary element in $\lim Im\mathbb{Q}$. But, then $e^{\pi x} = -1$ in the center of S^3, so it slides out and the denominators cancel out. It rests the doubt: which one is "faster," division by zero or commutation with a central element. However, all calculations done up to here are smoothly dependent on the parameters and the formula describes a smooth phenomenon. Indeed, one can readily see, applying high-school calculus, that the formula is essentially $\frac{\sin\theta}{\theta}$, which is analytic, [14]. So, commutation is faster than division by zero.

We can show using elementary topology that this map, call it $\beta : S^6 \longrightarrow S^3$, is homotopic to the commutator of quaternions, which also factors as a map $S^6 \longrightarrow S^3$ [16]. Up to now, we have proved with elementary means that the commutator of quaternions is homotopic to the analytic map β and its homotopy class generates the set of principal S^3 bundles over S^7, identified with $\pi_6 S^3$. That this group isomorphic to $\mathbb{Z}_{12}$ is a consequence of Serre's work [27], but it also follows through more elementary means as we will see.

3 Linear Algebra

Next, we employ the linear algebra and geometry of $Spin(n), 3 \leq n \leq 8$, and the exceptional Lie Group G_2 in relation with the commutator of quaternions β and in search of a homotopy between β^{12} and a constant. Recall that as mentioned above, such a homotopy would furnish explicit formulas describing exotic phenomena. This was our motivation.

The Cayley algebra $\mathbb{C}a$ is a nonassociative, noncommutative algebra on $\mathbb{R}^8 = \mathbb{Q} \oplus \mathbb{Q}$, the last division algebra on the $\mathbb{R}^{n'}s$. The usual definition of the Cayley product is: for $\begin{pmatrix} a \\ b \end{pmatrix}, \begin{pmatrix} c \\ d \end{pmatrix}$ in $\mathbb{Q} \oplus \mathbb{Q}$, we have

$$\begin{pmatrix} a \\ b \end{pmatrix}\begin{pmatrix} c \\ d \end{pmatrix} = \begin{pmatrix} ac - \overline{d}b \\ da + b\overline{c} \end{pmatrix} \in \mathbb{Q} \oplus \mathbb{Q}.$$

A convenient basis is $\begin{pmatrix} 1 \\ 0 \end{pmatrix} = e_0, \begin{pmatrix} i \\ 0 \end{pmatrix} = e_1, \begin{pmatrix} j \\ 0 \end{pmatrix} = e_2, \begin{pmatrix} k \\ 0 \end{pmatrix} = e_3, \begin{pmatrix} 0 \\ 1 \end{pmatrix} = e_4, \begin{pmatrix} 0 \\ i \end{pmatrix} = e_5, \begin{pmatrix} 0 \\ j \end{pmatrix} = e_6, \begin{pmatrix} 0 \\ k \end{pmatrix} = e_7$. Note that $e_0 = 1$, the unit element, $e_r^2 = -1$ and $e_r e_s = -e_s e_r$ for $r \neq s$, both ≥ 1. Conjugation $\overline{\begin{pmatrix} a \\ b \end{pmatrix}} = \begin{pmatrix} \overline{a} \\ -b \end{pmatrix}$ and we have for $\alpha, \beta \in \mathbb{C}a, \overline{\alpha\beta} = \overline{\beta}\overline{\alpha}, \alpha\overline{\alpha} = \overline{\alpha}\alpha = |\alpha|^2$, the Euclidean square length in $\mathbb{R}^8$. So, for $\alpha \neq 0, \alpha^{-1} = \overline{\alpha}\,|\alpha|^{-2}$ and the basis $\{e_r\}$ above is orthonormal with respect to the Euclidean metric on $\mathbb{R}^8$. Also, $|\alpha\beta| = |\alpha|\,|b|$ and left, resp., right multiplications **by unitary elements** are isometries with respect to the Euclidean scalar product: $L_\alpha(\eta) = \alpha\eta$ satisfies $|\alpha\eta| = |\eta|$ if $|\alpha| = 1$. Same with $R_\alpha(\eta) = \eta\alpha$.

The Triality Principle (T) $\forall A \in SO(8), \exists(B, C) \in SO(8) \times SO(8)$ a unique pair modulo common sign, i.e., (B, C) or $(-B, -C)$, such that $\forall\xi, \eta \in \mathbb{C}a$ we have $A(\xi\eta) = B(\xi)C(\eta)$.

This property was first described by Study at the beginning of 1900s and was formalized by Elie Cartan in the early 1920s as "duality" [5]. To prove it, note that if $\xi \in S^7 \subset \mathbb{C}a$ is a unit vector, then the reflection in $\mathbb{C}a = \mathbb{R}^8$ relative to the hyperplane perpendicular to ξ is given by $Refl_\xi(x) = -\xi\overline{x}\xi$. Note that $x = \frac{1}{2}(x + \overline{x}) + \frac{1}{2}(x - \overline{x})$ so $2\langle x, y\rangle = x\overline{y} + y\overline{x}$ and $x^2 = 2Re(x)x - |x|^2$. Now, test the formula for $x = \xi$ and for $x \perp \xi$. Since the formula describes an isometry, this ends the proof.

The Moufang Identities (M) In the 1930s, Emily Moufang in her doctoral thesis proved that the following identities hold: $\forall\alpha, x, y \in \mathbb{C}a$, one has

$$\alpha(xy)\alpha = (\alpha x)(y\alpha)$$
$$(\alpha x\alpha)y = \alpha[x(\alpha y)]$$
$$x(\alpha y\alpha) = [(x\alpha)y]\alpha$$

$\alpha(xy) = (axa)(\overline{\alpha}y)$
$(xy)\alpha = (x\overline{\alpha})(\alpha y\alpha)$
$\alpha(xy)\overline{\alpha} = (\alpha x\alpha^2)(\overline{\alpha}^2 y\overline{\alpha})$

If $(xp_1)(\overline{p_1}y) = (xp_2)(\overline{p_2}y)$, $\forall x, y \in \mathbb{C}a$, then $p_1 = \pm p_2$.

The proofs of these identities are relatively elementary and can be found in most algebra texts dealing with $\mathbb{C}a$.

The nonassociativity of $\mathbb{C}a$ can be seen from $(e_2e_4)e_7 = -e_1 \neq e_1 = e_2(e_4e_7)$. Given $x, y, z \in \mathbb{C}a$, the element $(xy)z - x(yz) \equiv [x, y, z] \in \mathbb{C}a$ is called the associator of the three elements. Analogous to the commutator of quaternions, the associator of Cayley numbers factors through to a map $A : S^{21} \longrightarrow S^7$. It turns out $[A] \in \pi_{21}S^7 \cong \mathbb{Z}_{120}$ and it should be interesting to know how is A^{120} homotopic to a constant map.

A corollary of a theorem of Artin says that any subalgebra of $\mathbb{C}a$ generated by two elements is associative.

What follows is described in [8–11].

Now, we remember (T). All triplets (A, B, C) that satisfy (T) form a subgroup of the Cartesian product $SO(8) \times SO(8) \times SO(8)$. Each element $A \in SO(8)$ is a product of an even number of reflections. Suppose $A = Refl_\xi \circ Refl_\eta$, then, using the appropriate Moufang identities, $A = Refl_\xi(-\eta(\overline{xy})\eta) = \xi(\overline{\eta}(xy)\overline{\eta})\xi = \xi[(\overline{\eta}x)(y\overline{\eta})]\xi = [\xi(\overline{\eta}x)][(y\overline{\eta})\xi] \equiv [L_\xi \circ L_{\overline{\eta}}(x)][R_\xi \circ R_{\overline{\eta}}(y)] = B(x)C(y)$ with $B = L_\xi \circ L_{\overline{\eta}}$, $C = R_\xi \circ R_{\overline{\eta}}$. One can show easily the pair $\pm(B, C)$ to be unique. For a product of an even number of reflections $A(x) = Refl_{u_{2r}} \circ \ldots \circ Refl_{u_1} = u_{2r}(\ldots u_2(\overline{u_1}\overline{x}\overline{u_1})u_2 \ldots)u_{2r}$, the negative sign and conjugation appear an even number of times and they cancel out. The appropriate Moufang identity implies $A(xy) = [u_{2r}(\ldots u_2(u_1x)\ldots)][(\ldots(y\overline{u_1})u_2\ldots)u_{2r}] = B(x)C(y)$.

We will show that this subgroup is isomorphic to $Spin(8)$, the universal (double) covering group of $SO(8)$ with projection $(A, B, C) \longmapsto A$. Note that the A and B uniquely determine C, so we claim that the triality triple provides an explicit representation of $Spin(8)$ in $SO(8) \times SO(8)$. It is immediately seen that the projection to A is a double covering group morphism onto $SO(8)$ and it is also connected: the path $(A(t), B(t), C(t))$, $t \in [0, \pi]$, where $A(t) = Refl_{e_1} \circ Refl_{e_1 \exp(t\pi e_1)}$, $B(t) = L_{e_1} \circ L_{e_1 \exp(t\pi e_1)}$, $C(t) = R_{e_1} \circ R_{e_1 \exp(t\pi e_1)}$, joins the point (I, I, I) to $(I, -I, -I)$. We still need to show that this group is simply connected. We leave this part for a collective proof: all $3 \leq n \leq 8$ together. For now, call this group $Spin(8)'$ subgroup of the Cartesian product of three copies of $SO(8)$.

Define $SO(8) \ni A \longmapsto \widetilde{A} \in SO(8)$ where $\widetilde{A}(x) = \overline{A(\overline{x})}$.

Proposition *If* $(A, B, C) \in Spin(8)'$, *then all the following triples also live in* $Spin(8)$*:* $(C, \widetilde{B}, A)$, $(\widetilde{A}, \widetilde{C}, \widetilde{B})$, $(\widetilde{C}, \widetilde{A}, B)$, $(\widetilde{B}, C, \widetilde{A})$, *and* $(B, A, \widetilde{C})$.

Proof We show only the first one, the proof of the rest is similar: $\forall x \in \mathbb{C}a$, $|x|^2 A(y) = A(\overline{x}(xy)) = B(\overline{x})C(xy)$. So, $|x|^2\, \overline{B(\overline{x})}A(y) = \overline{B(\overline{x})}B(\overline{x})C(xy) = |x|^2\, C(xy)$ because B is orthogonal. So, $C(xy) = \widetilde{B}(x)A(y)$ and $(C, \widetilde{B}, A) \in Spin(8)'$.

The center $ZSpin(8)' = \{(I, I, I), (I, -I, -I), (-I, I, -I), (-I - I, I)\} \cong \mathbb{Z}_2 \times \mathbb{Z}_2$ and the quotient of all automorphisms by the internal ones, i.e., the $ExtAut(Spin(8))$ is parametrized by the group of automorphisms of $\mathbb{Z}_2 \times \mathbb{Z}_2$, that is the permutation group of three elements, usually denoted by S_3. The six elements of S_3 correspond to the six external automorphisms of $Spin(8)'$. In detail, these are:

(1) id,
(2) $\delta(A, B, C) = (C, \widetilde{B}, A)$
(3) $\tau(A, B, C) = (\widetilde{A}, \widetilde{C}, \widetilde{B})$
(4) $\gamma(A, B, C) = \tau \circ \delta(A, B, C) = (\widetilde{C}, \widetilde{A}, B)$
(5) $\gamma^2(A, B, C) = (\widetilde{B}, C, \widetilde{A})$
(6) $\delta \circ \gamma(A, B, C) = (B, A, \widetilde{C})$.

The automorphism γ expresses the basic properties of triality and has order 3. Each one of the other two, δ and τ, has order two. Rigorously speaking, triality is represented by the group S_3, above, but it is usual to say that triality is $\{id, \gamma, \gamma^2\}$.

Lemma 3.1 *If* $A \in SO(7)$, *then* $\widetilde{A} = A$.

Proof $A(1) = 1$ and being orthogonal, A preserves $Re(\mathbb{C}a) \oplus Im(\mathbb{C}a)$, so, for $\alpha = \alpha_0 + \alpha_1 \in Re(\mathbb{C}a) \oplus Im(\mathbb{C}a)$ we get $\widetilde{A}(\alpha) = \overline{A(a_0 - a_1)} = \overline{A(a_0)} - \overline{A(a_1)} = A(a_0) + A(a_1) = A(\alpha)$.

Define: $Spin(7)' \subset Spin(8)'$ to be the subgroup of all $(A, B, C) \in Spin(8)'$ with $A \in SO(7)$. Similarly, $Spin(6)' = \{(A, B, C) \in Spin(7)' \mid A(e_1) = e_1\} = (A, B, C) \mid A \in SO(6)\}$, $Spin(5)' = \{\ldots A(1) = 1, A(e_1) = e_1, A(e_2) = e_2\}$, $Spin(4)'$ has $A(e_3) = e_3$ as well, i.e., $A \in SO(4)$ and finally $Spin(3)'$ has also $A(e_4) = e_4$, i.e., $A \in \begin{pmatrix} I_5 & 0 \\ 0 & SO(3) \end{pmatrix} \subset SO(8)$.

Note that $Spin(7)' = \{(A, B, \widetilde{B}) \in Spin(8)'\}$: because $\forall x \in \mathbb{C}a$, $|x|^2 = |x|^2 A(1) = A(\overline{x}x) = B(\overline{x})C(x)$ and B, C being orthogonal, from $|x|^{-2} B(\overline{x})C(x) = 1$ follows $C(x) = \widetilde{B}(x)$.

The group of all algebra automorphisms of $\mathbb{C}a$ is denoted by G_2, so $G_2 = \{A \in SO(8) \mid A(xy) = A(x)A(y)\}$. So, (A, A, A) and $(A, -A, -A)$ are the two triality triples corresponding to $A \in G_2$. Since any automorphism preserves the unit $= 1$, $A(1) = 1$ and $G_2 \subset SO(7)$. And, also $G_2 \subset Spin(7)'$.

Claim G_2 is the fixed point group of the automorphism γ.

Proof If $\gamma(A, B, C) = (A, B, C)$, then $(A, B, C) = (A, \widetilde{A}, \widetilde{A})$ with $A \in SO(7)$. So, $A = \widetilde{A}$ and $(A, B, C) = (A, A, A)$, $A \in G_2$. Clearly, γ fixes G_2.

Claim (1) $Fix(\tau) = Spin(7)'$, (2) $Fix(\delta) = \gamma(Spin(7)')$, (3) $Spin(7)' \cap \gamma(Spin(7)') = Spin(7)' \cap \gamma^2(Spin(7)' = \gamma(Spin(7)' \cap \gamma^2(Spin(7)') = G_2$. The proofs are immediate.

Proposition 1 *The following maps define principal bundles:*

(a) $Spin(7)' \dashrightarrow Spin(8)' \longrightarrow S^7$ *with* $(A, B, C) \mapsto A(1)$
(b) $Spin(6)' \dashrightarrow Spin(7)' \longrightarrow S^6$ *with* $(A, B, \widetilde{B}) \mapsto A(e_1)$
(c) $Spin(5)' \dashrightarrow Spin(6)' \longrightarrow S^5$ *with* $(A, B, \widetilde{B}) \mapsto A(e_2)$
(d) $Spin(4)' \dashrightarrow Spin(5)' \longrightarrow S^4$ *with* $(A, B, \widetilde{B}) \mapsto A(e_3)$
(e) $Spin(3)' \dashrightarrow Spin(4)' \longrightarrow S^3$ *with* $(A, B, \widetilde{B}) \mapsto A(e_4)$
(f) $G_2 \dashrightarrow Spin(8)' \longrightarrow S^7 \times S^7$ *with* $(A, B, C) \mapsto (A(1), B(1))$
(g) $G_2 \dashrightarrow Spin(7)' \longrightarrow S^7$ *with* $(A, B, \widetilde{B}) \mapsto B(1)$
(h) $G_2 \dashrightarrow \gamma(Spin(7)') \longrightarrow S^7$ *with* $(B, A, B) \mapsto B(1)$.

Proof We just show (h). The fiber over $1 = B(1)$ is G_2, because $B(x) = B(x \cdot 1) = A(x)B(1) = A(x)$. So, $A = B$ (in G_2). The rest are similar.

Time to show that our $Spin(k)'$ are simply connected and therefore equal to $Spin(k)$.

We have defined the Lie group epimorphism $f : Spin(8)' \longrightarrow SO(8)$, $f(A, B, C) = A$. So, $\ker(f) = \{(I, I, I), (I, -I, -I)\} \cong \mathbb{Z}_2$ and f is a double cover. Similarly for $k = 7, 6, 5, 4, 3$.

Claim $Spin(k)'$ is simply connected and the projection $c : Sp(1) = S^3 \longrightarrow SO(3)$ with $c(q)$ a matrix in $SO(3)$ sends the imaginary quaternion $x \in \lim Im\mathbb{Q} = \mathbb{R}^3$ to $qx\overline{q}$. And, c is an epimorphism of Lie groups with $Ker(c) = \{1, -1\}$, a double covering connected and simply connected.

Claim Since $\pi_1 S^3 = 0$, it is the universal cover of $SO(3)$ and therefore $Sp(1) \equiv S^3 = Spin(3)$. To show $Spin(4) = S^3 \times S^3$, take $S^3 \times S^3 \ni (p, q) \longmapsto F(p, q) \in SO(4)$ that maps the vector $v \in \mathbb{R}^4 \equiv \mathbb{Q}$ to $pv\overline{q}$. $F : S^3 \times S^3 \longrightarrow SO(4)$ is the universal (double) cover. Clearly, it is a group morphism with $\ker(F) = \{(1, 1), (-1, -1)\}$. Equality of dimensions implies that F is an epimorphism. We write $D = l_p \circ r_{\overline{q}}$, by abuse of notation and we are done.

Now, we show $Spin(4)' = S^3 \times S^3$. Take $A \in SO(4) \subset SO(8)$. From above, $\exists (p, q) \in S^3 \times S^3$, with $A = \begin{pmatrix} I_4 & 0 \\ 0 & l_p \circ r_{\overline{q}} \end{pmatrix}$. Take $B = \begin{pmatrix} r_{\overline{q}} & 0 \\ 0 & l_p \end{pmatrix}$, $C = \begin{pmatrix} l_q & 0 \\ 0 & l_p \end{pmatrix}$. Then, $Spin(4)' = \{(A, B, C), A, B, C$ as indicated$\}$. It is easy to verify using the multiplication rule for $\mathbb{C}a$ that this is a triality triple in $Spin(4)'$ and that the map from $S^3 \times S^3$ is a Lie group isomorphism. Therefore, $Spin(4)' = Spin(4)$. The exact homotopy sequence of $Spin(4)' \dashrightarrow Spin(5)' \longrightarrow S^4$ implies that $\pi_1 Spin(5)' = 0$ and so $Spin(5)' = Spin(5)$. Similarly, $Spin(6)' = Spin(6)$, $Spin(7)' = Spin(7)$, and $Spin(8)' = Spin(8)$.

Triality provides explicit identifications $Spin(5) \cong Sp(2)$, $Spin(6) \cong SU(4)$: Roughly speaking, we can represent $SU(4)$ in $SO(8)$ as all matrices A that commute with complex multiplication by i on $\mathbb{C}^4 = \mathbb{R}^8$. We replace this complex multiplication by L_{e_1}. $Spin(6) \ni (A, B, \widetilde{B}) \mapsto \gamma(A, B, \widetilde{B}) = (B, A, B) \in Spin(8)$. So, $B(e_1 x) = A(e_1)B(x) = e_1 B(x)$, i.e., $B \circ L_{e_1} = L_{e_1} \circ B$ and $B \in SU(4)$. A similar consideration shows that $Sp(2)$ is all matrices in $SO(8)$ that commute with two mutually anti-commuting linear complex structures: quaternionic multiplications by i, and by j on $\mathbb{Q}^2 = \mathbb{R}^8$. Replace these quaternionic multiplications by the Cayley ones L_{e_1}, L_{e_2}. We also get an isomorphism $Sp(2) \cong Spin(5)$. For details, see [9]. These, together with the lower spins, above, and $SU(2) = Sp(1) = S^3$ are the only identifications between elements of different infinite families of compact Lie groups, a fact known since E. Cartan in the 1920s through classification of their Lie Algebras. The novelty here is the unique argument and the formulas for the isomorphisms provided by triality.

Next, we consider two faithful representations of S^3 in G_2 and the resulting representation of $SO(4) \subset G_2$.

1. $S^3 \ni q \mapsto \begin{pmatrix} I_3 & 0 \\ 0 & l_q \end{pmatrix} = \phi(q) \in G_2$, 2. $S^3 \ni p \mapsto \begin{pmatrix} l_p \circ r_{\overline{p}} & 0 \\ 0 & r_{\overline{p}} \end{pmatrix} = \psi(p) \in G_2$,

3. $S^3 \times S^3 \ni (p,q) \mapsto \begin{pmatrix} l_p \circ r_{\overline{p}} & 0 \\ 0 & l_q \circ r_{\overline{p}} \end{pmatrix} = \Phi(p,q) \in G_2$, the last one has kernel $\{(1,1), (-1,-1)\} = \mathbb{Z}_2$ so it defines an inclusion of $SO(4)$. The verifications are a consequence of the definition of the Cayley product through two quaternions. Note that $\Phi(1,-1) = \begin{pmatrix} I_3 & 0 \\ 0 & -I_4 \end{pmatrix} = A$ and the image of Φ consists of all $X \in G_2$, that commute with A. Note that $A^2 = I_7$ and define the involutive inner automorphism of G_2, call it σ, by $\sigma(X) = AXA^{-1} = AXA$. As we saw, $Fix(\sigma) = \ker(\Phi)$, so σ factors through $G_2/SO(4) \longrightarrow G_2$ and the Cartan inclusion of the symmetric space in the group is $\widetilde{\sigma}([X]) = X\sigma(X^{-1}) = XAX^{-1}A^{-1} = [X, A]$, the commutator with A. The image is a totally geodesic inclusion into the group (following E. Cartan) and this symmetric space parametrizes the quaternionic subalgebras of $\mathbb{C}a$. This can be proved easily from the above and it was known to Cartan.

Another fact seems worth noting in this context: The inclusions ϕ, ψ above induce distinct morphisms in $\pi_3 S^3 = \mathbb{Z} \longrightarrow \mathbb{Z} = \pi_3 G_2 : \phi(1) = 1$ and $\psi(1) = 3$. This is a consequence of $G_2 \dashrightarrow SO(7) \to \mathbb{R}P^7$, $SO(7) \dashrightarrow SO(8) \longrightarrow S^7$, their exact homotopy sequences and elementary considerations, for example, the triality triple $\begin{pmatrix} l_q & 0 \\ 0 & I_4 \end{pmatrix}, \begin{pmatrix} l_q & 0 \\ 0 & I_4 \end{pmatrix}, \begin{pmatrix} I_4 & 0 \\ 0 & r_{\overline{q}} \end{pmatrix}$.

Next, we take a close look at G_2. From $e_0 = 1, e_1 = \binom{i}{0}, \ldots, e_7 = \binom{0}{k}$ and the multiplication rule of $\mathbb{C}a$, we deduce the multiplication table of the orthonormal basis $\{e_j\}$ of $\mathbb{C}a$.

-1	e_1	e_2	e_3	e_4	e_5	e_6	e_7
e_1	-1	e_3	$-e_2$	e_5	$-e_4$	$-e_7$	e_6
e_2	$-e_3$	-1	e_1	e_6	e_7	$-e_4$	$-e_5$
e_3	e_2	$-e_1$	-1	e_7	$-e_6$	e_5	$-e_4$
e_4				-1	e_1	e_2	e_3
e_5					-1	$-e_3$	e_2
e_6						-1	$-e_1$
e_7							-1

If A_j is the jth column of the matrix A in G_2, automorphisms of $\mathbb{C}a$, then the Cayley product $A_i A_j$ obeys the same rule as in the table above.

For S^3_q defined above, we have the principal bundle $S^3_q \dashrightarrow Spin(5) \longrightarrow S^7$ with projection $\pi(A, B, \widetilde{B}) = B(1)$. (Subgroup multiplies from the right). Routine proof. We can express the pullback diagram $Sp(2) = h^*(-h)$ at the beginning of the section. First note that the quotient $Spin(4) \dashrightarrow Spin(5) \longrightarrow S^4$ is

$(A, B, \widetilde{B}) \mapsto A(e_3) = (e_1\overline{B(1)})(\overline{B(1)}e_2) \in S^4 \subset \mathbb{R}^5 = span\{e_3, \ldots, e_7\} \subset \mathbb{R}^8 = \mathbb{C}a$ [9].

The Hopf map in terms of Cayley products, instead of quaternionic ones, was done in [10]. We reproduce it here:

For any orthonormal pair $(J, K) \in \lim Im\mathbb{C}a \times \lim Im\mathbb{C}a$, equivalently, for any element of the unit tangent bundle of S^6, define $\delta : S^7 \longrightarrow S^7$, by $\delta(\alpha) = (J\overline{\alpha})(\alpha K)$ and note that it is orthogonal to 1, J, and K. So, the image, being a unit vector, lives in the sphere S^4 specified above and the map $h(\alpha) = (e_1\overline{\alpha}\,)(\alpha e_2)$ is isomorphic, as a principal S^3 bundle, to the Hopf map described at the beginning.

Remark The map $-h$ has class in $\pi_7 S^4 \simeq \pi_7 S^7 + \pi_6 S^3 = \mathbb{Z} + \mathbb{Z}_{12}$, $[h] = 1 \pm \Sigma b$, where b generates $\pi_6 S^3$ and Σ is the suspension. The sign depends on the choice of orientation. The formula for h reflects the nonassociativity of $\mathbb{C}a$, just as the alternative formula $S^3 \ni q \mapsto qi\overline{q} \in S^2$ reflects the noncommutativity of quaternions. h is the invariant projection of the free S^3 action on S^7. $\begin{pmatrix} a \\ b \end{pmatrix} q = \begin{pmatrix} aq \\ bkq\overline{k} \end{pmatrix}$. There is an algebraic relation between three Hopf-type maps at different levels to be described later. For now, we note that the pullback diagram above is in triality/Cayley terms

$$\begin{array}{ccc} (A, B, \widetilde{B}) & \longmapsto & \widetilde{B}\,(e_4) \\ \downarrow & & \downarrow \\ B(1) & \longmapsto & A\,(e_3) \end{array}$$

. The proofs are in [9].

4 Infinitesimal Triality and $\widehat{G_2} \subset \widehat{SO}(7)$

Here, we denote with $\widehat{G}$ the Lie algebra of the Lie group G identified with $T_e G$. Consider the curve $\Gamma(t) = (A(t), B(t), C(t)) \in Spin(8)$, with $\Gamma(0) = (I, I, I)$ and $\Gamma'(0) = (A_0, B_0, C_0)$, take the derivative at $t = 0$ of $A(t)(\xi\eta) = B(t)(\xi) \cdot C(t)(\eta)$. We have

$$A_0(\xi\eta) = B_0(\xi) \cdot \eta + \xi \cdot C_0(\eta), \tag{1}$$

this is the infinitesimal version of Triality. It is convenient to write it as $\widehat{Spin(8)} = \{(X, X^\lambda, X^\rho) \in \widehat{SO(8)} \times \widehat{SO(8)} \times \widehat{SO(8)}\}$, where $X(\xi\eta) = X^\lambda(\xi)\cdot\eta + \xi\cdot X^\rho(\eta)$. So, $\widehat{Spin(7)} = \{(X, X^\lambda, \widetilde{X^\lambda}) \in \widehat{Spin(8)}\}$. The automorphisms γ, δ, and τ of $Spin(8)$ define, by linearity, Lie Algebra automorphisms of $\widehat{Spin(8)}$, for example, $d\gamma =$

$\widehat{\gamma}(X, X^\lambda, X^\rho) = (\widetilde{X^\rho}, \widetilde{X}, X^\lambda)$. It is not hard to show that the maps $\frac{1}{2}(id + \tau)$ and $\frac{1}{3}(id + \gamma + \gamma^2)$ are the Killing–Cartan projections of $\widehat{Spin(8)}$ onto $\widehat{Spin(7)}$ and $\widehat{G_2}$.

5 Generators of Some Homotopy Groups

Toda et al. [40] showed using triality that $\pi_7 SO(7) \simeq \mathbb{Z}$ and is generated by the conjugation of Cayley numbers: $S^7 \ni \alpha \mapsto C_\alpha = (L_a \circ R_{\overline{\alpha}}, L_\alpha \circ R_{\alpha^2}, L_{\overline{a}^2} \circ R_{\overline{\alpha}}) \equiv \Psi(\alpha) \in Spin(7)$ the triality triple of Cayley conjugation by α is a consequence of the appropriate Moufang identity. Note that the matrix B in the triple $(A, B, C) \in Spin(8)$ determines the triple, so we can say $B \in Spin(7)$. That is, $f_\alpha = L_a \circ R_{\alpha^2}$ generates $\pi_7 Spin(7) = \mathbb{Z}$. The identity $\alpha(xy) = (\alpha x \alpha)(\overline{\alpha} y)$ implies $\alpha \mapsto \Theta(\alpha) = (L_\alpha, L_\alpha \circ R_\alpha, L_{\overline{\alpha}}) \in Spin(8) \cong Spin(7) \times S^7$. The group $\pi_7 SO(7)$ is already stable due to the triviality of $Spin(8) \cong Spin(7) \times S^7$, and one can write easily the formulas for the isomorphism as well as their inverses [34]. Note that the classes $[\Theta]$ and $[\Psi]$ provide the spin versions of the homotopy generators in question.

Back to G_2 now we recall that any $D \in G_2$ is determined by $D(e_1) = x$, $D(e_2) = y$, and $D(e_4) = z$: its first, second, and fourth columns (the third one being 1st × 2nd). So, $\langle x, y \rangle = \langle z, x \rangle = \langle z, y \rangle = \langle z, xy \rangle = 0$. It follows that the action $G_2 \times S^6 \longrightarrow S^6$ with $(D, X) \mapsto DX$, X is a unit element in lim $Im(\mathbb{C}a)$, is transitive, and the orbit of e_1 covers the whole S^6.

Claim The isotropy of e_1 is a subgroup $H \subset G_2$ isomorphic to $SU(3)$. In fact, the homogeneous bundle $SU(3) \ldots G_2 \longrightarrow S^6$ is a reduction of the unit tangent bundle of S^6, namely $SO(6) \ldots SO(7) \longrightarrow S^6$.

The proof is relatively a routine one and we omit it, but for a few remarks: Consider the unitary Cayley number $\alpha \in S^7$ with $\alpha = \exp \frac{2\pi}{3} e_1 \in S^7$, so α lives in the parallel six sphere at 120° and $\alpha^3 = 1$. Consider $\Lambda = C_\alpha \equiv L_\alpha \circ R_{\overline{\alpha}} \in G_2$ because $\alpha^2 = \overline{\alpha}$. It fixes e_1 and can easily show that Λ commutes with all elements of H, i.e., $\Lambda \in Z(H)$, the center of H. Similarly, $C_{\alpha^2} \in Z(H)$ which is $\{I, \Lambda, \Lambda^2\}$. Note now that for any $\alpha \in S^6_{\frac{2\pi}{3}}$ the map f_α has values in G_2. In fact, it is an injection with image $S^6 \subset G_2$. Each such $a = \cos \frac{2\pi}{3} + \sin \frac{2\pi}{3} J$, for $J \in S^6_\pi$, i.e., $J^2 = -I$, L_J a linear complex structure in $Spin(8)$. Consider now the 7-cell $e^7 = \{\cos(t) + \sin(t) J,\ J^2 = -1,\ \frac{2\pi}{3} \leq t \leq \pi\} \subset S^7$, the unitary Cayley sphere. The map $Spin(7) \ni (A, B, \widetilde{B}) \mapsto B(1) \in S^7$ defines a principal bundle $G_2 \ldots Spin(7) \longrightarrow S^7$ and from the map Ψ above the projection to S^7 induces $\pi_7 Spin(7) \cong \mathbb{Z} \ni 1 \mapsto 3 \in \mathbb{Z} \cong \pi_7 S^7$. The homotopy diagram of this fibration says that $\partial(1) \neq 0 \in \pi_6 G_2$, since $1 \notin Image \pi_7 Spin(7)$, but $\partial(3) = 0$ and ∂ is an epimorphism, since $\pi_6 Spin(7) = \pi_6 Spin(8) = 0$ in the stable range. So, $\pi_6 G_2 \cong \mathbb{Z}_3$ and is generated by the class of $C_\alpha \mid S^6_{\frac{2\pi}{3}}$, by the definition of the boundary map.

Claim This map is the inclusion of the conjugate orbit of $\Lambda \in ZSU(3)$ into G_2 [11].

The proof of this is straightforward and is omitted. Note, however, that $(G_2, SU(3))$ is not a symmetric pair and that Λ has order 3 (not 2) and the inclusion of this orbit in the third roots of I in G_2 is not a totally geodesic submanifold of G_2; however, it is diffeomorphic to S^6. The "not totally geodesic" part is a result of computing the relevant Lie brackets to show that $\widehat{S^6} \equiv \widehat{SU(3)}^{\perp} \subset \widehat{G_2}$ is not a Lie Triple System. It is a minimal submanifold, though, being an isolated orbit of the conjugate action of G_2 on itself.

Now, we take a tour of elementary diagram chasing to indicate how to prove in a naive way that $\pi_3 S^3 \cong \mathbb{Z}_{12}$. Recall that $Spin(3) = Sp(1) = S^3$, $Spin(6) \cong SU(4)$, $Spin(5) \cong Sp(2)$, and $Spin(8)$ is diffeomorphic to $Spin(7) \times S^7$. The said diagram is the following:

$$\begin{array}{ccccccc} Spin(3) & \hookrightarrow & SU(3) & \hookrightarrow & G_2 & \hookrightarrow & Spin(7) \\ \vdots & & \vdots & & \vdots & & \vdots \\ Spin(5) & \hookrightarrow & Spin(6) & \hookrightarrow & Spin(7) & \hookrightarrow & Spin(8) \\ \downarrow & & \downarrow & & \downarrow & & \downarrow \\ S^7 & = & S^7 & = & S^7 & = & S^7 \end{array},$$

all projections onto S^7 are $(A, B, C) \mapsto B(1)$. From $SU(4) \hookrightarrow SU(5) \to S^9$, we get $\pi_7 SU(4) \cong \pi_7 SU = \mathbb{Z}$ and $\pi_6 SU(4) \cong \pi_6 SU = 0$. From the homotopy sequence of $SU(3) \hookrightarrow G_2 \to S^6$, we get

$\ldots \pi_7 G_2 \to \pi_7 S^6 \to \pi_6 SU(3) \to \pi_6 G_2 \to \pi_6 S^6 \to \ldots .$

Recall now that $\pi_7 S^6 \cong \mathbb{Z}_2$. So, if $\pi_7 G_2 = 0$, then $\pi_6 SU(3) \cong \mathbb{Z}_2 \oplus \mathbb{Z}_3 \cong \mathbb{Z}_6$. Now, the homotopy sequence of $G_2 \ldots Spin(7) \to S^7$, remembering that the homotopy class of $\alpha \mapsto C_\alpha$ generates $\pi_7 SO(7)$ [40], we get that the triality triple $(C_\alpha, L_\alpha \circ R_{\alpha^2}, L_{\overline{a^2}} \circ R_{\overline{\alpha}})$ generates $\pi_7 Spin(7)$. Projecting this generator to S^7 sends $\alpha \mapsto L_\alpha \circ R_{\alpha^2}(1) = \alpha^3$, so $\mathbb{Z} = \pi_7 Spin(7) = \mathbb{Z} \ni 1 \mapsto 3 \in \mathbb{Z} = \pi_7 S^7$ and this map is a monomorphism, so $\partial : \pi_8 S^7 = \mathbb{Z}_2 \to \pi_7 G_2$ is an epimorphism and the last group is either 0 or $\mathbb{Z}_2$. To show next that it is zero. This was calculated by Mimura [30] using less elementary means.

From $\pi_6 Spin(6) = \pi_6 SU(4) = 0$ and $\pi_7 Spin(6) = \pi_7 SU(4) = \mathbb{Z}$ in the exact sequence of $Spin(6) \cdots Spin(7) \to S^6$, we get

$\pi_7 Spin(6) = \mathbb{Z} \ni 1 \mapsto 2 \in \mathbb{Z} = \pi_7 Spin(7)$.

Now, use this information on the homotopy ladder of the two middle columns of the big diagram to follow $1 \in \mathbb{Z}$ around the square and get that the vertical image of this 1 is $6 \in \mathbb{Z} = \pi_7 S^7$ and therefore, $\pi_6 SU(3) \cong \mathbb{Z}_6$. Feed this $(1 \mapsto 2)$ information back into the homotopy ladder of the two left columns of the big diagram and get, again following the square around, that $\pi_7 Spin(5) = \mathbb{Z} \ni 1 \mapsto 2 \in \mathbb{Z} = \pi_7 Spin(6)$, vertically down to $12 \in \mathbb{Z} = \pi_7 S^7$, so $\pi_6 Spin(3) \equiv \pi_6 S^3 \cong \mathbb{Z}_{12}$, as promised to prove with an elementary argument. Remember that there is still one group pending, $\pi_7 G_2 = 0$ (and not $\mathbb{Z}_2$). Let φ be a generator of $\pi_8 S^7 = \mathbb{Z}_2$, then φ^3 also generates. Consider now the homotopy sequence of $G_2 \cdots Spin(7) \to S^7$, call the projection p and recall that $p(A, B, \widetilde{B}) = B(1)$. Let $S^8 \ni x \mapsto (A, B, \widetilde{B}) =$

$(C_{\phi(x)}, L_{\varphi(x)} \circ R_{\varphi(x)^2}, L_{\varphi(x)^{-2}} \circ R_{\varphi(x)^{-1}}) \in Spin(7)$ and this projects to $\varphi(x)^3 \in S^7$, whose class generates, i.e., the class of $p \circ Conj \circ \varphi$ generates $\pi_8 S^7 = \mathbb{Z}_2$ and p_* is onto $\pi_8 S^7$ and $\pi_7 G_2$ is squashed between two zero maps and is zero.

Now, look at $S^3_q \ldots G_2 \to V$, where V stands for $V_{7,2}$ the 2-orthonormal frames in Euclidean $\mathbb{R}^7 = Imaginary\ \mathbb{C}a$. This represents the two first columns of a matrix in G_2 (the third one is their Cayley product and the rest quotient out by the right action of the subgroup S^3_q). The homotopy sequence of this fibration implies (since $\pi_7 G_2 = 0$) that

$\partial : \pi_7 V \rightarrowtail \pi_6 S^3 \to \pi_6 G_2 \to \pi_6 V \ldots$ which gives us, after a moment's reflection, that $\pi_7 V = \mathbb{Z}_4$ and so $\mathbb{Z}_4 \rightarrowtail \mathbb{Z}_{12} \twoheadrightarrow \mathbb{Z}_3$. Of all these groups, we have nice maps generating them and the generator of the last one has its third power a constant. This may be a good point to start looking for a homotopy between $b\begin{pmatrix} x \\ w \end{pmatrix}^{12} = \left(\frac{w}{|w|} e^{\pi x} \frac{\overline{w}}{|w|}\right)^{12}$ and a constant. There is an algebraic section $\psi : \mathbb{Z}_3 \to \mathbb{Z}_{12}$, say, $1 \mapsto 4[b]$. A generator of $\pi_7 V$ can be $S^7 \ni \alpha \mapsto (\alpha e_1 \overline{\alpha}, \alpha e_2 \overline{\alpha}) \equiv H(\alpha) \in V$ as a consequence of $V = SO(7)/SO(5)$ and the [40] theorem that Cayley conjugation generates $\pi_7 SO(7)$. So, $\partial[H] = 3[b]$. A convenient generator of $\mathbb{Z}_4 \oplus \mathbb{Z}_3$, from this point of view, seems to be $\partial[H] \oplus \psi(1) = 3[b] \oplus \psi(1)$. Note that $\psi(1) = 4[b]$.

Thomas Püttmann, in his Habilitationsschrift [33], has done some remarkable work exhibiting formulas describing generators of geometrically relevant homotopy groups and some of the geometry behind the cancellations that determine the order of these groups. Between them is $\pi_7 Sp(2)$. It is a pity that he never published his manuscript. See also [35].

6 Hopf Maps

The classical formula for the three Hopf maps $S^3 \to S^2$, $S^7 \to S^4$, and $S^{15} \to S^8$ (the last one is not a principal bundle since S^7, the fiber, is not a group) is $\binom{X}{Y} \mapsto \binom{|X|^2 - |Y|^2}{2X\overline{Y}}$ and we already saw an alternative formula for the first one ($q \mapsto qi\overline{q}$). Such a formula, based on the nonassociativity of $\mathbb{C}a$, is $\mathbb{C}a \ni \alpha \mapsto (e_1\overline{\alpha})(\alpha e_2) \in S^4 \subset \mathbb{R}^5 = span\{e_3, \ldots, e_7\}$. In fact, one can use any pair of $(J, K) \in V$, in place of $(e_1 e_2)$, just as one can use any unitary vector $\lambda \in Imaginary\ \mathbb{Q}$, in place of i in the last case. The proof is elementary linear algebra [10]. Now, consider another Hopf-type map, which we used above. $H' : S^7 \times S^6 \to S^6$ with $H'(\alpha, J) = \alpha J \overline{\alpha}$ in the unitary $S^6 \subset \lim Im\mathbb{C}a$. We saw that for each fixed $J \in S^6$ (domain), say $e_1 \in S^6$, the map $\alpha \mapsto \alpha e_1 \overline{\alpha}$ generates $\pi_7 S^6 = \mathbb{Z}_2$. There three Hopf maps H

(2)

(above), H' and $h : S^7 \times V \to S^6$ with $h(\alpha, (J, K)) = (J\overline{\alpha})(\alpha K)$ satisfy the algebraic formula below, for all $\alpha \in S^7$, $m, n \in \mathbb{Z}$, $* = (e_1, e_2)$, the basepoint of $V = V_{7,2}$, defining $h(\beta, *) \equiv (e_1\overline{\beta})(\beta e_2)$:

$$h(\alpha^m, H(\alpha^n)) = H'(\alpha^n, h(\alpha^{m+3n}, *)). \tag{3}$$

For the proof, one uses the Moufang identities [10].

Example $(\alpha e_1\overline{\alpha})(\alpha e_2\overline{\alpha}) = \alpha h(\alpha^3)\overline{\alpha}$.

One can use this formula to look for an explicitly described (by formulas) trivial principal bundle $E(12)$ over S^7, and, equivalently, look for a formula for a generator of $\pi_7 Sp(2) = \mathbb{Z}$ in the following sense: The pullback of $E(1) = Sp(2)$ by the 12th power map from S^7 is trivial and so it has a section. This is equivalent to producing a map $\beta : S^7 \to S^7$ such that $-h \circ \beta(\alpha) = h(\alpha^{12})$ and the matrix with columns α^{12} and $\beta(\alpha)$ represents a generator of $\pi_7 Sp(2)$. Inversely, suppose we are given a map $g : S^7 \to Sp(2)$ with $[g] = 1 \in \mathbb{Z} = \pi_7 Sp(2)$. It is immediate that the columns of the matrix $g(\alpha)$ both have degree 12 or (-12). Then, g provides a section of the pullback of the bundle $Sp(2)$ over its first column. Consequently, it provides a bundle isomorphism $S^7 \times S^3 \cong g_1^*(Sp(2))$. So, the homotopy classes of their classifying maps are equal or $0 = [g_1] \in \mathbb{Z}_{12}$. But, the classifying map of all these bundles is the boundary map $\partial : \pi_7 S^7 \to \pi_6 S^3$, and we have seen that it is the commutator of quaternions.

At the end of [10], there is a sketch of a possible path to obtaining a formula for a homotopy between the 12th power of the commutator of quaternions and a constant. There is also a neat formula for the Cartan inclusion of the symmetric space $\Lambda : G_2/SO(4) \hookrightarrow G_2$. This inclusion is the conjugate orbit of the matrix $\Sigma = \begin{pmatrix} I_3 & 0 \\ 0 & -I_4 \end{pmatrix}$ a square root of I and the Cartan inclusion is $[A] \mapsto A\Sigma A^{-1} = L_{A_3} \circ L_{A_2} \circ L_{A_1}$, where A_j is the jth column of any matrix A in the given class. One can use this combined with the principal bundle $SO(3) \cdots V_{7,2} \to G_2/SO(4)$ with projection s to define $\phi = \Lambda \circ s \circ H$ and try to produce a generator of $\pi_7 Sp(2)$ following the suggestions of [10]. Püttmann in his work mentioned above gave a formula for a generator of $\pi_7 Sp(2)$ with column maps of degree 12 but not related to the Cayley power.

7 The Geometry of the Commutator and Exotic Phenomena

In 1956, John Milnor showed there are exotic 7-spheres: smooth 7-manifolds, homeomorphic but not diffeomorphic to the usual S^7. The first examples are linear (non-principal) S^3 bundles over S^4. In 1962, [29] classified all manifolds homeomorphic to S^n for $n \geq 5$, and [21] constructed an invariant, based on cohomological data, capable of distinguishing differential structures on 7-, 11-, and 15-dimensional manifolds. The homeomorphic spheres denoted by Σ^n may be constructed as a topological quotient of a union of two closed discs glued along their boundaries by a diffeomorphism, which is not isotopic to the identity: $\Sigma^n = D^n \cup_\sigma D^n$ and $\sigma : S^{n-1} = \partial D^n \to \partial D^n = S^{n-1}$ is a diffeomorphism of degree one, i.e., orientation preserving, but σ cannot be continuously deformed to $id_{S^{n-1}}$ through diffeomorphisms. This means that σ and $id_{S^{n-1}}$ belong to a different element of $\pi_0 Diff^+ S^{n-1}$. Composition of diffeomorphisms goes through to define a group structure on $\pi_0 Diff^+ S^{n-1}$, denoted by Γ_n, the group of homotopy n-spheres. This is an Abelian group isomorphic to the h-cobordism classes of comotopy n-spheres under the connected sum operation: #, usually denoted by Θ_n. So, $\Gamma_n \cong \Theta_n$. The isomorphism is given by using the diffeomorphism σ. For details, see [28]. For $n \geq 5$, every homotopy sphere is homeomorphic to S^n, a result due to Smale.

In 1972, Gromoll and Meyer [24] constructed, using geometry, a $\Sigma^7 \in \Gamma_7 \cong \mathbb{Z}_{28}$ and using the Eells–Kuiper invariant showed that it is a generator.

8 The G-M Sphere

The Gromoll–Meyer sphere is the quotient of the following free action $\star : S^3 \times Sp(2) \to Sp(2) : q \star \begin{pmatrix} a & c \\ b & d \end{pmatrix} = \begin{pmatrix} qa\overline{q} & qc \\ qb\overline{q} & qd \end{pmatrix}$; it is easy to check that it consists of a subgroup product by diagonal (q, q) from the left and subgroup product by diagonal $(\overline{q}, 1)$ from the right. It is free and so we get a principal bundle with a compact 7-manifold as a quotient. Using the fact that diagonal inclusion (q, q) induces $\pi_3 S^3 = \mathbb{Z} \ni 1 \mapsto 2 \in \pi_3 Sp(2) = \mathbb{Z}$ and the diagonal inclusion $(\overline{q}, 1)$ induces $1 \mapsto -1$, we conclude from the homotopy exact sequence that the quotient is 3-connected. Elementary algebraic topology now implies that the seven-dimensional quotient is a homology sphere. Smale's work [38] implies that it is homeomorphic to the 7-sphere and we denote it by Σ^7_1. The basic feature was that it inherits a metric of nonnegative sectional curvature by Riemannian submersion from the bi-invariant metric of $Sp(2)$, since the subgroup action is by isometries from each side. So, Σ^7_1 has $K \geq 0$. Duran uses the diagram

$$\begin{array}{ccccc} & & S^3 & & \\ & & \vdots & \star & \\ S^3 & \cdots & Sp(2) & \to & \Sigma^7_1 \\ & \bullet & \downarrow & & \\ & & S^7 & & \end{array} \tag{4}$$

and notes that the fibers of $\star$ and $\bullet$ through all elements of $SO(2) \subset Sp(2)$ (all entries are real) coincide, the action of $Sp(2)$ from the left is by $\bullet$-bundle isomorphisms, and that the subgroup multiplication by diagonal $(1, \overline{p})$ from the right commutes with the $\star$-action as does $SO(2)$ from the left. We saw at the beginning that the bi-invariant metric of $Sp(2)$ does not project to the round metric on S^7 and followed Duran's idea to use the Kaluza–Klein process to put a Left Invariant metric on $Sp(2)$, call it g, whose quotient metric on $S^7$is the Euclidean $K \equiv 1$. The $\bullet$-fibers are totally geodesic in the Duran metric. It follows that g is right invariant by diagonal $(r, s) \in S^3 \times S^3$. So, the metric g also (as well as the bi-invariant one) projects to a $K \geq 0$ Riemannian metric on Σ_1^7 with isometry group $S^1 \times S^3$. Furthermore, if we denote with $[A]$ the $\star$projection to Σ_1^7 we get that any unit geodesic γ_Σ of Σ_1^7 starting from $[I]$ reaches $[-I]$ at time π and returns to $[I]$ at time 2π. To prove the last statement, lift γ_Σ to a $\star$-horizontal geodesic $\gamma \in (Sp(2), g)$ from I. Note that γ is $\bullet$-horizontal too, since $H_\star(I) = H_\bullet(I)$. So, γ projects to some geodesic γ_S from 1 to -1 at time π and is back at time 2π, since S^7 has $K = 1$. So, γ cuts the $\bullet$-fibers at times π, respectively, 2π. But, the fibers of $\bullet$ and $\star$ coincide at all points of $SO(2)$ and γ_Σ satisfies the conditions.

It follows that Σ_1^7 with the induced metric (from g) has this Blaschke property at all points of $SO(2)$, since this subgroup of $Sp(2)$ acts by isometries on Σ_1^7 preserving geodesics, etc.

We reproduced Duran's formula (above) for unit horizontal, rel. $\bullet$ or $\star$, geodesics of $(Sp(2), g)$ from I. From the formula follows $\gamma_X(2\pi) = \begin{pmatrix} 1 & 0 \\ 0 & \frac{w}{|w|}e^{2\pi x}\frac{\overline{w}}{|w|} \end{pmatrix}$ and the tangent vector $\gamma_X'(t) = \begin{pmatrix} -\sin(t) + \cos(t)x & -(\cos(t) + \sin(t)x)e^{tx}\overline{w} \\ -\cos(t)w & -\sin(t)we^{tx}\overline{w} \end{pmatrix}$ and $\gamma_X'(\pi) = \begin{pmatrix} -x & e^{\pi x}\overline{w} \\ w & 0 \end{pmatrix}$. Putting all this information together with the Blaschke property at $[I]$ and $[-I]$, Duran concludes that Σ_1^7 is the union of two closed 7-discs made out of geodesics of length $\pi/2$ the first one from $[I]$ and the second from $[-I]$, with their boundary 6-spheres identified by the diffeomorphism $\sigma\binom{x}{w} = \binom{[b\binom{x}{w}]x[b\binom{x}{w}]^{-1}}{[b\binom{x}{w}]w[b\binom{x}{w}]^{-1}}$, i.e., acts on each component x, w, through the projection $S^3 \to SO(3)$ then "(3 × 3) matrix on 3 vector." Note that $\mathbb{R}e(w)$ remains unchanged by σ, but still σ depends on it through b. Remember that $b\binom{x}{w} = \frac{w}{|w|}e^{\pi x}\frac{\overline{w}}{|w|}$ is the Blakers–Massey element, essentially the commutator of quaternions, whose homotopy class in $\pi_6 S^3$ generates this group isomorphic to $\mathbb{Z}_{12}$.

Although there does not exist linear algebra based on Cayley numbers, due to their nonassociativity, the formulas for $b : S^6 \to S^3$ and $\sigma \in Diff^+S^6$ above generalize (just replace quaternionic coordinates with Cayley ones) to give a generator of $\pi_{14}S^7$ as well as exotic diffeomorphisms of S^{14} [14].

An elementary calculation shows that $\sigma^n = \sigma \circ \sigma \circ \cdots \circ \sigma$ is just

$$\binom{x}{w} \mapsto \binom{\left[b\binom{x}{w}\right]^n x \left[b\binom{x}{w}\right]^{-n}}{\left[b\binom{x}{w}\right]^n w \left[b\binom{x}{w}\right]^{-n}}$$

and also for negative values of n. Also, $b \circ \sigma = b$, i.e., b is σ invariant. Although there is no analog to the Gromoll–Meyer construction of Σ_1^7 to the 15-dimensional case, one can redo the arguments in a different way and extend some of the results above to $\Sigma^{15} = D^{15} \cup_\sigma D^{15}$ an exotic 15-sphere, generating a subgroup of index 2 in $\Gamma_{15} \cong \mathbb{Z}_2 \oplus \mathbb{Z}_{8.128}$ with the operation $\#_1^k \Sigma^{15} = D^{15} \cup_{\sigma^k} D^{15}$. As many Σ^7 are linear (non-principal) S^3 bundles over S^4 so are many Σ^{15}'s linear S^7 bundles over S^8. It would be interesting to know if one can replace two, or even all three, of the spheres with exotic ones in the Hopf fibration $S^7 \cdots S^{15} \to S^8$. By the way, $\Gamma_8 \cong \mathbb{Z}_2$. Some progress in this direction was done in Llohann Dalagnol Sperança's PhD thesis at IMECC under the supervision of Duran [39].

9 Exotic Involutions

In [1] was proved using Cerf's work [6] that $\rho = \alpha \circ \sigma$ is a free, exotic involution on S^6, for α the antipodal involution $\alpha\binom{x}{w} = \binom{-x}{-w}$. It is elementary to show free and "involution" using the σ-invariance of b. Note too that $\alpha \circ \sigma = \sigma^{-1} \circ \alpha$. Now, replace quaternions for Cayley numbers. All arguments involved take place in $span\{1, X, W, XW\} \subset \mathbb{C}a$, which is associative, so they generalize to the Cayley case. If $\mathbb{R}e(w) = 0$, i.e., $\binom{x}{w} \in S^5 \subset Imaginary\mathbb{Q} \times Imaginary\mathbb{Q}$, a unit vector, then the restriction of ρ to S^5 and, respectively, to S^{13} is still a free involution. It is shown that the four free involutions are exotic, i.e., not isotopic to the antipodal map of the relevant sphere, so each of the respective quotient manifolds is an exotic real projective space $\mathbb{R}P^n$, $n = 5, 6, 13, 14$. It follows from topological classifications of involutions on S^5 and S^{13} that the quotients S^5/ρ and S^{13}/ρ are not even homeomorphic to the corresponding standard projective spaces [32]. It is quite instructive to follow the drawing showing the steps for the case $n = 5$ as it is easy to picture in $\mathbb{R}^3$, which we used to picture in the plane since high school. The only thing one has to remember is that conjugation by a unit quaternion q on $Imaginary\ \mathbb{Q} \equiv \mathbb{R}^3$ is a rotation about the axis defined by $Imaginary\ \mathbb{Q}$. So, conjugating $Imaginary\ \mathbb{Q}$ by $b\binom{x}{w}$ is a rotation about the axis defined by $\frac{w}{|w|}\frac{x}{|x|}\frac{\overline{w}}{|w|}$. The degenerate cases, $x = 0$ or $w = 0$, offer no difficulty. Before we close this section, we remark on the action $A : \mathbb{Z}_2 \times Diff^+(S^m) \to Diff^+(S^m)$ defined by $(-1) \cdot h = \alpha \circ h \circ \alpha^{-1}$. This action preserves the group structure of $Diff^+$, which is composition. It is easily seen to descend to an action on $\pi_0 Diff^+(S^m) \equiv \Gamma_{m+1}$, by permuting the connected components of $Diff^+(S^m)$. In particular, $A(n) = -n$ on Γ_7. It is true also that $\alpha \circ \sigma^k \equiv \rho_k$ is also a free involution on each of our chosen S^m. One can show, employing action A some known results, like: [31]: Every orientation reversing diffeomorphism of S^6 is isotopic to a free involution. In this context, it seems interesting to know if there exist orientation preserving diffeomorphisms f, respectively, g (isotopic to σ^{14}) with $A(f) = f$, respectively, $g^2 = id_{S^6}$. The inverse involution $B(f) = f^{-1}$ may help in this context: note that it follows from the relations above that $A(\sigma^k) = B(\sigma^k)$ for all $k \in \mathbb{Z}$. That is, the powers of σ

are contained in the subset of $Diff^+(S^6)$, where the orbits of the actions A and B coincide. Are there other elements in this set? Do there exist orientation preserving diffeos η of S^6, with isotopy class $4 \in \mathbb{Z}_{28}$ and $\eta^7 = id_{S^6}$? Such elements of finite order could help express the structure of Γ_7. (Just imagine a map $\varphi : S^6 \to S^3$ with $[\varphi] = 1 \in \mathbb{Z}_{12} = \pi_6 S^3$ with φ^{12} =constant).

10 Non-cancellation Phenomena

In [25] are given examples of three manifolds M, N, R, such that, M and N are not homotopy equivalent, but $M \times R$ is diffeomorphic to $N \times R$; in other words, you cannot cancel R in the equation $M \times R = N \times R$. They called this a "non-cancellation phenomenon." In his PhD thesis at IMECC, Barros [2] showed some examples of existence related to S^3 bundles over S^4 and S^7. See also [3]. At that time, we could not produce any explicit formula. Such a formula became possible only after Duran's work [1] as I will describe below. But, first let us look at some non-explicitly described examples. Let Σ^7 be any homotopy 7-sphere. It follows using moderately easy Topology (Σ^7 is framed cobordant to S^7, $\Sigma^7 \times S^3$ is framed cobordant to $S^7 \times S^3$, the obstruction to them being diffeomorphic lives in the Wall group $L_{10}(0)$, which is zero, 10 being the dimension of the product). The same is true for the bundle of orthonormal frames $SO(8) = S^7 \times SO(7) \cong \Sigma^7 \times SO(7) = \Sigma O(8)$. Here, $L_{28}(0) = 0$. So, there are free actions and exotic quotients $S^3 \cdots S^7 \times S^3 \to \Sigma^7$ for each $\Sigma^7 \in \Gamma_7$. Analogously, $SO(7) \cdots SO(8) \to \Sigma^7$. These are the bundles of orthonormal frames of the 7-spheres, that are all parallelizable. De Sapio [20] showed that the total spaces of the bundles of orthonormal frames of all homotopy n-spheres are diffeomorphic to $SO(n+1)$. It is an elementary fact that for any two Riemannian metrics on a compact manifold, the bundles of orthonormal frames are isomorphic. So, there are, at least, as many exotic free actions $SO(n) \times SO(n+1) \to SO(n+1)$ as many exotic structures exist on S^n. I know of no exact formula describing such an action. However, there is an exact description of two actions r_1, r_2 : $(\mathbb{Z}_2 \times S^3) \times (S^6 \times S^3) \to (S^6 \times S^3)$, such that the two actions are not differentiably conjugate, i.e., there is no diffeomorphism $f : S^6 \times S^3 \to S^6 \times S^3$, such that $g \star f(z) = f(g \bullet z)$, for all $g \in (\mathbb{Z}_2 \times S^3)$ and all $z \in (S^6 \times S^3)$; but $r_1$and r_2 restricted to $\{1\} \times S^3$ are differentiably conjugate and also, restricted to $\mathbb{Z}_2 \times id_{S^3}$ are, again, differentiably conjugate. Here, we use $\mathbb{Z}_2 \equiv \{-1, 1\}$. The actions are the consequence of the $\bullet$, respectively, $\star$ actions of S^3 on $Sp(2)$ restricted to the trivialization over each of the two S^6 equators of S^7, respectively, Σ_1^7. The action of $\mathbb{Z}_2$ is the restriction of multiplication by $-I$ on the same sets. The trivialization is done through Duran's formula for $\gamma_{\binom{x}{w}}\binom{\pi}{2}$. The two actions are $r_2\left((1, p), \left(\binom{x}{w}, q\right)\right) = \left(\binom{x}{w}, pq\right)$ and $r_2\left((-1, p), \left(\binom{x}{w}, q\right)\right) = \left(\binom{-x}{-w}, pq\overline{b}\binom{x}{w}\right)$; the other action is $r_1\left((1, p), \left(\binom{x}{w}, q\right)\right) = \left(\binom{px\overline{p}}{pw\overline{p}}, q\overline{p}\right)$ and $r_2\left((-1, p), \left(\binom{x}{w}, q\right)\right) =$

$\left(\rho_{-1}\binom{x}{w}, qb\binom{x}{w}\right)$. If r_1 and r_2 were differentiably conjugate, then $S^6/\rho(\mathbb{Z}_2)$ would be diffeomorphic to $\mathbb{R}P^6$, which is not. The restrictions of r_1, r_2 to $(1 \times S^3)$ coincide. The restrictions to $(-1 \times S^3)$ are conjugate to each other by the involutive diffeomorphism $F\left(\binom{x}{w}, q\right) = \left(\binom{qx\overline{q}}{qw\overline{q}}, \overline{q}\right)$.

11 An Infinite Family of Gromoll–Meyer Spheres

One can imitate the Gromoll–Meyer action of S^3 on the pullback of the principal bundle (denoted by E_1) $S^3 \cdots Sp(2) \rightarrow S^7$ by the nth Cayley power of S^7 [16]. Let $\binom{\cos(t)+\sin(t)x}{\sin(t)w} = \alpha \mapsto \alpha^n = \binom{\cos(nt)+\sin(nt)x}{\sin(nt)w} = \alpha^n$ with $x \in Im\ \mathbb{Q}, w \in \mathbb{Q}, |x|^2 + |w|^2 = 1$, be the nth power map $\psi_n(\alpha) = \alpha^n$. Call E_n the pullback of E_1 by ψ_n. The ten-dimensional manifold representing the total space of E_n, denoted by the same symbol E_n, consists of all $(\beta, \gamma) \in S^7 \times S^7 \subset \mathbb{Q} \times \mathbb{Q}$ with $<< \psi_n(\beta), \gamma >>= 0$. Here, $<< \eta, \theta >>= \overline{\eta}^{tr}\theta$ is the standard Hermitian product on $\mathbb{Q} \times \mathbb{Q}$. The free action of the fiber S^3 is on the vector θ from the right and the free left action by $q \in S^3$, imitating $\star$, the Gromoll–Meyer action is $q \star (\eta, \theta) = (q\eta\overline{q}, q\theta)$. Remember that η, θ have two quaternionic coordinates, say, $\eta = \binom{a}{b}$ so $q\eta\overline{q} = \binom{qa\overline{q}}{qb\overline{q}}$ and $q\theta = \binom{qc}{qd}$. The same reasoning applied to the free quotient of $Sp(2) = E_1$ by the Gromoll–Meyer action shows that the (free again) quotient of E_n by the $\star$action of S^3 (just described) is again a 7-manifold homeomorphic to S^7, denoted now by Σ^7_n. It turns out quite naturally that Σ^7_n represents the $n(mod\ 8)$ element of $\mathbb{Z}_{28} = \Gamma_7$. There is equivariance too: Let $\mathbb{Z}_2 \times \mathbb{Z}_2$ be the subgroup $\begin{pmatrix} \pm 1 & 0 \\ 0 & \pm 1 \end{pmatrix} \subset Sp(2)$. For all n, E_n admits a smooth action, denoted by $\bullet$, by the group $\mathbb{Z}_2 \times \mathbb{Z}_2 \times S^3$ that commutes with the free $\star$ action. As follows: the matrix $B \in \mathbb{Z}_2 \times \mathbb{Z}_2$ acts as $B \bullet (\eta, \theta) = (B\eta, B\theta)$ and $p \in S^3$ acts as $(\eta, \theta\overline{p})$. The induced effective action on Σ^7_n is by $\mathbb{Z}_2 \times \mathbb{Z}_2 \times SO(3)$. On $\Sigma^7_0 \equiv S^7$, this is the linear action $(B, \pm p)\binom{a}{b} = \binom{\pm a}{\pm pb\overline{p}}$.

It is shown that for all n even Σ^7_n is equivariantly homeomorphic to S^7 with the above linear $\mathbb{Z}_2 \times \mathbb{Z}_2 \times SO(3)$ action, while all Σ^7_{2m+1} are equivariantly homeomorphic to Σ^7_1, the Gromoll–Meyer sphere, with respect to the same action. The fixed point sets are spheres in all even spheres. But, in the odd spheres, there are three-dimensional fixed point sets with fundamental groups $\mathbb{Z}_2$ and $\mathbb{Z}_3$. In particular, the subsets of E_n with $\eta \in Im\ \mathbb{Q} \times Im\ \mathbb{Q}$ project to invariant submanifolds $\Sigma^5_n \subset \Sigma^7_n$. These are $\mathbb{Z}_2 \times \mathbb{Z}_2 \times SO(3)$ diffeomorphic to S^5 for n even and to the Brieskorn sphere W^5_3 if n is odd. Moreover, the sphere Σ^5_n is minimal for every $(\pm 1) \times SO(3)$ invariant metric on Σ^7_n. The invariant Σ^5_n is dual to the invariant circle Σ^1_n, the quotient of $\begin{pmatrix} \cos\theta & -\sin\theta \\ \sin\theta & \cos\theta \end{pmatrix}$ in E_n. Moreover, all points of the invariant circle $\Sigma_n^1 \subset \Sigma_n^7$ have the wiedersehen property, Σ_n^1 and Σ_n^5 have constant distance $\frac{\pi}{2}$ and the obvious (from this geometry) map $\Sigma_n^1 \star \Sigma_n^5 \to \Sigma_n^7$ is a homeomorphism. This invariant geodesic join structure is used in the proofs of the previously mentioned results. A very broad generalization of these results was recently released [23], showing, between other results, that all exotic 7-spheres admit $SO(3)$ invariant metrics with nonnegative sectional curvature.

12 Homotopy Revisited

We can feed the Duran procedure in the commutator homotopy problem [17]. This article is an overview of much of the work described here. Essentially, we apply the Duran horizontal lifting of Euclidean geodesics in the diagram right after claim 4 above. Naturally, we have to endow the Lie groups $Sp(2)$, $SU(4)$, and $Spin(7)$ with left invariant metrics that are also right invariant under subgroups: $Sp(1) \times Sp(1) \subset Sp(2)$, $U(3) \subset SU(4)$, and $G_2 \subset Spin(7)$. These metrics on the total spaces induce the $K \equiv 1$ metric on the base S^7 by Riemannian submersion. Lifting horizontally the Euclidean geodesics of S^7 from the identity elements of the Lie groups up to the corresponding fibers over the south pole, we get maps from $S^6 = unit(T_1 S^7) \to Fiber$. These fibers are diffeomorphic to the subgroups $Sp(1)$, $SU(3)$, resp., G_2. The maps, denoted by $b : S^6 \to Sp(1)^-$, $\phi : S^6 \to SU(3)^-$, and $\chi : S^6 \to G_2^-$, represent generators of the relevant π_6. Each two of these maps are homotopic to each other in the bigger fiber. We can see this "unfolding" of b to ϕ and χ as follows. Start from a geodesic γ_v of S^7 from the north pole with initial unitary tangent vector $v \in S^6$ and let γ_v^{Sp} and γ_v^{SU} denote the horizontal lifts to $Sp(2)$, respectively, $SU(4)$. Both lifts project to the same curve γ_v, so the bundle inclusion above implies $\gamma_v^{Sp}(t) \in \gamma_v^{SU}(t) \cdot \begin{pmatrix} 1 & 0 \\ 0 & SU(3) \end{pmatrix}$ for all $t \in \mathbb{R}$. The homotopy $H(t, v) = \gamma_v^{SU}(\pi) \cdot \gamma_v^{SU}(t)^{-1} \cdot \gamma_v^{Sp}(t)$ attains values in $SU(3)^-$ and we have $H(0, v) = \gamma_v^{SU}(\pi) = \phi(v)$ and $H(\pi, v) = \gamma_v^{Sp}(\pi) = b(v)$. So, H is a homotopy between b and ϕ with values in $SU(3)^-$. Similarly, we can describe homotopies between b, ϕ, χ with values in G_2^-. To see the cancellations in homotopy, we employ algebraic identities and ad hoc exercises [10], [33] together with the above homotopies. As a result, [33] presents a homotopy in $SU(3)$ between b^6 and a constant.

In [18], it is shown that any homotopy between b^{12} and the constant map $S^6 \longrightarrow \{1\} \in S^3$ cannot be $SO(3)$ equivariant. This explains a little the difficulty of producing such a formula: there is a braking of symmetry. This is a consequence of a description in a general equivariant setting of the results in [1] and also in [19].

References

1. U. Abresch, C.E.Duran, T. Püttmann, A. Rigas: Wiedersehen metrics and exotic involutions on Euclidean spheres. J.Reine Angew. Math. **605**, 1–21 (2007).
2. T. E. Barros, Fenómenos de não cancelamento relacionados a S^3 fibrados. Tese de Doutorado, IMECC, Unicamp (1997) and Correction for the paper "S^3bundles and exotic actions" Bull. Soc. Math. France **129**(4), 543–545 (2001).
3. T. E. Barros, A Rigas: The role of commutators in a non - cancellation phenomenon. Math. Jour. Okayama Univ., **43,** 73–93 (2001).
4. A. Borel, J.-P. Serre: Groupes de Lie et puissances reduites de Steenrod, Amer. J. Math. **75**, 409–448 (1953).
5. E. Cartan: Le principe de dualité et la theorie des groups simples et semi-simples. Bull. Sci. Math. **49**, 361–374 (1925).
6. J. Cerf: La stratification naturelle des espaces de fonctions differentiables réeles et le théorème de la pseudo-isotopie. Inst. Hautes Études Sci. Publ. Math. **39**, 5–173 (1970).
7. J. Cheeger, D. Gromoll: On the structure of complete manifolds of non negative curvature. Ann. of Math. **96**, 413–443 (1972).
8. L. M. Chaves: Resultados sobre a geometria dos fibrados. Tese de Doutorado, IMECC, Unicamp (1992).
9. L. M. Caves, A. Rigas: From the triality viewpoint. Note di Matematica **18**, no. 2, 155–163 (1998).
10. L. M. Chaves, A. Rigas: Hopf maps and triality. Math. Jour. Okayama Univ. **38**, 197–208 (1996).
11. L. M. Chaves, A. Rigas: On a conjugate orbit of G_2. Math. Jour. Okayama Univ. **33**, 155–161 (1991).
12. C. E. Duran: Pointed Wiedersehen metrics on exotic spheres and diffeomorphisms of S^6. Geom. Dedicata **88**, 199–210 (2001).
13. Dan A. Agüero Cerna: On the geometry and topology of the commutator of unit quaternions. Master's Dissertation, IMECC, Unicamp (2016).
14. C.E. Duran, A. Mendoza, A. Rigas: Bakers-Massey elements and exotic diffeomorphisms of S^6 and S^{14}. Trans. Am. Math. Soc. **356** (12), 5025–5034 (2004).
15. O. Dearricot, F. Galaz-García, L. Kennard, C. Searle, G Weigart, W. Ziller: Geometry of manifolds with non negative sectional curvature. Lecture notes in Mathematics **2110**.
16. C.E. Duran, T. Püttmann, A. Rigas: An infinite family of Gromoll-Meyer spheres. Archiv der Math. (printed ed.) **95**, 269–282 (2010).
17. C.E. Duran, T. Püttmann, A. Rigas: Some Geometric formulas and cancellations in Algebraic and Differential Topology. Matem. Contemp. **28**, 133–149 (2005).
18. C.E. Duran, A. Rigas: Equivariant homotopy and deformations of diffeomorphisms. Differ. Geo. Appl. **27**(2), 206–211 (2009).
19. C.E. Duran, A. Rigas, L. Dallagnol Sperança: Bootstrapping Ad-equivariant maps, diffeomorphisms and involutions. Matem. Contemp. **35**, 27–39 (2008).
20. R. De Sapio: Manifolds homeomorphic to sphere bundles over spheres. Bull. Amer. Math. Soc. **75**(1), 59–63 (1969).
21. J. Eells, N. Kuiper: An invariant for certain smooth manifolds. Annali Mat. Pura e Appl. **60**, 93–110 (1962).
22. l. Florit, W. Ziller: Non negatively curved Euclidean submanifolds in codimension two. Comm. Math. Helv. (Printed ed.) **91**, 629–651 (2016).
23. S. Goette, M. Kerin, K. Shankar: Highly connected 7-manifolds and non negative sectional curvature. Preprint (2017).
24. D. Gromoll, W. Meyer: An exotic sphere with non negative sectional curvature. Ann. of Math. **100**, 401–406 (1974).
25. P. Hilton, G. Mislin, J. Roitberg: Sphere bundles over spheres and non cancellation phenomena. J. London Math. Soc. **6**, 15–23 (1972).

26. I.M. James: On H-spaces and their homotopy groups. Quart. Jour. Math. Oxford Ser. (2) **11**, 161–179 (1960).
27. J.-P. Serre: Cohomologie modulo 2 des complexes d' Eilenberg MacLane. Comm. Math. Helv. **27**, 198–232 (1953).
28. A. Kosinski: Differentiable Manifolds. Dover Publications (2007).
29. M. A.a Kervaire, J. Milnor: Groups of homotopy spheres. Ann.of Math. **77**, 504–537 (1963).
30. M. Mimura: Homotopy groups of Lie groups of low rank. Jour. of Math. Kyoto Univ., **6**(2), 131–176 (1967).
31. B. Mann, E. Miller: The construction of the Kervaire sphere by means of an involution. Mich. Math. J. **27**, 301–308 (1980).
32. W. Oledzki: Exotic involutions of low dimensional spheres and the eta invariant. Tohoku Math. J. **52**, 173–198 (2000).
33. T. Püttmann: Einige Homotopiegruppen der klassischen Gruppen aus geometrischer Sicht. Habilitationsschrift, Ruhr-Universität Bochum (2004).
34. T. Püttmann, A. Rigas: When is $\mathbb{R}P^n \times Spin(n)$ diffeomorphic to $S^n \times SO(n)$ and how. Math. J. Okayama Univ. **45**, 111–115 (2003).
35. T. Püttmann, A. Rigas: Presentations of the first homotopy groups of the unitary groups. Comment. Math. Helv. **78**(3), 648–662 (2003).
36. A. Rigas: Riemannian metrics of non negative sectional curvature on stable vector bundles over spheres. Ph.D. thesis, University of Chicago (1974) and Geodesic generators of $\pi_n(O)$, $\pi_{n+1}(BO)$. J. Diff. Geom. **13**, 527–545 (1978).
37. A. Rigas: S^3 fibrados e ações exóticas. Tese de Livre docência, IMECC, Unicamp (1983) and S^3 bundles and exotic actions. Bull. Soc. Math. de France **112**, 69–92 (1984).
38. S. Smale: Generalized Poincaré's conjecture in dimensions greater than four. Ann. of Math. **74**, 391–406 (1961).
39. L.Dallagnol Sperança: Geometria e Topologia de Cobordos. Tese de Doutorado, IMECC, Unicamp, (2012).
40. H. Toda, Y. Saito, I. Yokota: Note on the generator of $\pi_7 SO(n)$. Mem. Coll. Sci. Univ. Kyoto, Ser. A, XXX, Math. No. 3, 227–230 (1957).
41. C.T.C. Wall: Classification problems in differential topology, VI. Topology **6**, 273–296 (1967).
42. W. Ziller: Examples of Riemannian manifolds with non negative sectional curvature. Metric and comparison Geometry. Surv. Diff. Geom., ed. K.Grove and J. Cheeger. International Press, 63–102, (2007).

Life in the Rindler Reference Frame: Does a Uniformly Accelerated Charge Radiate? Is There a Bell 'Paradox'? Is Unruh Effect Real?

Waldyr A. Rodrigues Jr. and Jayme Vaz Jr.

Abstract The determination of the electromagnetic field generated by a charge in hyperbolic motion is a classical problem for which the majority view is that the Liénard-Wiechert solution which implies that the charge radiates is the correct one. However we analyze in this paper a less known solution due to Turakulov that differs from the Liénard-Wiechert one and which according to him does not radiate. We prove his conclusion to be wrong. We analyze the implications of both solutions concerning the validity of the Equivalence Principle. We analyze also two other issues related to hyperbolic motion, the so-called Bell's "paradox" which is as yet source of misunderstandings and the Unruh effect, which according to its standard derivation in the majority of the texts is a correct prediction of quantum field theory. We recall that the standard derivation of the Unruh effect does not resist any tentative of any rigorous mathematical investigation, in particular the one based in the algebraic approach to field theory which we also recall. These results make us to align with some researchers who also conclude that the Unruh effect does not exist.

1 Introduction

There are some problems in Relativity Theory that are continuously source of controversies, among them we discuss in this paper: (a) the problem of determining if a uniformly accelerated charge does or does not radiate[1]; (b) the so-called Bell's paradox; and (c) the Unruh effect.[2]

[1]This problem is important concerning one of the formulations of the Equivalence principle.

[2]We call the reader's attention that the references quoted in this paper are far from complete, so we apologize for papers not quoted.

W. A. Rodrigues Jr. · J. Vaz Jr. (✉)
Institute of Mathematics, Statistics and Scientific Computing, University of Campinas, Campinas, SP, Brazil
e-mail: walrod@ime.unicamp.br; vaz@ime.unicamp.br

C. Lavor, F. A. M. Gomes (eds.), *Advances in Mathematics and Applications*,
https://doi.org/10.1007/978-3-319-94015-1_13

In order to throw some light on the controversies we discuss in detail in Sect. 2 the concept of (right and left) Rindler reference frames, Rindler observers, and a chart naturally adapted to a given Rindler frame. These concepts are distinct and thus represented by different mathematical objects and having this in mind is a necessary condition to avoid misunderstandings, both of mathematical and of physical nature.

In Sect. 3 we analyze Bell's "paradox" that even having a trivial solution seems to not have been understood for some people even recently for it is confused with another distinct problem which if one does not pay the required attention seems analogous to the one formulated by Bell.

In Sect. 4 we discuss at length the problem of the electromagnetic field generated by a charge in hyperbolic motion. First we present the classical Liénard-Wiechert solution, which implies that an observer at rest in an inertial reference frame observes that the charge radiates. Next we analyze (accepting that the Liénard-Wiechert solution is correct) if an observer comoving with the charge detects or no radiation. We argue with details that contrary to some views it is possible for a real observer living in a real laboratory[3] in hyperbolic motion to detect that the charge is radiating. Our conclusion is based (following [41]) on a careful analysis of different concepts of *energy* that are used in the literature, the one defined in an inertial reference frame and the other in the Rindler frame. In particular, we discuss in detail the error in Pauli's argument.

But now we ask: Is it necessary to accept the Liénard-Wiechert solutions as the true one describing the electromagnetic field generated by a charge in hyperbolic motion? To answer that question we analyzed the Turakulov [57] solution to this problem, which consisting in solving the wave equation for the electromagnetic potential in a special systems of coordinates where the equation gets separable. We have verified that Turakulov solution (which differs form the Liénard-Wiechert one) is correct (in particular, by using the Mathematica software). Turakulov claims that in his solution the charge does not radiate. However, we prove that his claim is wrong, i.e., we show that as in the case of the Liénard-Wiechert solution an observer comoving with the charge can detect that it is emitting radiation.

In Sect. 5 we discuss, taking into account that it seems a strong result the fact that a charge at rest in the Schwarzschild spacetime does not radiate [41], what the results of Sect. 4 implies for the validity or not of one of the forms in which Equivalence principle is presented in many texts.

Section 6 is dedicated to the Unruh effect. We first recall the standard presentation (emphasizing each one of the hypothesis used in its derivation) of the supposed fact that Rindler observers are living in a thermal bath with a Planck spectrum with temperature proportional to its local proper acceleration and thus such radiation may *excite* detectors on board. Existence of the Unruh radiation and Rindler particles seems to be the majority view. However, we emphasize that rigorous mathematical analysis of standard procedure (which is claimed to predict

[3]This, of course, means that the laboratory (whatever its mathematical model) [9] must have finite spatial dimensions as determined by the observer at any instant of its propertime.

the Unruh effect) done by several authors shows clearly that such a procedure contains several inconsistencies. These rigorous analyses show that the Unruh effect does not exist, although it may be proved that detectors in hyperbolic notion can get excited, although the energy for that process comes from the source accelerating the detector and it is not (as some claim) due to fluctuations of the Minkowski vacuum. We recall in Appendix 2 a (necessarily resumed) introduction to the algebraic approach to quantum theory as applied to the Unruh effect in order to show how much we can trust each one of the suppositions used in the standard derivation of the Unruh effect. Detailed references are given at the appropriate places.

Section 7 presents our conclusions and in Appendix 1 we present our conventions and some necessary definitions of the concepts of reference frames, observers, instantaneous observers, and naturally adapted charts to a given reference frame.

2 Rindler Reference Frame

A proper understanding of almost any problem in Relativity theory requires that we know (besides the basics of differential geometry[4]) exactly the meaning and the precise mathematical representation of the concepts of: (a) reference frames and their classification; (b) a naturally adapted chart to a given reference frame; (c) observers; and (d) instantaneous observers. The main results necessary for the understanding of the present paper and some other definitions are briefly recalled in Appendix 1.[5] It is essential to have in mind that most of the possible reference frames used in Relativity theory are *theoretical instruments*, i.e., they are not physically realizable as a material system. This is particularly the case of the right and left Rindler reference frames and respective observers that we introduce next.

Let $\sigma : I \to M, s \mapsto \sigma(s)$ a timelike curve in M describing the motion of an accelerated observer (or an accelerated particle) where s is the proper time along σ. The coordinates of σ in ELP gauge (see Appendix 1) are

$$x^{\mu}_{\sigma}(s) = \boldsymbol{x}^{\mu} \circ \sigma\ (s) \tag{1}$$

and for motion along the $x^3 = z$ axis it is

$$(x^{o}_{\sigma})^2 - (x^{3}_{\sigma})^2 = -\frac{1}{\mathrm{a}^2_{\sigma}}, \tag{2}$$

[4]Basics of differential geomety may be found in [12, 18, 20, 36]. Necessary concepts concerning Lorentzian manifods may be found in [39, 50].

[5]More details may be found in [23, 45].

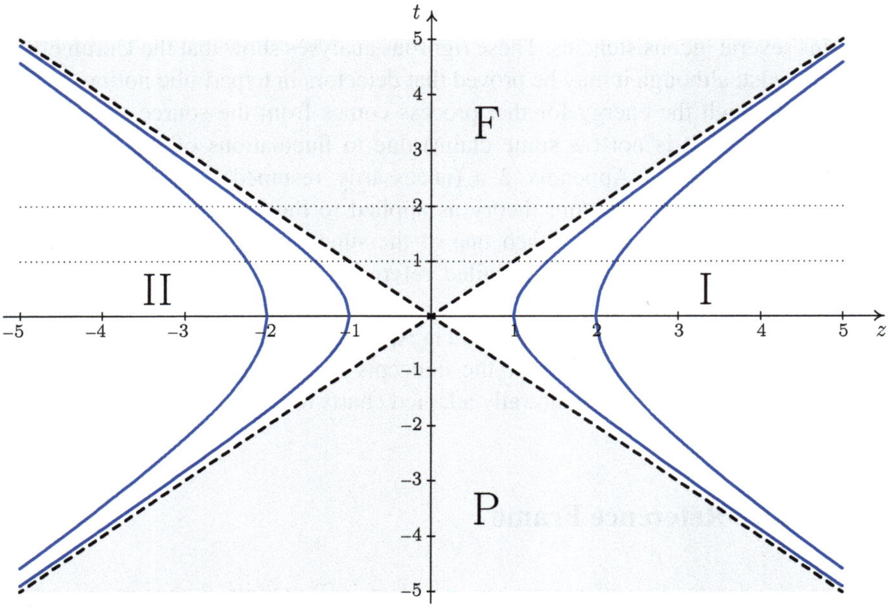

Fig. 1 Some integral lines of the right ***R*** and left ***L*** Rindler reference frames

where a_σ is a real constant for each curve σ. In Fig. 1 we can see two curves σ and σ' for which $\frac{1}{\mathrm{a}_\sigma} = 1$ and $\frac{1}{\mathrm{a}_{\sigma'}} = 2$. To understand the meaning of the parameter a_σ in Eq. (2) we write

$$x^0_\sigma(s) = \frac{1}{\mathrm{a}_\sigma}\sinh(\mathrm{a}_\sigma s), \quad x^3_\sigma(s) = \frac{1}{\mathrm{a}_\sigma}\cosh(\mathrm{a}_\sigma s). \tag{3}$$

The unit velocity vector of the observer is

$$\boldsymbol{v}_\sigma(s) = \sigma_*(s) := v^\mu(s)\frac{\partial}{\partial x^\mu} = \cosh(\mathrm{a}_\sigma s)\frac{\partial}{\partial t} + \sinh(\mathrm{a}_\sigma s)\frac{\partial}{\partial \mathrm{z}}.$$

Now, the acceleration of σ is

$$\boldsymbol{a}_\sigma = \frac{d}{ds}\sigma_*(s) = \mathrm{a}_\sigma\left(\sinh(\mathrm{a}_\sigma s)\frac{\partial}{\partial t} + \cosh(\mathrm{a}_\sigma s)\frac{\partial}{\partial z}\right)\Big|_\sigma \tag{4}$$

and of course, $\boldsymbol{a}_\sigma \cdot \boldsymbol{v}_\sigma = 0$ and $\boldsymbol{a}_\sigma \cdot \boldsymbol{a}_\sigma = -\mathrm{a}^2_\sigma$.

2.1 Rindler Coordinates

Introduce first the regions I, II, F, and P of Minkowski spacetime

$$\mathscr{I} = \{(t, x, y, z) \mid -\infty < t < \infty, -\infty < x < \infty, , -\infty < y < \infty, 0 < z < \infty\}, \tag{5}$$

and two coordinate functions $(\boldsymbol{x}^0, \boldsymbol{x}^1, \boldsymbol{x}^2, \boldsymbol{x}^3)$ and $(\boldsymbol{x}'^0,\boldsymbol{x}'^1,\boldsymbol{x}'^2,\boldsymbol{x}'^3)$ covering such regions. For $\mathfrak{e} \in M$ it is $\{\boldsymbol{x}^0(\mathfrak{e}) = x^0 = t, \boldsymbol{x}^1(\mathfrak{e}) = x, \boldsymbol{x}^2(\mathfrak{e}) = y, \boldsymbol{x}^3(\mathfrak{e}) = z\}$ and $\{\boldsymbol{x}'^0(\mathfrak{e}) = \mathrm{t}, \boldsymbol{x}'^1(\mathfrak{e}) = \mathrm{x} = x, \boldsymbol{x}'^2(\mathfrak{e}) = \mathrm{y} = y, \boldsymbol{x}'^3(\mathfrak{e}) = \mathrm{z}\}$ with[6]

$$\begin{aligned} &\mathrm{z} = \pm\sqrt{z^2 - t^2}, \quad \mathrm{t} = \tanh^{-1}\left(\frac{t}{z}\right), \quad |z| \geq |t|, \\ &x^0 = t = \mathrm{z}\sinh \mathrm{t}, \quad x^3 = z = \mathrm{z}\cosh \mathrm{t} \quad \text{in region I}, \\ &x^0 = t = -\mathrm{z}\sinh \mathrm{t}, \quad x^3 = z = -\mathrm{z}\cosh \mathrm{t} \quad \text{in region II} \end{aligned} \tag{6}$$

and

$$\begin{aligned} &\mathrm{z} = \pm\sqrt{t^2 - z^2}, \quad \mathrm{t} = \tanh^{-1}\left(\frac{z}{t}\right), \quad |t| \geq |z|, \\ &x^0 = t = \mathrm{z}\cosh \mathrm{t}, \quad x^3 = z = \mathrm{z}\sinh \mathrm{t} \quad \text{in region F}, \\ &x^0 = t = -\mathrm{z}\cosh \mathrm{t}, \quad x^3 = z = -\mathrm{z}\sinh \mathrm{t} \quad \text{in region P}. \end{aligned} \tag{7}$$

The right Rindler *reference frame* $\boldsymbol{R} \in \sec T\mathrm{I}$ has support in region I and is defined by

$$\begin{aligned} &\boldsymbol{R} = \frac{z}{\sqrt{z^2 - t^2}}\frac{\partial}{\partial t} + \frac{t}{\sqrt{z^2 - t^2}}\frac{\partial}{\partial z} = \frac{1}{\mathrm{z}}\frac{\partial}{\partial \mathrm{t}}, \\ &z > 0; \quad |z| \geq t. \end{aligned} \tag{8}$$

The left reference Rindler frame $\boldsymbol{L} \in \sec T\mathrm{II}$ is defined by

$$\begin{aligned} &\boldsymbol{L} = \frac{z}{\sqrt{z^2 - t^2}}\frac{\partial}{\partial t} + \frac{t}{\sqrt{z^2 - t^2}}\frac{\partial}{\partial z} = \frac{1}{\mathrm{z}}\frac{\partial}{\partial \mathrm{t}}, \\ &z < 0: \quad |z| \geq t. \end{aligned} \tag{9}$$

[6]Of course the coordinates (t, x, y, z) cover all M but the coordinates (t, x, y, z) do not cover all M, they are singular in origin.

Then, we see that in $\mathrm{I} \subset M$, $(t, x^1, x^2, \mathrm{z})$ as defined in Eq. (6) are a *naturally adapted coordinate system* to $\boldsymbol{R}$ [(nacs|$\boldsymbol{R}$)] and $\boldsymbol{L}$ [(nacs|$\boldsymbol{L}$)]. With D being the Levi-Civita connection of $\boldsymbol{g}$, the acceleration vector field associated with $\boldsymbol{R}$ is

$$\boldsymbol{a} = D_{\boldsymbol{R}}\boldsymbol{R} = \frac{1}{\mathrm{z}}\frac{\partial}{\partial \mathrm{z}}. \tag{10}$$

Also,

$$\boldsymbol{a}_\sigma = \frac{d}{ds}\sigma_*(s) = \mathrm{a}_\sigma \left.\frac{\partial}{\partial \mathrm{z}}\right|_\sigma \tag{11}$$

i.e., $\boldsymbol{a}_\sigma = D_{\boldsymbol{R}}\boldsymbol{R}|_\sigma = \left.\frac{1}{\mathrm{z}}\frac{\partial}{\partial \mathrm{z}}\right|_\sigma = \mathrm{a}_\sigma \frac{\partial}{\partial \mathrm{z}}\big|_\sigma$. Moreover, recall that since σ is clearly an integral line of the vector field $\boldsymbol{R}$, it is $\boldsymbol{v}_\sigma = \boldsymbol{R}|_\sigma$.

Remark 1 Note that in Eq. (8) (respectively Eq. (9)) it is necessary to impose $z > 0$ (respectively, $z < 0$), this being the reason for having defined the right and left Rindler reference frames.

2.2 Decomposition of $D\boldsymbol{R}$

Recall that the Minkowski metric field $\boldsymbol{g} = \eta_{\mu\nu}dx^\mu \otimes dx^\nu$reads in Rindler coordinates (in region I)

$$\begin{aligned} \boldsymbol{g} = g_{\mu\nu}dx^\mu \otimes dx^\nu &= \mathrm{z}^2 d\mathrm{t} \otimes d\mathrm{t} - d\mathrm{x} \otimes d\mathrm{x} - dy \otimes dy - dz \otimes dz \\ &= \eta_{\mathbf{ab}}\boldsymbol{\gamma}^{\mathbf{a}} \otimes \boldsymbol{\gamma}^{\mathbf{b}} \end{aligned} \tag{12}$$

where $\{\boldsymbol{\gamma}^0, \boldsymbol{\gamma}^0, \boldsymbol{\gamma}^2, \boldsymbol{\gamma}^3\} = \{zdt, dx, dy, dz\}$ is an orthonormal coframe for $T^*\mathrm{I}$ which is dual to the orthonormal frame $\{e_0, e_1, e_2, e_3\} = \{\boldsymbol{R} = \frac{1}{\mathrm{z}}\frac{\partial}{\partial \mathrm{t}}, \frac{\partial}{\partial \mathrm{x}}, \frac{\partial}{\partial \mathrm{y}}, \frac{\partial}{\partial \mathrm{z}}\}$ for $T\mathrm{I}$. We write

$$D_{\frac{\partial}{\partial \mathrm{x}^\nu}} d\mathrm{x}^\mu = -\Gamma^{\mu\cdot\cdot}_{\cdot\nu\iota} d\mathrm{x}^\iota, \qquad D_{e_{\mathbf{b}}}\boldsymbol{\gamma}^a = -\boldsymbol{\Gamma}^{\mathbf{a}\cdot\cdot}_{\cdot\mathbf{bc}}\boldsymbol{\gamma}^c \tag{13}$$

and keep in mind that it is $\boldsymbol{\Gamma}^{\mathbf{a}\cdot\mathbf{c}}_{\cdot\mathbf{b}\cdot} = -\boldsymbol{\Gamma}^{\mathbf{c}\cdot\mathbf{a}}_{\cdot\mathbf{b}\cdot}$ (and of course, $\Gamma^{\mu\cdot\cdot}_{\cdot\nu\iota} = \Gamma^{\mu\cdot\cdot}_{\cdot\iota\nu}$)

Define the 1-form field (physically equivalent to $\boldsymbol{R}$)

$$R = \boldsymbol{g}(\boldsymbol{R},\ ) = R_\mu dx^\mu = \mathrm{z}dx^0 = \boldsymbol{\gamma}^0. \tag{14}$$

Then, as well known[7] DR has the invariant decomposition

$$DR = a \otimes R + \omega_R + \varkappa + \frac{1}{3}\mathfrak{E}h, \tag{15}$$

with

$$\begin{aligned}
a &:= \boldsymbol{g}(\boldsymbol{a},\ ),\\
\boldsymbol{\omega}_R &:= \omega_{\mu\nu}dx^\mu \otimes dx^\nu = \frac{1}{2}\left(R_{\sigma;\tau} - R_{\tau;\sigma}\right)h^\sigma_\mu h^\tau_\nu dx^\mu \otimes dx^\nu\\
\varkappa &:= \varkappa_{\mu\nu}dx^\mu \otimes dx^\nu = \left[\frac{1}{2}\left(R_{\sigma;\tau} + R_{\tau;\sigma}\right)h^\sigma_\mu h^\tau_\nu - \frac{1}{3}\mathfrak{E}h_{\sigma\tau}h^\sigma_\mu h^\tau_\nu\right]dx^\mu \otimes dx^\nu\\
\mathfrak{E} &:= \operatorname{div} \boldsymbol{R} = R^\mu_{;\mu} = \delta R\\
h &:= \left(g_{\mu\nu} - R_\mu R_\nu\right)dx^\mu \otimes dx^\nu
\end{aligned} \tag{16}$$

where a and ω are, respectively, the (form) *acceleration* and the *rotation tensor* (or vortex) of R, $\varkappa$ is the shear tensor of R, and $\mathfrak{E}$ is the *expansion ratio* of R.

Now, $d\boldsymbol{\gamma}^0 = dz \wedge dx^0 = \frac{1}{z}\boldsymbol{\gamma}^3 \wedge \boldsymbol{\gamma}^0$ and thus $\boldsymbol{\gamma}^0 \wedge d\boldsymbol{\gamma}^0 = 0$ which implies that $\boldsymbol{\omega}_R = 0$. See Appendix 1 and details in [45].

This means that the Rindler reference frame $\boldsymbol{R}$ is locally synchronizable, but since R is not an exact differential $\boldsymbol{R}$ is *not* proper time synchronizable, something that is obvious once we look at Fig. 1 and see that for each time $t > 0$ of the inertial reference frame $\boldsymbol{I} = \partial/\partial t$ the Rindler observers following paths σ and σ' (which have of course, different proper accelerations) have also different speeds, so their clocks (according to an inertial observer) tic-tac at different ratios.

2.3 Constant Proper Distance Between σ and σ'

We can easily verify using the orthonormal coframe introduced above that since $d\boldsymbol{\gamma}^{\mathbf{i}} = 0, \mathbf{i} = 1, 2, 3$ it is $\boldsymbol{\Gamma}^{\mathbf{i}}_{\mathbf{ab}} = \boldsymbol{\Gamma}^{\mathbf{i}}_{\mathbf{ba}}$ for $\mathbf{i} = 1, 2, 3$ and $\mathbf{a}, \mathbf{b} = 0, 1, 2, 3$ and also from the form of $d\boldsymbol{\gamma}^0$ we realize that $\boldsymbol{\Gamma}^{\mathbf{0}\cdot\cdot}_{\cdot\mathbf{00}} = \boldsymbol{\Gamma}^{\mathbf{0}\cdot\mathbf{0}}_{\cdot\mathbf{0}\cdot} = -\boldsymbol{\Gamma}^{\mathbf{0}\cdot\mathbf{0}}_{\cdot\mathbf{0}\cdot} = 0$. Thus,

$$\mathfrak{E} = \delta R = -\boldsymbol{\gamma}^{\mathbf{a}} \lrcorner D_{e_{\mathbf{a}}}(\boldsymbol{\gamma}^{\mathbf{0}}) = \boldsymbol{\Gamma}^{\mathbf{0}\cdot\cdot}_{\cdot\mathbf{ab}}\boldsymbol{\gamma}^{\mathbf{a}} \lrcorner \boldsymbol{\gamma}^{\mathbf{b}} = \eta^{\mathbf{ab}}\boldsymbol{\Gamma}^{\mathbf{0}\cdot\cdot}_{\cdot\mathbf{ab}} = -\boldsymbol{\Gamma}^{\mathbf{a}\cdot\mathbf{0}}_{\cdot\mathbf{a}\cdot} = \boldsymbol{\Gamma}^{\mathbf{a}\cdot\cdot}_{\cdot\mathbf{a0}} = 0 \tag{17}$$

and we realize that each observer following an integral line of $\boldsymbol{R}$, say σ_1 will maintain a *constant proper distance* to any of its neighbor observers which are following a different integral line of $\boldsymbol{R}$.

[7] See, e.g., [45].

Of course, proper distance between an observer following σ and another one following σ' is operationally obtained in the following way: Using Rindler coordinates at an event, say $\mathfrak{e}_1 = (0, 0, 0, \mathrm{z}_1)$ the observer following σ sends a light signal to σ' (in the direction $\boldsymbol{e}_3$) which arrives at the σ' worldline at the event $\mathfrak{e}_2 = (\mathrm{t}_2, 0, 0, \mathrm{z}_1 + \ell)$ where it is immediately reflected back to σ arriving at event $\mathfrak{e}_3 = (\mathrm{t}_3, 0, 0, \mathrm{z}_1)$. So, the total coordinate time for the two-way trip of the light signal is t_3 and immediately we get (from the null geodesic equation followed by the light signal)

$$\mathrm{t}_2 = \ln\left(1 + \frac{\ell}{\mathrm{z}_1}\right),$$
$$\mathrm{t}_3 - \mathrm{t}_2 = \ln\left(1 + \frac{\ell}{\mathrm{z}_1}\right) \tag{18}$$

and thus

$$\mathrm{t}_3 = 2\ln\left(1 + \frac{\ell}{\mathrm{z}_1}\right). \tag{19}$$

Now, the observer at σ evaluates the total proper time for the total trip of the signal, it is $\mathrm{z}_1 t_3$. The *proper distance* is by definition

$$d_{\sigma\sigma'} := \frac{1}{2}\mathrm{z}_1\mathrm{t}_3 = \mathrm{z}_1 \ln\left(1 + \frac{\ell}{\mathrm{z}_1}\right). \tag{20}$$

Equation (20) shows that proper distance and coordinate distance are different in a Rindler reference frame.

Remark 2 A look at Fig. 1 shows immediately that inertial observers in $\boldsymbol{I} = \partial/\partial t$ will find that the distance between σ and σ' is shortening with the passage of t time. It is opportune to take into account that despite the fact that the Rindler coordinate times for the going and return paths are equal (the coordinate time being equal to proper time in σ) measured by the inertial observers are different and indeed as it is intuitive the return path is realized in a shorter inertial time.

Remark 3 Of course, if $\boldsymbol{R} = \frac{1}{\mathrm{z}}\partial/\partial t$ is physically realizable by a rocket with the constraint that, e.g., $\mathrm{z}_1 \leq \mathrm{z} \leq (\mathrm{z}_1+\ell)$ then it needs to have a very special propulsion system, with its rear accelerating faster than the front. We do not see how such a rocket could be constructed.[8]

[8]Note that the original Rindler reference frame $\boldsymbol{R}$ for which $(0 < z < \infty)$ is only supposed to be a theoretical construct, it obviously cannot be realized by any material system.

3 Bell 'Paradox'

In [3] it is proposed the following question:

> Three small spaceships, A, B, and C, drift freely in a region of spacetime remote from other matter, without rotation and without relative motion, with B and C equidistant from A (Fig. 1).
>
> On reception of a signal from A the motors of B and C are ignited and they accelerate gently (Fig. 2)
>
> Let ships B and C be identical, and have identical acceleration programmes. Then (as reckoned by an observer at A) they will have at every moment the same velocity, and so remain displaced one from the other by a fixed distance. Suppose that a fragile thread is tied initially between projections from B to C (Fig. 3). If it is just long enough to span the required distance initially, then as the rockets speed up, it will become too short, because of its need to FitzGerald contract, and must finally break. It must break when at a sufficiently high velocity the artificial prevention to the natural contraction imposes intolerable stress.

Then Bell continues saying:

> Is this really so? This old problem came up for discussion once in the CERN canteen. A distinguished experimental physicist refused to accept that the thread would break, and regarded my assertion, that indeed it would, as a personal misinterpretation of special relativity. We decided to appeal to the CERN Theory Division for arbitration, and made a (not very systematic) canvas of opinion in it. there emerged a clear consensus that the tread would **not** break.
>
> Of course many people who give this wrong answer at first get the right answer on further reflection.

Recently Motl [35] wrote a note saying that Bell did not understand Special Relativity since the correct answer to his question is the CERN majority (first sight) view. Now, reading Motl's article one arrives at the conclusion that he did not understand correctly the *formulation* of Bell's problem. Indeed, the problem that

Fig. 2 Figure 1 in Bell [3] (adapted)

Fig. 3 Figure 2 in Bell [3] (adapted)

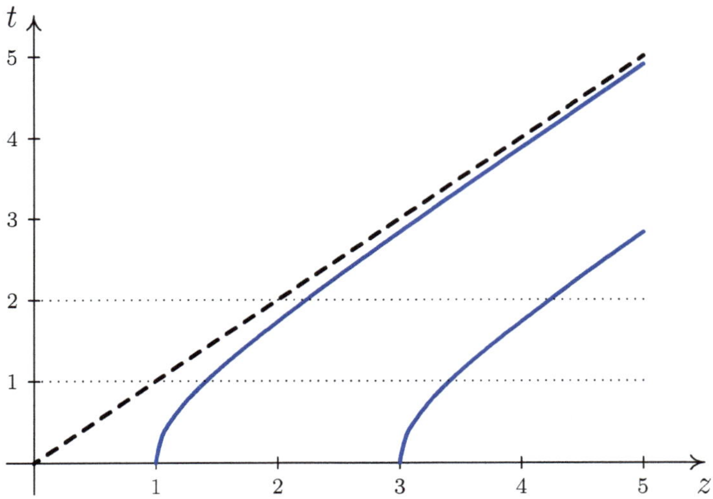

Fig. 4 Spacetime diagram for Bell's question with ships B (tick line on the left) and C (tick line on the right) having the same acceleration relative to the inertial observer A

is correctly analyzed in [35] was the one in each ships B and C are modelled as two distinct observers following two different integral lines of the Rindler reference frame $\boldsymbol{R}$ introduced in the previous section.

It is quite obvious to any one who reads Sect. 1 that in this case (which is *not* the Bell's one) B and C did not have the *same* acceleration programme as seem by observer A (represented by a particular integral line of the inertial frame $\boldsymbol{I} = \partial/\partial t$ the t axis in Fig. 4).

In the case of Bell's question ships B and C are modelled (as a first approximation) as observers, i.e., as the timelike curves

$$t_B^2 - x_B^2 = -\frac{1}{a_B^2},$$

$$t_C^2 - (x_C - d)^2 = -\frac{1}{a_C^2} = -\frac{1}{a_B^2},$$

where to illustrate the situation we draw Fig. 4 with $a_B = 1$ and $d = 2$. It is absolutely clear from Fig. 4 that the distance between B and C any instant $t > 0$ as determined by the inertial observer is the same as it was at $t = 0$, when B and C start accelerating with the same accelerating programme.

A trivial calculation similar to the one in Sect. 2.3 above shows that proper distance between B and C as determined by B (or C) is *increasing* with the coordinate time t used by these observers which are modelled as integral lines of the Rindler reference frame $\boldsymbol{R}$. As a consequence of this fact we arrive at the conclusion that the thread cannot go during the acceleration period to its *natural* Lorentz deformed configuration and thus will break.

Bell's problem illustrates that bodies subject to special acceleration programs do not go to their Lorentz deformed configuration immediately. After the acceleration programme ends the body will acquire adiabatically its Lorentz deformed configuration. More on this issue is discussed in [44].

4 Does a Charge in Hyperbolic Motion Radiate?

4.1 The Answer Given by the Liénard-Wiechert Potential

It is usually assumed (see, e.g., [26, 30, 31, 40–42] that the electromagnetic potential $A = A_\mu(x)dx^\mu \in \sec T^*M$ generated by a charged particle in hyperbolic motion with world line given by $\sigma : \mathbb{R} \to M, s \mapsto \sigma(s)$, with parametric equations given by Eq. (3) and electric current given $J = J_\mu(x(s))dx^\mu\big|_\sigma = eV_\mu(s))dx^\mu\big|_\sigma \sec T^*M$ where

$$v^\mu(s) := \frac{d}{ds}x^\mu \circ \sigma(s), \quad v := (v^0, \mathbf{v}) = \left(\frac{1}{\sqrt{1-\mathbf{v}^2}}, 0, 0, \frac{\mathbf{v}^i}{\sqrt{1-\mathbf{v}^2}}\right), \tag{21}$$

$$J_\mu(x) = e\int ds v_\mu(s)\delta^{(4)}(x' - x \circ \sigma(s)) \tag{22}$$

is given by the solution of the differential equation

$$\Box A_\mu = J_\mu \tag{23}$$

through the well-known formula

$$A_\mu(x) = e\int d^4x D_r(x - x')J_\mu(x') \tag{24}$$

where $D_r(x - x')$ is the retarded Green function[9] given by

$$\begin{aligned} D_r(x - x') &= \frac{1}{2\pi}\theta(x^0 - x'^0)\delta^{(4)}[(x - x')^2] \\ &= \frac{\theta(x^0 - x'^0)}{4\pi \mathrm{R}}\delta(x^0 - x'^0 - \mathrm{R}) \end{aligned} \tag{25}$$

[9] I.e., a solution of $\Box D_r(x - x') = \delta^{(4)}(x - x')$.

with from the light cone constraint in Eq. (25)

$$\mathrm{R} = |\mathbf{x} - \mathbf{x}(\sigma(s)| = \left| x^0 - x^0(s) \right|. \tag{26}$$

Thus using Eq. (25) in Eq. (24) gives the famous Liénard-Wiechert formula, i.e.,

$$A_\mu(x) = \frac{e}{4\pi} \left. \frac{v_\mu(s)}{v \cdot [x - x(\sigma(s))]} \right|_{s=s_0} \tag{27}$$

and putting $\gamma = 1/\sqrt{1 - \mathbf{v}^2}$, we have

$$v \cdot [x - x(\sigma(s))] = \gamma \mathrm{R}(1 - \mathbf{v} \bullet \mathbf{n}) \tag{28}$$

and thus

$$A^0(t, x) = \frac{e}{4\pi} \left. \frac{1}{(1 - \mathbf{v} \bullet \mathbf{n})\mathrm{R}} \right|_{\mathrm{ret}}, \quad \mathbf{A}(t, x) = \frac{e}{4\pi} \left. \frac{\mathbf{v}}{(1 - \mathbf{v} \bullet \mathbf{n})\mathrm{R}} \right|_{\mathrm{ret}} \tag{29}$$

where ret means that the value of the bracket must be calculated at the instant $x^0(s_0) = x^0 - \mathrm{R}$.

We also have for the components of the field $F = dA \in \sec \bigwedge^2 t^* M$

$$F_{\mu\nu}(x) = \frac{e}{4\pi} \frac{1}{v \cdot [x - x(\sigma(s))]} \frac{d}{ds} \left[\frac{[x - x_\sigma(s)]_\mu v_\nu - [x - x_\sigma(s)]_\nu v_\mu}{v \cdot [x - x(\sigma(s))]} \right]_{\mathrm{ret}} \tag{30}$$

and taking into account that $[x - x_\sigma(s)] = (\mathrm{R}, \mathrm{R}\mathbf{n})$, $\quad v_\mu = (\gamma, -\gamma \mathbf{v})$ and putting $\overset{\bullet}{\mathbf{v}} = d\mathbf{v}/dt$ it is

$$\frac{dv_\mu}{ds} = \gamma^2 \left(\gamma^2 \mathbf{v} \bullet \dot{\mathbf{v}}, - \left(\dot{\mathbf{v}} + \gamma^2 \mathbf{v}(\mathbf{v} \bullet \dot{\mathbf{v}}) \right) \right) \tag{31}$$

and

$$\frac{d}{ds}[v \cdot (x - x(\sigma(s))] = -1 + (x - x(\sigma(s)))_\alpha \frac{dv^\alpha}{ds} \tag{32}$$

and thus we get

$$\mathbf{E}(t, \mathbf{x}) = \frac{e}{4\pi} \left[\frac{(\mathbf{n} - \mathbf{v})}{\gamma^2 (1 - \mathbf{v} \bullet \mathbf{n})^3 \mathrm{R}^2} \right]_{\mathrm{ret}} + \frac{e}{4\pi} \left[\frac{\mathbf{n} \times [(\mathbf{n} - \mathbf{v}) \times \dot{\mathbf{v}}}{\gamma^2 (1 - \mathbf{v} \bullet \mathbf{n})^3 \mathrm{R}} \right]_{\mathrm{ret}}, \tag{33}$$

$$\mathbf{B}(t, x) = \mathbf{n} \times \mathbf{E}(t, \mathbf{x}). \tag{34}$$

Since

$$\mathbf{n} \times [(\mathbf{n} - \mathbf{v}) \times \dot{\mathbf{v}} = (\mathbf{n} \bullet \dot{\mathbf{v}})(\mathbf{n} - \mathbf{v}) - \mathbf{n} \cdot (\mathbf{n} - \mathbf{v})\dot{\mathbf{v}} \tag{35}$$

we see that for the hyperbolic motion where $\mathbf{v}$ is parallel to $\dot{\mathbf{v}}$ and

$$\mathbf{v}(t) = a_\sigma \frac{t}{\sqrt{1 + a_\sigma^2 t^2}} \hat{\mathbf{e}}_3,$$

$$\dot{\mathbf{v}}(t) = a_\sigma \frac{1}{(1 + a_\sigma^2 t^2)^{3/2}} \hat{\mathbf{e}}_3$$

the Liénard-Wiechert potential implies in a radiation field, i.e., a field that goes in the infinity (radiation zone) as $1/R$.

In Jackson's book [26] (page 667) one can read that when a charge is accelerated in a reference frame where its speed is $|\mathbf{v}| \ll 1$, the Poynting vector associated with the field given by Eqs. (33) and (34) is

$$\mathbf{S} = \mathbf{E} \times \mathbf{B} = |\mathbf{E}|\, \mathbf{n} \tag{36}$$

and the power irradiated per solid angle is [26]

$$\frac{dP}{d\Omega} = \frac{e^2}{(4\pi)^2} (\mathbf{n} \times \dot{\mathbf{v}}) \tag{37}$$

Thus the total instantaneous irradiated power (for a nonrelativistic accelerated charge) is

$$P = \frac{2}{3} \frac{e^2}{4\pi} |\dot{\mathbf{v}}|^2, \tag{38}$$

a result known as Larmor formula.

The correct formula valid for arbitrary speeds and with $P^\mu = mV^\mu$ (as one can verify after some algebra) is

$$\begin{aligned} P &= -\frac{2}{3} \frac{1}{4\pi} \frac{e^2}{m^2} \left(\frac{dP_\mu}{ds} \frac{dP^\mu}{ds} \right) \\ &= \frac{2}{3} \frac{1}{4\pi} e^2 \gamma^6 \left[|\dot{\mathbf{v}}|^2 - (\mathbf{v} \times \dot{\mathbf{v}})^2 \right]. \end{aligned} \tag{39}$$

Remark 4 Equation (37) shows that the radiated power in a linear accelerator is, of course, bigger for electrons than for, e.g., protons. However, as commented by Jackson [26] even for electrons in a linear accelerator with typical gain of 50 MeV/m the radiation loss is completely negligible. In the case of circular accelerators like synchrotrons since the momentum $\mathbf{p} = \gamma m \mathbf{v}$ changes in direction rapidly we can show that the radiated power (predicted from the Liénard-Wiechert potential) is

$$P = \frac{2}{3}\frac{1}{4\pi}\frac{e^2}{m^2}\gamma^2\omega^2 \left|\mathbf{p}\right|^2 \tag{40}$$

where ω is the angular momentum of the charged particle. This formula fits well the experimental results.

4.2 Pauli's Answer

In this section we use the same parametrization as before for the coordinates of the charged particle in hyperbolic motion. Let $\mathfrak{e}$ (see Fig. 5) be an arbitrary observation point with coordinates $x = (x^0 = t, x^1, x^2, x^3 = z)$. In what follows for simplicity of writing we denote the expression for the Lenard-Wiechert potential (Eq. (27)) as

$$A_\mu(x) = \frac{e}{4\pi}\frac{v_\mu(s)}{v \cdot [x - x(\sigma(s))]}, \tag{41}$$

but we cannot forget that at the end of our calculations we must put $s = s_0$. We have, explicitly for the velocity of the particle (moving in the x^3-direction with $a_\sigma = 1$)

$$v^0(s) = \cosh s, \quad v^3(s) = \sinh s \tag{42}$$

and so

$$\begin{aligned} v \cdot [x - x(\sigma(s))] &= x^0 \cosh s - x^3 \sinh s = \mathrm{x}^3 \sinh(s - x^0) \\ &= \mathrm{z} \sinh(s - t). \end{aligned} \tag{43}$$

Then, we have

$$A^0(x) = \frac{e}{4\pi}\frac{\cosh s}{\mathrm{z}\sinh(s - \mathrm{t})}, \quad A^3(x) = \frac{e}{4\pi}\frac{\sinh s}{\mathrm{z}\sinh(s - \mathrm{t})} \tag{44}$$

which are Eqs. (249) in Pauli's book [43].

Pauli's argument for saying that a charge in hyperbolic motion does not radiate is as follows:

(i) Consider the inertial reference frame $\boldsymbol{I}'$ where the charge is momentarily at rest at the instant $(x^0_{\mathfrak{e}'} - \mathrm{R}) = t_0$. This is the time coordinate (in the coordinates of the inertial frame $\boldsymbol{I}$) of the event $\mathfrak{e}_0$ in Fig. 5.

A naturally adapted coordinate system for the reference frame $\boldsymbol{I}'$ is ($\mathfrak{v} = |\mathbf{v}|$)

$$
\begin{aligned}
x'^0 &= t_0 + \gamma(x^0 - \mathfrak{v}x^3),\\
x'^3 &= z_0 + \gamma(x^3 - \mathfrak{v}x^0),\\
x'^1 &= x^1, \quad x'^2 = x^2
\end{aligned}
\tag{45}
$$

and

$$
\begin{aligned}
\frac{\partial x'^0}{\partial x^0} &= \gamma = \cosh s, & \frac{\partial x'^0}{\partial x^3} &= -\sinh s,\\
\frac{\partial x'^3}{\partial x^0} &= -\gamma\mathfrak{v} = -\sinh s, & \frac{\partial x'^3}{\partial x^3} &= \cosh s,
\end{aligned}
\tag{46}
$$

from where it follows that the components of the potential A in the new coordinates $\{x'^\mu\}$ are

$$
A'^0(x') = \frac{e}{4\pi}\frac{1}{\mathrm{z}\sinh(s - \mathrm{t})}, \quad A'^3(x') = 0. \tag{47}
$$

As a consequence of Eq. (47) it follows that the magnetic field $\mathbf{B}'$ as measured in the reference frame $\boldsymbol{I}'$ is null, thus the Poynting vector in this frame $\mathbf{S}' = \mathbf{E}' \times \mathbf{B}' = 0$ and thus (according to Pauli) an observer instantaneously at rest at event $\mathfrak{e}_0$ with respect to the charge will detect no radiation.

(ii) To conclude his argument Pauli considers a second inertial reference frame $\breve{\boldsymbol{I}}$ where the events $\mathfrak{o}$ and $\mathfrak{e}'$ are simultaneous and where $\mathfrak{e}'$ is an event on the world line of another observer at rest in the $\boldsymbol{R}$ frame which supposedly will receive—if it exists—the radiation field emitted by the charge at event $\mathfrak{e}_0$ (see Fig. 5). A naturally adapted coordinate system to $\breve{\boldsymbol{I}}$ is

$$
\begin{aligned}
\breve{x}^0 &= \breve{\gamma}(x^0 - \breve{\mathfrak{v}}x^3),\\
\breve{x}^1 &= x^1, \quad \breve{x}^2 = x^2,\\
\breve{x}^3 &= \breve{\gamma}(x^3 - \breve{\mathfrak{v}}x^0),
\end{aligned}
\tag{48}
$$

with

$$
\breve{\mathfrak{v}} = \sinh \mathrm{t}/\cosh \mathrm{t}, \quad \breve{\gamma} = (1 - \breve{\mathfrak{v}}^2)^{-1/2} = \cosh \mathrm{t}. \tag{49}
$$

A trivial calculation gives

$$
\breve{A}^0(\breve{x}) = \frac{e}{4\pi}\frac{\coth(s - \mathrm{t})}{\sqrt{(\breve{x}^3)^2 - (\breve{x}^0)^2}}, \quad \breve{A}^3(\breve{x}) = \frac{e}{4\pi}\frac{1}{\sqrt{(\breve{x}^3)^2 - (\breve{x}^0)^2}}. \tag{50}
$$

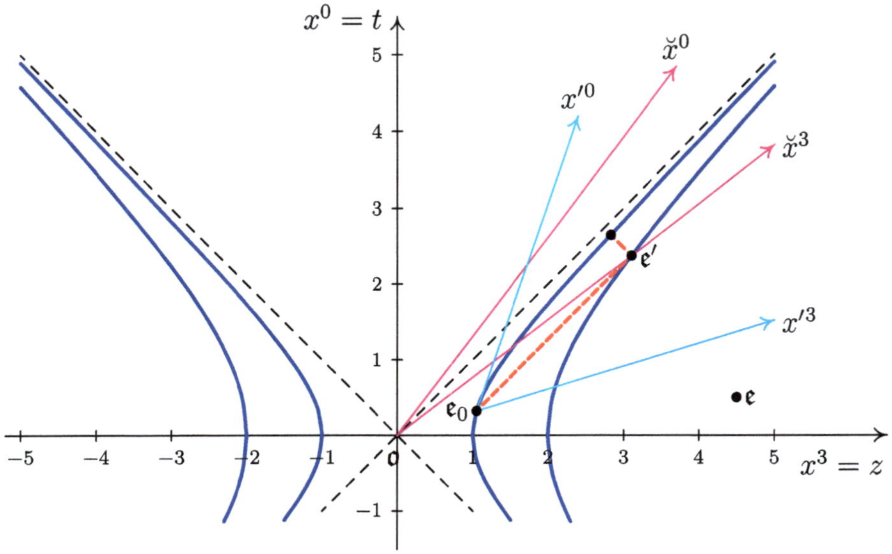

Fig. 5 Graphic for presenting Pauli's argument

and since $\breve{\mathbf{B}} = (F_{32}, F_{13}, F_{21}) = 0$ it follows that the Poynting vector $\breve{\mathbf{S}} = \breve{\mathbf{E}} \times \breve{\mathbf{B}} = 0$. Thus an instantaneous observer $(\mathfrak{e}', \breve{\boldsymbol{I}}_{\mathfrak{e}'})$ in the $\breve{\boldsymbol{I}}$ frame momentously at rest relative to instantaneous observer $(\mathfrak{e}', \boldsymbol{R}_{\mathfrak{e}'})$ observer in the $\boldsymbol{R}$ frame at the considered event will also not detect any radiation emitted from $\mathfrak{e}_0$.

4.2.1 Calculation of Components of the Potentials in the *R* Frame

Using an obvious notation we write the components of the electromagnetic potential in the $\boldsymbol{R}$ frame as $A(\boldsymbol{x}'(\mathfrak{e})) = (\mathrm{A}^0(\mathrm{t}, \mathrm{z}), 0, 0, -\mathrm{A}^3(\mathrm{t}, \mathrm{z}))$ and we have

$$\begin{aligned} \mathrm{A}_0 &= \frac{\partial x^0}{\partial \mathrm{x}^0} A_0 + \frac{\partial x^3}{\partial \mathrm{x}^0} A_3 = \frac{e}{4\pi} \coth(\mathrm{t}-s)|_{s=s_0}\,, \\ \mathrm{A}_3 &= \frac{\partial x^0}{\partial \mathrm{x}^3} A_0 + \frac{\partial x^3}{\partial \mathrm{x}^3} A_3 = -\frac{e}{4\pi \mathrm{z}} \tanh(s-\mathrm{t})|_{s=s_0}\,. \end{aligned} \tag{51}$$

So,

$$\overrightarrow{\mathrm{E}}(\mathrm{t}, \mathrm{z}) := (0, 0, \mathrm{F}_{03}(\mathrm{t}, \mathrm{z})), \quad \overrightarrow{\mathrm{B}}(\mathrm{t}, \mathrm{z}) = 0, \tag{52}$$

$$\mathrm{F}_{03}(\mathrm{t}, \mathrm{z}) = \frac{\partial}{\partial \mathrm{t}} \mathrm{A}_3(\mathrm{t}, \mathrm{z})\Big|_{s=s_0} - \frac{\partial}{\partial \mathrm{z}} \mathrm{A}_0(\mathrm{t}, \mathrm{z})\Big|_{s=s_0} \tag{53}$$

and again the Poynting vector $\overrightarrow{\mathrm{E}} \times \overrightarrow{\mathrm{B}}$ is null. So, by Paul's argument the observers at rest in the $\boldsymbol{R}$ frame will detect no radiation.

4.3 Is Pauli Argument Correct?

In order to evaluate if Pauli's argument is correct we recall that the Liénard-Wiechert potential $A \in \sec \bigwedge^1 T^*M$ by construction is in Lorenz gauge, i.e., $\delta A = 0$ and moreover it satisfies the homogeneous wave equation for all spacetime points outside the worldline of the accelerated charge, i.e.,

$$\lozenge A = -d\delta A - \delta d A = -\delta d A = 0 \tag{54}$$

where $\lozenge$ is the Hodge Laplacian, and δ is the Hodge coderivative. Since $F = dA \in \sec \bigwedge^2 T^*M$ and

$$\lozenge F = -d\delta d A - \delta d d A = -d\delta d A = 0 \tag{55}$$

it follows that the electromagnetic field satisfies also a wave equation.

Remark 5 Well, it is a common practice to call an electromagnetic field satisfying the wave equation a electromagnetic wave. So, despite the fact that $\overrightarrow{\mathrm{B}} = 0$ observers outside the worldline of the accelerated charge (and living in the same accelerated laboratory) will perceive a pure electric wave.

In our case

$$F = \mathrm{F}_{03} d\mathrm{x}^0 \wedge d\mathrm{x}^3 \tag{56}$$

and the energy momentum tensor of the electromagnetic field

$$\mathbf{T} = \mathrm{T}_{\mu\nu} d\mathrm{x}^\mu \otimes d\mathrm{x}^\nu \in \sec T_0^2 M \tag{57}$$

in the coordinates $\{\mathrm{x}^\mu\}$ (naturally adapted to the Rindler frame $\boldsymbol{R}$) has only the following non-null component:

$$\mathrm{T}^{00}(\mathrm{t}, \mathrm{z}) = \frac{1}{2} |\mathrm{F}_{03}(\mathrm{t}, \mathrm{z})|^2 . \tag{58}$$

So an observer, following the worldline σ' with $\mathrm{z} = \mathrm{z}_0 =$ constant $(\mathrm{z} > 1)$ will detect a pseudo-energy density "wave" passing through the point where he is located. Moreover, if this observer carries with him an electric charge say e' he will certainly detect that his charge is acted by the electromagnetic field with a (1-form) force

$$\mathfrak{F} = e' v_{\sigma'} \lrcorner F = v^0_{\sigma'} \mathrm{F}_{03} d\mathrm{x}^3 \tag{59}$$

and he certainly will need more pseudo energy or better more Minkowski energy (fuel in his rocket) to maintain his charge (with mass m') at constant acceleration than the energy that he would have to use to maintain at a constant acceleration a particle with mass m' and null charge.

Also, since the energy arriving at the σ' worldline must be coming from energy radiated by the charge following σ, an observer maintaining the charge e (of mass m) at constant acceleration will expend more Minkowski energy than the one necessary for maintaining at a constant acceleration a particle with mass m and null charge.

4.4 The Rindler (Pseudo) Energy

It is a well-known fact that outside the worldline σ of the accelerating charge the electromagnetic energy-momentum tensor has null divergence, i.e., satisfy

$$D \cdot \mathbf{T} = \mathbf{0} \tag{60}$$

where D is the Levi-Civita connection of $\boldsymbol{g}$. Since $\boldsymbol{K} = \frac{\partial}{\partial \mathrm{t}}$ is a Killing vector field for the metric $\boldsymbol{g}$ as it is obvious looking at the representation of $\boldsymbol{g}$ in terms of the coordinates $\{\mathrm{x}^\mu\}$ adapted to the $\boldsymbol{R} = \frac{1}{\mathrm{z}} \boldsymbol{K}$ frame we have that the current

$$\mathcal{J}_R = \mathrm{K}^\nu \mathrm{T}_{\nu\mu} d\mathrm{x}^\mu \tag{61}$$

is conserved, i.e.,

$$\underset{\boldsymbol{g}}{\delta} \mathcal{J}_R = -\boldsymbol{\partial} \lrcorner \mathcal{J}_R = -\frac{1}{\sqrt{-\det \boldsymbol{g}}} \frac{\partial}{\partial \mathrm{x}^\mu} \left(\sqrt{-\det \boldsymbol{g}} \mathrm{K}^\nu \mathrm{T}^\mu_\nu \right) = 0. \tag{62}$$

Then, of course, the *scalar* quantity[10]

$$\mathcal{E} = \int_{\Sigma'} \star \mathcal{J}_R \tag{63}$$

is a conserved one. However, take notice that differently of the case of the similar current calculated with the Killing vector field $\partial/\partial t$ it does not qualify as the zero component of a momentum covector (*not* covector field). See details in [47].

[10] If $N \subset M$ is the region where $\mathcal{J}_R$ has support, then $\partial N = \Xi + \Xi' + \digamma$ where Ξ and Ξ' are spacelike surfaces and $\mathcal{J}_R$ is null in $\digamma$ (spatial infinity).

In our case we have

$$\frac{\partial}{\partial x^{\mu}}\left(zT_{0}^{\mu}\right)=0 \tag{64}$$

Consider the accelerating charge following the σ worldline (for which $z = 1$ and $s = t$) surrounded by a 2-dimensional sphere Σ_t of constant radius $r = \mathfrak{R}$ at time t. Now, from propertime $s_1 = t_1$ to propertime $s_2 = t_2$ the surface Σ_t moves producing a world tube in Minkowski spacetime.

Since

$$\frac{\partial}{\partial x^{0}}\left(zT_{0}^{0}\right)=-\frac{\partial}{\partial x^{i}}\left(zT_{0}^{i}\right) \tag{65}$$

the quantity $\mathscr{E}(t_1 \mapsto t_2)$ given by

$$\begin{aligned}\mathscr{E}(t_1 \mapsto t_2) &= \int_{t_1}^{t_2} dt \iiint r^2 \sin\theta dr d\theta d\varphi \frac{\partial}{\partial t}\left(zT_{0}^{0}\right)\\ &= -\int_{t_1}^{t_2} dt \iiint r^2 dr d\Omega \frac{\partial}{\partial x^{i}}\left(zT_{0}^{i}\right)\\ &= -\int_{t_1}^{t_2} dt \iint \left(zT_{0}^{i}\right) n_i\, \mathfrak{R}^2 d\Omega \end{aligned} \tag{66}$$

(where $\{r, \theta, \varphi\}$ are polar coordinates associated with $\{x^1, x^2, x^3\}$ and n_i are the components of the normal vector to Σ_t) is null since $T_0^i = 0$.

Thus if the observer following σ (of course, at rest relative to the accelerating charge) decides to call $\mathscr{E}(t_1 \mapsto t_2)$ the energy radiated by the charge he will arrive at the conclusion that he did not see any radiated energy.

But of course, $\mathscr{E}(t_1 \mapsto t_2)$ is not the extra Minkowski energy (calculated above) necessary for the observer to maintain the charge at constant acceleration. Parrott [42] quite appropriately nominates $\mathscr{E}(t_1 \mapsto t_2)$ the *pseudo-energy*, other people as authors of [14] call it Rindler energy.

Conclusion 1 *What seems clear at least to us is that whereas any one can buy Minkowski energy (e.g., in the form of fuel) for his rocket no one can buy the "magical" Rindler energy.*

4.5 The Turakulov Solution

In a paper published in the *Journal of Geometry and Physics* [56] Turakulov presented a solution for the problem of finding the electromagnetic field of a charge in uniformly accelerated motion by directly solving the wave equation for the potential $A \in \sec \bigwedge^1 T^*M$ using a separation of variables method instead of using the Liénard-Wiechert potential used in the previous discussion. Since this solution is not well known we recall and analyze it here with some details.

Turakulov started his analysis with the coordinates (t, x, y, z) introduced in Sect. 2 and proceeds as follows. In the t = constant Euclidean semi-spaces he introduced[11] *toroidal* coordinates (u, v, φ) by

$$\mathrm{z} = \frac{a \sinh u}{\cosh u + \cos v}, \quad \rho = \frac{a \sin v}{\cosh u + \cos v},$$

$$u = \tanh^{-1}\left(\frac{2a\mathrm{z}}{\mathrm{z}^2 + \rho^2 + a^2}\right), \quad v = \tanh^{-1}\left(\frac{2a\mathrm{z}}{\mathrm{z}^2 + \rho^2 - a^2}\right). \tag{67}$$

(where $\rho = +\sqrt{x^2 + y^2}$) and also introduced their pseudo Euclidean generalizations for the other domains, i.e.,

$$\mathrm{z} = \frac{a \sin u}{\cos u + \cos v}, \quad \rho = \frac{a \sin v}{\cos u + \cos v},$$

$$u = \tan^{-1}\left(\frac{2a\mathrm{z}}{-\mathrm{z}^2 + \rho^2 + a^2}\right), \quad v = \tan^{-1}\left(\frac{2a\mathrm{z}}{-\mathrm{z}^2 + \rho^2 - a^2}\right). \tag{68}$$

Let σ be the world line a uniformly accelerate charge, as we know it corresponds to z =constant and thus the surfaces u = constant forms a family of spheres defined by the equation

$$(z - a \coth u_0) + \rho^2 = a \sinh^{-1} u \tag{69}$$

involving the charge. The Minkowski metric in region I and II using the coordinates (t, u, v, ρ) reads

$$\boldsymbol{g} = \left(\frac{a}{\cosh u + \cos v}\right)^2 \left(\sinh^2 u \, dt \otimes dt - du \otimes du - dv \otimes dv - \sin^2 v d\varphi \otimes d\varphi\right) \tag{70}$$

and for regions F and P it is

$$\boldsymbol{g} = \left(\frac{a}{\cosh u + \cos v}\right)^2 \left(- \sin^2 u \, dt \otimes dt + du \otimes du - dv \otimes dv - \sin^2 v d\varphi \otimes d\varphi\right). \tag{71}$$

[11] Toroidal coordinates (also called bispherical coordinates) is discussed in Section 10.3 in volume II of the classical book by Morse and Feshbach [34].

As we know the potential A^T in the Lorenz gauge $\delta A^T = 0$ satisfies the wave equation $\delta d A^T = 0$ Then supposing (as usual) that the potential is tangent to the integral lines of $\boldsymbol{R}$ we can write[12]

$$A^T = \Theta(u, v)dt \tag{72}$$

and the general solution of the wave equation is

$$\Theta(u, v) = \alpha_0(\cosh u - 1) + \textstyle\sum_{n=1}^{\infty} \alpha_n \sinh u \frac{d}{du} P_n(\cosh u) P_n(\cos v), \tag{73}$$

where P_n are Legendre polynomials and $\alpha_{0,}\alpha_n$ are constants. The field of a charge is simply specified only by the first term with $\alpha_0 = e$ the value of the charge generating the field. Thus, if the charge is at $u = \infty$ we have for regions I and II and P and F

$$A^T_{\mathrm{I,II}} = e(\cosh u - 1)dt, \quad A^T_{\mathrm{P,F}} = e(\cos u - 1)dt. \tag{74}$$

In terms of the coordinates (t, x, y, z), writing $A^T = A^T_\mu dx^\mu$ we have the following solution valid for all regions[13]:

$$\begin{aligned} A^T_0 &= -\frac{z}{z^2 - t^2}\left(\frac{t^2 - \rho^2 + z^2 - a^2}{\Lambda_+ \Lambda_-} - 1\right), \\ A^T_3 &= \frac{t}{z^2 - t^2}\left(\frac{t^2 - \rho^2 + z^2 - a^2}{\Lambda_+ \Lambda_-} - 1\right), \\ A^T_1 &= A^T_2 = 0, \\ \Lambda_\pm(t, x, y, z) &= \sqrt{(\sqrt{z^2 - t^2} \pm a)^2 + x^2 + y^2}. \end{aligned} \tag{75}$$

From these formulas we infer that

$$F^T = F_{tu} dt \wedge du = -e \sinh u dt \wedge du \tag{76}$$

and thus an observer comoving with the charge will see only an "electric field" which for him is in the u-direction and the *pseudo* energy evaluated beyond a given sphere $u = u_0$ of radius $\boldsymbol{r}$ is

[12]Here the value of the charge is $e/4\pi = 1$.

[13]We have verified using the Mathematica software that indeed A_0 and A_3 satisfy the wave equation. Note that there is are signal misprints in the formulas for A_0 and A_3 in [56] and the modulus $\sqrt{|z^2 - t^2|}$ in those formulas is not necessary.

$$\mathscr{E} = \frac{e^2}{2\boldsymbol{r}}. \tag{77}$$

Thus, Turakulov concludes as Pauli did that there is no radiation. But is his conclusion correct?

4.5.1 Does the Turakulov Solution Imply that a Charge in Hyperbolic Motion Does Not Radiate?

Recall that in Sect. 4.3 we showed that supposing that the Liénard-Wiechert solution is the correct one then Pauli's argument is incorrect since an observer following another integral line of $\boldsymbol{R}$ will see an electric "wave" (recall Eq. (58)). We now make the same analysis as the one we did in the case of the Turakulov solution in order to find the correct answer to our question. We first explicitly calculate the electric and magnetic fields in the inertial frame $\boldsymbol{I} = \partial/\partial t$. We have

$$E_x = \frac{8a^2xz}{\Lambda_+^3\Lambda_-^3}, \quad E_y = \frac{8a^2yz}{\Lambda_+^3\Lambda_-^3},$$

$$E_z = \frac{-4a^2[x^2+y^2+a^2-z^2+t^2]}{\Lambda_+^3\Lambda_-^3},$$

$$B_x = \frac{8a^2yt}{\Lambda_+^3\Lambda_-^3}, \quad B_y = \frac{-8a^2xt}{\Lambda_+^3\Lambda_-^3}, \quad B_z = 0. \tag{78}$$

The Poincaré invariants of the Turakulov solution $I_1 := \mathbf{E}^2 - \mathbf{B}^2$ and $I_2 := \mathbf{E} \bullet \mathbf{B}$ are

$$I_1 = \frac{16a^4}{\Lambda_+^6\Lambda_-^6}[(x^2+y^2-z^2+t^2)^2 + 4(x^2+y^2)(z^2+t^2)], \quad I_2 = 0. \tag{79}$$

This shows that an inertial observer at rest at (x, y, z) will detect a *time dependent* electromagnetic field configuration passing through his observation point. Of course, it is *not* a null field, but it certainly qualifies as an electromagnetic wave. And what is important for our analysis is that the field carries energy and momentum from the accelerating charge to the point (x, y, z).

Indeed, consider a charge q at rest in the Rindler frame following an integral line σ' of $\boldsymbol{R}$ with constant Rindler coordinates $(\mathrm{t}, \mathrm{x} = \mathrm{x}_0, \mathrm{y} = \mathrm{y}_0\ \mathrm{z} = \mathrm{z}_0)$ and thus with inertial coordinates $(t, \mathrm{x}_0, y = \mathrm{y}_0, z = \sqrt{\mathrm{z}_0^2 + t^2})$.

As determined by the inertial observer the density of *real* energy and the Poynting vector arriving from the uniformly accelerated charge moving along the z-axis of the inertial frame to where the charge q is located are:

$$\frac{1}{2}(\mathbf{E}^2+\mathbf{B}^2) = \frac{1}{2}\mathring{\Lambda}_+^{-6}\mathring{\Lambda}_-^{-6}\Big(128(\mathrm{x}_0^2+\mathrm{y}_0^2)t^2 + 64a^4(\mathrm{x}_0^2+\mathrm{y}_0^2)\mathrm{z}_0^2$$
$$+16a^4(\mathrm{x}_0^2+\mathrm{y}_0^2+a^2-\mathrm{z}_0^2)^2\Big),$$
$$\mathbf{S} = \mathbf{i}\frac{32a^4\mathrm{x}_0}{\mathring{\Lambda}_+^6\mathring{\Lambda}_-^6}(\mathrm{x}_0^2+\mathrm{y}_0^2+a^2-\mathrm{z}_0^2)t + \mathbf{j}\frac{-32a^4\mathrm{y}_0}{\mathring{\Lambda}_+^6\mathring{\Lambda}_-^6}(\mathrm{x}_0^2+\mathrm{y}_0^2+a^2-\mathrm{z}_0^2)t$$
$$+\mathbf{k}\frac{64a^4\sqrt{z_0^2+t^2}}{\mathring{\Lambda}_+^6\mathring{\Lambda}_-^6}(\mathrm{x}_0^2+\mathrm{y}_0^2)t,$$
$$\mathring{\Lambda}_\pm = \sqrt{(\mathrm{z}_0\pm a)^2+\mathrm{x}_0^2+\mathrm{y}_0^2} \tag{80}$$

Thus, we see that indeed there is a flux of *real* energy and momentum arriving at the charge q located at $(t, \mathrm{x}_0, y = \mathrm{y}_0, z = \sqrt{\mathrm{z}_0^2+t^2})$.

Moreover, the Lorentz force $\mathbf{F}_L$ acting on the charge q (according to the inertial observer) is

$$\mathbf{F}_L = q\mathbf{E} + \mathbf{q}\mathbf{v}_{\sigma'}\times\mathbf{B}, \tag{81}$$

depends on t and is doing work on the charge q. So, an observer comoving with the charge q will need to expend more *real* energy to carry this charge than to carry a particle with zero charge.

More important: since the energy arriving at the charge q is the one produced by the charge e generating the field we arrive at the conclusion, as in the case of the Pauli solution that an observer carrying the charge e will speed more energy (fuel of its rocket) than when it carries a particle with zero charge.

Remark 6 We already observed in [32] that the use of the retarded Green's function may result in non-sequitur solutions in some cases. Most important is the fact that in [58] it is observed that the Green's function for a massless scalar field is the integral $(\omega = k_0)$

$$G(x,x') = \frac{1}{(2\pi)^4}\int d^3\mathbf{k}\int d\omega\frac{e^{-i(\omega(t-t')-\mathbf{k}\cdot(\mathbf{x}-\mathbf{x}'))}}{\mathbf{k}^2-\omega^2} \tag{82}$$

and the evaluation of the integral is done in all classical presentations in the complex ω-plane and thus its result depends, as is well known from the path of integration chosen. But, contrary to what is commonly accepted this is not necessary for the integrand is not singular. This can be shown as follows. Recalling that G depends only on

$$\tau^2 - \mathfrak{r}^2 = (t-t')^2 - (\mathbf{x}-\mathbf{x}')^2$$

we can choose a coordinate system where $(\mathbf{x}-\mathbf{x}')^2 = 0$ for the point under consideration, Then, introducing the coordinates

$$\begin{aligned}\varkappa &= \omega^2 - \mathbf{k}^2, \quad \xi = \tanh^{-1}(|\mathbf{k}|/\omega),\\ \omega(t-t') - \mathbf{k}\cdot(\mathbf{x}-\mathbf{x}') &= \varkappa\varsigma\cosh\xi\end{aligned} \tag{83}$$

Eq. (83) becomes after some algebra

$$G(\tau,\mathfrak{r}) = \frac{1}{4\pi^3}\int d\varkappa \int d\xi \int d\theta \int d\varphi \sin\theta \sinh^3\xi\varkappa^2 e^{i\varkappa\varsigma\cosh\xi}. \tag{84}$$

This important result obtained in [58] shows explicitly that it is possible to evaluate the Green's function without introducing the "famous" $i\varepsilon$ prescription! Turakulov also observed that putting $\lambda = \varkappa\varsigma$ the Eq. (84) gives

$$G(\tau,\mathfrak{r}) = \frac{\pi^2}{\varsigma^2}\int d\lambda\lambda \int d\xi \sinh^2\xi e^{i\lambda\cosh\xi}. \tag{85}$$

The conclusion is thus that integration only predetermines the factor $1/\varsigma^2$ and it is now possible to select any path of integration in the complex plane, which means that the retarded Green's function is create by inserting a non-existence singularity into the integrand!

Moreover, in it is shown in [58] that the use of the retarded Green's function produces problems with energy-conservation when, e.g., a charge is accelerated in an external potential. Finally we observe that in [57] it is shown that when there are infinitesimally small changes of the acceleration there is emission of radiation.

5 The Equivalence Principle

Consider first the statements (a) and (b):

(a) an observer (say Mary) living in a small constantly accelerated reference frame (e.g., a "small" world tube, with non-transparent walls of the reference frame $\boldsymbol{R}$) following an integral line σ of the $\boldsymbol{R}$ frame and for which $D_{\boldsymbol{R}}\boldsymbol{R}|_\sigma = \boldsymbol{a}_{|\sigma}$;

(b) an observer (say John) living in a "small" reference frame, (e.g., a "small" world tube, with non-transparent walls of the reference frame $\boldsymbol{Z}$ in a Lorentzian spacetime structure $(M, \mathbf{g}, \mathbf{D}, \tau_{\mathbf{g}}, \uparrow)$ modelling a gravitational field (generated by some energy-momentum distribution) in General Relativity theory and such that $\mathbf{D}_{\boldsymbol{Z}}\boldsymbol{Z}|_\lambda = \boldsymbol{a}_{|\lambda} = \boldsymbol{a}_{|\sigma}$.

Then a common formulation of the *Equivalence Principle*[14] says that Mary or John cannot with *local*[15] experiments determine if she(he) lives in a uniformly accelerated frame in Minkowski spacetime or in the gravitational field modelled by $(M, \mathbf{g}, \mathbf{D}, \tau_{\mathbf{g}}, \uparrow)$.

Now, as well known (since long ago) and as proved rigorously (under well determined conditions) in [42] a charge in a static gravitational field in General Relativity theory does not radiate if it follows an integral line of a reference frame like $\boldsymbol{Z}$ in (b). An observer comoving with the charge will see only an electric field and thus will see no radiation since the Poynting vector is null.

Does this imply that the Equivalence Principle holds for local experiments with charged matter?

Well, if we accept that the Liénard-Wiechert solution is the correct one, then the answer from the analysis given in the previous section is *no* (see also, [30, 31, 42]. In particular Parrot's argument is the following: since there is no radiation in the true gravitational field an observer at rest in the Schwarzschild spacetime following a worldline λ will spend the same amount of "energy" to maintain at constant acceleration $\boldsymbol{a}_{|\lambda} = \boldsymbol{a}_{|\sigma}$ a particle with mass m and null charge and one with mass m and charge $e \neq 0$.

Since we already know that in the $\boldsymbol{R}$ frame it is clear that an observer σ will spend *different* amounts (of Minkowski) energy to maintain at constant acceleration $\boldsymbol{a}_{|\lambda} = \boldsymbol{a}_{|\sigma}$ a particle with mass m and null charge and one with mass m and charge $e \neq 0$.

Of course, even supposing that the Liénard-Wiechert solution is the correct one many people do not agree with this conclusion and some of the arguments of the opposition are discussed in [42].

Remark 7 From our point of view we think necessary to comment that Parrot' s argument would be a really strong one only if the concept of energy (and momentum) would be well defined in General Relativity, which is definitively not the case [45–47]. However, take notice that the quantity defined as "energy" by Parrot (the zero component of current of the form given by Eq. (61), where in this case $\boldsymbol{K}$ is a timelike Killing vector field for the Schwarzschild metric is not the component of any energy-momentum covector field, it looks more as the concept of energy in Newtonian physics. Anyway, the quantity of the pseudo " energy" necessary to carry a particle in uniformly accelerated motion will certainly be different in the two cases of a charged and a non-charged particle. In our opinion what is necessary is to construct an analysis of the problem charge in a gravitational

[14] A thoughtful discussion of the Equivalence Principle and the so-called Principle of Local Lorentz Invariance is given in [44].

[15] Of course, by local mathematicians means a (4-dimensional) open set U of the appropriate spacetime manifold. So, by doing experiments in U observers will detect using a gradiometer tidal force fields (proportional to the Riemann curvature tensor) if at rest in $\boldsymbol{Z}$ in a real gravitational field and will not detect any tidal force field if living in $\boldsymbol{R}$ in Minkowski spacetime. For more details see, e.g., [38, 44].

theory where energy-momentum of a system can be defined and is a conserved quantity [45, 46].

On the other hand, if we accept that Turakulov solution as the correct one then again the Equivalence Principle is violated and for the same reason than in the case of the Liénard-Wiechert solution as discussed in Sect. 4.5.1.

So, which solution, Liénard-Wiechert or Turakulov is the correct one?

An answer can be given to the above question only with a clever experiment and for the best of our knowledge no such experiment has been done yet.

6 Some Comments on the Unruh Effect

6.1 Minkowski and Fulling-Unruh Quantization of the Klein-Gordon Field

(**u1**) To discuss the Unruh effect it is useful to introduce coordinates such that the solution of the Klein-Gordon equation in these variables becomes as simple as possible. A standard choice is to take $(\mathfrak{t},\mathfrak{x},\mathfrak{y},\mathfrak{z})$ and $(\mathfrak{t}',\mathfrak{x}',\mathfrak{y}',\mathfrak{z}')$ for regions I and II defined by[16]

$$
\begin{aligned}
&\mathfrak{t}=\frac{1}{a}\tanh^{-1}\left(\frac{t}{z}\right), \quad \mathfrak{z}=\frac{1}{2a}\ln[a(z^2-t^2)], \quad \mathfrak{x}=x, \quad \mathfrak{y}=y\\
&t=\frac{1}{a}\exp(a\mathfrak{z})\sinh(a\mathfrak{t}), \quad z=\frac{1}{a}\exp(a\mathfrak{z})\cosh(a\mathfrak{t}), \quad |z|\geq t, \;\; z>0,\\
&\mathfrak{t}'=\frac{1}{a}\tanh^{-1}\left(\frac{t}{z}\right), \quad \mathfrak{z}'=\frac{1}{2a}\ln[a^2(z^2-t^2)], \quad \mathfrak{x}'=x, \quad \mathfrak{y}'=y.,\\
&t=\frac{1}{a}\exp(a\mathfrak{z}')\sinh(a\mathfrak{t}'), \quad z=-\frac{1}{a}\exp(a\mathfrak{z}')\cosh(a\mathfrak{t}'), \quad |z|\geq t, \;\; z<0,\\
&\qquad \mathfrak{t},\mathfrak{z}\in(-\infty,\infty), \qquad a\in\mathbb{R}^+.
\end{aligned}
\tag{86}
$$

Take notice that in regions I and II the coordinates $\mathfrak{t}$ and $\mathfrak{z}$ are respectively timelike and spacelike and in region II the decreasing of $\mathfrak{t}$ corresponds to the increase of t.

The Minkowski metric in these coordinates (and in the regions I and II) reads

$$
\begin{aligned}
&\boldsymbol{g}=\exp(2a\mathfrak{z})d\mathfrak{t}\otimes d\mathfrak{t}-d\mathfrak{x}\otimes d\mathfrak{x}-d\mathfrak{y}\otimes d\mathfrak{y}-\exp(2a\mathfrak{z})d\mathfrak{z}\otimes d\mathfrak{z}=\eta_{\mathbf{ab}}\mathfrak{g}^{\mathbf{a}}\otimes\mathfrak{g}^{\mathbf{b}},\\
&\mathfrak{g}^{\mathbf{0}}=\exp(a\mathfrak{z})d\mathfrak{t}, \quad \mathfrak{g}^{\mathbf{1}}=d\mathfrak{x}, \quad \mathfrak{g}^{\mathbf{2}}=d\mathfrak{y}, \quad \mathfrak{g}^{\mathbf{3}}=\exp(a\mathfrak{z})d\mathfrak{z}.
\end{aligned}
\tag{87}
$$

[16]Note that $(\mathfrak{t},\mathfrak{z})$ differs form the coordinates (t, z) introduced in Sect. 2.

(u2) The right and left Rindler reference frames are represented by

$$\boldsymbol{R} = \frac{1}{\exp(a\mathfrak{z})}\partial/\partial\mathfrak{t}, \quad t \in (-\infty, \infty), \quad |z| \geq t, \quad z > 0,$$
$$\boldsymbol{L} = \frac{1}{\exp(a\mathfrak{z})}\partial/\partial\mathfrak{t}, \quad t \in (-\infty, \infty), \quad |z| \geq t, \quad z < 0. \tag{88}$$

and they are *not* Killing vector fields.[17]

Consider the integral line, say σ of $\boldsymbol{R}$ given by $\mathfrak{x}, \mathfrak{y} =$ constant and $\mathfrak{z} = \mathfrak{z}_0 =$ constant. We immediately find that its proper acceleration is

$$a_\sigma = 1/\sqrt{g_{00}(\mathfrak{z}_0)}. \tag{89}$$

(u3) However, the vector fields

$$\boldsymbol{I} = \partial/\partial t,$$
$$\boldsymbol{Z}_{\mathrm{I}} = \partial/\partial\mathfrak{t}, \text{ with } t \in (-\infty, \infty), |z| \geq t \text{ and } z > 0,$$
$$\boldsymbol{Z}_{\mathrm{II}} = \partial/\partial\mathfrak{t}, \text{ with } t \in (-\infty, \infty), |z| \geq t \text{ and } z < 0, \tag{90}$$

are Killing vector fields, i.e., $\mathscr{L}_{\partial/\partial t}\boldsymbol{g} = \mathscr{L}_{\boldsymbol{Z}_{\mathrm{I}}}\boldsymbol{g} = \mathscr{L}_{\boldsymbol{Z}_{\mathrm{II}}}\boldsymbol{g} = 0$. The inertial reference frame $\boldsymbol{I}$ besides being locally synchronizable is also propertime synchronizable, i.e., $\boldsymbol{g}(\boldsymbol{I},\ ) = dt$ and the fields $\boldsymbol{Z}_{\mathrm{I}}$ and $\boldsymbol{Z}_{\mathrm{II}}$ although does not qualify as reference frames (according to our definition) play an important role for our considerations of the Unruh effect. The reason is that both fields in the regions where they have support are such that

$$Z_{\mathrm{I}} = \boldsymbol{g}(\boldsymbol{Z}_{\mathrm{I}},\ ) = \exp(2a\mathfrak{z})d\mathfrak{t}, \text{ with } t \in (-\infty, \infty), |z| \geq t \text{ and } z > 0,$$
$$Z_{\mathrm{II}} = \boldsymbol{g}(\boldsymbol{Z}_{\mathrm{II}},\ ) = \exp(2a\mathfrak{z})d\mathfrak{t}, \text{ with } t \in (-\infty, \infty), |z| \geq t \text{ and } z > 0. \tag{91}$$

Thus the field $\boldsymbol{I}$ can be used to foliate all M as $M = \cup_t(\mathbb{R}\times\Sigma(t))$ where $\Sigma(t) \simeq \mathbb{R}^3$ is a Cauchy surface. Moreover, the field Z_{I} (respectively Z_{II}) can be used to foliate region I (respectively region II) as $\mathrm{I} = \cup_\mathfrak{t}(\mathbb{R}\times\Sigma_{\mathrm{I}}(\mathfrak{t}))$ (respectively $\mathrm{II} = \cup_\mathfrak{t}(\mathbb{R}\times\Sigma_{\mathrm{II}}(\mathfrak{t}))$) where $\Sigma_{\mathrm{I}}(\mathfrak{t}) \simeq \Sigma_{\mathrm{I}}$ and $\Sigma_{\mathrm{II}}(\mathfrak{t}) \simeq \Sigma_{\mathrm{II}}$ are Cauchy surfaces.

We now briefly describe how the Unruh effect for a complex Klein-Gordon field is presented in almost all texts[18] dealing with the issue.

---

[17] This can easily be verified taking into account that $\mathscr{L}_{\boldsymbol{R}}\boldsymbol{g} = 2\eta_{\mathbf{ab}}\mathscr{L}_{\boldsymbol{R}}\mathfrak{g}^{\mathbf{a}} \otimes \mathfrak{g}^{\mathbf{b}}$ and recalling that if $R = \boldsymbol{g}(\boldsymbol{R},\ ) = \mathfrak{g}^{\mathbf{0}}$ we may evaluate [45] as $\mathscr{L}_{\boldsymbol{R}}\mathfrak{g}^{\mathbf{a}} = d(\mathfrak{g}^{\mathbf{0}} \cdot \mathfrak{g}^{\mathbf{a}}) + \mathfrak{g}^{\mathbf{0}} \lrcorner d\mathfrak{g}^{\mathbf{a}}$.

[18] E.g., in [14, 17, 25, 52, 53, 59, 63]. The presentations eventually differ in the use of other coordinate systems.

(u4) Let $\phi \in \sec(\mathbb{C}\otimes\bigwedge^0 T^*M)$. Our departure point is to first solve the Klein-Gordon equation

$$-\delta d\phi + \mu^2\phi = 0 \tag{92}$$

valid for all M, in the global naturally adapted coordinates (in ELP gauge) to $\boldsymbol{I}$ and next to solve it in regions I and II using the coordinates defined in Eq. (86) (and then extend this new solution for all M). In the first case we use the $t = 0$ as Cauchy surface to given initial data. In the second case we use the $\mathfrak{t} = 0$ Cauchy surface to give initial data (see below).

The positive energy solutions will be called *Minkowski modes* for the first case and *Fulling-Unruh modes* for the second case (i.e., the solutions in regions I and II). In order to simplify the writing of the formulas that follows we introduce the notations

$$\phi_M(x) = \phi_M(t, x, y, z),\ \ \phi_{\mathrm{I}}(\mathfrak{l}) = \phi_{\mathrm{I}}(\mathfrak{t}, \mathfrak{x}, \mathfrak{y}, \mathfrak{z}),\ \ \phi_{\mathrm{II}}(\mathfrak{l}') = \phi_{\mathrm{II}}(\mathfrak{t}, \mathfrak{x}, \mathfrak{y}, \mathfrak{z}),$$
$$k \cdot x = k_\alpha x^\alpha,\ \ \omega_{\mathbf{k}} = k_0 = +\sqrt{\mathbf{k}^2 + \mu^2},\ \ k \cdot k = (k_0)^2 - \mathbf{k}^2 = \mu^2,\ \ \mathbf{k}^2 = \mathbf{k} \bullet \mathbf{k},$$
$$\mathbf{q} = (k_1, k_2),\ \ \mathbf{r} = (x^1, x^2) = (x, y) \text{ and },$$
$$\mathbf{q} \bullet \mathbf{r} = k_1 x^1 + k_2 x^2,\ \ \nu = +\sqrt{\mathbf{q}^2 + \mu^2}. \tag{93}$$

Observing that in region II the timelike coordinate $\mathfrak{t}'$ decreases when t increases we have that the elementary modes (of positive energy) which are solutions of the Klein-Gordon equation in the three regions:

$$\phi_{M\mathbf{k}}(x) = [(2\pi)^3 2\omega_{\mathbf{k}})]^{-1/2} e^{-ik\cdot x},$$
$$\phi_{\mathrm{I}\nu\mathbf{q}}(\mathfrak{l}) = [(2\pi)^2 2\nu)]^{-1/2} F_{\mathrm{I}\nu\mathbf{q}}(\mathfrak{z}) e^{-i(\nu\mathfrak{t} - \mathbf{q}\bullet\mathbf{r})},$$
$$\phi_{\mathrm{I}\nu\mathbf{q}}(\mathfrak{l}') = [(2\pi)^2 2\nu)]^{-1/2} F_{\mathrm{II}\nu\mathbf{q}}(\mathfrak{z}) e^{+i(\nu\mathfrak{t}' + \mathbf{q}\bullet\mathbf{r})}, \tag{94}$$

with

$$F_{\mathrm{I}\nu\mathbf{q}}(\mathfrak{z}) = (2\pi^{-1})^{1/2} C_{\mathrm{I}\mathbf{q}} \frac{1}{\Gamma(i\nu)} \left(\frac{\nu}{2a}\right)^{i\nu} K_{i\nu}(\nu\mathfrak{z}),$$
$$F_{\mathrm{II}\nu\mathbf{q}}(\mathfrak{z}') = (2\pi^{-1})^{1/2} C_{\mathrm{II}\mathbf{q}}(a) \frac{1}{\Gamma(i\nu)} \left(\frac{\nu}{2a}\right)^{i\nu} K_{i\nu}(\nu\mathfrak{z}'), \tag{95}$$

where $C_{\mathrm{I}\mathbf{q}}$ are arbitrary "phase factor," Γ is the gamma function, and $K_{i\nu}$ are the modified Bessel functions of second kind.

Remark 8 Before we continue it is important to emphasize that the concept of energy defined in regions I and II is indeed the pseudo-energy concept that we discussed in previous section.

(u5) We use the positive frequencies in standard way in order to construct Hilbert spaces $\mathscr{H}$, $\mathscr{H}_{\mathrm{I}}$, and $\mathscr{H}_{\mathrm{II}}$ by defining the well-known scalar products for the spaces of positive energy-solutions. This is done by introducing the spaces of square integrable functions $\mathscr{K}_M$, $\mathscr{K}_{\mathrm{I}}$, and $\mathscr{K}_{\mathrm{II}}$, respectively, of the forms

$$\begin{aligned} \Phi_M(x) &= \int d^3\mathbf{k}[a(\mathbf{k})\phi_{M\mathbf{k}}(x) + \bar{a}^*(\mathbf{k})\phi^*_{M\mathbf{k}}(x)] \\ \Phi_{\mathrm{I}}(\mathfrak{l}) &= \int_0^\infty d\nu \int d^2\mathbf{q}[b_{\mathrm{I}\nu}(\mathbf{q})\phi_{\mathrm{I}\nu\mathbf{q}}(\mathfrak{l}) + \bar{b}^*_{\mathrm{I}\nu}(\mathbf{q})\phi^*_{\mathrm{I}\nu\mathbf{q}}(\mathfrak{l})] \\ \Phi_{\mathrm{II}}(\mathfrak{l}') &= \int_0^\infty d\nu \int d^2\mathbf{q}[b_{\mathrm{II}\nu}(\mathbf{q})\phi_{\mathrm{II}\nu\mathbf{q}}(\mathfrak{l}') + \bar{b}^*_{\mathrm{II}\nu}(\mathbf{q})\phi^*_{\mathrm{II}\nu\mathbf{q}}(\mathfrak{l}')] \end{aligned} \tag{96}$$

where $a, b_{\mathrm{I}\nu}, b_{\mathrm{II}\nu}, \bar{a}, \bar{b}_{\mathrm{I}\nu}, \bar{b}_{\mathrm{II}\nu}$ are arbitrary square integrable functions (elements of $\mathfrak{L}(\mathbb{R}^3)$).

Take notice that $\hat{\phi}_{\mathrm{I}} + \hat{\phi}_{\mathrm{II}}$ can be extended to all M by extending $\phi_{\mathrm{I}\nu\mathbf{q}}$ and $\phi_{\mathrm{II}\nu\mathbf{q}}(\mathfrak{l})$ to all M.

Now, we construct in the space of these functions the usual inner products ($J = M, \mathrm{I}, \mathrm{II}$)

$$\langle \Phi_J, \Psi_J \rangle_J = i \int_\Sigma d\Sigma n^a (\Phi_J^* \frac{\partial}{\partial x_J^a} \Psi_J - \Phi_J \frac{\partial}{\partial x_J^a} \Psi_J^*) \tag{97}$$

where $J = M, \mathrm{I}, \mathrm{II}$ and x_J^a denotes the appropriate variables for each domain and finally we construct as usual the Hilbert spaces $\mathscr{H}$, $\mathscr{H}_{\mathrm{I}}$, and $\mathscr{H}_{\mathrm{II}}$ by completion of the respective $\mathscr{K}$ spaces and n^a are the components of the normal to the spacelike surface Σ.

In particular, choosing Σ to be hypersurface $t = 0$ for the Minkowski modes and $\mathfrak{t} = 0$ for the Rindler modes we have

$$\begin{aligned} &\langle \phi_{M\mathbf{k}}, \phi_{M\mathbf{k}'} \rangle_M = \delta(\mathbf{k} - \mathbf{k}'), \quad \langle \phi^*_{M\mathbf{k}}, \phi^*_{M\mathbf{k}'} \rangle_M = -\delta(\mathbf{k} - \mathbf{k}'), \\ &\langle \phi_{\mathrm{I}\nu\mathbf{q}}, \phi_{\mathrm{I}\nu'\mathbf{q}'} \rangle_{\mathrm{I}} = \delta(\nu - \nu')\delta(\mathbf{q} - \mathbf{q}'), \quad \langle \phi_{\mathrm{I}\nu\mathbf{q}}, \phi_{\mathrm{I}\nu'\mathbf{q}'} \rangle_{\mathrm{I}} = -\delta(\nu - \nu')\delta(\mathbf{q} - \mathbf{q}'), \\ &\langle \phi_{\mathrm{II}\nu\mathbf{q}}, \phi_{\mathrm{I}\nu'\mathbf{q}'} \rangle_{\mathrm{II}} = \delta(\nu - \nu')\delta(\mathbf{q} - \mathbf{q}'), \quad \langle \phi_{\mathrm{II}\nu\mathbf{q}}, \phi_{\mathrm{II}\nu'\mathbf{q}'} \rangle_{\mathrm{II}} = -\delta(\nu - \nu')\delta(\mathbf{q} - \mathbf{q}'), \\ &\langle \phi_{M\mathbf{k}}, \phi^*_{M\mathbf{k}'} \rangle_M = 0, \quad \langle \phi_{\mathrm{I}\nu\mathbf{q}}, \phi^*_{\mathrm{I}\nu'\mathbf{q}'} \rangle_{\mathrm{I}} = 0, \quad \langle \phi_{\mathrm{II}\nu\mathbf{q}}, \phi^*_{\mathrm{II}\nu'\mathbf{q}'} \rangle_{\mathrm{II}} = 0. \end{aligned} \tag{98}$$

(u6) From $\mathscr{H}$, $\mathscr{H}_{\mathrm{I}}$, and $\mathscr{H}_{\mathrm{II}}$ we construct the Fock-Hilbert space $\mathscr{F}(\mathscr{H})$, $\mathscr{F}(\mathscr{H}_{\mathrm{I}})$ and $\mathscr{F}(\mathscr{H}_{\mathrm{II}})$ which describe all possible physical states of the quantum fields

$$\hat{\phi}_M(x) = \int d^3\mathbf{k} \left[\mathbf{a}(\mathbf{k})\, \phi_{M\mathbf{k}} + \bar{\mathbf{a}}^\dagger(\mathbf{k})\, \phi^*_{M\mathbf{k}} \right], \tag{99a}$$

$$\hat{\phi}_{\rm I}(\mathfrak{l}) = \int_0^\infty d\nu \int d^2\mathbf{q} \left[\mathbf{b}_{\rm I\nu}(\mathbf{q})\,\phi_{{\rm I}_\nu\mathbf{q}}(\mathfrak{l}) + \bar{\mathbf{b}}^\dagger_{\rm I\nu}(\mathbf{q})\,\phi^*_{{\rm I}_\nu\mathbf{q}}(\mathfrak{l})\right], \tag{99b}$$

$$\hat{\phi}_{\rm II}(\mathfrak{l}') = \int_0^\infty d\nu \int d^2\mathbf{q} \left[\mathbf{b}_{\rm II\nu}(\mathbf{q})\,\phi_{{\rm II}_\nu\mathbf{q}}(\mathfrak{l}') + \bar{\mathbf{b}}^\dagger_{\rm II\nu}(\mathbf{q})\,\phi^*_{{\rm I}_\nu\mathbf{q}}(\mathfrak{l}')\right], \tag{99c}$$

which are operator valued distributions acting, respectively, on $\mathscr{F}(\mathscr{H})$, $\mathscr{F}(\mathscr{H}_{\rm II})$ $\mathscr{F}(\mathscr{H})$ and where the $\mathbf{a}$, $\mathbf{a}^\dagger$, $\mathbf{b}_{\rm I\nu}$, $\mathbf{b}^\dagger_{\rm I\nu}$ and $\mathbf{b}_{\rm II\nu}$, $\mathbf{b}^\dagger_{\rm II\nu}$ (respectively $\bar{\mathbf{a}}$, $\bar{\mathbf{a}}^\dagger$, $\bar{\mathbf{b}}_{\rm I\nu}$, $\bar{\mathbf{b}}^\dagger_{\rm I\nu}$ and $\bar{\mathbf{b}}_{\rm II\nu}$, $\bar{\mathbf{b}}^\dagger_{\rm II\nu}$) are destruction and creation operators for *positive* (respectively negative) charged particles. We have for the *non-null* commutators:

$$\begin{aligned}
[\bar{\mathbf{a}}(\mathbf{k}), \bar{\mathbf{a}}^\dagger(\mathbf{k}')] &= [\mathbf{a}(\mathbf{k}), \mathbf{a}^\dagger(\mathbf{k}')] = \delta(\mathbf{k}-\mathbf{k}'),\\
[\bar{\mathbf{b}}_{\rm I\nu}(\mathbf{q}), \bar{\mathbf{b}}_{\rm I\nu'}(\mathbf{q}')] &= [\mathbf{b}_{\rm I\nu}(\mathbf{q}), \mathbf{b}_{\rm I\nu'}(\mathbf{q}')] = \delta(\nu-\nu')\delta(\mathbf{q}-\mathbf{q}'),\\
[\bar{\mathbf{b}}_{\rm II\nu}(\mathbf{q}), \bar{\mathbf{b}}_{\rm II\nu'}(\mathbf{q}')] &= [\mathbf{b}_{\rm II\nu}(\mathbf{q}), \mathbf{b}_{\rm II\nu'}(\mathbf{q}')] = \delta(\nu-\nu')\delta(\mathbf{q}-\mathbf{q}').
\end{aligned} \tag{100}$$

We suppose that we have a second quantum field construction for all Minkowski spacetime (with eigenfunctions properly extended for all domains) once we choose as the one-particle Hilbert space $\mathscr{H}_{\rm II} \oplus \mathscr{H}_{\rm I}$. Now, take notice that [63]

$$\mathscr{F}(\mathscr{H}_{\rm II} \oplus \mathscr{H}_{\rm I}) \simeq \mathscr{F}(\mathscr{H}_{\rm II}) \otimes \mathscr{F}(\mathscr{H}_{\rm I}). \tag{101}$$

(u7) The Minkowski vacuum and the vacua for regions I, II are defined, respectively, by the states $|0\rangle_M \in \mathscr{F}(\mathscr{H})$, $|0\rangle_{\rm I} \in \mathscr{F}(\mathscr{H}_{\rm I})$, $|0\rangle_{\rm II} \in \mathscr{F}(\mathscr{H}_{\rm II})$ such that

$$\begin{aligned}
&\mathbf{a}(\mathbf{k})\,|0\rangle_M = \bar{\mathbf{a}}(\mathbf{k})\,|0\rangle_M = 0\ \forall\mathbf{k},\\
&\mathbf{b}_{\rm I\nu}(\mathbf{q})|0\rangle_{\rm I} = \bar{\mathbf{b}}_{\rm I\nu}(\mathbf{q})|0\rangle_{\rm I} = 0, \text{ and } \mathbf{b}_{\rm II\nu}(\mathbf{q})\,|0\rangle_{\rm II} = \bar{\mathbf{b}}_{\rm II\nu}(\mathbf{q})\,|0\rangle_{\rm II} = 0,\ \forall\mathbf{q},\nu.
\end{aligned} \tag{102}$$

The respective particle number operators for modes $\mathbf{k}$, Iν, and IIν are $N_\mathbf{k} = \mathbf{a}^\dagger(\mathbf{k})\,\mathbf{a}(\mathbf{k})$, $\bar{N}_\mathbf{k} = \mathbf{a}^\dagger(\mathbf{k})\,\mathbf{a}(\mathbf{k})$, $N_{\rm I\nu\mathbf{q}} = \mathbf{b}^\dagger_{\rm I\nu}(\mathbf{q})\,\mathbf{b}_{\rm I\nu}(\mathbf{q})$, $\bar{N}_{\rm I\nu\mathbf{q}} = \mathbf{b}^\dagger_{\rm I\nu}(\mathbf{q})\,\mathbf{b}_{\rm I\nu}(\mathbf{q})$ and $N_{\rm II\nu\mathbf{q}} = \mathbf{b}^\dagger_{\rm II\nu}(\mathbf{q})\,\mathbf{b}_{\rm II\nu}(\mathbf{q})$, $\bar{N}_{\rm II\nu\mathbf{q}} = \mathbf{b}^\dagger_{\rm II\nu}(\mathbf{q})\,\mathbf{b}_{\rm II\nu}(\mathbf{q})$. Of course,

$$\begin{aligned}
&{}_M\langle 0|N_\mathbf{k}|0\rangle_M = 0,\ {}_{\rm I}\langle 0|N_{\rm I\nu\mathbf{q}}|0\rangle_{\rm I} = 0,\ {}_{\rm II}\langle 0|N_{\rm II\nu\mathbf{q}}|0\rangle_{\rm II} = 0,\\
&{}_M\langle 0|\bar{N}_\mathbf{k}|0\rangle_M = 0,\ {}_{\rm I}\langle 0|\bar{N}_{\rm I\nu\mathbf{q}}|0\rangle_{\rm I} = 0,\ {}_{\rm II}\langle 0|\bar{N}_{\rm II\nu\mathbf{q}}|0\rangle_{\rm II} = 0.
\end{aligned} \tag{103}$$

(u8) In some presentations it is supposed that the quantum field in regions I + II obtained through the above quantization procedures can be described by

$$\hat{\phi}_{\rm I} + \hat{\phi}_{\rm II} \tag{104}$$

acting on $\mathscr{F}(\mathscr{H}_{\mathrm{II}} \oplus \mathscr{H}_{\mathrm{I}})$. However, here we suppose that the quantum field $\hat{\phi}'$ in regions I + II is described by an "entangled field" made from $\hat{\phi}_{\mathrm{I}}(x)$ and $\hat{\phi}_{\mathrm{II}}(x)$ acting on $\mathscr{F}(\mathscr{H}_{\mathrm{II}}) \otimes \mathscr{F}(\mathscr{H}_{\mathrm{I}})$, i.e., described by

$$\hat{\phi}' = \mathbf{1}_{\mathrm{II}} \otimes \hat{\phi}_{\mathrm{I}} + \hat{\phi}_{\mathrm{II}} \otimes \mathbf{1}_{\mathrm{I}} \tag{105}$$

acting (see Eq. (101)) on the Fock-Hilbert space $\mathscr{F}(\mathscr{H}_{\mathrm{II}}) \otimes \mathscr{F}(\mathscr{H}_{\mathrm{I}})$.

Moreover, it is taken as obvious that (see e.g., [59]) it is not necessary to analyze what happens in regions F and P.

6.2 "Deduction" of the Unruh Effect

(u9) As it is well known the delta functions in Eqs. (98) and (100) lead to problems and so to continue the analysis it is usual to introduce in the Hilbert spaces[19] $\mathscr{H}$, $\mathscr{H}_{\mathrm{I}}$, and $\mathscr{H}_{\mathrm{II}}$ countable basis, which we denote in Fourier space by

$$f_{\mathbf{m},\mathbf{l},\varrho}(\mathbf{k}) = \varrho^{-\frac{3}{2}} \exp\left(-\frac{2\pi i \mathbf{k} \bullet \mathbf{l}}{\varrho}\right) \chi_{[(|\mathbf{m}|-1/2)\varrho,(|\mathbf{m}|+1/2)\varrho]}(\mathbf{k}), \tag{106}$$

where $\varrho \in \mathbb{R}^{+}$ (has inverse length dimension) and χ_S is the characteristic function of the set S.[20] The functions $f_{\mathbf{m},\mathbf{l},\varrho}(\mathbf{k})$ are localized in Fourier space around[21] $\mathbf{m} = (m_1, m_2, m_3)$ and have wave number vector $\mathbf{l} = (\ell_1, \ell_2, \ell_3)$, and thus in $\mathbb{R}^3$ they are localized around $\mathbf{l}$ with wave number vector $\mathbf{m}$. We immediately have that[22]

$$\int d\mathbf{k} f^{*}_{\mathbf{m},\mathbf{l},\varrho}(\mathbf{k}) f_{\mathbf{m}',\mathbf{l}',\varrho}(\mathbf{k})$$
$$:= \frac{1}{\varrho^3}\delta_{\mathbf{m}\mathbf{m}'} \prod_i \int_{(m_i-1/2)\varrho}^{(m_i+1/2)\varrho} dk_i \exp\left(-\frac{2\pi i k_i(\ell_i - \ell'_i)}{\varrho}\right) = \delta_{\mathbf{m}\mathbf{m}'}\delta_{\ell\ell'} \tag{107}$$

and

---

[19]Note that $\mathscr{H}$, $\mathscr{H}_{\mathrm{I}}$, and $\mathscr{H}_{\mathrm{II}}$ are isomorphic to $\mathfrak{L}^2(\mathbb{R}^3)$.

[20]For each $\mathbf{m} = (m_1, m_2, m_3)$ it is $S = \{(x^1, x^2, x^3) \mid (m_i - 1/2)\varrho < x^i < (m_i + 1/2)\varrho,\ x^i \in \mathbb{R}, i = 1, 2, 3\}$.

[21]The $m_i, \ell_i \in \mathbb{Z}$, $i = 1, 2, 3$.

[22]Take notice that in the term $\exp\left(-\frac{2\pi i k_i(\ell_i - \ell'_i)}{\varrho}\right)$ in Eq. (108) $k_i \ell_i$ does not mean that we are summing in the indice i.

$$\begin{aligned}&\sum\nolimits_{\mathbf{l}\in\mathbb{Z}^3} f_{\mathbf{m},\mathbf{l},\varrho}(\mathbf{k}) f_{\mathbf{m},\mathbf{l},\varrho}(\mathbf{k}') = \chi_{[(|\mathbf{m}|-1/2)\varrho,(|\mathbf{m}|+1/2)\varrho]}(\mathbf{k})\delta(\mathbf{k}-\mathbf{k}'),\\ &\sum\nolimits_{\mathbf{l},\mathbf{m}\in\mathbb{Z}^3} f_{\mathbf{m},\mathbf{l},\varrho}(\mathbf{k}) f_{\mathbf{m},\mathbf{l},\varrho}(\mathbf{k}') = \delta(\mathbf{k}-\mathbf{k}'). \end{aligned} \tag{108}$$

(u10) Now, in the Hilbert spaces $\mathscr{H}$, $\mathscr{H}_{\mathrm{I}}$, and $\mathscr{H}_{\mathrm{II}}$ we construct the *positive frequencies* solutions of the Klein-Gordon equation, i.e.,

$$\begin{aligned}\Phi_{M,\mathbf{m},\mathbf{l},\varrho}(x) &= \int d^3\mathbf{k} f_{\mathbf{m},\mathbf{l},\varrho}(\mathbf{k})\phi_{M\mathbf{k}}(x),\\ \Phi_{\mathrm{I},\mathbf{m},\mathbf{l},\varrho}(\mathfrak{l}) &= \int_0^\infty d\nu \int d^2\mathbf{q} f_{\mathbf{m},\mathbf{l},\varrho}(\mathbf{k})\phi_{\mathrm{I}\nu\mathbf{q}}(\mathfrak{l}),\\ \Phi_{\mathrm{II},\mathbf{m},\mathbf{l},\varrho}(\mathfrak{l}') &= \int_0^\infty d\nu \int d^2\mathbf{q} f_{\mathbf{m},\mathbf{l},\varrho}(\mathbf{k})\phi_{\mathrm{II}\nu\mathbf{q}}(\mathfrak{l}'). \end{aligned} \tag{109}$$

We have

$$\begin{aligned}&\langle \Phi_{M,\mathbf{m},\mathbf{l},\varrho}, \Phi_{M,\mathbf{n},\mathbf{l}',\varrho}\rangle_M = \delta_{\mathbf{m}\mathbf{m}'}\delta_{\ell\ell'},\quad \langle \Phi^*_{M,\mathbf{m},\mathbf{l},\varrho}, \Phi^*_{M,\mathbf{n},\mathbf{l}',\varrho}\rangle_M = -\delta_{\mathbf{m}\mathbf{m}'}\delta_{\ell\ell'},\\ &\langle \Phi_{\mathrm{I},\mathbf{m},\mathbf{l},\varrho}, \Phi_{\mathrm{I},\mathbf{m}',\mathbf{l}',\varrho}\rangle_{\mathrm{I}} = \delta_{\mathbf{m}\mathbf{m}'}\delta_{\ell\ell'},\quad \langle \Phi^*_{\mathrm{I},\mathbf{m},\mathbf{l},\varrho}, \Phi^*_{\mathrm{I},\mathbf{m}',\mathbf{l}',\varrho}\rangle_{\mathrm{I}} = -\delta_{\mathbf{m}\mathbf{m}'}\delta_{\ell\ell'},\\ &\langle \Phi_{\mathrm{II},\mathbf{m},\mathbf{l},\varrho}, \Phi_{\mathrm{II},\mathbf{m}',\mathbf{l}',\varrho}\rangle_{\mathrm{II}} = \delta_{\mathbf{m}\mathbf{m}'}\delta_{\ell\ell'},\quad \langle \Phi^*_{\mathrm{II},\mathbf{m},\mathbf{l},\varrho}, \Phi^*_{\mathrm{II},\mathbf{m}',\mathbf{l}',\varrho}\rangle_{\mathrm{II}} = -\delta_{\mathbf{m}\mathbf{m}'}\delta_{\ell\ell'},\\ &\langle \Phi_{M,\mathbf{m},\mathbf{l},\varrho}, \Phi^*_{M,\mathbf{n},\mathbf{l}',\varrho}\rangle_M = 0,\quad \langle \Phi_{\mathrm{I},\mathbf{m},\mathbf{l},\varrho}, \Phi^*_{\mathrm{I},\mathbf{m}',\mathbf{l}',\varrho}\rangle_{\mathrm{I}} = 0,\\ &\qquad\qquad \langle \Phi_{\mathrm{II},\mathbf{m},\mathbf{l},\varrho}, \Phi^*_{\mathrm{II},\mathbf{m}',\mathbf{l}',\varrho}\rangle_{\mathrm{II}} = 0 \end{aligned} \tag{110}$$

and so

$$\begin{aligned}\phi_{M\mathbf{k}}(x) &= \sum\nolimits_{\mathbf{l},\mathbf{m}\in\mathbb{Z}^3} f_{\mathbf{m},\mathbf{l},\varrho}(\mathbf{k})\Phi_{M,\mathbf{m},\mathbf{l},\varrho}(x),\\ \phi_{\mathrm{I}\nu\mathbf{q}}(\mathfrak{l}) &= \sum\nolimits_{\mathbf{l},\mathbf{m}\in\mathbb{Z}^3} f_{\mathbf{m},\mathbf{l},\varrho}(\mathbf{k})\Phi_{\mathrm{I},\mathbf{m},\mathbf{l},\varrho}(\mathfrak{l}),\\ \phi_{\mathrm{II}\nu\mathbf{q}}(\mathfrak{l}') &= \sum\nolimits_{\mathbf{l},\mathbf{m}\in\mathbb{Z}^3} f_{\mathbf{m},\mathbf{l},\varrho}(\mathbf{k})\Phi_{\mathrm{II},\mathbf{m},\mathbf{l},\varrho}(\mathfrak{l}'). \end{aligned} \tag{111}$$

The field operators are then written as

$$\hat{\phi}_M(x) = \sum\nolimits_{\mathbf{l},\mathbf{m}\in\mathbb{Z}^3}\left[\mathbf{a}_{\mathbf{m},\mathbf{l},\varrho}\phi_{M,\mathbf{m},\mathbf{l},\varrho}(x) + \bar{\mathbf{a}}^\dagger\phi^*_{M,\mathbf{m},\mathbf{l},\varrho}(x)\right], \tag{112a}$$

$$\hat{\phi}_{\mathrm{I}}(\mathfrak{l}) = \sum\nolimits_{\mathbf{l},\mathbf{m}\in\mathbb{Z}^3}\left[\mathbf{b}_{\mathrm{I}\mathbf{m},\mathbf{l},\varrho}\phi_{\mathrm{I}_\nu\mathbf{q}}(\mathfrak{l}) + \bar{\mathbf{b}}^\dagger_{\mathrm{I}\mathbf{m},\mathbf{l},\varrho}\phi^*_{\mathrm{I}_\nu\mathbf{q}}(\mathfrak{l})\right], \tag{112b}$$

$$\hat{\phi}_{\mathrm{II}}(\mathfrak{l}') = \sum\nolimits_{\mathbf{l},\mathbf{m}\in\mathbb{Z}^3}\left[\mathbf{b}_{\mathrm{II}\nu}\phi_{\mathrm{II}_\nu\mathbf{q}}(\mathfrak{l}) + \bar{\mathbf{b}}^\dagger_{\mathrm{II}\nu}\phi^*_{\mathrm{I}_\nu\mathbf{q}}(\mathfrak{l})\right], \tag{112c}$$

with

$$\mathbf{a}_{\mathbf{m},\mathbf{l}_{,\varrho}} = \int d^3\mathbf{k} f^*_{\mathbf{m},\mathbf{l},\varrho}(\mathbf{k})\mathbf{a}(\mathbf{k}),$$

$$\mathbf{b}_{\mathrm{I}\mathbf{m},\mathbf{l}_{,\varrho}} = \int_0^\infty d\nu \int d^2\mathbf{q} f^*_{\mathbf{m},\mathbf{l},\varrho}(\mathbf{k})\mathbf{b}_{\mathrm{I}\nu}(\mathbf{q}),$$

$$\mathbf{b}_{\mathrm{II}\mathbf{m},\mathbf{l}_{,\varrho}} = \int_0^\infty d\nu \int d^2\mathbf{q} f^*_{\mathbf{m},\mathbf{l},\varrho}(\mathbf{k})\mathbf{b}_{\mathrm{II}\nu}(\mathbf{q}) \tag{113}$$

and analogous equations for the operators $\bar{\mathbf{a}}_{\mathbf{m},\mathbf{l}_{,\varrho}}$, $\bar{\mathbf{b}}_{\mathrm{I}\mathbf{m},\mathbf{l}_{,\varrho}}$ and $\bar{\mathbf{b}}_{\mathrm{II}\mathbf{m},\mathbf{l}_{,\varrho}}$. The non-null commutators are

$$[\mathbf{a}_{\mathbf{m},\mathbf{l}_{,\varrho}}, \mathbf{a}^\dagger_{\mathbf{m}',\mathbf{l}'_{,\varrho}}] = \delta_{\mathbf{m}\mathbf{m}'}\delta_{\mathbf{l}\mathbf{l}'}, \ [\mathbf{b}_{\mathrm{J}\mathbf{m},\mathbf{l}_{,\varrho}}, \mathbf{b}_{\mathrm{J}'\mathbf{m}',\mathbf{l}'_{,\varrho}}] = \delta_{\mathrm{J}\mathrm{J}'}\delta_{\mathbf{m}\mathbf{m}'}\delta_{\mathbf{l}\mathbf{l}'},$$
$$[\mathbf{b}_{\mathrm{J}\mathbf{m},\mathbf{l}_{,\varrho}}, \mathbf{b}_{\mathrm{J}\mathbf{m}',\mathbf{l}'_{,\varrho}}] = \delta_{\mathrm{J}\mathrm{J}'}\delta_{\mathbf{m}\mathbf{m}'}\delta_{\mathbf{l}\mathbf{l}'} \tag{114}$$

with J = I, II (and analogous equations involving the operators $\bar{\mathbf{a}}_{\mathbf{m},\mathbf{l}_{,\varrho}}$, $\bar{\mathbf{b}}_{\mathrm{I}\mathbf{m},\mathbf{l}_{,\varrho}}$ and $\bar{\mathbf{b}}_{\mathrm{II}\mathbf{m},\mathbf{l}_{,\varrho}}$). Of course,

$$_M\langle 0|\mathbf{a}_{\mathbf{m},\mathbf{l}_{,\varrho}}\mathbf{a}^\dagger_{\mathbf{m}',\mathbf{l}'_{,\varrho}}|0\rangle_M = 1, \ _\mathrm{I}\langle 0|\mathbf{b}_{\mathbf{m},\mathbf{l}_{,\varrho}}\mathbf{b}^\dagger_{\mathbf{m}',\mathbf{l}'_{,\varrho}}|0\rangle_\mathrm{I} = 1, \ _\mathrm{II}\langle 0|\mathbf{b}_{\mathbf{m},\mathbf{l}_{,\varrho}}\mathbf{b}^\dagger_{\mathbf{m}',\mathbf{l}'_{,\varrho}}|0\rangle_\mathrm{II} = 1 \tag{115}$$

and analogous equations involving the operators $\bar{\mathbf{a}}_{\mathbf{m},\mathbf{l}_{,\varrho}}$, $\bar{\mathbf{b}}_{\mathrm{I}\mathbf{m},\mathbf{l}_{,\varrho}}$ and $\bar{\mathbf{b}}_{\mathrm{II}\mathbf{m},\mathbf{l}_{,\varrho}}$.

(u11) The Fulling-Rindler vacuum $|0\rangle_F := |0\rangle_\mathrm{II} \otimes |0\rangle_\mathrm{I} \in \mathscr{F}(\mathscr{H}')$ is then defined by

$$\mathbf{1}_\mathrm{II} \otimes \mathbf{b}_{\mathrm{I}\mathbf{m},\mathbf{l}_{,\varrho}}|0\rangle_F = \mathbf{1}_\mathrm{II} \otimes \bar{\mathbf{b}}_{\mathrm{I}\mathbf{m},\mathbf{l}_{,\varrho}}|0\rangle_F = 0,$$
$$\mathbf{b}_{\mathrm{II}\mathbf{m},\mathbf{l}_{,\varrho}} \otimes \mathbf{1}_\mathrm{I}|0\rangle_F = \bar{\mathbf{b}}_{\mathrm{II}\mathbf{m},\mathbf{l}_{,\varrho}} \otimes \mathbf{1}_\mathrm{I}|0\rangle_F = 0. \tag{116}$$

(u12) Let $\hat{\phi}_{M,\mathrm{I+II}}$ be the representation in $\mathscr{F}(\mathscr{H}_\mathrm{II}) \otimes \mathscr{F}(\mathscr{H}_\mathrm{I})$ of the restriction of the field $\hat{\phi}_M$ given by Eq. (99a) to regions I + II. It is a well-known fact [21] that the Minkowski quantization of the Klein-Gordon field and the Unruh quantization producing $\hat{\phi}'$ are *not* unitary equivalent.[23]

Anyhow, it is supposed that we can identify

$$\mathscr{F}(\mathscr{H})|_{\mathscr{H}'} = \mathscr{F}(\mathscr{H}') = \mathscr{F}(\mathscr{H}_\mathrm{I}) \otimes \mathscr{F}(\mathscr{H}_\mathrm{II}) \tag{117}$$

and writing

$$\hat{\phi}_{M,\mathrm{I+II}} = \mathbf{1}_\mathrm{II} \otimes \hat{\phi}_{M,\mathrm{I}} + \hat{\phi}_{M,\mathrm{II}} \otimes \mathbf{1}_\mathrm{I}$$

[23] See Appendix 2 to know how this result is obtained in the algebraic approach to quantum theory.

we thus put

$$\hat{\phi}_{M,\mathrm{I+II}} = \hat{\phi}'. \tag{118}$$

(u13) Under these conditions the relation between those representations is supposed to be given by the well-known Bogolubov transformations which express the operators $\mathbf{b}, \mathbf{b}^\dagger$ as functions of the operators $\mathbf{a}, \mathbf{a}^\dagger$. We have $(\mathrm{J} = \mathrm{I}, \mathrm{II})$

$$\begin{aligned}
\mathbf{b}_{\mathbf{Jm,l}_{,\varrho}} &= \sum\nolimits_{\mathbf{l,m}\in\mathbb{Z}^3} \mathbf{a}_{\mathbf{m,l}_{,\varrho}} \Xi_{\mathbf{Jm,l,m',l'},\varrho} + \bar{\mathbf{a}}^\dagger_{\mathbf{m,l}_{,\varrho}} \Upsilon_{\mathbf{Jm,l,m',l'},\varrho}, \\
\bar{\mathbf{b}}_{\mathbf{Jm,l}_{,\varrho}} &= \sum\nolimits_{\mathbf{l,m}\in\mathbb{Z}^3} \mathbf{a}^\dagger{}_{\mathbf{m,l}_{,\varrho}} \Upsilon_{\mathbf{Jm,l,m',l'},\varrho} + \bar{\mathbf{a}}_{\mathbf{m,l}_{,\varrho}} \Xi_{\mathbf{Jm,l,m',l'},\varrho}.
\end{aligned} \tag{119}$$

The explicit calculation of the operators $\mathbf{b}_{\mathbf{Jm,l}_{,\varrho}}$ and $\bar{\mathbf{b}}_{\mathbf{Jm,l}_{,\varrho}}$ is done by first evaluating $\Xi_{\mathbf{Jm,l,m',l'},\varrho}$ and $\Upsilon_{\mathbf{Jm,l,m',l'},\varrho}$. The well-known result is [54]

$$\begin{aligned}
&\Xi_{\mathbf{Jm,l,m',l'},\varrho} \\
&= \int_0^\infty d\nu \int_{-\infty}^{\infty} dp_1 \int\int\int\int dk_1 dk_2 dp_2 dp_3 [f^*_{m_1,\ell_1,\varrho}(\nu) f_{m'_1,\ell'_1,\varrho}(p_1) \\
&\quad \times f^*_{m_2,\ell_2,\varrho}(k_1) f^*_{m_3,\ell_3,\varrho}(k_2) f_{m_2,\ell_2,\varrho}(p_2) f_{m_3,\ell_3,\varrho}(p_3) \Xi_{\mathrm{J}\nu,\mathbf{pk}}
\end{aligned} \tag{120}$$

(with analogous expression for $\Upsilon_{\mathbf{Jm,l,m',l'},\varrho}$ where $\Xi_{\mathrm{J}\nu,\mathbf{pk}}$ is substituted by $\Upsilon_{\mathrm{I}\nu,\mathbf{pk}}$) with

$$\begin{aligned}
\Xi_{\mathrm{I}\nu,\mathbf{pk}} &= \frac{1}{2\pi}\delta(p_1 - k_1)\delta(k_2 - p_2) e^{\frac{\pi\nu}{2}} \left|\Gamma(i\nu)\right| \left(\frac{\nu}{\omega_{\mathbf{k}}}\right)^{\frac{1}{2}} \left(\frac{\omega_{\mathbf{k}} + p_3}{\omega_{\mathbf{k}} - p_3}\right)^{\frac{i\nu}{2}}, \\
\Upsilon_{\mathrm{I}\nu,\mathbf{pk}} &= \frac{1}{2\pi}\delta(p_1 - k_1)\delta(k_1 - p_1) e^{-\frac{\pi\nu}{2}} \left|\Gamma(i\nu)\right| \left(\frac{\nu}{\omega_{\mathbf{k}}}\right)^{\frac{1}{2}} \left(\frac{\omega_{\mathbf{k}} + p_3}{\omega_{\mathbf{k}} - p_3}\right)^{\frac{i\nu}{2}}
\end{aligned} \tag{121}$$

Next $\mathbf{b}_{\mathbf{Jm,l}_{,\varrho}}$ and $\bar{\mathbf{b}}_{\mathbf{Jm,l}_{,\varrho}}$ are approximated for the case where ϱ is very small and such that $\varrho m_3 \approx 1$ by the corresponding $\mathbf{b}_{\mathrm{J}\nu}(\mathbf{q})$. We have that

$$\nu \mapsto \nu_{m_3} : m_3\varrho, \quad \omega_{\mathbf{k}} \mapsto \omega_{\mathbf{m}'} := \sqrt{\varrho^2 \sum_i (m'_i)^2 + \mu^2} \tag{122}$$

and thus using this approximation we write

$$\Xi_{\mathbf{Im},\mathbf{l},\mathbf{m}',\mathbf{l}',\varrho} = \frac{\varrho}{\sqrt{2\pi}}\Theta\left(m_3+\frac{1}{2}\right)\delta_{m_1,m_1'}\delta_{\ell_1',0}\delta_{m_2,m_2'}\delta_{m_3',0}\delta_{\ell_2,\ell_2'}\delta_{\ell_3,\ell_3'}$$
$$\times\frac{1}{\sqrt{\omega_{\mathbf{n}}}}\frac{1}{\sqrt{1-e^{-2\pi\nu_{m_3}}}}\left(\frac{\omega_{\mathbf{m}'}+m_3'\varrho}{\omega_{\mathbf{m}'}-m_3'\varrho}\right)^{\frac{i\nu_{m_3}}{2}},$$
$$\Upsilon_{\mathbf{Im},\mathbf{l},\mathbf{m}',\mathbf{l}',\varrho} = \frac{\varrho}{\sqrt{2\pi}}\Theta\left(m_3+\frac{1}{2}\right)\delta_{m_1,m_1'}\delta_{\ell_1',0}\delta_{m_2,-m_2'}\delta_{m_3',0}\delta_{\ell_2,-\ell_2'}\delta_{\ell_3,-\ell_3'}$$
$$\times\frac{1}{\sqrt{\omega_{\mathbf{n}}}}\frac{1}{\sqrt{1-e^{-2\pi\nu_{m_3}}}}\left(\frac{\omega_{\mathbf{m}'}+m_3'\varrho}{\omega_{\mathbf{m}'}-m_3'\varrho}\right)^{\frac{i\nu_{m_3}}{2}}. \tag{123}$$

where the errors $\triangle\Xi_{\mathbf{Im},\mathbf{l},\mathbf{m}',\mathbf{l}',\varrho}$ and $\triangle\Upsilon_{\mathbf{Im},\mathbf{l},\mathbf{m}',\mathbf{l}',\varrho}$ are estimated to be of order ϱ.

Denoting by $|0,\mathrm{II},\mathrm{I}\rangle_M$ the restriction of the Minkowski vacuum state $|0\rangle_M$ to the region II + I we have putting $\nu_{m_3} = \nu_j/a$ that, e.g., the expectation value of particles of type $\mathbf{b}^\dagger_{\mathbf{Im},\mathbf{l}_{,\varrho}}$ in the state $|0,\mathrm{II},\mathrm{I}\rangle_M$ is:

$$_M\langle 0,\mathrm{II},\mathrm{I}|\mathbf{1}_{\mathrm{II}}\otimes\mathbf{b}^\dagger_{\mathbf{Im},\mathbf{l}_{,\varrho}}\mathbf{b}_{\mathbf{Im},\mathbf{l}_{,\varrho}}|0,\mathrm{II},\mathrm{I}\rangle_M$$
$$=\frac{\varrho^2}{2\pi}\delta_{\ell_1 0}\;{}_M\langle 0,\mathrm{II},\mathrm{I}||0,\mathrm{II},\mathrm{I}\rangle_M\frac{1}{e^{2\pi\nu_j/a}-1}\sum\nolimits_{j\in\mathbb{Z}}\frac{1}{\omega_j} \tag{124}$$

Equation (126) shows that even if we suppose that $_M\langle 0,\mathrm{I},\mathrm{II}||0,\mathrm{II},\mathrm{I}\rangle_M = {}_M\langle 0|0\rangle_M = 1$, the vector $\mathbf{b}_{\mathbf{Im},\mathbf{l}_{,\varrho}}|0,\mathrm{II},\mathrm{I}\rangle_M \in \mathscr{F}(\mathscr{H}_{\mathrm{I}})\otimes\mathscr{F}(\mathscr{H}_{\mathrm{II}})$ has not a finite norm, thus showing that the procedure we have been using until now is not a mathematical legitimate one.

(u14) Nevertheless, taking the above approximation for the Bogolubov transformation as a good one for at least a region where $\varrho m_3 \approx 1$, the state $|0,\mathrm{II},\mathrm{I}\rangle_M$ is written

$$|0,\mathrm{II},\mathrm{I}\rangle_M = \Omega^{-1}\exp\left\{\sum\nolimits_{j,m_1}e^{-2\pi\nu_{m_1}}\left(\left(\mathbf{b}^+_{\mathbf{IIm},\mathbf{l}_{,\varrho}}\right)^{n_j}\otimes\mathbf{1}_{\mathrm{I}}\right.\right.$$
$$\left.\left.+\mathbf{1}_{\mathrm{II}}\otimes\left(\mathbf{b}^+_{\mathbf{Im},\mathbf{l}_{,\varrho}}\right)^{n_j}\right)\right\}|0\rangle_{\mathrm{II}}\otimes|0\rangle_{\mathrm{I}}$$
$$=\Omega^{-1}\prod_j\sum\nolimits_n e^{-\pi n\nu_j/a}|\check{n}_j\rangle_{\mathrm{II}}\otimes|\check{n}_j\rangle_{\mathrm{I}}, \tag{125}$$

where Ω is a normalization constant and $|\check{n}_j\rangle_{\mathrm{J}} = |\check{n}_j\rangle_{\mathrm{J}} + |0\rangle_{\mathrm{J}}$, J = I, II.

(u15) Using the fact that regions I and II are causally disconnected, i.e., observers following integral lines of ***R*** can only detect *right* Rindler particles it is supposed that these observers can only describe (according to standard quantum mechanics prescription) the state of the Minkowski quantum vacuum by a mixed state [63], i.e., a density matrix obtained by tracing over the states of the region II the pure state density matrix $\hat{\rho} = |0, \mathrm{I}, \mathrm{II}\rangle_M \langle 0, \mathrm{I}, \mathrm{II}|_M$. The result is

$$\hat{\rho}_{\mathrm{I}} = \mathrm{tr}_{\mathrm{I}}(\hat{\rho}) = \Omega^{-1} \prod_j \sum\nolimits_n e^{-2\pi n \nu_j / a} |n_j\rangle_{\mathrm{I}} \otimes {}_{\mathrm{I}}\langle n_j|, \tag{126}$$

which looks like a thermal spectrum with temperature parameter $a/2\pi$.

Remark 9 Take notice that for an observer following the worldline σ with $\mathfrak{z}$ =constant in region I the local temperature of the thermal radiation is [59]

$$T(\mathfrak{z}) = \frac{1}{\sqrt{g_{00}(\mathfrak{z})}} \frac{a}{2\pi} \tag{127}$$

and thus $T(\mathfrak{z})\sqrt{g_{00}(\mathfrak{z})}$ is a constant. This is extremely important for otherwise thermodynamical equilibrium (according to Tolman's version [55]) would not be possible in the $\boldsymbol{R}$ frame.

(u16) Given Eq. (126) since $n\nu_j$ is the value of the *pseudo energy* in the $|n_j\rangle_{\mathrm{I}}$ state and since $\hat{\rho}_{\mathrm{I}}$ looks like a thermal density matrix $\rho_T = e^{-H/T}$ it is claimed that:

The Minkowski vacuum in region I is seem by observers living there as a thermal bath at temperature $a/2\pi$ of the so-called Rindler particles, which can excite well designed detectors [24, 25, 49, 53, 59, 60, 63]. *Even more, it is claimed, (e.g., in* [53]*) that the Rindler particles are irradiated from the boundary of the region I (which is supposed to be "analogous" to the horizon of a blackhole which is supposed to radiate due to the so-called Hawking effect).*

(u17) The fact is that a rigorous mathematical analysis of the problem, based on the algebraic approach to field theory[24] (which for completeness, we recall in Appendix 2), it is possible to show that the hypothesis given by Eq. (117) and thus Eq. (124) are *not* correct. Indeed, there we recall that strictly speaking the density matrix $\hat{\rho}$ and thus $\hat{\rho}_{\mathrm{I}}$ are meaningless. Also, many people have serious doubts if Fulling-Rindler vacuum $|0\rangle_F := |0\rangle_{\mathrm{II}} \otimes |0\rangle_{\mathrm{I}}$.can be physically realizable. These arguments are, in our opinion stronger ones and the reader is invited to at least give a look in Appendix 2 (where the main references on original papers dealing with the issue of the algebraic approach to the Unruh effect may be found) in order to have an idea of the truth of what has just been stated.

(u18) As it is the case of the problem of the electromagnetic field generated by a charge in hyperbolic motion, there are several researchers that are convinced that the Unruh effect does *not* exist.

[24] First applied to the Unruh effect problem in [27].

Besides the inconsistencies recalled in Appendix 2 several others are discussed, e.g., in [1, 9, 13, 19]. The most important one in our opinion has been realized in [19] where it is shown that both in the conventional approach and in the algebraic approach to quantum field theory it is impossible to perform the quantization of Unruh modes in Minkowski spacetime. Authors claim (and we agree with them) that Unruh quantization in a Rindler frame implies setting a *boundary condition* for the quantum field operator which changes the topological properties and symmetry group of the spacetime (where the Rindler reference frame has support) and leads to a field theory in the two disconnected regions I and II. They concluded that the Rindler effect does not exist.

(**u19**) Despite this fact, in a recent publication [11] authors that pertain to the majority view (i.e., those that believe in the existence of the thermal radiation) state:

> Then, instead of waiting for experimentalists to perform the experiment, we use standard classical electrodynamics to anticipate its output and show that it reveals the presence of a thermal bath with temperature T_U in the accelerated frame. Unless one is willing to question the validity of classical electrodynamics, this must be seen as a virtual observation of the Unruh effect.

Well, authors of [11] also believe that a charge in hyperbolic motion radiates, and that the correct solution to the problem is the one given by the Liénard-Wiechert potential. But what will be of the statement that we cannot doubt classical electrodynamics if turns out that the Turakulov solution is the correct one (i.e., experimentally confirmed)?

Another important question is the following one: does a detector following an integral line of $\boldsymbol{R}$ get excited?

(**u20**) Several thoughtful analyses of the problem done from the point of view of an inertial reference frame show that the detector gets excited. This is discussed in [14] and a very simple model of a detector showing that the statement is correct may be found in [37]. But, of course, it is necessary to leave clear that this excitation energy can only come from the source that maintains the detector accelerated and it is not an excitation due to fluctuations of the zero point of the field as claimed, e.g. in [1].

7 Conclusions

There are some problems in Relativity Theory that are source of controversies since a long time. One of them has to do with the question if a charge in uniformly accelerated motion radiates. This problem is important, in particular, in its connection with one of the forms of the Equivalence Principle. In this paper we recalled that there are two different solutions for the electromagnetic field generated by a charge in hyperbolic motion, the Liénard-Wiechert (LW) one (obtained by the retarded Green function) and the less known one discovered by Turakulov in 1994 (and which we have verified to be correct, in particular using the software Mathematica). According to the LW solution the charge radiates and claim that an observer comoving with the charge does not detect any radiation is shown to be wrong. This is done by analyzing the different concepts of energy used by people that claims that no radiation is detected. Turakulov claims in [56] that his solution implies that there is no radiation. However, we have proved that he is also wrong, the reason being essentially the same as in the case of the Liénard-Wiechert solution. On the other hand, we recalled that a charge at rest in Schwarzschild spacetime does not radiate. Thus, if the LW or the Turakulov solution is the correct one, then experiment with charges may show that the Equivalence Principle is false.

Another problem which we investigate is the so-called Bell's "paradox." We discussed it in detail since it is, as yet, a source of misunderstandings.

Finally, we briefly recall how the so-called Unruh effect is obtained in almost all texts using some well ideas of quantum field theory. We comment that this standard approach seems to imply that an observer in hyperbolic motion is immersed in a thermal bath with temperature proportional to its proper acceleration. Acceptance that this is indeed the case is almost the majority view among physicists. However, the fact is that the standard approach does not resist a rigorous mathematical analysis, in particular when one uses the algebraic approach to quantum field theory. Thus as it is the case with the problem of determining the electromagnetic field of a charge in hyperbolic motion there are dissidents of the majority view. Having studied the arguments of several papers we presently agree with [9, 19] that there is no Unruh effect. However, it is not hard to show that a detector in hyperbolic motion on the Minkowski vacuum gets excited, but the energy producing such excitation, contrary to some claims (as, e.g., in [1]) does not come from the fluctuations of the zero point field, but comes from the source pushing the charge.

Appendix 1: Some Notations and Definitions

(**a1**) Let M be a four-dimensional, real, connected, paracompact, and non-compact manifold. We recall that a *Lorentzian manifold* as a pair $(M, \mathbf{g})$, where $\mathbf{g} \in \sec T_2^0 M$ is a Lorentzian metric of signature $(1, 3)$, i.e., $\forall \mathfrak{e} \in M$, $T_x M \simeq T_{\mathfrak{e}}^* M \simeq \mathbb{R}^4$. Moreover, $\forall x \in M$, $(T_x M, \mathbf{g}_x) \simeq \mathbb{R}^{1,3}$, where $\mathbb{R}^{1,3}$ is the Minkowski *vector* space We define a *Lorentzian spacetime* M as pentuple $(M, \mathbf{g}, \boldsymbol{D}, \tau_{\mathbf{g}}, \uparrow)$, where $(M, \mathbf{g}, \tau_{\mathbf{g}}, \uparrow))$ is an *oriented* Lorentzian manifold (oriented by $\tau_{\boldsymbol{g}}$) and *time* oriented[25] by $\uparrow$, and D is the Levi-Civita connection of $\boldsymbol{g}$. Let $U \subseteq M$ be an open set covered, say, by coordinates (y^0, y^1, y^2, y^3). Let $U \subseteq M$ be an open set covered by coordinates $\{x^\mu\}$. Let $\{e_\mu = \partial_\mu\}$ be a coordinate basis of $T\mathscr{U}$ and $\{\boldsymbol{\vartheta}^\mu = dx^\mu\}$ the dual basis on $T^*\mathscr{U}$, i.e., $\boldsymbol{\vartheta}^\mu(\partial_\nu) = \delta_\nu^\mu$. If $\mathbf{g} = g_{\mu\nu}\boldsymbol{\vartheta}^\mu \otimes \boldsymbol{\vartheta}^\nu$ is the metric on $T\mathscr{U}$ we denote

[25] Please, consult, e.g., [45].

by $\mathtt{g} = g^{\mu\nu}\boldsymbol{\partial}_\mu \otimes \boldsymbol{\partial}_\nu$ the metric of $T^*\mathscr{U}$, such that $g^{\mu\rho}g_{\rho\nu} = \delta^\mu_\nu$. We introduce also $\{\boldsymbol{\partial}^\mu\}$ and $\{\boldsymbol{\vartheta}_\mu\}$, respectively, as the reciprocal bases of $\{e_\mu\}$ and $\{\boldsymbol{\vartheta}_\mu\}$, i.e., we have

$$\mathbf{g}(\boldsymbol{\partial}_\nu, \boldsymbol{\partial}^\mu) = \delta^\mu_\nu, \quad g(\boldsymbol{\vartheta}^\mu, \boldsymbol{\vartheta}_\nu) = \delta^\mu_\nu. \tag{128}$$

(a2) Call $(M \simeq \mathbb{R}^4, \boldsymbol{g}, D, \tau_{\boldsymbol{g}}, \uparrow)$ the *Minkowski spacetime structure*. When $M \simeq \mathbb{R}^4$ there is (infinitely) global charts. Call (x^0, x^1, x^2, x^3) the coordinates of one of those charts. These coordinates are said to be in *Einstein-Lorentz-Poincaré* (ELP) gauge. In these coordinates

$$\boldsymbol{g} = \eta_{\mu\nu} dx^\mu \otimes dx^\nu \text{and } \mathtt{g} = \eta^{\mu\nu} \frac{\partial}{\partial x^\mu} \otimes \frac{\partial}{\partial x^\nu} \tag{129}$$

where the matrix with entries $\eta_{\mu\nu}$ and also the matrix with entries $\eta^{\mu\nu}$ are diagonal matrices $\mathrm{diag}(1, -1, -1, -1)$.

(a3) In a general Lorentzian structure if $\mathbf{Q} \in \sec TU \subset \sec TM$ is a time-like vector field such that $\boldsymbol{g}(\mathbf{Q}, \mathbf{Q}) = 1$, then there exist, in a coordinate neighborhood U, three space-like vector fields $\boldsymbol{e}_\mathbf{i}$ which together with $\mathbf{Q}$ form an orthogonal moving frame for $x \in U$ [12, 45].

(a4) A *moving frame* at $x \in M$ is a basis for the tangent space $T_x M$. An orthonormal (moving) frame at $x \in M$ is a basis of orthonormal vectors for $T_x M$.

(a5) An *observer* in a general Lorentzian spacetime is a future pointing time-like curve $\sigma : \mathbb{R} \supset I \to M$ such that $\boldsymbol{g}(\sigma_*, \sigma_*) = 1$. The timelike curve σ is said to be the worldline of the observer.

(a6) An *instantaneous observer* is an element of TM, i.e., a pair $(x, \mathscr{Q})$, where $x \in M$, and $\mathscr{Q} \in T_x M$ is a future pointing unit timelike vector. $\mathrm{Span}\mathscr{Q} \subset T_x M$ is the *local time axis* of the observer and $\mathscr{Q}^\perp$ is the *observer rest space.*

(a7) Of course, $T_x M = \mathrm{Span}\mathscr{Q} \oplus \mathscr{Q}^\perp$, and we denote in what follows $\mathrm{Span}\mathscr{Q} = T$ and $\mathscr{Q}^\perp = H$, which is called the *rest space* of the instantaneous observer. If $\sigma : \mathbb{R} \supset I \to M$ is an observer, then $(\sigma u, \sigma_* u)$ is said to be the local observer at u and write $T_{\sigma u} M = T_u \oplus H_u \ , \ u \in I$.

(a8) The *orthogonal projections* are the mappings

$$\mathbf{p}_u = T_{\sigma u} M \to H_u \ , \ \mathbf{q}_u : T_{\sigma u} M \to T_u. \tag{130}$$

Then if $\mathbf{Y}$ is a vector field over σ then $\mathbf{pY}$ and $\mathbf{qY}$ are vector fields over σ given by

$$(\mathbf{pY})_u = \mathbf{p}_u(\mathbf{Y}_u), \ (\mathbf{qY})_u = \mathbf{q}_u(\mathbf{Y}_u). \tag{131}$$

(a9) Let $(x, \mathscr{Q})$ be an instantaneous observer and $\mathbf{p}_x : T_x M \to H$ the orthogonal projection. The *projection tensor* is the symmetric bilinear mapping $\mathbf{h} : \sec(TM \times TM) \to \mathbb{R}$ such that for any $\mathbf{U}, \mathbf{W} \in T_x M$ we have:

$$\mathbf{h}_x(\mathbf{U}, \mathbf{W}) = \boldsymbol{g}_x(\mathbf{pU}, \mathbf{pW}) \tag{132}$$

Let $\{x^\mu\}$ be coordinates of a chart covering $U \subset M$, $x \in U$ and $\alpha_{\mathcal{Q}} = \boldsymbol{g}_x(\mathcal{Q}, \ )$. We have the properties:

(a)	$\mathbf{h}_X = \boldsymbol{g}_X - \alpha_{\mathcal{Q}} \otimes \alpha_{\mathcal{Q}}$	(133)
(b)	$\mathbf{h}\vert_{\mathcal{Q}^\perp} = \boldsymbol{g}_x\vert_{\mathcal{Q}^\perp}$	
(c)	$\mathbf{h}(\mathcal{Q}, \ ) = 0$	
(d)	$\mathbf{h}(\mathbf{U}, \ ) = \boldsymbol{g}(\mathbf{U}, \ ) \Leftrightarrow \ \boldsymbol{g}(\mathbf{U}, \mathcal{Q}) = 0$	
(e)	$\mathbf{p} = h^\mu_\nu \left.\frac{\partial}{\partial x^\mu}\right\vert_x \otimes dx^\nu\vert_x$	
(f)	$\mathrm{trace}(h^\mu_\nu \left.\frac{\partial}{\partial x^\mu}\right\vert_x \otimes dx^\nu\vert_x) = -3$	

The result quote in (**a3**) together with the above definitions suggests to introduce the following notions:

(**a10**) A *reference frame* for $U \subseteq M$ in a spacetime structure $(M, \boldsymbol{g}, D, \tau_{\boldsymbol{g}}, \uparrow)$ is a time-like vector field which is a section of TU such that each one of its integral lines is an observer.

(**a11**) Let $\mathbf{Q} \in \sec TM$ be a reference frame. A chart in $U \subseteq M$ of an oriented atlas of M with coordinate functions $(\boldsymbol{y}^\mu)$ and coordinates $(\boldsymbol{y}^0(\mathfrak{e}) = y^0, \boldsymbol{y}^1(\mathfrak{e}) = y^1, \boldsymbol{y}^2(\mathfrak{e}) = y^2, \boldsymbol{y}^3(\mathfrak{e}) = y^3)$ such that $\partial/\partial y^0 \in \sec TU$ is a timelike vector field and the $\partial/\partial y^i \in \sec TU$ $(i = 1, 2, 3)$ are spacelike vector fields is said to be a possible naturally adapted coordinate chart to the frame $\mathbf{Q}$ *(*denoted $(nacs - \mathbf{Q})$ in what follows*)* if the space-like components of $\mathbf{Q}$ are null in the natural coordinate basis $\{\partial/\partial x^\mu\}$ of TU associated with the chart. We also say that (y^0, y^1, y^2, y^3) are naturally adapted coordinates to the frame $\mathbf{Q}$.

Remark 10 It is crucial, in order to avoid misunderstandings, to have in mind that most of the reference frames used in the formulation of physical theories are theoretical objects, i.e., a reference frame does not need to have material support in the region where it has mathematical support.

(**a12**) Reference frames in Lorentzian spacetimes can be *classified* according to the decomposition of $D\mathbf{Q}$ and according to their *synchronizability*. Details may be found in [45]. We analyze in detail the nature of the right Rindler reference frame in Sect. 2. Here we only recall that $\mathbf{Q}$ is locally synchronizable if its rotation tensor ω (coming form the decomposition of $Q = \boldsymbol{g}(\mathbf{Q})$), is such that $\omega \wedge d\omega = 0$, and we can show $\omega = 0 \Longleftrightarrow Q \wedge dQ = 0$. Also, $\mathbf{Q}$ is synchronizable if besides being irrotational also there exist a function H on U and a timelike coordinate, say u (part of a naturally adapted coordinate system to $\mathbf{Q}$) such that $Q = Hdu$. Finally, $\mathbf{Q}$ is said to be propertime synchronizable if $Q = du$.

(**a13**) We also used in the main text the following conventions:

$$\boldsymbol{g}(\boldsymbol{A},\boldsymbol{B}) = \boldsymbol{A} \cdot \boldsymbol{B}, \quad \mathrm{g}(C, D) = C \cdot D,$$
$$\boldsymbol{A},\boldsymbol{B} \in \sec TM, \quad C, D \in \sec \bigwedge{}^1 T^*M. \tag{134}$$

and the scalar product of Euclidean vector fields is denoted by $\bullet$.

(**a14**) Moreover, d and δ denote the differential and Hodge codifferential operators acting on sections of $\bigwedge T^*M$ and $\lrcorner$ denotes the left contraction operator of form fields [45].

Appendix 2: $C^\star$ Algebras and the Unruh "Effect"

The reason for including this Appendix in this paper is for the interested reader to have an idea of how much he can *trust* the standard approach recalled in the main text which results in the claim that Rindler observers live in a thermal bath. The algebraic approach to quantum field theory is based on $C^\star$-algebras[26] which are now briefly recalled.

(**b1**) Let then be $\mathscr{A}$ a C^*-algebra over $\mathbb{C}$ whose some of its elements may be associated to the observables[27] (associated to the quantum field $\hat{\phi}$). We recall that a *representation* of a C^*-algebra is a linear mapping

$$f : \mathscr{A} \to \mathscr{B}(\mathfrak{H}), \;\; A \mapsto f(A), \;\; f(A^\star) = f(A)^\dagger. \tag{135}$$

where $\mathscr{B}(\mathfrak{H})$ is an algebra of bounded linear operators on a Hilbert space $\mathfrak{H}$. The observables are associated with elements $A = A^*$, where * denotes the *involution* operation in $\mathscr{A}$, i.e., $\mathscr{A}\mathscr{A}^\star = 1$ and † denotes the Hermitian conjugate in $\mathscr{B}(\mathfrak{H})$

(**b2**) A representation $(f, \mathfrak{H})$ of $\mathscr{A}$ is said *faithful* if $f(A) = 0$ if $A = 0$ and $(f, \mathscr{H})$ is irreducible if the only closed subspaces of $\mathfrak{H}$ invariant under f are $\{0\}$ and $\mathfrak{H}$.

(**b3**) Let $\mathfrak{L} \subset \mathfrak{H}$ be a non-zero closed subspace of invariant under f. Let $\hat{\mathbf{P}}_{\mathscr{L}}$ be the *orthogonal projection* operator on $\mathscr{L}$. A *subrepresentation* of $f_{\mathscr{L}}$ is the mapping

$$f_{\mathscr{L}} : \mathscr{A} \to \mathscr{B}(\mathfrak{H}), \quad A \mapsto f(A)\hat{\mathbf{P}}_{\mathscr{L}}. \tag{136}$$

(**b3**) Two representations, say $(f_1, \mathfrak{H}_1)$ and $(f_2, \mathfrak{H}_2)$ of $\mathscr{A}$ are said to be unitarily equivalent if there exists an isomorphism $\mathbf{U} : \mathfrak{H}_1 \to \mathfrak{H}_2$, such that

$$\mathbf{U} f_1(\mathscr{A})\mathbf{U}^{-1} = f_2(\mathscr{A}). \tag{137}$$

(**b4**) A *state* on $\mathscr{A}$ is a mapping

[26]For a succint presentation of C^*-algebras, enough for the understanding of the following see, e.g., [16]. There the reader will find the main references on the algebraic (and axiomatic) approach to quantum field theory. Also, the reader who wants to know all the details concerning the algebraic approach to the Unruh effect must study the texts quoted below which has been heavily used in the writing of this Appendix 2.

[27]I.e., the self-adjoints elements of $\mathscr{A}$.

$$\omega : \mathscr{A} \to \mathbb{R},$$
$$\omega(1) = 1, \quad \omega(A^{\star}A) \geq 0, \forall A \in \mathscr{A}. \tag{138}$$

(**b5**) A *pure state* ω on $\mathscr{A}$ is one that cannot be written as a non-trivial convex linear combination other states. On the other hand, a *state* ω on $\mathscr{A}$ is said to be *mixed* if it can be written as a non-trivial convex linear combination other states.

(**b6**) It is important to recall that a result (theorem) due to Gel'fand, Naimark, and Segal (GNS) [22, 48] establishes that for any ω on $\mathscr{A}$ there always exists a representation $(f_\omega, \mathfrak{H}_\omega)$ of $\mathscr{A}$ and $\Phi_\omega \in \mathfrak{H}_\omega$ (usually called a *cyclic vector*) such that $f_\omega(\mathscr{A})\Phi_\omega$ is dense in $\mathfrak{H}_\omega$ and

$$\omega(A) = \langle \Phi_\omega | f_\omega(A) | \Phi_\omega \rangle. \tag{139}$$

Moreover the GNS result warrants that up to unitary equivalence, $(f_\omega, \mathfrak{H}_\omega)$ is the unique *cyclic* representation of $\mathscr{A}$.

(**b7**) The *folium* $\mathfrak{F}(\omega)$ of ω on $\mathscr{A}$ is the set of all abstract states that can be expressed as density matrices on the Hilbert space of the GNS representation determined by $\mathfrak{H}_\omega$.

(**b8**) Given states ω_1, ω_2 on $\mathscr{A}$ they are said *quasi-equivalent* if and only if $\mathfrak{F}(\omega_1) = \mathfrak{F}(\omega_2)$. The states ω_1, ω_2 on $\mathscr{A}$ are said to be *disjoint* if $\mathfrak{F}(\omega_1) \cap \mathfrak{F}(\omega_2) = \varnothing$.

(**b9**) It is possible to show that:

(i) *Any irreducible representation has no proper subrepresentations and in this case if ω_1 and ω_2 are pure states, quasi-equivalence reduces to unitary equivalence and disjointness reduces to non-unitary equivalences;*
(ii) *When ω_1 and ω_2 are mixed states they in general are not quasi-equivalent or disjoint.*

This happens when, e.g., ω_1 has disjoint representations and one of then is unitarily equivalent to ω_2.

(**b10**) For our considerations it is important to recall the following result [8]:

The states ω_1 and ω_2 are disjoint if and only if the GNS representation of $f_{\omega_1+\omega_2}$ determined by $\omega_1 + \omega_2$ satisfies

$$(f_{\omega_1+\omega_2}, \mathfrak{H}_{\omega_1+\omega_2}) = (f_{\omega_1} \oplus f_{\omega_2}, \mathfrak{H}_{\omega_1} \oplus \mathfrak{H}_{\omega_2}), \tag{140}$$

i.e., the direct sum of the representations f_{ω_1} and f_{ω_2}. Elements of $\mathfrak{H}_{\omega_1+\omega_2}$ are *denoted by*

$$|\Phi_{\omega_1+\omega_2}\rangle = |\Phi_{\omega_1}\rangle \oplus |\Phi_{\omega_2}\rangle \tag{141}$$

(**b11**) To continue the presentation it is necessary to use a particular C^*-algebra, namely the Weyl algebra[28] $\mathscr{A}_W(M)$ which encodes (see, e.g., [10]), in particular an exponential version of the canonical commutation relations for the Klein-Gordon field used in the analysis of the Unruh effect in this paper. Use of the Weyl algebras is opportune because in a version appearing in [29] it leads to a net of algebras $\{\mathscr{A}(U)\}$ where if $U \subset M$ is an open set of compact closure which qualifies as a globally hyperbolic spacetime structure $(U, \boldsymbol{g}|_U, D|_U, \tau_{\boldsymbol{g}}|_U, \uparrow)$ then if $U \subset U' \subset M$ it is $\mathscr{A}(U) \subset \mathscr{A}(U')$.

(**b12**) It is also necessary to know the following result [6–8]:

Let $\boldsymbol{Z} \in \sec TU$ *where* U *qualifies as a globally hyperbolic spacetime which is foliated with Cauchy surfaces*[29] $\Sigma(u)$. *Let* $n \in \sec TM$ *be the unit normal to* Σ, *a member of the foliation. Only if for some* $\varepsilon \in R$, Z *satisfies*

$$\boldsymbol{Z} \cdot \boldsymbol{Z} \geq \varepsilon \boldsymbol{Z} \cdot \boldsymbol{n} \geq \varepsilon^2 \tag{142}$$

there exists a procedure that associates with Σ *a so-called quasi-free state* ω_Σ *on* $A(M)$.

(**b13**) *Quasi-free states* are the ones for which the n-point functions of quantum field theory are determined by the two point functions and their importance here lies in the fact that it can be shown that the GNS representation of ω_Σ has a natural Fock-Hilbert space structure $\mathscr{F}(\Sigma)$ where ω_Σ is represented by the vacuum state $|0\rangle_\Sigma \in \mathscr{F}(\Sigma)$. Thus, ω_Σ qualifies as a candidate for the vacuum state.

Remark 11 Note that if we take $\boldsymbol{Z}$ equal to $\boldsymbol{I}$ since it is irrotational (and a Killing vector field), it can be used to foliate M and for $\boldsymbol{I}$ Eq. (142) is satisfied. Then we naturally can construct ω_M on $\mathscr{A}$ representing the state $|0\rangle_M \in \mathscr{F}(\mathscr{H})$. Also, if we take $\boldsymbol{Z} = \boldsymbol{Z}_{\mathrm{I}}$ *or* $\boldsymbol{Z} = \boldsymbol{Z}_{\mathrm{II}}$ (as defined in Eqs. (90)) since these fields besides being Killing vector fields are also irrotational, they can be used to foliate regions I and II where the respective Cauchy surfaces are of course, spacelike surfaces orthogonal respectively to $\boldsymbol{Z}_{\mathrm{I}}$ and $\boldsymbol{Z}_{\mathrm{Ii}}$. In these cases, Eq. (142) is violated near the " horizon" and it is not possible to construct[30] ω_{I} on $\mathscr{A}(\mathrm{I})$ and ω_{II} on $\mathscr{A}(\mathrm{II})$.These states are the ones associate with the vacuum states $|0\rangle_{\mathrm{I}}$ and $|0\rangle_{\mathrm{II}}$ described above.

(**b14**) We have now the fundamental result:

The states $\omega_M|_{\mathscr{A}(\mathrm{I})}$ *(respectively* $\omega_M|_{\mathscr{A}(\mathrm{II})}$*) and* ω_{I} *(respectively* ω_{II}*) are disjoint.*

(**b15**) To understand what is the meaning of this statement it is necessary to recall the definition of a *von Neumann algebra* [62].(denoted W^*-algebra). It is a special type of a C^*-algebra of bounded operators on a Hilbert space that is closed in the weak operator topology and contains the identity operator.

---

[28]Also called Symplectic Clifford Algebra [15, 64].

[29]u is a parameter indexing the foliation.

[30]The states ω_{I} on $\mathscr{A}(\mathrm{I})$ and ω_{II} on $\mathscr{A}(\mathrm{II})$ are called Boulware vacuum states[5].

(**b16**) What is important for us here is that if $\mathscr{A}$ is a C^*-algebra identified with the space of bound operators $\mathfrak{B}(\mathscr{H})$ of an appropriate Hilbert space then $\mathscr{A}$ is a W^*-algebra if and only if

$$\mathscr{A} = \mathscr{A}'', \tag{143}$$

where $\mathscr{A}'$ denotes the so-called *commutant* of $\mathscr{A}$, i.e., the set of operators that commute with all elements of $\mathscr{A}$. Of course, $\mathscr{A}''$ denotes the commutant of the commutant and is called bicommutant.

(**b17**) Given a representation $(f, \mathfrak{H})$ of $\mathscr{A}$ we denote $f''(\mathscr{A})$ the so-called *double commutant* of $f(\mathscr{A})$. It is called the von Neumann algebra and denoted $W_f(\mathscr{A})$. If the commutant $f'(\mathscr{A})$ is an Abelian algebra $W_f(\mathscr{A})$ is called type **I** and it is the case given von Neumann theorem that if ω is a state on $\mathscr{A}$ then $W_f(\mathscr{A})$ can be identified with $\mathfrak{B}(\mathscr{H}_\omega)$ for a GNS representation $(f_\omega, \mathscr{H}_\omega)$.

(**b18**) A *factorial state* ω on $\mathscr{A}$ (and their GNS representation $\Phi_\omega \in \mathscr{H}_\omega$) is one for which the only multiples of the identity are elements of $W_{f\omega}(\mathscr{A}) \cap W_{f\omega}(\mathscr{A})'$.

(**b19**) A *normal state* ω on $\mathscr{A}$ (and their GNS representation $\Phi_\omega \in \mathscr{H}_\omega$) is one whose canonical extension to a state $\breve{\omega} \in W_{f\omega}(\mathscr{A})$ is countably additive.

(**b20**) Von Neumann algebras can also be of types [4] **II** and **III**. Type **III** are important for the sequel and it is one where factors are factors that do not contain any nonzero finite projections at all.

(**b21**) Given these definitions it is possible to show the following results concerning C^*-algebras:

(**b21a**) If f and f' are non-degenerate representations of $\mathscr{A}$, then they are quasi-equivalent if and only if there is a *-isomorphism

$$\begin{aligned} \boldsymbol{i} &: W_f(\mathscr{A}) \to W_{f'}(\mathscr{A}), \\ \boldsymbol{i}(f(A)) &= f'(A) \end{aligned} \tag{144}$$

(**b21b**) The representations f and f' are quasi-equivalent if and only if f has no subrepresentation disjoint from f' and vice versa.

(**b21c**) A representation of a $\mathscr{A}$ is factorial if and only if every subrepresentation of f is quasi-equivalent to f'.

From (**b21a**) it follows (see, e.g., [10]) that $f_{\omega_{\mathrm{I}}}$ (respectively $f_{\omega_{\mathrm{II}}}$) and $f_{\omega_M|_{\mathscr{A}(\mathrm{I})}}$ (respectively $f_{\omega_M|_{\mathscr{A}(\mathrm{II})}}$) are not isomorphic since $W_{f_{\omega\mathrm{I}}}(\mathscr{A})$ (respectively $W_{f_{\omega\mathrm{I}}}(\mathscr{A})$) is a von Neumann algebra of type **I** whereas $W_{f_{\omega_M|_{\mathscr{A}(\mathrm{I})}}}(\mathscr{A})$ (respectively $W_{f_{\omega_M|_{\mathscr{A}(\mathrm{II})}}}(\mathscr{A})$) is a von Neumann algebra of type **III** [2].

(**b22**) It is the case that in general not to be quasi-equivalent does not implies being disjoint, but in our particular case ω_{I} (respectively ω_{II}) is a pure state which is irreducible and as such has no non-trivial representation. Also, $\omega_M|_{\mathscr{A}(\mathrm{I})}$ (respectively $\omega_M|_{\mathscr{A}(\mathrm{II})}$) is factorial and (c) implies that it is equivalent to each one of its subrepresentation. Finally, from (**a**) it follows that $f_{\omega_{\mathrm{I}}}$ (respectively $f_{\omega_{\mathrm{II}}}$)

and $f_{\omega_M|_{\mathscr{A}(\mathrm{I})}}$ (respectively $f_{\omega_M|_{\mathscr{A}(\mathrm{II})}}$) is disjoint if and only if they are not quasi-equivalent.

Now, what does it mean that f_{ω_I} *(respectively* f_{ω_II}*) and* $f_{\omega_M|_{\mathscr{A}(\mathrm{I})}}$ *(respectively* $f_{\omega_M|_{\mathscr{A}(\mathrm{II})}}$*) is disjoint?*

(**b23**) Recall, e.g., that what ω_M has to say about region I is given by $\omega_M|_{\mathscr{A}(\mathrm{I})}$ and from what we already recalled above cannot be represented by a density matrix in the representation f_{ω_I}, in particular for any representation on $\mathscr{A}(\mathrm{I})$.This happens because it is impossible to write $\mathscr{A}(M)$ as a tensor product $\mathscr{A}' \otimes \mathscr{A}(\mathrm{I})$ for some $\mathscr{A}'$. This result is called *expressive incompleteness.*

(**b24**) Despite expressive incompleteness we have the following result by Verch [61]:

On $U \subset I \subset M$ (which is open and of compact closure) *let* $f_{\omega_M} | \mathscr{A}(U)$ be the GNS representation constructed from ω_M restrict to the image $\omega_M|\mathscr{A}(U)$ under f_{ω_M} of $\mathscr{A}(U)$ (and completing in the natural topology of $\mathscr{H}_{\omega_M}$) and analogous construct[31] $\omega_\mathrm{I}|\mathscr{A}(U)$ the image of ω_I under $f_{\omega_\mathrm{I}}|\mathscr{A}(U)$[32]. *Then,* $f_{\omega_M} | \mathscr{A}(U)$ and $f_{\omega_\mathrm{I}}|\mathscr{A}(U)$ *are quasi-equivalent.*

(**b25**) The result presented in (**b24**) is the only one that would permit legitimately to physicists to talk about ω_M and ω_I as being quasi equivalents, for indeed as already recalled f_{ω_M} and f_{ω_I} are indeed disjoint representations of the algebra of observables $\mathscr{A}$ and thus not unitarily equivalents.

(**b26**) Anyway, the above result implies that only if we do measurements on observables of the algebra $\mathscr{A}$ in regions of non-compact closure can distinguish the representations f_{ω_M} and f_{ω_I}.

(**b27**) Finally one can ask the question: is $f_{\omega_M|_{\mathscr{A}(U)}}$ and $f_{\omega_M|_{\mathscr{A}(U)}}$ where again $U \subset I \subset M$ (open and of compact closure) quasi equivalent?

The answer to this question is (for the best of our knowledge) *not* known and this is another hindrance that makes one to affirm that no convincing theoretical proof that the Unruh effect is a real effect exists.

(**b28**) In the standard "deduction" (Sect. 6.1) of the Unruh effect it is claimed that the uniformly accelerated observer detects a thermal bath. Supporters that the effect is a real one try to endorse their claim by using the notion of KMS states[33] (which as well known generalizes the notion of equilibrium state) [6–8, 28, 33]. In fact, Sewell [51] argues that the restriction of the Minkowski vacuum ω_M to region I, i.e., $\omega_M|_{\mathscr{A}(\mathrm{I})}$ $(=\omega_M|_\mathrm{I})$ can be formulated as an algebraic state on $\mathscr{A}_\mathrm{I}$ which satisfies the KMS condition at temperature $\beta^{-1} = a/2\pi$ relative to the notion of time translation defined by vector field $\mathbf{Z}_\mathrm{I} = \partial/\partial \mathrm{t}$ (which then generates the one-parameter group of automorphism $a_{u=\mathrm{t}}$). However, it is necessary to have in mind

[31]Please, do not confuse $\omega_\mathrm{I}|\mathscr{A}(U)$ with $\omega_\mathrm{I}|_{\mathscr{A}(U)}$.

[32]The states $\omega_M|\mathscr{A}(U)$ and $\omega_\mathrm{I}|\mathscr{A}(U)$ are quasi free-Hadamard states, i.e., states for which

[33]Recall that a KMS state is an algebraic state (ζ_u, β) on $\mathscr{A}$ where $\zeta_u : \mathscr{A} \to \mathscr{A}$ one parameter group of automorphisms and $0 \leq \beta < \infty$ such that the condition $\omega(A\zeta_{u\beta}B) = \omega(BA)$. It is a basic result that a state satisfying the KMS condition at t act as a thermal reservoir, in the sense that any finite system coupled to it reaches thermal equilibrium at "temperature" $T = \beta^{-1}$.

that the proof that $\omega_M|_{\mathrm{I}}$ is a KMS state does not imply that it is a thermal bath of Rindler particles. The assumption that it is is only a suggestive one. The reason for that statement is that as commented in the main text a detector can indeed be excited when in uniform accelerated motion, but the excitation energy does *not* come from the *pseudo energy* of any hypothetical thermal bath, but from the *real* energy (as inferred from an inertial reference frame) of the source accelerating the device.

References

1. Arageorgis, A., Earman, J. and Ruetsch, L.,Fulling Non-uniqueness and the Unruh Effect: A Primer on Some Aspects of Quantum Field Theory, *Philos. of Sci.*. **70**, 164–202 (2003).
2. Araki, H., Type of the von Neumann Algebra Associated to Free Field, *Prog. Theor. Phys.* **32**, 956–965 (1964).
3. Bell, J. S., *Speakable and Unspeakable in Quantum Mechanics*, Cambridge Univ. Press, Cambridge, 1987.
4. Blackadar, B.,*Operator Algebras*, Springer, Heildelberg, 2005. revised text at http://wolfweb.unr.edu/homepage/bruceb/Cycr.pdf.
5. Boulware, D., Quantum Field Theory in Schwarzschild and Rindler Spaces, *Phys. Rev. D* **11**, 1404- (1975).
6. Bratelli, O, Kishimoto, A.,and Robinson, D. W., Stability Properties and the KMS Condition, *Comm. in Math. Phys.* **61**, 209–328 (1978)
7. Bratelli, O and Robinson, D. W., *Operator Algebras and Quantum Statistical Mechanics I*, Springer-Verlag, New York,1979.
8. Bratelli, O and Robinson, D. W., *Operator Algebras and Quantum Statistical Mechanics II*, Springer-Verlag, New York,1996.
9. Buchholz, D. and Verch, R., Macroscopic Aspects of the Unruh Effect, *Class.Quant.Grav.* **32**, 245004, (2015) [arXiv:1412.5892v4[gr-qc]]
10. Clifton, R., and Halvorson, H., Are Rindler Quanta Real?, in S. M. Cristensesn (ed.), *Quantum Theory of Gravity*, pp. 66–77, Adam Hilger, Bristol, 1984.
11. Cozzella,G., Landulfo, A.G.S., Matsas, G. E. A., and Vanzella, D. A. T., *Virtual Observation of the Unruh Effect*. [arXiv:1701.03446v1 [gr-qc]]
12. Choquet-Bruhat, Y., DeWitt-Morette, C. and Dillard-Bleick, M., *Analysis, Manifolds and Physics* (revised edition), North Holland Publ. Co., Amsterdam, 1982
13. Colosi, D. and Rätzel, D., The Unruh Effect in General Boundary Quantum Field Theory, *Symm. Integ. Geom: Meth. and Appl.* **9**, 019 (2013). [http://dx.doi.org/10.3842/SIGMA.2013.019]
14. Crispino, L. C. B., Higuchi, A. and Matsas, G. E. A., The Unruh Effect and its Applications, *Rev. Mod. Phys.* **80**, 786–838 (2008).
15. Crumeyrolle, A., *Orthogonal and Symplectic Clifford Algebras. Spinor Structures*, Kluwer Acad. Pub., Dordrecht, 1990.
16. David, F., *A Short Introduction to the Quantum Formalism*. [arXiv;1211-5627v1 {math-ph]]
17. Davies, P. C. W., Particles Do Not Exist, in Christensen, S.M. (ed.), *Quantum Theory of Gravity*, pp 66–77, Adam Hilger, Bristol, 1984.
18. Dodson, C. T. J. and Poston, T., *Tensor Geometry* (second edition), Springer Verlag, Berlin, Heidleberg, New York, 1991.
19. Fedotov, A. M.,Mur, V.D., Narozhny, N. B., Belinski, V.A. and Karnakov, B. M, Quantum Field Aspect of the Unruh Problem, *Phys. Lett. A* **254**, 126–132 (1999).
20. Frankel, T., *The Geometry of Physics*, Cambridge Univ. Press, Cambridge, 1997.
21. Fulling, S. A., *Aspects of Quantum Field Theory on Curved Spacetime*, Cambridge Uni. Press, Cambridge, 1989.

22. Gel'fand, I. M, and Naimark, M. A., On Embedding of Normed RIngs into the Ring of Operators in Hilbert Space, *Math. Sb* **12**, 197–213 (1943).
23. Giglio, J. F. T. and Rodrigues, W. A. Jr., Locally Inertial Reference Frames in Lorentzian and Riemann- Cartan Spacetimes, *Ann. der Physik*. **502**, 302–310 (2012).
24. Ginsburg, V. L. and Frolov, V. P.,Vacuum in Homogeneous Gravitational Field and Excitation of a Uniformly Accelerate Detector, *Sov. Phys Uspecki* **30**, 1073–1095 (1987).
25. Grib, A,. A., Mamayev, S. G., and Mostepanenko, V. M., *Vacuum Quantum Effects in Strong Fields*, Friedmann Laboratory Publishing, St. Petersburg, 1994.
26. Jackson, J. D., *Classical Electrodynamics*, (third edition), J. Wiley &Sons, New York, 1998.
27. Kay, B. S., The Double-Wedge Algebra for Quantum Fields on Schwarzschild and Minkowski Spacetimes, *Com. Math. Phys.* **100**, 57–81 (1985).
28. Kubo, Statistical-Mechanical Theory of Irreversible Processes. I. General Theory and Simple Applications to Magnetic and Conduction Problems", *J. Phys. Soc. of Japan* **12**, 570–586 (1957).
29. Kay, B. S. and Wald, R. M., Theorems on the Uniquenessand Thermal Properties of Stationary, Nonsingular, Quasi Free States on Spacetimes with a Bifircucate Killing Horizon, *Phys. Reports* **207**, 1709–1714 (1991).
30. Lyle, S. N., *Uniformly Accelerating Charges. A Treat to the Equivalence Principle*, Fundamental.Theories of Physics. **158**, Springer, Heidelberg, 2008.
31. Lyle, S. N., *Self-Force and Inertia. Old Light on New Ideas*, Lecture. Notes in Physics. **796,** Springer, Heidelberg, 2010.
32. Maiorino, J. E. and Rodrigues, W. A. Jr., Maxwell Theory is Still a Source of Surprises, in Dvoeglazov, V. V. (ed.), *Photon: Old Problems in Light of New Ideas,* Nova Sci. Publ., Inc. New York, 2000.
33. Martin, P. C.; Schwinger, J., Theory of Many-Particle Systems. I, *Phys. Rev.* **115**, 1342–1373 (1959).
34. Morse,,P. M. and Feshbach, *Methods of Theoretical Physics*, Part II, McGraw-Hill Book Co., Inc., New York,1953.
35. Motl, L., John Bell Actually Misunderstood Relativity,Too, http://motls.blogspot.com.br/2015/05/john-bell-actually-misunderstood.html
36. Nakahara, M., *Geometry, Topology and Physics*, Inst. Physics Publ., Bristol and Philadelphia, 1990.
37. Nöth, M., *The Unruh Effect Without Observer*, M.Sc. Thesis, Ludwig-Maximilians-Universität München, 2016. http://www.mathematik.uni-muenchen.de/~bohmmech/theses/Noeth_Markus_MA.pdf
38. Ohanian, H. and Ruffini, R., *Gravitation and Spacetime* (second edition), W.W.Norton &Co., New York, 1994.
39. O'Neill, B., *Semi-Riemannian Geometry With Applications to Relativity*, Academic Press, San Diego, 1983.
40. Panofski, W. K. H.,and Philiips, M., *Classical Eletricity and Magnetism,* Addison-Wesley Publ. Co., Reading MA, 1969.
41. Parrott, S., *Relativistic Electrodynamics and Differential Geometry*, Springer-Verlag, New York, 1987
42. Parrott, S., Radiation from an Uniformily Accelerated Charge and the Equivalence Principle, *Found. Phys.* **32**, 407–440(2002)[arXiv:gr-qc:9303025v8]
43. Pauli, W., *Theory of Relativity*, Dover edition, Dover Publ. Inc., New York, 1981.
44. Rodrigues, W. A. Jr., and Sharif, M., Equivalence Principle and the Principle of Local Lorentz Invariance, *Found. Phys.* **31**, 1785–1806 (2001).
45. Rodrigues, W. A. Jr. and Capelas de Oliveira, E., *The Many Faces of Maxwell, Dirac and Einstein Equations. A Clifford Bundle Approach* (second revised and enlarged edition). Lecture Notes in Physics **922**,Springer, Dordrechet, 2016.
46. Rodrigues, W. A. Jr. and Wainer, S. A.,Equations of Motion and Energy-Momentum 1-Forms for the Coupled Gravitational, Maxwell and Dirac Fields, *Adv, Appl.Clifford Algebras* (2016). https://doi.org/10-1---7/s0006016--0679-5. [arXiv: 1601.04878[math-ph]]

47. Rodrigues, W. A. Jr. and Wainer, S. A., Notes on Conservation Laws, Equations of Motion and Particle Field in Lorentzian and Teleparallel de Sitter Spacetime Structures, *Adv. Math. Phys.* **2016**, 5465263 (2016) [arXiv:1505.02935[math-ph]]
48. Segal, I, Irreducible Representaions of Operator Algebras, *Bull. Amer. Math. Soc.* **53**, 73–88 (1947).
49. Sciama, D. W., Candela, P., and Deutch, D., Quantum Fild Theory, Horizons and Thermodynamics, *Adv. in Phys.* **30**, 327–366 (1981)
50. Sachs, R. K. and Wu, H., *General Relativity for Mathematicians*, Springer-Verlag, New York, Heidelberg,Berlin, 1977.
51. Swell, G. l., Quantum Fields on Manifolds: PCT and Gravitationally Induced Thermal States, *Ann. Phys.* **141**, 201–224 (1982)
52. Socolovsy, M., Rindler Space, Unruh Effect and Hawking Temperature, *Ann. Fondation L. de Broglie* **30**, 1–49 (2014).
53. Susskind, L. and Lindsey, J., *An Introduction to Black Holes, Information and the String Theory Revolution. The Holographic Universe*, World Scientidfic, Singapore 2005.
54. Takagi, S.,Vacuum Noise and Stress Induced by Uniform Accelerating Hawking-Unruh Effect in Rindler Manifold of Arbitrary Dimension, *Prog. Theor. Phys.* (Suppl.) **88**, 1–142 (1986)
55. Tolman, R., Rela, tivity, *Thermodynamics and Cosmology,* Dover Publ., Inc., New York, 1987 (first published by Oxford Univ. Press, Oxford,1934).
56. Turakulov, Z. Ya., Electromagnetic Field of a Charge Moving with Constant Acceleration, *J. Geom. Phys* **14**, 305–308 (1994).
57. Turakulov, Z. Ya., Infinitesimal Radiation Phenomena, *Turkish. J. Phys.* **19**, 1567–1573 (1995).
58. Turakulov, Z. Ya..,Geometric Theory of Radiation, in Dowling, J. P. (ed.), *Electron Theory and Quantum Electrodynamics*, NATO ASI SERIES. Series B: Physics vol. 358, pp. 321–325, Plenum Press, New York, 1997.
59. Unruh, W. H., Notes on Blackhole Evaporation, *Phys. Rev. D* **14**, 870–892 (1976)
60. Unruh, W. H., and Wald, R. M., What Happens When an Accelerating Observers Detects a Rindler Particle, *Phys. Rev. D* **29**, 1047–1056 (1984).
61. Verch, R., Local Definitetness, Primitivity, and Quasi Equivalence of Quasi- Free Hadamard-vQunatum States in Curved Spacetime, *Comm. Math. Phys.* **160**, 507–536 (1994).
62. von Neumann, J. Zur Algebra der Funktional Operationen und Theorie der Normalen Operatoren", *Math. Ann.* **102**, 370–427 (1930).
63. Wald, R. M., *Quantum Field Theory in Curved Spacetime and Black-Hole Thermodynamics*, The Univ. Chicago Press, Chicago, 1994.
64. Oziewicz, Z., Sitarczyk, Cz, Parallel Treatment of Riemannian and Symplectic Clifford Algebras, in Artibano, M., Boudet, R. Helmstetter, J. (eds.), *Clifford Algebras and their Applications in Mathematical Physics*, pp.83–96, Kluwer Acad. Publ., Dordrecht, 1989.

Flag Type of Semigroups: A Survey

Luiz A. B. San Martin

Abstract In this chapter, we present an overview of the theory of semigroups in semi-simple Lie groups and its applications to dynamical systems, control systems, and random dynamical systems. A great deal of the results to be surveyed appeared first in Ph.D. theses by students of the Department of Mathematics of IMECC.

The piece of semigroup theory to be discussed here was constructed with the purpose of understanding semigroups with non-empty interior in semi-simple Lie groups. A characteristic of this theory is that it is built upon the actions of the semigroups on the flag manifolds of the Lie groups. These actions contain crucial information about the semigroups due to the strong structural properties of the semi-simple Lie groups.

The concept of *flag type* of a semigroup emerges as a synthesis of several results about the actions of the semigroups on the flag manifolds. This concept gives a classification of semigroups via block decompositions, much like the Jordan form of matrices. More than that, it provides key information about the structure of the semigroups in the semi-simple Lie groups. The results to be surveyed in this chapter exploit the concept of flag type to describe properties of the semigroups as well as to get applications to control and dynamical systems.

1 Control Sets and Flag Type

Let $S \times X \to X$ be an action of a semigroup S in the topological space X. A control set for the action is a subset $D \subset X$ such that $D \subset \mathrm{cl}\,(Sx)$ for all $x \in D$ and D is maximal with this property. The control set is invariant if $\mathrm{cl} D = \mathrm{cl}\,(Sx)$ for all $x \in D$.

The control sets are building blocks to the construction of the orbits of the action of a semigroup. These sets appeared in the literature of control systems, and hence

L. A. B. San Martin (✉)
Institute of Mathematics, Statistics and Scientific Computing, University of Campinas, Campinas, SP, Brazil
e-mail: smartin@ime.unicamp.br

C. Lavor, F. A. M. Gomes (eds.), *Advances in Mathematics and Applications*,
https://doi.org/10.1007/978-3-319-94015-1_14

the name. The invariant control sets were first considered in Arnold–Kliemann [6] as supports of invariant (stationary) measures for diffusion processes. The control sets were exhaustively studied in Colonius–Kliemann [20] that established their main dynamical properties.

In the paper Arnold–Kliemann–Oeljeklaus [7], the action of a semigroup S of matrices on the projective space $\mathbb{P}^d$ emerges as a tool to study Lyapunov exponents of linear stochastic equations. In [7], it is proved that there is a unique invariant control set under the accessibility assumption that the orbits of S have non-empty interior. Uniqueness of the invariant control set is relevant to understand the invariant measures for the stochastic processes defined by the differential equations.

Some results of [7] were extended to projective bundles in San Martin–Arnold [59]. The results of these two papers [7, 59] are the starting point of the development of the results to be surveyed in this paper.

In [61], the invariant control sets were studied in a much larger setup obtaining the following crucial characterization.

Theorem 1.1 ([61]) *Let G be a connected and noncompact semi-simple Lie group and $P_\Theta \subset G$ a parabolic subgroup, so that $\mathbb{F}_\Theta = G/P_\Theta$ is a flag manifold of G. If $S \subset G$ is a semigroup with* $\mathrm{int}S \neq \emptyset$*, then the action of S on $\mathbb{F}_\Theta$ has a unique invariant control set C_Θ.*

Moreover, C_Θ is the closure of the open set $(C_\Theta)_0$ whose elements are the attractor fixed points $\mathrm{att}_\Theta(h)$ *with h running through the regular elements in* $\mathrm{int}S$.

In the group $G = \mathrm{Sl}(d, \mathbb{R})$, a parabolic subgroup P_Θ is a subgroup of block upper diagonal matrices where Θ is a set of indices that tells the sizes of the diagonal blocks of the matrices in P_Θ. If these sizes are $k_1, \ldots, k_s$, then $\mathbb{F}_\Theta = G/P_\Theta$ identifies with the manifold of flags $(V_1 \subset \cdots \subset V_s)$ of subspaces of $\mathbb{R}^d$ with dimensions $\dim V_i = k_1 + \cdots + k_i$. Clearly, the projective space $\mathbb{P}^{d-1}$ and the Grassmannians $\mathrm{Gr}_k(d)$ are included among the flag manifolds. A matrix $h \in \mathrm{Sl}(d, \mathbb{R})$ is regular if it is diagonalizable with (real) distinct eigenvalues. This way, the second part of the above theorem when applied to $\mathbb{F}_\Theta = \mathbb{P}^{d-1}$ means that $(C_\Theta)_0$ is the set of eigenspaces associated to the highest eigenvalues of the regular matrices $h \in \mathrm{int}S$. In a Grassmannian $\mathbb{F}_\Theta = \mathrm{Gr}_k(d)$, the attractor $\mathrm{att}_\Theta(h)$ of a regular h is the sum of the eigenspaces associated to the k largest eigenvalues. Thus, $(C_\Theta)_0 \subset \mathrm{Gr}_k(d)$ is made of these k-dimensional subspaces again with h running through $\mathrm{int}S$.

The full description of the control sets in the flag manifolds is done in [75]. To state it, we let $\mathscr{W}$ be the Weyl group of G. If $h \in G$ is a regular element, then its action in the maximal flag manifold $\mathbb{F}$ has exactly $|\mathscr{W}|$ fixed points which we denote by $\mathrm{fix}(h, w)$ with $w \in \mathscr{W}$. In this notation, $\mathrm{fix}(h, 1)$ is the only attractor fixed point of h.

The control sets are characterized in terms of the fixed points $\mathrm{fix}(h, w)$ with h regular in $\mathrm{int}S$. The core D_0 of a control set D is defined by $D_0 = \{x \in D : \mathrm{int}Sx \cap \mathrm{int}S^{-1}x \neq \emptyset\}$. The control set is said to be effective if $D_0 \neq \emptyset$. In this case,

D_0 is open and dense in D. The core of the (unique) invariant control set, mentioned in Theorem 1.1, is called the attractor set of S. The repeller set of S is the attractor set of S^{-1}.

Theorem 1.2 ([75]) *Let G be a connected and noncompact semi-simple Lie group and $S \subset G$ a semigroup with* $\text{int}S \neq \emptyset$. *Let $\mathbb{F}$ be the maximal flag manifold of G. For each $w \in \mathcal{W}$, the set*

$$(D_w)_0 = \{\text{fix}\,(h, w) \in \mathbb{F} : h \in S^{\text{reg}}\}$$

is the core of a control set D_w, where S^{reg} is the set of regular elements in $\text{int}S$. *The control sets D_w, $w \in \mathcal{W}$, exhaust the effective control sets for S in $\mathbb{F}$. (D_1 is the unique invariant control set.)*

In a partial flag manifold $\mathbb{F}_\Theta$, the control sets are given by $\pi_\Theta(D_w)$, $w \in \mathcal{W}$, where $\pi_\Theta : \mathbb{F} \to \mathbb{F}_\Theta$ is the canonical projection.

In $G = \text{Sl}(d, \mathbb{R})$, we have $\mathbb{F} = \text{Sl}(d, \mathbb{R}) / P$ where P is the subgroup of upper triangular matrices and $\mathbb{F}$ is identified to be the manifold of complete flags of subspaces. Also, $\mathcal{W}$ is the group of permutations of $\{1, \ldots, d\}$.

A regular element h is diagonalizable in basis $\beta = \{v_1, \ldots, v_d\}$ with the eigenvalues ordered so that $\lambda_1 > \cdots > \lambda_d$. Then, $\text{fix}\,(h, w) = (V_1 \subset \cdots \subset V_d)$ where $V_i = \langle v_{w(1)}, \ldots, v_{w(i)} \rangle$. So that the control sets of S in $\mathbb{F}$ recover in some extent the bases diagonalizing the regular elements $h \in \text{int}S$.

The map $w \mapsto D_w$ defined by the above theorem is not in general injective. The results that give the full picture of this map are proved with the assumption that G has finite center which ensures that the K component of the Iwasawa decomposition is compact. This compactness enables to prove the following theorem about a transitive action of S on a flag manifold of G. This theorem is a basic tool to prove forthcoming results.

Theorem 1.3 *Suppose that G has finite center and let $S \subset G$ be a semigroup with* $\text{int}S \neq \emptyset$. *If S acts transitively on some flag manifold $\mathbb{F}_\Theta$ of G, then $S = G$.*

The next result gives an algebraic characterization of the level sets of $w \mapsto D_w$.

Theorem 1.4 ([75]) *Suppose that G has finite center. Then, there exists a subgroup $\mathcal{W}_S \subset \mathcal{W}$ such that $D_{w_1} = D_{w_2}$ if and only if $\mathcal{W}_S w_1 = \mathcal{W}_S w_2$. The subgroup $\mathcal{W}_S$ is parabolic in the sense that there is a set $\Theta(S)$ of simple reflections such that $\mathcal{W}_S$ is the subgroup $\mathcal{W}_{\Theta(S)}$ generated by the reflections in $\Theta(S)$.*

For $G = \text{Sl}(d, \mathbb{R})$, the simple reflections in the permutation group $\mathcal{W}$ are the permutations $(i, i+1)$. Hence, the set $\Theta(S)$ of simple reflections determines a partition of $\{1, \ldots, d\}$. In turn, the partition yields a parabolic subgroup $P_{\Theta(S)}$ given by block upper triangular matrices where the sizes of the diagonal blocks are equal to the partition elements. The parabolic subgroup $P_{\Theta(S)}$ gives rise to the flag manifold $\mathbb{F}_{\Theta(S)} = G/P_{\Theta(S)}$. The same relationship between a subset of simple reflections, a parabolic subgroup, and a flag manifold holds in general.

Definition 1.5 The flag type (or parabolic type) of a semigroup ($\text{int}S \neq \emptyset$) is given either by the set $\Theta(S)$ of simple reflections such that $\mathscr{W}_S = \mathscr{W}_{\Theta(S)}$or by the parabolic subgroup $P_{\Theta(S)}$ or the flag manifold $\mathbb{F}_{\Theta(S)} = G/P_{\Theta(S)}$.

The following results were proved also in [75]. They characterize the flag type in terms of the geometry of the control sets and the Jordan decompositions of the elements in the interior of S.

Theorem 1.6 *Assume that G has finite center. Let $\mathbb{E}$ be a flag manifold and denote by $\pi : \mathbb{F} \to \mathbb{E}$ the canonical projection from the maximal flag manifold and by $C_{\mathbb{E}}$ the invariant control set of S in $\mathbb{E}$. Then, $\mathbb{E} = \mathbb{F}_{\Theta(S)}$ (the flag type of S) if and only if the following two conditions hold:*

1. *$C = \pi^{-1}(C_{\mathbb{E}})$, where C is the invariant control set in $\mathbb{F}$.*
2. *$C_{\mathbb{E}}$ is contractible by every $h \in S^{\text{reg}}$, that is, $h^n C_{\mathbb{E}}$ shrinks to a point as $n \to \infty$.*

$\mathbb{F}_{\Theta(S)}$ is minimal among the flag manifolds that satisfy the first property and maximal with the second property.

Theorem 1.7 *Assume that G has finite center. Take $g \in \text{int}S$ and write its Jordan decomposition as the product of commuting elements $g = mhn$ where m is elliptic, h hyperbolic, and n unipotent. Then, the attractor set of h in $\mathbb{F}_{\Theta(S)}$ is a fixed point. Conversely, there exists a hyperbolic element $h \in \text{int}S$ such that $\mathbb{F}_{\Theta(S)}$ is the flag manifold which is minimal with the property that the attractor fixed point set of h reduces to a point.*

For a matrix $g \in \text{Sl}(d, \mathbb{R})$, the component h in the Jordan decomposition $g = mhn$ is a diagonal matrix $h = \text{diag}\{a_1, \ldots, a_d\}$ (in some basis) where $a_i = |\lambda_i|$ with λ_i running through the eigenvalues of g. If $a_1 \geq a_2 \geq \cdots \geq a_d$, then the attractor set of h in $\mathbb{F}_\Theta$ reduces to a fixed point if and only if the partition of $\{1, \ldots, d\}$ given by the multiplicities of a_i refines the partition of the parabolic subgroup P_Θ. Hence, Theorem 1.7 says that the Jordan block decomposition of any $g \in \text{int}S$ is a refinement of the block decomposition of $P_{\Theta(S)}$ (w.r.t. different bases) and there is a diagonalizable $h \in \text{int}S$ with exactly the same block decomposition as $P_{\Theta(S)}$ (if the eigenvalues of h are ordered decreasingly). For the particular case, when $\mathbb{F}_{\Theta(S)}$ is the projective space $\mathbb{P}^{d-1}$, the theorem shows that any $g \in \text{int}S$ has a principal (real) eigenvalue. This is one of the statements of the classical Perron–Frobenius theorem that considers the semigroup S_W of matrices leaving invariant a cone $W \subset \mathbb{R}^d$, whose flag type is $\mathbb{P}^{d-1}$. Thus, a particular instance of Theorem 1.7 is a generalization of Perron–Frobenius theorem.

There is a natural partial ordering between the control sets saying that $D_1 < D_2$ if D_2 is attained from D_1 by the action of the semigroup, that is, if there are $x \in D_1$, $y \in D_2$, and $g \in S$ such that $gx = y$. A related concept is the domain of attraction $\mathscr{A}(D)$ of a control set which is defined by

$$\mathscr{A}(D) = \{x : \exists g \in S, gx \in D\}.$$

For example, in a flag manifold, the invariant control set C of S is maximal w.r.t this ordering and its domain of attraction is the whole flag manifold. On the other hand, still in a flag manifold, the core C^* of the invariant control of S^{-1} is a control set of S which is minimal w.r.t. the partial order.

In [68], the ordering between the control sets D_w were given algebraically in terms of the Bruhat–Chevalley order in the Weyl group which is defined from the way $w \in \mathscr{W}$ is generated by simple reflections.

Theorem 1.8 ([68]) *For $w \in \mathscr{W}$, let D_w denote the control set in the maximal flag manifold $\mathbb{F}$ as given by Theorem 1.2. Let $w_1, w_2 \in \mathscr{W}$. Then, the following statements are equivalent:*

1. $D_{w_1} \leq D_{w_2}$.
2. *There exists $w \in \mathscr{W}$ such that $w_1 \geq w$ and $w \in \mathscr{W}_S w_2$, that is, $D_w = D_{w_2}$.*

The domain of attraction of the control sets is given in [68] in terms of the Schubert cells. In order to describe these domain of attractions, it is developed in [68] an alternative way to write down the Schubert cells.

Let α_i be a simple root and denote by s_i the reflection w.r.t. α_i. Denote by $P_i = P_{\{\alpha_i\}}$ the parabolic subgroup defined by $\Theta = \{\alpha_i\}$ and by $\mathbb{F}_i = G/P_i$ the associated flag manifold. If $\pi_i : \mathbb{F} \to \mathbb{F}_i$ is the fibration from the maximal flag manifold $\mathbb{F}$ and $A \subset \mathbb{F}$, then we write

$$\gamma_i(A) = \pi_i^{-1}\pi_i(A)$$

for the union of the fibers $\pi_i : \mathbb{F} \to \mathbb{F}_i$ through A. It is proved in [68] that a Schubert cell in $\mathbb{F}$ has the form $\gamma_1 \cdots \gamma_n\{b\}$ for suitable $b \in \mathbb{F}$ where $\gamma_1 \cdots \gamma_n$ is associated to a reduced expression $w = s_1 \cdots s_n$ of $w \in \mathscr{W}$ as a product of simple reflections.

Theorem 1.9 ([68]) *Let $C^* = D(w_0)$ be the minimal control set where w_0 is the principal involution of $\mathscr{W}$ (that is, the element of largest length). Then, for any $w \in \mathscr{W}$ the domain of attraction $\mathscr{A}(D_w)$ of D_w is given by*

$$\mathscr{A}(D_w) = \gamma_1 \cdots \gamma_n \left(C^*\right). \tag{1}$$

Here, the sequence $\gamma_1, \ldots, \gamma_n$ comes from a reduced expression

$$w_0 w = s_n \cdots s_1.$$

Examples of control sets and flag types of semigroups were produced in several places (see [9, 9, 24, 32, 74, 77]).

Apart from the above results, the flag type of a semigroup S encodes several geometric and algebraic properties of S that will be described later.

In [43, 86], there is an application of the flag type to study control sets on the adjoint orbits $\mathrm{Ad}(G)H$ where H is such that $h = \exp H$ is hyperbolic. (It is proved in [26] that these orbits are diffeomorphic to the cotangent bundles of the flag manifolds.) The fact to be remarked are:

1. As happens to the flag manifolds the control sets are given by the set of fixed points of the regular elements in int$S$. This permits to parametrize them by the Weyl group.
2. There are no invariant control sets unless $S = G$.
3. Different control sets are not related by their ordering, that is, a control set is not reached by the action of the (proper) semigroup from another control set.

2 Topological Properties

An open subsemigroup of a Lie group may have two kinds of connected components, namely components K such that $K^2 \cap K = \emptyset$ and components that are themselves subsemigroups (e.g., the semigroup $(2, 3) \cup (4, 6) \cup (6, +\infty) \subset \mathbb{R}$ has the two types of components).

The following result proved in [50] describes the semigroup components in terms of control sets. It is the first of the topological results to be recalled here.

Theorem 2.1 ([50]) *Let G be a noncompact semi-simple Lie group and $S \subset G$ an open semigroup. Denote by C^+ (respectively, C^-) the attractor set (respec., repeller set, that is, the attractor set of S^{-1}) of S in the maximal flag manifold.*

Given a pair of connected components K_1 of C^+ and K_2 of C^-, there exists a unique semigroup component Γ of S such that K_1 and K_2 are, respectively, the attractor and repeller sets of Γ. Therefore, S has $\text{card}\left(C^+\right) \cdot \text{card}\left(C^-\right)$ semigroup components.

In [54, 55], the connected components of a semigroup S are related to the algebraic property of reversibility, that is, to the set of $g \in G$ such that $S \cap gS$ is not empty.

Another topological result goes back to a classical theorem of Cartan saying that the manifold underlying a Lie group is the product of a compact Lie group by a Euclidian space $\mathbb{R}^n$. This result implies that the bulk of the topology of a Lie group is concentrated in a compact Lie group. The next result shows that this happens with semigroups as well with the proviso that the semigroup S is infinitesimally generated, that is, $S = \langle \exp W \rangle$ where W is a cone in the Lie algebra $\mathfrak{g}$ of the group. Such condition is needed in the proofs to get homotopies with the aid of 1-parameter semigroups.

Theorem 2.2 ([2, 78]) *Let $S = \langle \exp W \rangle$ be an infinitesimally generated semigroup with non-empty interior in a noncompact semi-simple Lie group G. Let $\mathbb{F}_{\Theta(S)} = G/P_{\Theta(S)}$ be the flag type of S and denote by $K_{\Theta(S)}$ the maximal compact subgroup of the Levi component of $P_{\Theta(S)}$. Then, there exists a coset $gK_{\Theta(S)}$ that is a deformation retract of S. Hence, the homotopy groups of S and $K_{\Theta(S)}$ are the same.*

In $\text{Sl}(d, \mathbb{R})$, a group $K_{\Theta(S)}$ has the form $\text{SO}(k_1) \times \cdots \times \text{SO}(k_s)$ where $k_1, \ldots, k_s$ are the sizes of the block diagonal matrices defining $P_{\Theta(S)}$. Hence, an open semigroup $S \subset \text{Sl}(d, \mathbb{R})$ has the homotopy type of a product of special orthogonal groups.

An example is the homotopy type of the semigroup $S = \mathrm{Sl}^{+}(d, \mathbb{R}) \subset \mathrm{Sl}(d, \mathbb{R})$ of matrices with nonnegative entries. Its flag type is the projective space $\mathbb{P}^{d-1} = \mathbb{F}_{\Theta(S)}$ which is associated to the partition $\{1\} \cup \{2, \ldots, d\}$ of $\{1, \ldots, d\}$. Hence, $K_{\Theta(S)} = \mathrm{SO}(d-1)$ that has the same homotopy type as $\mathrm{Sl}^{+}(d, \mathbb{R})$.

3 Maximal Semigroups

A maximal subsemigroup of a group is a semigroup which is not properly contained in a proper semigroup. A general result says that a subsemigroup S of a topological group G with $\mathrm{int} S \neq \emptyset$ is contained in a maximal semigroup (see Hilgert–Hofmann–Lawson [28]). The concept of flag type for semigroups in semi-simple Lie groups suggests a refinement of the notion of maximality, namely a Θ-maximal semigroup is a semigroup S ($\mathrm{int} S \neq \emptyset$) with flag type Θ which is not properly contained in a semigroup with the same flag type. If Θ is the complement of a singleton (which means that $\mathbb{F}_{\Theta}$ is a minimal flag manifold), then a Θ-maximal semigroup is maximal in G. So that in semi-simple Lie groups maximality of semigroups with non-empty interior is a particular case of Θ-maximality.

In [71], the Θ-maximal (and hence the maximal) semigroups were characterized with the aid of the so-called $\mathscr{B}$-convex sets in the flag manifolds.

Any flag manifold $\mathbb{F}_{\Theta}$ is in duality to another flag manifold $\mathbb{F}_{\Theta^*}$ whose points parametrize the open Bruhat cells of $\mathbb{F}_{\Theta}$ and conversely. For example, the projective space $\mathbb{P}^{d-1}$ is dual to the Grassmannian $\mathrm{Gr}_{d-1}(d)$ in the sense that a subspace $V \in \mathrm{Gr}_{d-1}(d)$ defines the open Bruhat cell of lines in $\mathbb{P}^{d-1}$ that are transversal to V. This way, a dual of a set $C \subset \mathbb{F}_{\Theta}$ is the set $C^* \subset \mathbb{F}_{\Theta^*}$ given by $x \in \mathbb{F}_{\Theta^*}$ such that the open cell defined by x contains C. The same way one defines the dual $D^* \subset \mathbb{F}_{\Theta}$ of a subset $D \subset \mathbb{F}_{\Theta^*}$.

A $\mathscr{B}$-convex set $C \subset \mathbb{F}_{\Theta}$ is defined to be a set such that $C = (C^*)^*$. With these notions, we have the following characterization of the Θ-maximal semigroups.

Theorem 3.1 ([71]) *A semigroup S with $\mathrm{int} S \neq \emptyset$ is Θ-maximal if and only if there exists a $\mathscr{B}$-convex set $C \subset \mathbb{F}_{\Theta}$ with $C = \mathrm{cl}(\mathrm{int} C)$ such that S is the compression semigroup of C, that is,*

$$S = \{g \in G : gC \subset C\}.$$

In this case, C is the invariant control set of S. The semigroup S is maximal if and only if it is Θ-maximal and $\mathbb{F}_{\Theta}$ is a minimal flag manifold.

A $\mathscr{B}$-convex set may be quite wild. For example, if $G = \mathrm{Sl}(2, \mathbb{R})$, then the only flag manifold is the projective line $\mathbb{P}^1$ which is self-dual. Since an open Bruhat cell is the complement of a point, any proper subset is $\mathscr{B}$-convex. Hence, the compression semigroup of any proper subset $C = \mathrm{cl}(\mathrm{int} C) \subset \mathbb{P}^1$ is a maximal subsemigroup of $\mathrm{Sl}(2, \mathbb{R})$ (a similar picture holds in the groups with real rank 1). On the other side, it is proved in [71] that the connected $\mathscr{B}$-convex subsets of a projective space $\mathbb{P}^{d-1}$

(viewed as flag manifold of $\mathrm{Sl}(d, \mathbb{R})$) are those subsumed by a pointed convex cone $W \subset \mathbb{R}^d$. Hence, the compression semigroup S_W of the double cone $W \cup -W$ is maximal in $\mathrm{Sl}(d, \mathbb{R})$.

4 Integration on Semigroups

In this section, the flag type is applied to measured theoretic questions in semi-simple Lie groups. The first question goes back to the path breaking paper by Furstenberg [25] that generalizes to semi-simple Lie groups the classical representation of harmonic functions in the disk known as Poisson space. Afterwards, we look at the behavior of the Helgason–Laplace transform of the indicator function of a semigroup and the so-called moment Lyapunov exponent associated to a probability measure. These questions are clarified by the flag type of a semigroup.

4.1 Poisson Spaces

If ν is a probability measure in a group G, a function f on G is called ν-harmonic if

$$f(g) = \int_G f(gh)\, d\nu(h).$$

The Poisson space for ν is a compact G-space Π with a ν-invariant probability measure μ (the Poisson measure) such that there exists a one-to-one correspondence between bounded (respectively, left uniformly continuous) ν-harmonic functions on G and measurable (respectively, continuous) functions on Π. The Poisson formula performs the bijective relation: given a left uniformly continuous bounded ν-harmonic function, there exists a unique continuous function $\hat{f}$ on Π such that

$$f(g) = \int_\Pi \hat{f}(gx)\, d\mu(x).$$

Let $G = KAN$ be an Iwasawa decomposition of a semi-simple group G, with the corresponding decomposition of the Lie algebra $\mathfrak{g} = \mathfrak{k} \oplus \mathfrak{a} \oplus \mathfrak{n}$. Let M be the subgroup of G which is the centralizer of A in K. If ν is absolutely continuous, the main results in Furstenberg [25] establish that: (1) A Poisson space of G is a homogeneous space of the form $\Pi_\nu = G/M_\nu AN$, where the M_ν is a subgroup of M containing its identity component M_0, and hence Π_ν is a covering of the maximal flag manifold $\mathbb{F} = G/MAN$ of the group G; (2) If the identity of the group belongs to the interior of the support of some convolution power ν^k, then the Poisson space is the maximal flag manifold $\mathbb{F} = G/MAN$ itself, also called Furstenberg boundary. This condition implies that the semigroup $S = S_\nu$ generated by the support of ν coincides with G.

In [34, 40], it was considered the case where the semigroup S_ν is proper but has non-empty interior. The key point is to identify the subgroup M_ν of M such that $\Pi_\nu = G/M_\nu AN$. As in other cases, the idea is to get M_ν through the action of S_ν on homogeneous spaces of G. Here, however, the right places are not the flag manifolds of G; that is, the flag type of S_ν is not enough to obtain M_ν.

As in other instances, assume that G has finite center so that K and M are compact subgroups. The homogeneous space G/M_0AN is compact as well and the canonical fibration $\pi : G/M_0AN \to \mathbb{F} = G/MAN$ is a covering with a finite number of leaves. By compactness, there are invariant control sets of S_ν in G/M_0AN, possibly more than one although all of them are projected by π to the unique invariant control set $C \subset \mathbb{F}$.

Theorem 4.1 *Let D be an invariant control set in G/M_0AN. Define*

$$M(D) = \{m \in M/M_0 : D \cdot m = D\}$$

where $D \cdot m$ stands for the right action of M/M_0 on G/M_0AN. Then, $M_\nu = \pi_0^{-1}(M(D))$ where $\pi_0 : M \to M/M_0$ is the canonical fibration.

The following theorem says in particular that in case S_ν is connected the parabolic type of S_ν completely determines M_ν.

Theorem 4.2 *Suppose that the invariant control set $C \subset \mathbb{F}$ is connected (this happens if S_ν is itself connected). Then, $M_\nu = M \cap P^0_{\Theta(S_\nu)}$ where $\Theta(S_0)$ is the flag type of S_ν and $P^0_{\Theta(S_\nu)}$ is the identity component of $P_{\Theta(S_\nu)}$.*

4.2 Characteristic Function

Let G be a noncompact connected semi-simple Lie group with finite center and $S \subset G$ a semigroup with non-empty interior. The question addressed here is the convergence of the Helgason–Laplace spherical transform of S, that is, integrals over S of the type

$$I_S(\lambda, u) = \int_S e^{\lambda(\mathsf{a}(g,u))} dg \tag{2}$$

where dg is the Haar measure of G. To write the integrand, we take an Iwasawa decomposition $G = KAN$ and let $\mathfrak{a}$ be the Lie algebra of A. In the integral, the parameters are $\lambda \in \mathfrak{a}^*$ and $u \in K$ while the function $\mathsf{a}(g, u) = \log h \in \mathfrak{a}$ if $gu = khn \in KAN$ is the Iwasawa decomposition of gu.

The integral $I_S(\lambda, u)$ is called the *characteristic function* of S by analogy with the characteristic function I_W of a cone $W \subset \mathbb{R}^d$ which is the classical Laplace transform of the indicator function of W. The characteristic function I_W of a cone is extensively used in the statistic literature since it yields an "exponential model."

As happens usually to Laplace transforms, the integral (2) may be $+\infty$ for some values of (u, λ). This poses the question of determining the domain of convergence of $I_S(\lambda, u)$.

Such domain of convergence is provided in [2, 78] in terms of the flag type of S. The domain is divided into two pieces, the $\lambda \in \mathfrak{a}^*$ and the $u \in K$ components.

As to the K component, it is noted that the integrand $e^{\lambda(\mathsf{a}(g,u))}$ as a function of $u \in K$ factors to a function on the full flag manifold $\mathbb{F} = K/M$ since it is invariant by right multiplication of u by any m in the subgroup M. This permits to write, for $g \in G$ and $x \in \mathbb{F}$, $\mathsf{a}(g, x) = \mathsf{a}(g, u)$ where $x = ux_0$ and x_0 is the origin of $\mathbb{F} = K/M$.

This way, $I_S(\lambda, u)$ can be seen as a function defined in $\mathfrak{a}^* \times \mathbb{F}$ and the domain of convergence of the K-component is actually determined by a subset of the maximal flag manifold $\mathbb{F}$.

In order to state the domain of convergence for the radial component λ, define for a subset Θ the partial chamber

$$\left(\mathfrak{a}_\Theta^*\right)^+ = \{\gamma \in \mathfrak{a}_\Theta^* : \forall \alpha \in \Sigma,\ \langle \alpha, \gamma \rangle > 0\}$$

which is an open cone in the annihilator $\mathfrak{a}_\Theta^*$ of Θ and put

$$\mathscr{C}_\Theta^+ = \bigcap_{\alpha \in \Pi^+ \setminus \langle \Theta \rangle^+} \left(d_\Theta D_\Theta \alpha + 2\rho_\Theta + \left(\mathfrak{a}_\Theta^*\right)^+\right)$$

where $d_\Theta = \dim \mathbb{F}_\Theta$, $D_\Theta = \dim \mathbb{F} - \dim \mathbb{F}_\Theta$, and $\rho_\Theta(H) = \frac{1}{2}\mathrm{tr}\left(\mathrm{ad}(H)_{|\mathfrak{n}_\Theta^+}\right)$ if $H \in \mathfrak{a}$. Since $\mathscr{C}_\Theta^+$ is the intersection of a finite number of translates of $\left(\mathfrak{a}_\Theta^*\right)^+$, it is an open cone in $\mathfrak{a}_\Theta^*$ as well.

The following convergence theorem on the flag type $\mathbb{F}_{\Theta(S)}$ of S is one of the main results of [39, 81].

Theorem 4.3 ([78]) *Let $S \subset G$ be a proper semigroup with* $\mathrm{int} S \neq \emptyset$ *and flag type $\Theta = \Theta(S)$. Then,*

$$I_S(\lambda, x) = \int_S e^{\lambda(\mathsf{a}(g,x))} dg \tag{3}$$

converges for any $\lambda \in -\mathscr{C}_{\Theta(S)}^+$ and x in the core C_0 of the invariant control set C of S in $\mathbb{F}$.

In case $\Theta(S) = \emptyset$ and $\mathbb{F}_\Theta$ is the maximal flag manifold, we have convergence if $\lambda + 2\rho$ belongs to the Weyl chamber $-(a^*)^+$.

If $S = G$, then $I_S(\lambda, x) = +\infty$ for all (λ, x).

This theorem can be improved by the remark that $I_S(\lambda, x) \leq I_T(\lambda, x)$ if T is a semigroup containing S. Hence, $I_S(\lambda, x)$ converges if x belongs to the attractor set of T which may be larger than C_0. In [2, 78], a $\Theta(S)$-maximal semigroup $T \supset S$

is defined as a compression semigroup of a $\mathscr{B}$-convex set $D \supset C_{\Theta(S)}$. Namely, D is the complement of the union of the domains of attraction of the control sets in $\mathbb{F}_{\Theta(S)}$ different from the invariant control set $C_{\Theta(S)}$. It can be proved that D is $\mathscr{B}$-convex (see Sect. 3 above) and hence its compression semigroup T is indeed $\Theta(S)$-maximal.

It is a classical fact that Laplace transforms are analytic functions. For the Helgason–Laplace transform (3), it is proved in [78] the following partial result on the smoothness of $I_S(\lambda, x)$.

Theorem 4.4 *If $\Theta(S) = \emptyset$, then $I_S(\lambda, x)$ is analytic as a function of x in its domain for fixed $\lambda \in -\mathscr{C}^+$. If the flag type $\Theta(S) \neq \emptyset$, then $I_S(\lambda, x)$ is $\mathscr{C}^k$ with k becoming larger as the size of λ increases.*

Example As an example of a characteristic function, let $S = \mathrm{Sl}^+(d, \mathbb{R}) \subset \mathrm{Sl}(d, \mathbb{R})$ be the semigroup of matrices with nonnegative entries. Its flag type $\mathbb{F}_{\Theta(S)}$ is the projective space $\mathbb{P}^{d-1}$ and C_0 is the interior of the subset subsumed by the positive orthant $\mathbb{R}^d_+ \subset \mathbb{R}^d$.

Thanks to the fact that the group $A \subset \mathrm{Sl}(d, \mathbb{R})$ of diagonal matrices with positive entries is contained in $\mathrm{Sl}^+(d, \mathbb{R})$ and acts transitively in C_0 it is possible to compute explicitly the characteristic function of S. It is given by the analytic function defined on C_0 by

$$I_S(s\lambda_1, [x]) = d^{s/2} I_S(s\lambda_1, [x_0]) (x_1 \cdots x_d)^{1/d}$$

if $x = (x_1, \ldots, x_d) \in \left(\mathrm{int}\mathbb{R}^d_+\right) \cap S^{d-1}$. In this expression, d is the dimension, $x_0 = (1, \ldots, 1)$, and λ_1 is the generator of $\mathfrak{a}^*_{\Theta(s)}$ which is the linear map that associates to a diagonal matrix its first eigenvalue.

4.3 Moment Lyapunov Exponents

Let G be a noncompact semi-simple Lie group and μ a probability measure on G. Denote by S_μ the semigroup generated by the support $\mathrm{supp}\mu$ of μ and assume that $\mathrm{int}S_\mu \neq \emptyset$.

Take Iwasawa decompositions $\mathfrak{g} = \mathfrak{k} \oplus \mathfrak{a} \oplus \mathfrak{n}$ and $G = KAN$. For $g \in G$, write $g = ue^{H(g)}n \in KAN$ with $H(g) \in \mathfrak{a}$. The map $\sigma(g, k) = H(gk)$, $g \in G$, $k \in K$ descends to a map say $\mathsf{a} : G \times \mathbb{F} \to \mathfrak{a}$ where $\mathbb{F}$ is the maximal flag manifold of G. It satisfies the cocycle property

$$\mathsf{a}(gh, y) = \mathsf{a}(g, hy) + \mathsf{a}(h, y).$$

Take $\lambda \in \mathfrak{a}^*$. The λ-moment Lyapunov exponent $\gamma_\lambda(x)$ in the "direction" of $x \in \mathbb{F}$ is defined by

$$\gamma_\lambda(x) = \lim_{n\to+\infty} \sup \frac{1}{n} \log \int_G e^{\lambda \mathsf{a}(g,x)} \mu^{*n}(dg)$$

where μ^{*n} is the nth convolution power. (For this definition to make sense, it must be assumed that the integrals are finite which holds if $\int_G e^{\lambda \mathsf{a}(g,x)} \mu(dg) < \infty$ for all λ. For instance, if suppμ is compact, then the integrals are finite.)

The following results relate the moment Lyapunov exponents with the flag type of S_μ.

Theorem 4.5 ([70]) *Suppose that $S_\mu = G$. Then, for every $\lambda \in \mathfrak{a}^*$ and $x \in \mathbb{F}$, there exists $p < 0$ such that*

$$\gamma_{p\lambda}(x) > 0.$$

Now, let $\mathbb{F}_\Theta$ be the flag type of S_μ (assuming that S_μ is proper). Denote by Σ the simple system of roots and by Φ the corresponding set of dominant weights. Also, let Φ_Θ be the dominant weights corresponding to Θ, that is,

$$\Phi \setminus \Phi_\Theta = \{\omega \in \Phi : \forall \alpha \in \Theta,\ \langle \alpha, \omega \rangle = 0\}.$$

The cone generated by $\Phi \setminus \Phi_\Theta$ is the closure of the partial chamber $\left(\mathfrak{a}_\Theta^*\right)^+$ given by $\lambda \in \mathfrak{a}^*$ such that $\langle \lambda, \alpha \rangle > 0$ if $\alpha \in \Sigma \setminus \Theta$ and $\langle \lambda, \beta \rangle = 0$ if $\beta \in \Theta$.

Theorem 4.6 ([70]) *Let Θ be such that $\mathbb{F}_\Theta$ is the flag type of S_μ. Take $\lambda \in$ cl$\left(\mathfrak{a}_\Theta^*\right)^+$. Then, there exists $x \in \mathbb{F}$ such that*

$$\gamma_{p\lambda}(x) \leq 0$$

for all $p < 0$.

The following theorem is a partial converse to the previous one and implies the first theorem.

Theorem 4.7 ([70]) *Let Θ be such that $\mathbb{F}_\Theta$ is the flag type of S_μ. Take λ in the subspace spanned by Θ. Then, for all $x \in \mathbb{F}$, there exists $p < 0$ such that*

$$\gamma_{p\lambda}(x) > 0.$$

5 Controllability and Transitive Actions

In a semi-simple noncompact Lie group G with finite center, a proper semigroup S with int$S \neq \emptyset$ cannot act transitively on any flag manifold of G (see Theorem 1.3 above). As proved in [69], after a preparation in [76], proper semigroups do not act transitively in almost all homogeneous spaces of G.

Theorem 5.1 ([69]) *Let G be a semi-simple noncompact Lie group with finite center, $L \subset G$ a closed subgroup, and $S \subset G$ a subsemigroup with* $\text{int}S \neq \emptyset$. *Then, the following two conditions are necessary for S to act transitively in G/L.*

1. *The action of L on $\mathbb{F}_{\Theta(S)}$ is minimal, that is, every orbit of L in $\mathbb{F}_{\Theta(S)}$ is dense.*
2. *There exists a sequence $g_n \in L$ which is contractive in $\mathbb{F}_{\Theta(S)}$; that is, there exists an open Bruhat cell σ such that $g_n\sigma$ shrinks to a point as $n \to \infty$.*

If S is the compression semigroup of its invariant control set $C_{\Theta(S)}$, then the conditions are sufficient as well.

A kind of subgroup that satisfies both conditions of this theorem is a lattice $L \subset G$, which is a discrete subgroup such that G/L admits a finite G-invariant measure. In such a homogeneous space, Poincaré's recurrence theorem permits to show that every semigroup S with $\text{int}S \neq \emptyset$ acts transitively in G/L.

On the other hand in [76], there is the example of $\text{Sl}(2n, \mathbb{R})/\text{Sp}(n, \mathbb{R})$ where a proper semigroup $S \subset \text{Sl}(2n, \mathbb{R})$ acts transitively if the flag type of S is the projective space $\mathbb{P}^{d-1}$. The subgroup $\text{Sp}(n, \mathbb{R}) \subset \text{Sl}(2n, \mathbb{R})$ is one of the few subgroups L with a finite number of connected components such that there are semigroups acting transitively in G/L. For a subgroup L with a finite number of connected components, there are the following strong restrictions so that it satisfies the conditions of Theorem 5.1.

Theorem 5.2 ([69]) *Suppose that L has a finite number of connected components and satisfies the two conditions of Theorem 5.1 for a flag manifold $\mathbb{F}_\Theta$. Let L_0 be the identity component of L. Then,*

1. *L_0 also satisfies the conditions;*
2. *L_0 is reductive, noncompact, and acts transitively in $\mathbb{F}_\Theta$;*
3. *$\mathbb{F}_\Theta$ is a flag manifold of L_0.*

In the thesis [31], one can find several pairs (L, G) of noncompact semi-simple Lie groups with $L \subset G$ that have a common flag manifold.

One of the motivations to look at transitive actions of semigroups is the controllability problem for control systems. Let

$$\dot{g} = X(g) + \sum_{i=1}^{m} u_i Y_i(g) \tag{4}$$

be a control system in G where X and $Y_1, \ldots, Y_m$ are right invariant vector fields and $u_i : \mathbb{R}_+ \to \mathbb{R}$ are control functions such that $u_1(t) Y_1 + \cdots + u_m(t) Y_m$ assume values in a certain subset U of the Lie algebra $\mathfrak{g}$ of G. The controllability properties of (4) are described by the semigroup S generated by the exponentials e^{tZ} with $Z \in X + U$ and $t \geq 0$. This semigroup has non-empty interior if and only if X and $Y_1, \ldots, Y_m$ generate the Lie algebra $\mathfrak{g}$ of G.

The control system (4) is said to be controllable if $S = G$. The above results on the transitivity of S allows to study controllability by looking at transitive actions on homogeneous spaces. For instance, (4) is controllable if and only if its control semigroup S acts transitively on some (and hence all) flag manifold of G. Instead of a flag manifold, one can take any homogeneous space not falling into the conditions of Theorem 5.1.

A complete picture of the controllability in $\mathrm{Sl}(2, \mathbb{R})$ was obtained in [8, 14] by analyzing the action of the control semigroup on the projective line $\mathbb{P}^1$. A sample result is given as follows.

Theorem 5.3 *Suppose that $X, Y \in \mathfrak{sl}(2, \mathbb{R})$ generate $\mathfrak{sl}(2, \mathbb{R})$. Then, the control system $\dot{g} = X(g) + uY(g)$, $u \in [-\rho, \rho]$ is controllable if and only if the segment $\{X + uY : u \in [-\rho, \rho]\}$ meets the open double cone*

$$\{Z \in \mathfrak{sl}(2, \mathbb{R}) : \det Z > 0\}.$$

Other results for rank 1 Lie groups appear in [51] for the group $\mathrm{SO}(1, n)$.

A method that emerges from the existence of the flag type comes from the fact that if S is a proper semigroup then the invariant control set $C_{\Theta(S)}$ is contractible in the flag type $\mathbb{F}_{\Theta(S)}$. In [5, 82, 83], this fact was exploited to get the next result. In its statement, we denote by $G(\alpha)$ the subgroup generated by $\exp \mathfrak{g}_{\pm\alpha}$. For instance, if $G = \mathrm{Sl}(d, \mathbb{R})$ or $\mathrm{Sl}(d, \mathbb{C})$, then a $G(\alpha)$ is a subgroup of matrices leaving invariant a subspace $\langle e_i, e_j \rangle$ spanned by two basic vectors and which is the identity in the subspace spanned by the remaining basic vectors. In this case, $G(\alpha)$ is isomorphic to $\mathrm{Sl}(2, \mathbb{R})$.

Theorem 5.4 *Let G be a connected simple Lie group with Lie algebra $\mathfrak{g}$ and $S \subset G$ a semigroup with* $\mathrm{int} S \neq \emptyset$. *Then, $S = G$ if there is a root α with $G(\alpha) \subset S$ in the following cases:*

1. *$\mathfrak{g}$ is complex.*
2. *$\mathfrak{g} = \mathfrak{sl}(l + 1, \mathbb{R})$.*
3. *$\mathfrak{g} = \mathfrak{sp}(l, \mathbb{R})$ and α is a long root.*
4. *$\mathfrak{g}$ is the split real form associated to G_2 and α is a short root.*

The proof of this theorem consists in showing that in any flag manifold $\mathbb{F}_\Theta$ there is a compact orbit of $G(\alpha)$ that must be contained in the invariant control set of S but is not contractible in $\mathbb{F}_\Theta$ permitting to conclude that S must be the whole group G. In case $\mathfrak{g}$ is complex, the compact $G(\alpha)$-orbit is a 2-sphere so the second homotopy group $\pi_2(\mathbb{F}_\Theta)$ is worked out to check noncontractibility. In the other cases, the orbit in question is a circle and the fundamental group shows up.

In [83], the same technique was applied to other subgroups besides $G(\alpha)$.

The choice of the subgroup $G(\alpha)$ in Theorem 5.4 was inspired by the following result by Jurdjevic–Kupka [36].

Theorem 5.5 *Suppose that A and B are $d \times d$ trace zero matrices such that*

1. *$B = \mathrm{diag}\{b_1, \ldots, b_d\}$ with $b_i - b_j \neq b_r - b_s$ if $(i, j) \neq (r, s)$ and*
2. *$A = (a_{ij})$ satisfies $a_{1n}a_{n1} < 0$.*

Then, the control system $\dot{g} = Ag + uBg$ is controllable in $\mathrm{Sl}(d, \mathbb{R})$ *if A and B generates the Lie algebra $\mathfrak{sl}(d, \mathbb{R})$.*

This theorem was generalized to semi-simple Lie groups in Jurdjevic–Kupka [37] and motivated several papers in the 1980s containing improvements and generalizations (see El Assoudi-Gauthier-Kupka [23]).

The relationship between these classical results and Theorem 5.4 is that in the very first step of the proof of Theorem 5.5 it is shown that the control semigroup contains a subgroup $G(\alpha)$, namely the $\mathrm{Sl}(2,\mathbb{R})$ subgroup of the subspace spanned by the basic vectors e_1 and e_d. Hence, the proof of Theorem 5.5 reduces to Theorem 5.4. It should be said that the methods of [5, 82, 83] are global hence Theorem 5.4 is not restricted to infinitesimally generated semigroups as are the needs of Theorem 5.5.

In the same vein as Theorem 5.5, it was proved in [62] the following result for discrete-time control systems.

Theorem 5.6 *Consider the discrete-time control system*

$$g_{n+1} = e^{A+uB} g_n \tag{5}$$

in $\mathrm{Sl}(d,\mathbb{R})$ *with* A *and* B *trace zero* $d \times d$ *matrices. Suppose that* $B = \mathrm{diag}\{\lambda_1, \ldots, \lambda_d\}$ *with* $\lambda_1 > \cdots > \lambda_d$. *If* $A = (a_{ij})$, *denote by* A^B *the matrix which is zero on the diagonal and whose entries* $i, j, i \neq j$ *are given by* $\frac{a_{ij}}{\lambda_j - \lambda_i}$. *Assume that* A *and* B *generates* $\mathfrak{sl}(d,\mathbb{R})$ *and that*

$$(-1)^{k+1} a_k^+(A^B) a_k^-(A^B) > 0$$

for all $k = 1, \ldots, d$ *where* $a_k^+(A^B)$ *(respectively,* $a_k^-(A^B)$*) is the upper right (respectively, lower left)* $k \times k$ *minor of* A^B.

Then, (5) is controllable, that is the semigroup generated by e^{A+uB}, $u \in \mathbb{R}$, *is the whole* $\mathrm{Sl}(d,\mathbb{R})$.

This theorem was extended to the symplectic group in [17, 19].

The papers [52, 53] study the existence of cones invariant by bilinear control systems in $\mathbb{R}^d$ (that are the same as invariant systems in the group $\mathrm{Gl}(d,\mathbb{R})$). This kind of question was posed by Sachkov [58] where it is conjectured that a bilinear control system is not controllable if and only if there is a cone $W \subset \mathbb{R}^d$ invariant by the system. It is shown in [52, 53] that this conjecture is not true. Furthermore, necessary and sufficient conditions for the existence of invariant cones are given in terms of the flag type of the semigroup generated by the control system.

6 Dynamical Systems

Let ϕ_t ($t \in \mathbb{Z}$ or $\mathbb{R}$) be a continuous flow on a metric space (X, d). For $\varepsilon, T > 0$, an (ε, T)-chain (or pseudo-orbit) of ϕ_t is given by a finite sequence of points $x_1, \ldots, x_k$ and times $t_1, \ldots, t_{k-1} \geq T$ such that $d(\phi_{t_i}(x_i), x_{i+1}) < \varepsilon$, $i = 1, \ldots, k-1$. A subset $M \subset X$ is chain transitive if for any pair of points $x, y \in M$ there are (ε, T)-chains starting in one of the points and ending at the other for all $\varepsilon, T > 0$. Chain transitive sets that are maximal by set inclusion are used as tools to build Morse decompositions in the context of Conley theory (see Conley [21]).

In [16, 46–48], and in the thesis [44], semigroup theory was applied to study maximal chain transitive sets. The idea is to "close chains" with continuous maps yielding to the definition of the semigroups $S_{\varepsilon,T}$, $\varepsilon, T > 0$, which are the semigroups generated by continuous maps that are ε-close (in their domains) to some ϕ_t, $t > T$. The semigroups $S_{\varepsilon,T}$ were named shadowing semigroups. The principles relating the shadowing semigroups to the chain transitive sets are summarized in the following items:

1. There exists an (ε, T)-chain starting at $x \in X$ and ending at $y \in X$ if and only if $y \in S_{\varepsilon,T}x$;
2. The maximal chain transitive sets of the flow ϕ_t are obtained by shrinking a family of control sets, say $D_{\varepsilon,T}$, of the semigroups $S_{\varepsilon,T}$ as $\varepsilon \to 0$ and $T \to +\infty$.

These facts are true under suitable assumptions on the space X. In [16], shadowing semigroups $S_{\varepsilon,T}$ are taken in the local group of local homeomorphisms of X and the above facts are proved with a local transitivity hypothesis. These results are generalized in [44, 46, 47] in two directions, namely the shadowing semigroups are taken in the whole space of continuous functions and more importantly the theory is developed in Hausdorff topological spaces.

The idea of the shadowing semigroups was used in [15, 19] in a different context, namely for the study of the chain control sets of semigroup actions.

The following theorem is an application of the shadowing semigroup method combined with the description of the control sets on the flag manifolds discussed in Sect. 1. In the next statement, we use the notation $\mathrm{fix}\,(h, w)$ for a connected component of the fixed point set of the hyperbolic element $h \in G$ in a flag manifold $\mathbb{F}_\Theta$. These components are parametrized by the elements w of the Weyl group $\mathcal{W}$ in such a way that $\mathrm{fix}\,(h, 1)$ is the only attractor fixed point set.

Theorem 6.1 ([16, 48]) *Let G be a noncompact semi-simple Lie group and ϕ_t a flow on $X \times G$ which is right invariant, that is, $\phi_t(x, gh) = \phi_t(x, g)h$ so that $\phi_t(x, g) = (\theta_t(x), \rho(t, x)g)$ where θ_t is a flow on X and ρ is a cocycle over θ with values in G.*

If $\mathbb{F}_\Theta$ is a flag manifold of G, then we have a flow ψ_t on $X \times \mathbb{F}_\Theta$ given by $\psi(x, f) = (\theta_t(x), \rho(t, x)f)$. Suppose that the flow θ_t is chain transitive. Then, there exists a hyperbolic element $h_\phi \in G$ and a continuous map $\sigma : X \to \{gh_\phi g^{-1} : g \in G\}$ such that the maximal chain transitive sets of ψ (Morse components) are given by

$$\mathcal{M}_w = \bigcup_{x\in X} \{x\} \times g_x \mathrm{fix}\left(h_\phi, w\right) g_x^{-1}$$

where $g_x \in G$ is such that $\sigma(x) = g_x h_\phi g_x^{-1}$.

For instance, in $G = \mathrm{Sl}(d, \mathbb{R})$, a hyperbolic element h_ϕ is a diagonal matrix with positive entries. If $\mathbb{F}_\Theta$ is the projective space $\mathbb{P}^{d-1}$, then a fixed point component fix (h_ϕ) is the set of lines contained in an eigenspace of h_ϕ. Thus, the maximal chain transitive sets on $X \times \mathbb{P}^{d-1}$ (Morse components) are obtained from a Whitney decomposition of the bundle $X \times \mathbb{R}^d$. This is the content of the theorem by Selgrade [84] (see also [20]). Hence, Theorem 6.1 generalizes Selgrade's theorem with an independent proof.

In [16, 48], the above theorem is worked out in the more general setting of a right invariant flow in a principal bundle $Q \to X$ with structural group G (not necessarily the trivial one $X \times G$). Also, in [48], there are results for semiflows which bring some subtleties related to the invariant subsets.

Proposition – Definition 6.2 *Let h_ϕ be as in Theorem 6.1. Then, there exists a unique flag manifold $\mathbb{F}_{\Theta(\phi)}$ which is maximal with the property that the attractor fixed point set* fix $(h_\phi, 1)$ *of h_ϕ in $\mathbb{F}_{\Theta(\phi)}$ is a singleton. This flag manifold is called the flag type for the Morse decomposition of ϕ_t.*

Now, let $G = KAN$ be an Iwasawa decomposition of G. If $\phi_t(x, g) = (\theta_t(x), \rho(t, x) g)$ is a right invariant flow on $X \times G$ write for $x \in X$ and $u \in K$,

$$\rho(t, x) u = k_t(x, u) a_t(x, u) n_t(x, u) \in KAN.$$

The component $k_t(x, u)$ is a flow that is understood via the induced flows on the flag bundles $X \times \mathbb{F}_\Theta$. The characteristic exponents are given by the asymptotics of the component $a_t(x, u)$. The mapping $a_t(x, u)$ is a multiplicative cocycle (with values in A) that factors to a cocycle over $X \times \mathbb{F}$ where $\mathbb{F}$ is the maximal flag manifold. Its logarithm $\mathsf{a}(t, x, u) = \log a_t(x, u)$ is an additive cocycle. The limits

$$\lambda(x, u) = \lim_{t \to \infty} \frac{1}{t} \mathsf{a}(t, x, u) \tag{6}$$

are the (vector valued) Lyapunov exponents of the flow.

Related to the Morse decomposition, it was introduced by Colonius–Kliemann [20] the concept of Morse spectrum set that measures the growth ratio along chains. The Morse spectrum $\Lambda_{\mathrm{Mo}}(\mathcal{M}) \subset \mathfrak{a}$ over a chain component $\mathcal{M}$ of the flow is defined by evaluating the cocycle $\mathsf{a}(t, x, u)$ along chains in $\mathcal{M}$ taking into account the jumps of the chains. In [42, 79], the concept of Morse spectra of [20] was extended to the vector valued cocycle $\mathsf{a}(t, x, u)$. Their main properties were derived in the light of the Morse decomposition on the flag bundles and the flag type of a flow.

Theorem 6.3 ([42, 79]) *For $w \in \mathcal{W}$, let $\mathcal{M}_w$ be the Morse component in $X \times \mathbb{F}$ as in Theorem 6.1 where $\mathbb{F}$ is the maximal flag manifold. Write $\mathcal{M}^+ = \mathcal{M}_1$ for the attractor component. Then, the Morse spectra $\Lambda_{\mathrm{Mo}}(\mathcal{M}_w) \subset \mathfrak{a}$ satisfy the following properties:*

1. *For every* $w \in \mathscr{W}$, *we have*

$$\Lambda_{\mathrm{Mo}}\left(\mathscr{M}_{w}\right) = w^{-1}\Lambda_{\mathrm{Mo}}(\mathscr{M}^{+}),$$

so that the whole Morse spectra is read off from the spectrum of the attractor component.
2. *The spectra* $\Lambda_{\mathrm{Mo}}\left(\mathscr{M}^{+}\right)$ *of the attractor component is invariant by the subgroup* $\mathscr{W}_{\phi} \subset \mathscr{W}$ *that fixes* $H_{\phi} = \log h_{\phi}$.
3. $\Lambda_{\mathrm{Mo}}\left(\mathscr{M}^{+}\right)$ *is contained in the interior of* $\bigcup_{w\in\mathscr{W}_{\phi}} \mathrm{cl}\mathscr{C}_{w}$, *where* $\mathscr{C}_{w}$ *is the Weyl chamber associated to* w.
4. $\Lambda_{\mathrm{Mo}}\left(\mathscr{M}^{+}\right)$ *intercepts the closure of every chamber* $\mathscr{C}_{w}$, $w \in \mathscr{W}_{\phi}$.
5. *Different Morse spectra do not overlap. (This fact is not true for linear flows on vector bundles as shown in [20], Example 5.5.11. The point here is that the vector bundle Morse spectra are images under linear maps of the vector valued spectra. Overlappings of the images may occur.)*

The limits (6) (that is, the vector valued Lyapunov exponents) are analyzed in [3, 41] where it is offered an analogous of the classical multiplicative ergodic theorem of Oseledets. For this theorem, one must have in advance a probability measure ν on the base space which is invariant by the flow on X. Then, it is proved that for ν-almost every $x \in X$ the limit (6) exists for every point in the fiber over x. If ν is an ergodic measure, then there is a flag type $\Theta_{\mathrm{Ly}}(\nu)$ describing the decomposition of the fibers given by the level sets of the Lyapunov exponents.

About the spectra, there are also the following results:

1. In [4, 41], there are necessary and sufficient conditions ensuring that the flag type $\Theta_{\mathrm{Ly}}(\nu)$ given by the multiplicative ergodic theorem equals the flag type Θ_{Mo} coming from the Morse decomposition. In general, $\Theta_{\mathrm{Ly}}(\nu) \subset \Theta_{\mathrm{Mo}}$. Equality between the two flag types means that the measurable Oseledets decomposition is well behaved in the sense that it has a continuous extension to all of the base space.
2. The differentiable dependence of the Lyapunov exponents as a function of the flow is treated in [24, 85] where the flow ϕ on a principal bundle Q is perturbed as $\phi\gamma$ with γ varying in the gauge group $\mathscr{G}(Q)$ of Q. It is proved that if ω belongs to the annihilator of Θ_{Mo}, then $\omega\left(\Lambda_{\mathrm{Ly}}\right)$ depends differentiable of γ, recalling that the $\mathscr{G}(Q)$ has the structure of a Banach Lie group (usually infinite dimensional). This is a generalization of a result by Ruelle [56] that is proved for a continuous linear flow leaving invariant a cone. One of the main achievements of [24, 85] is to put in evidence the Morse decomposition as a tool to look at the differentiability properties of the Lyapunov exponents.

Acknowledgements Supported by CNPq grant no. 303755/09-1, FAPESP grant no. 2012/18780-0, and CNPq/Universal grant no. 476024/2012-9.

References

1. Adriano João da Silva. Invariance entropy for control systems on Lie groups and homogeneous spaces. Thesis, Unicamp 2014.
2. Alexandre José Santana. Homotopia de semigrupos. Thesis, Unicamp 2000.
3. Alves, L.A.; San Martin, L. A. B.: *Multiplicative ergodic theorem on flag bundles of semi-simple Lie groups*. Discrete and Continuous Dynamical Systems. Series A, **33**, (2013) 1247–1273.
4. Alves, L.A.; San Martin, L. A. B.: *Conditions for equality between Lyapunov and Morse decompositions*. Ergodic Theory & Dynamical Systems, v. 36, p. 1007–1036, 2016.
5. Ariane Luzia dos Santos. Controlabilidade de sistemas de controle em grupos de Lie simples e a topologia das variedades flag. Thesis, Unicamp 2011.
6. Arnold, L., and W. Kliemann: *Qualitative theory of stochastic systems*. Probabilistic Analysis and Related Topics, vol. **3**, A. T. Barucha-Reid, Ed., Academic Press (1983) 1–79.
7. Arnold, L., W. Kliemann, and E. Oeljeklaus: *Lyapunov exponents of linear stochastic systems*. In: Lyapunov Exponents, L. Arnold, and V. Wihstutz, Eds., LNM (Springer) 1186 (1986).
8. Ayala, V.; San Martin, L. A. B.: *Controllability of two-dimensional bilinear systems: restricted controls and discrete-time*. Proyecciones, v. 18, n.2, p. 207–223, 1999.
9. Ayala, V.; Kliemann, W.; San Martin, L. A. B.: *Control sets and total positivity*. Semigroup Forum, v. 69, n.1, p. 113–140, 2004.
10. Ayala, V.; Ribeiro Jr., R and San Martin, L. A. B.: *Controllability on* Sl(2, $\mathbb{C}$) *with restricted controls*. SIAM J. Control Optim, v. 52, number 4, 2548–2567, 2014.
11. W.M. Boothby: *A transitivity problem from control theory*. J. Differential Equations 17: 296–307, 1975.
12. Boothby, W.M. and E.N. Wilson: *Determination of the transitivity of bilinear systems*. SIAM J. on Control Optim. **17** (1979), 212–221.
13. Braga Barros, C. J.; San Martin, L. A. B.: *On the action of semigroups in fiber bundles*. Matemática Contemporânea, Brasil, v. 13, p. 1–19, 1997.
14. Braga Barros, C. J.; Ribeiro, J. ; Rocio, O. G.; San Martin, L. A. B.: *Controllability of two-dimensional bilinear systems*. Proyecciones, v. 15, n.2, p. 111–139, 1996.
15. Braga Barros, C. J.; San Martin, L. A. B.: *Chain control sets for semigroup actions*. Computational and Applied Mathematics, v. 15, n.3, p. 257–276, 1996.
16. Braga Barros, C. J.; San Martin, L. A. B.: *Chain transitive sets for flows on flag bundles*. Forum Mathematicum, v. 19, p. 19–60, 2007.
17. Braga Barros, C. J.; San Martin, L. A. B.: *Controllability of discrete-time control systems on the symplectitc group*. Systems & Control Letters, v. 42, n.2, p. 95–100, 2001.
18. Braga Barros, C. J.; San Martin, L. A. B.: *On the number of control sets on projective Spaces*. Systems & Control Letters, v. 29, p. 21–26, 1996.
19. Carlos José Braga Barros. Conjuntos controláveis e conjuntos controláveis por cadeias para a ação de semigrupos. Thesis, Unicamp 1995.
20. Colonius, F. and W. Kliemann: The dynamics of control. Birkhäuser, Boston (2000).
21. Conley C.: *Isolated invariant sets and the Morse index*. CBMS Regional Conf. Ser. in Math., **38**, American Mathematical Society, (1978).
22. Duistermaat, J.J.; J.A.C. Kolk and V.S. Varadarajan: *Functions, flows and oscillatory integrals on flag manifolds and conjugacy classes in real semisimple Lie groups*. Compositio Mathematica, **49** (1983), 309–398.
23. El Assoudi R.; Gauthier, J. P.; Kupka I. K.: *On subsemigroups of semisimple Lie groups*. Ann. Inst. H. Poincaré Anal. Non Linéaire, 13(1):117–133, 1996.
24. Ferraiol, T.; San Martin, L. A. B.: *Differentiability of Lyapunov exponents*. To appear.
25. Furstenberg H. – A Poisson formula for semi-simple Lie groups. *Annals of Maths.*, **77** (1963), 335–386.
26. Gasparim, E.; Grama, L.; San Martin, L. A. B.: *Adjoint Orbits and Cotangent Bundles of Flag Manifolds*. To appear.

27. Guivarc'h, Y. and A. Raugi: *Frontière de Furstenberg, propriétés de contraction et théorèmes de convergence*. Z.W. **69** (1985), 187–242.
28. Hilgert, J., K. H. Hofmann, and J. D. Lawson: Lie Groups, Convex Cones and Semigroups, Oxford University Press, 1989.
29. Hilgert, J. and K.-H. Neeb: Lie semigroups and their applications. Springer Lecture Notes in Mathematics **1552**. Springer-Verlag (1993).
30. Humphreys, J. E.: Reflection groups and Coxeter groups. Cambridge Studies in Advanced Mathematics, **29**. Cambridge University Press (1990).
31. Janete de Paula Ferrareze. Transitividade de semigrupos em variedades homogêneas. Thesis, Unicamp 2012.
32. João Ribeiro Gonçalves Filho. Cones e semigrupos. Thesis, Unicamp 2001.
33. K. D. Johnson: *The structure of parabolic subgroups*. J. Lie Theory **14** (2004), 287–316.
34. Jorge Nicolás Lopez. Espaços de Poisson-Furstenberg e medidas invariantes em grupos de Lie semi-simples. Thesis, Unicamp 2005.
35. Josiney Alves de Souza. Ações de semigrupos: recorrência por cadeias em fibrados e compactificações de Ellis. Thesis, Unicamp 2008.
36. Jurdjevic V.; Kupka I.: *Control systems subordinated to a group action: accessibility*. J. Differential Equations, 39(2):186–211, 1981.
37. Jurdjevic V.; Kupka I.: *Control systems on semisimple Lie groups and their homogeneous spaces*. Ann. Inst. Fourier (Grenoble), 31(4):vi, 151–179, 1981.
38. Kliemann, W.: *Recurrence and invariant measures for degenerate diffusions*. Ann. Probab. **15** (1987), 690–707.
39. Laércio José dos Santos. Semigrupos gerados por classes laterais e funções características de semigrupos. Thesis, Unicamp 2007.
40. Lopez, J. N.; Ruffino, P. R. C.; San Martin, L. A. B.: *Poisson spaces for proper semigroups of semi-simple Lie groups*. Stochastics and Dynamics, v. 7, p. 273–297, 2007.
41. Luciana Aparecida Alves. Expoentes de Lyapunov e Morse em fibrados flag. Thesis, Unicamp 2010.
42. Lucas Conque Seco Ferreira. Expoentes de Morse vetoriais e semifluxos em fibrados flag. Thesis, Unicamp 2007.
43. Marcos André Verdi. Conjuntos de controle em órbitas adjuntas e compactificações ordenadas de semigrupos. Thesis, Unicamp 2007.
44. Mauro Moraes Alves Patrão. Semifluxos em fibrados flag e seus semigrupos de sombreamento. Thesis, Unicamp 2006.
45. Osvaldo Germano do Rocio. Semigrupos Discretos em Grupos Solúveis. Thesis, Unicamp 1995.
46. Patrão, M.: *Morse decomposition of semiflows on topological spaces*. J. Dyn. Diff. Eq., **19** (2007), 181–198.
47. Patrão, M. M. A.; San Martin, L. A. B.: *Semiflows on topological spaces: Chain transitivity and shadowing semigroups*. Journal of Dynamics and Differential Equations, v. 19, p. 155–180, 2007.
48. Patrão, M. M. A.; San Martin, L. A. B.: *Morse decomposition of semiflows on fiber bundles*. Discrete and Continuous Dynamical Systems. Series A, v. 17, p. 561–587, 2007.
49. Patrão, M. M. A.; San Martin, L. A. B.; Seco, Lucas: *Conley indexes and stable sets for flows on flag bundles*. Dynamical Systems, v. 24, p. 249–276, 2009.
50. Rocio, O. G.; San Martin, L. A. B.: *Connected components of open semigroups in semi-simple Lie groups*. Semigroup Forum, v. 69, n.1, p. 1–29, 2004.
51. Richard Manuel Mamani Troncoso. Aspectos da teoria de semigrupos em grupos de Lie semi-simples e aplicações. Thesis, Unicamp 1999.
52. Rocio, O. G.; San Martin, L. A. B.; Santana, A. J.: *Invariant cones and convex sets for bilinear control systems and parabolic type of semigroups*. Journal of Dynamical and Control Systems, v. 12, n.3, p. 419–432, 2006.
53. Rocio, O. G. ; San Martin, L. A. B. ; Santana, A. J.: *Invariant cones for semigroups in transitive Lie groups*. Journal of Dynamical and Control Systems, v. 14, p. 559–569, 2008.

54. Ronan Antonio dos Reis. Ações de Semigrupos em Espaços Homogêneos. Thesis, Unicamp 2004.
55. Reis, Ronan A. ; San Martin, Luiz A.B.: *Reversibility properties of semigroup actions on homogeneous spaces*. Semigroup Forum, v. 84, p. 472–486, 2012.
56. Ruelle, D.: *Analycity Properties of the Characteristic Exponents of Random Matrix Products*. Adv. Math. **32** (1979), 68–80.
57. Ruppert, W. A. F.: *On open subsemigroups of connected groups*. Semigroup Forum, **39** (1989), 347–362.
58. Y.L. Sachkov: *On invariant orthants of bilinear systems*. J. Dynam. Control Systems, vol. 4, **1** (1998), 137–147.
59. San Martin, L. A. B.; Arnold, L.: *A control problem related to the Lyapunov spectrum of stochastic flows*. Computational and Applied Mathematics, **5** (1986), 31–64.
60. San Martin, L.: *Controllability of families of measure preserving vector fields*. Systems & Control Letters **8** (1987), 459–462.
61. San Martin, L.A.B.: *Invariant control sets on flag manifolds*. Math. of Control, Signals, and Systems, **6** (1993), 41–61.
62. San Martin, L.A.B.: *On global controllability of discrete-time control systems*. Math. of Control, Signals, and Systems, **8** (1995), 279–297.
63. San Martin, L. A. B.: *Nonexistence of invariant semigroups in affine symmetric spaces*. Mathematische Annalen, v. 321, p. 587–600, 2001.
64. San Martin,L.A.B.: *Control sets and semigroups in semi-simple Lie groups*. In Semigroups in algebra, geometry and analysis. Gruyter Verlag (1994).
65. San Martin, L.A.B.: *Orders and domains of attraction of control sets on flag manifolds*. J. of Lie Theory,
66. San Martin, L.A.B.: *Nonreversibility of subsemigroups of semi-simple Lie groups*. Semigroup Forum 44: 376–387, 1992.
67. San Martin, L. A. B.: *On global controllability of discrete-time control systems*. Mathematics of Control, Signals and Systems, v. 8, p. 279–297, 1995.
68. San Martin, L. A. B.: *Order and domains of attraction of control sets in flag manifolds*. Journal of Lie Theory, v. 8, n.2, p. 335–350, 1998.
69. San Martin, L. A. B.: *Homogeneous spaces admitting transitive semigroups*. Journal of Lie Theory, **8** (1998), 111–128.
70. San Martin, L. A. B.: *Moment Lyapunov exponents and semigroups in semi-simple Lie groups*. To appear.
71. San Martin, L. A. B.: *Maximal semigroups in semi-simple Lie groups*. Transactions of the American Mathematical Society, v. 353, p. 5165–5184, 2001.
72. San Martin, L.A.B.: Álgebras de Lie. Ed. Unicamp, second edition (2010).
73. San Martin, L.A.B.: Grupos de Lie. Ed. Unicamp, (2016).
74. San Martin, L. A. B.: *A family of maximal noncontrollable Lie wedges with empty interior*. Systems & Control Letters, v. 43, n.1, p. 53–57, 2001.
75. San Martin, L.A.B. and P. A. Tonelli: *Semigroup actions on homogeneous spaces*. Semigroup Forum **50** (1995), 59–88.
76. San Martin, L.A.B. and P. A. Tonelli: *Transitive actions of semigroups in semi-simple Lie groups*. Semigroup Forum, **58** (1999), 142–151.
77. San Martin, L. A. B.; Ribeiro, J.: *The compression semigroup of a cone is connected*. Portugaliae Mathematica, v. 60, p. 305–317, 2003.
78. San Martin, L. A. B.; Santana, A. J.: *Homotopy type of Lie semigroups in semi-simple Lie groups*. Monatshefte fur Mathematik, v. 136, n.2, p. 151–173, 2002.
79. San Martin, L. A. B.; Seco, Lucas: *Morse and Lyapunov spectra and dynamics on flag bundles*. Ergodic Theory & Dynamical Systems, v. 30, p. 893–922, 2009.
80. San Martin, L. A. B.; Santos, L. J.: *Semigroups in symmetric Lie groups*. Indagationes Mathematicae, v. 18, p. 135–146, 2007.
81. San Martin, L. A. B.; Laércio J. Santos: *Characteristic Functions of Semigroups in Semi-simple Lie Groups*. To appear.

82. Santos, A.L. ; San Martin, L. A. B.: *Controllability of control systems on complex simple Lie groups and the topology of flag manifolds*. Journal of Dynamical and Control Systems, v. 19, p. 157–171, 2013.
83. Santos, A. L.; San Martin, L. A. B.: *A method to find generators of a semi-simple Lie group via the topology of its flag manifolds*. Semigroup Forum. To appear.
84. Selgrade, J.: *Isolate invariant sets for flows on vector bundles*. Trans. Amer. Math. Soc., **203** (1975), 259–390.
85. Thiago Fanelli Ferraiol. *Diferenciabilidade dos expoentes de Lyapunov*. Thesis, Unicamp 2012.
86. Verdi, M. A.; Rocio, O. G.; San Martin, L. A. B.: *Semigroup Actions on Adjoint Orbits*. Journal of Lie Theory, v. 22, p. 931–948, 2012.
87. Vinberg, E. B.: *Invariant convex cones and orderings on Lie groups*. Funct. Anal. and Appl. **14** (1980), 1–13.
88. Zhang Cunhong. *Semigrupos assintóticos e semiálgebricos*. Thesis, Unicamp 2002.

Generic Singularities of 3D Piecewise Smooth Dynamical Systems

Marco Antonio Teixeira and Otávio M. L. Gomide

Abstract The aim of this paper is to provide a discussion on the current directions of research involving typical singularities of $3D$ nonsmooth vector fields. A brief survey of known results is also presented.

We describe the dynamical features of a fold–fold singularity in its most basic form and we give a complete and detailed proof of its local structural stability (or instability). In addition, classes of all topological types of a fold–fold singularity are intrinsically characterized. Such proof essentially follows from some lines laid out by Colombo, García, Jeffrey, Teixeira, and others and it offers a rigorous mathematical treatment under clear and crisp assumptions and solid arguments.

One should highlight that the geometric–topological methods employed lead us to the mathematical understanding of the dynamics around a T-singularity. This approach lends itself to applications in generic bifurcation theory. It is worth to say that such subject is still poorly understood in higher dimension.

1 Introduction

Certain aspects of the theory of nonsmooth vector fields (piecewise smooth vector fields) have been mainly motivated by the study of vector fields near the boundary of a manifold. Concerning this topic, many authors provided results and techniques which have been very useful in piecewise smooth systems. It is worthwhile to cite in the 2-dimensional case works from Andronov et al., Peixoto, and Teixeira (see [1, 18, 23]) and in higher dimensions the works from Sotomayor and Teixeira, Vishik, and Percell (see [19, 22, 31]). In particular, in [31], Vishik provided a classification of generic points lying in the boundary of a manifold, using techniques from Theory of Singularities.

M. A. Teixeira (✉) · O. M. L. Gomide
Institute of Mathematics, Statistics and Scientific Computing, University of Campinas, Campinas, SP, Brazil
e-mail: teixeira@ime.unicamp.br

C. Lavor, F. A. M. Gomes (eds.), *Advances in Mathematics and Applications*,
https://doi.org/10.1007/978-3-319-94015-1_15

Many papers have contributed to the analysis and generic classification of singularities of 2D Filippov systems (Kuznetsov et al., Guardia et al., and Kozlova among others, see [12, 14, 15]). Specifically with respect to the fold–fold singularity, we point Ekeland (see [6]) and Teixeira (see [25]). Regarding the n-dimensional problem, we point out the work from Colombo and Jeffrey (see [9]) which analyzes an n-dimensional family having a two-fold singularity, nevertheless the generic classification for $n > 2$ is much more complicated and still poorly understood.

As far as we know, the first approach where a generic 3D fold–fold singularity was studied was offered by Teixeira in [24] where one finds a discussion on some features of the first return mapping defined around this singularity. Maybe due to this fact, the invisible fold–fold singularity is known as T-singularity.

In [10], Filippov provided a mathematical formalization of the theory of piecewise smooth vector fields. In the last chapter of [10], Filippov studied generic singularities in $3D$ piecewise smooth systems, and a systematic mathematical analysis of the behavior around a fold–fold singularity was officially arisen. However, most of the proofs were only roughly sketched and would require a better explanation and interpretation. In particular, the proofs of the results concerning the fold–fold singularity were obscure and unfinished. Many works appeared lately trying to explain it (see [7, 8, 11, 21, 27]).

In [27], Teixeira established necessary conditions for the structural stability of the fold–fold singularity and he proved that it is not a generic property. Nevertheless, the case of the invisible fold–fold point having a hyperbolic first return map was not understood. He also provided results concerning asymptotic stability.

In [7, 8, 11], Jeffrey et al. also studied the problem of classification of the structural stability around a fold–fold singularity. More specifically, in [11], the authors studied the behavior of a 2-parameter semi-linear model $Z_{\alpha,\beta}$ having a T-singularity at $Z_{0,0}$. By studying the first return map explicitly, they have found countably many curves γ_k in a region of the parameter space, where the topological type β_k of a system in γ_k satisfies $\beta_k \neq \beta_l$ provided $k \neq l$. Moreover, they predict the existence of classes of structural stability between the curves γ_k in this region.

Guided by these results, we show that in the region of the parameter space considered in [11], a general Filippov system Z having a T-singularity at p always has a first return map with complex eigenvalues. It brings several consequences to the behavior of Z around p; in particular, it produces a foliation of this region in the parameter space depending on the argument of the eigenvalues of Z, such that two systems in different leaves are not topologically equivalent near the T-singularity, which means that there is no class of stability in this region of parameters. It provides a negative answer to the questions raised in [11] concerning to the validity of the results for general Filippov systems around a T-singularity.

A 3D-fold–fold singularity is an intriguing phenomenon that has no counterparts in smooth systems, and the complete characterization of the local structural stability of a 3D-nonsmooth system around an elliptic fold–fold singularity has been an open problem over the last 30 years. In this work, we believe that all existing mathematical gaps were filled up and the precise statement of results and proofs were well established.

It is worth to mention that the methods and techniques used in this paper provide a solution from a geometric–topological point of view. In addition, we present a generic and qualitative characterization of a fold–fold singularity, in order to clarify any fact concerning the generality of the results.

2 Setting the Problem

In what follows, we summarize a rough overall description of the basic concepts and results in order to set the problem.

2.1 Filippov Systems

For simplicity, let M be a connected bounded region of $\mathbb{R}^3$ and let $f : M \to \mathbb{R}$ be a smooth function having 0 as a regular value, therefore $\Sigma = f^{-1}(0)$ is a compact embedded codimension one submanifold of M which splits it in the sets $M^{\pm} = \{p \in M; \pm f(p) > 0\}$.

Denote the set of germs of vector fields of class $\mathscr{C}^r$ at Σ by χ^r. Endow χ^r with the $\mathscr{C}^r$ topology and consider $\Omega^r = \chi^r \times \chi^r$ with the product topology.

If $Z = (X, Y) \in \Omega^r$, then a **nonsmooth vector field** is defined in some neighborhood V of Σ in M as follows:

$$Z(p) = F(p) + \operatorname{sgn}(f(p))G(p), \tag{1}$$

where $F(p) = \frac{X(p)+Y(p)}{2}$ and $G(p) = \frac{X(p)-Y(p)}{2}$.

Definition 1 The **Lie derivative** of f in the direction of the vector field $X \in \chi^r$ at $p \in \Sigma$ is defined by $Xf(p) = X(p) \cdot \nabla f(p)$. The **tangency set** of X with Σ is given by $S_X = \{p \in \Sigma;\ Xf(p) = 0\}$.

If $X_1, \ldots, X_n \in \chi^r$, the higher order Lie derivatives are defined as:

$$X_n \ldots X_1 f(p) = X_n(X_{n-1} \ldots X_1 f)(p),$$

that is, $X_n \ldots X_1 f(p)$ is the Lie derivative of the smooth function $X_{n-1} \ldots X_1 f$ in the direction of the vector field X_n at p. In particular, $X^n f(p)$ denotes the Lie derivative $X_n \ldots X_1 f(p)$, where $X_i = X$, for $i = 1, \ldots, n$.

If $Z = (X, Y) \in \Omega^r$, then the switching manifold Σ generically splits into three distinct open regions (Fig. 1):

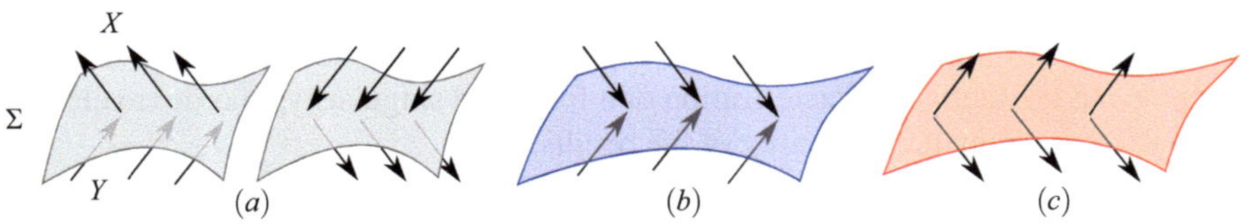

Fig. 1 Regions in Σ: Σ^c in (**a**), Σ^{ss} in (**b**), and Σ^{us} in (**c**)

- Crossing Region: $\Sigma^c = \{p \in \Sigma;\ Xf(p)Yf(p) > 0\}$;
- Stable Sliding Region: $\Sigma^{ss} = \{p \in \Sigma;\ Xf(p) < 0,\ Yf(p) > 0\}$;
- Unstable Sliding Region: $\Sigma^{us} = \{p \in \Sigma;\ Xf(p) > 0,\ Yf(p) < 0\}$.

Consider the **sliding region** of Z as $\Sigma^s = \Sigma^{ss} \cup \Sigma^{us}$.

The tangency set of Z will be referred as $S_Z = S_X \cup S_Y$. Notice that Σ is the disjoint union $\Sigma^c \cup \Sigma^{ss} \cup \Sigma^{us} \cup S_Z$.

The concept of solution of Z follows Filippov's convention. More details can be found in [10, 12, 30].

We highlight that the local solution of $Z = (X, Y) \in \Omega^r$ at a point $p \in \Sigma^s$ is given by the **sliding vector field**:

$$F_Z(p) = \frac{1}{Yf(p) - Xf(p)} \left(Yf(p)X(p) - Xf(p)Y(p)\right). \tag{2}$$

Remark 1 Notice that F_Z is a vector field tangent to Σ^s. The singularities of F_Z in Σ^s are called **pseudo-equilibria** of Z.

Definition 2 If $p \in \Sigma^s$, the **normalized sliding vector field** is defined by:

$$F_Z^N(p) = Yf(p)X(p) - Xf(p)Y(p). \tag{3}$$

Remark 2 If R is a connected component of Σ^{ss}, then F_Z^N is a re-parameterization of F_Z in R, and they have exactly the same phase portrait. If R is a connected component of Σ^{us}, then F_Z^N is a (negative) re-parameterization of F_Z in R, then they have the same phase portrait, but the orbits are oriented in opposite direction.

If $Z = (X, Y) \in \Omega^r$, consider all the integral curves of X in M^+, all the integral curves of Y in M^-, and the integral curves of F_Z in Σ^s. In this work, any oriented piecewise smooth curve passing through q is considered as a solution of Z through q.

2.2 Σ-Equivalence

An orbital equivalence relation is defined in $\Omega^r(M)$ as follows:

Definition 3 Let $Z_0, Z \in \Omega^r$ be two germs of nonsmooth vector fields. We say that Z_0 is **topologically equivalent** to Z at p if there exist neighborhoods U and V of p in M and an order-preserving homeomorphism $h : U \to V$ such that it carries orbits of Z_0 onto orbits of Z, and it preserves Σ, i.e., $h(\Sigma \cap U) = \Sigma \cap V$.

The concept of local structural stability at a point $p \in \Sigma$ is defined in the natural way.

Definition 4 $Z_0 \in \Omega^r$ is said to be **Σ-locally structurally stable** if Z_0 is locally structurally stable at p, for each $p \in \Sigma$.

Denote the space of germs of nonsmooth vector fields $Z \in \Omega^r$ which are Σ-locally structurally stable by Σ_0.

2.3 *Reversible Mappings*

We introduce concepts which will be useful throughout this work. More details can be found in [38, 40].

Definition 5 A germ of an **involution** at 0 is a $\mathscr{C}^r$ germ of a diffeomorphism $\varphi : \mathbb{R}^2 \to \mathbb{R}^2$ such that $\varphi(0) = 0$, $\varphi^2(x, y) = (x, y)$, and $\det[\varphi'(0, 0)] = -1$.

The set of all germs of involutions at 0 is denoted by I^r and it is endowed with the $\mathscr{C}^r$ topology. Consider $W^r = I^r \times I^r$ endowed with the product topology.

Definition 6 Let $\varphi = (\varphi_0, \varphi_1)$, $\psi = (\psi_0, \psi_1) \in W^r$ be two pairs of involutions at 0. Then, φ and ψ are said to be **topologically equivalent** at 0 if there exists a germ of a homeomorphism $h : (\mathbb{R}^2, 0) \to (\mathbb{R}^2, 0)$ which satisfies $h\varphi_0 = \psi_0 h$ and $h\varphi_1 = \psi_1 h$, simultaneously.

The local structural stability of a pair of involutions in W^r is defined in the natural way. The proof of the next theorem can be found in [24] as well as more details about involutions.

Theorem 1 *A pair of involutions (φ, ψ) is locally and simultaneous structurally stable at 0 if and only if 0 is a hyperbolic fixed point of the composition $\varphi \circ \psi$. Moreover, the structural stability in the space of pairs of involutions is not a generic property.*

3 Generic Singularities

In this section, we present the classification of the generic points of Σ.

Definition 7 Let $Z = (X, Y) \in \Omega^r$, a point $p \in \Sigma$ is said to be a **tangential singularity** of Z if $Xf(p)Yf(p) = 0$ and $X(p), Y(p) \neq 0$.

Definition 8 Let $Z = (X, Y) \in \Omega^r$, a point $p \in \Sigma$ is said to be a Σ-**singularity** of Z if p is either a tangential singularity or a pseudo-equilibrium of F_Z. Otherwise, it is said to be a **regular–regular** point of Z

Definition 9 Let $Z = (X, Y) \in \Omega^r$. A tangential singularity $p \in \Sigma$ is said to be **elementary** if it satisfies one of the following conditions:

(FR) - $Xf(p) = 0$, $X^2 f(p) \neq 0$, and $Yf(p) \neq 0$ (resp. $Xf(p) \neq 0$, $Yf(p) = 0$, and $Y^2 f(p) \neq 0$). In this case, p is said to be a **fold–regular** (resp. regular–fold) point of Σ.
(CR) - $Xf(p) = 0$, $X^2 f(p) = 0$, $X^3 f(p) \neq 0$, and $Yf(p) \neq 0$ (resp. $Xf(p) \neq 0$, $Yf(p) = 0$, $Y^2 f(p) = 0$, and $Y^3 f(p) \neq 0$), and $\{df(p), dXf(p), dX^2 f(p)\}$ (resp. $\{df(p), dYf(p), dY^2 f(p)\}$) is a linearly independent set. In this case, p is said to be a **cusp–regular** (resp. regular–cusp) point of Σ.
(FF) - If $Xf(p) = 0$, $X^2 f(p) \neq 0$, $Yf(p) = 0$, $Y^2 f(p) \neq 0$, and $S_X \pitchfork S_Y$ at p. In this case, p is said to be a **fold–fold** point of Σ.

Definition 10 Define Ξ_0 as the set of all germs of nonsmooth vector fields $Z \in \Omega^r$ such that, for each $p \in \Sigma$, either p is a regular–regular point of Z or p is an elementary tangential singularity.

From [31], we derive the following result:

Proposition 1 *Ξ_0 is an open dense set of Ω^r.*

In order to classify Σ_0, we assume, without loss of generality, that p is either a regular–regular point or an elementary tangential singularity.

The next step is devoted to characterize the locally structurally stable systems at generic singularities.

Lemma 3.1 *Let $Z = (X, Y) \in \Omega^r$ and assume that R is a connected component of Σ^s. Then:*

1. *The sliding vector field F_Z is of class $\mathscr{C}^r$ and it can be smoothly extended beyond the boundary of R.*
2. *If $p \in \partial R$ is a fold–regular point of Z, then F_Z is transverse to ∂R at p.*
3. *If $p \in \partial R$ is a cusp–regular point of Y, then F_Z has a quadratic contact with ∂R at p.*

This result is proved in [27]. It is a very useful tool to construct topological equivalences.

Theorem 2 *Let $Z = (X, Y) \in \Omega^r$, then:*

1. *Z is locally structurally stable at a regular–regular point $p \in \Sigma$ if and only if $p \in \Sigma^c$ or $p \in \Sigma^s$ and, in the second case, p is either a regular point or a hyperbolic singularity of F_Z.*
2. *Z is locally structurally stable at any fold–regular singularity $p \in \Sigma$.*
3. *Z is locally structurally stable at any cusp–regular singularity $p \in \Sigma$.*

The proof of this result can be found in [10, 12].

4 Statement of the Main Results

Define the following subsets of Ω^r:

- $\Sigma(G)$: $Z \in \Omega^r$ such that each point $p \in \Sigma$ is either a tangential singularity or a regular–regular point.
- $\Sigma(R)$: $Z \in \Omega^r$ such that for each regular–regular point $p \in \Sigma$ of Z we have either $p \in \Sigma^c$ or $p \in \Sigma^s$ and, in the second case, p is either a regular point or a hyperbolic singularity of F_Z;
- $\Sigma(H)$: $Z \in \Omega^r$ such that for each visible fold–fold point $p \in \Sigma$, the normalized sliding vector field F_Z^N has no center manifold in Σ^s.
- $\Sigma(P)$: $Z \in \Omega^r$ such that for each invisible–visible point $p \in \Sigma$, the normalized sliding vector field F_Z^N is either transient in Σ^s or it has a hyperbolic singularity at p. Moreover, if ϕ_X is the involution associated to Z, then it satisfies:

  1. $\phi_X(S_Y) \pitchfork S_Y$ at p;
  2. F_Z^N and $\phi_X^* F_Z^N$ are transversal at each point of $\Sigma^{ss} \cap \phi_X(\Sigma^{us})$;
  3. $\phi_X(S_Y) \pitchfork F_Z^N$ in a neighborhood of p.

- $\Sigma(E)$: $Z \in \Omega^r$ such that for each T-singularity $p \in \Sigma$, the first return map ϕ_Z associated to Z has a fixed point at p of type saddle with both local invariant manifolds $W^{u,s}_{loc}$ contained in Σ^c.

Remark 3 If Z has a visible–invisible fold–fold singularity at p, then the roles of X and Y in the condition $\Sigma(P)$ are interchanged.

The main result of this work is the following theorem.

Theorem 3 *$Z \in \Omega^r$ is locally structurally stable at a T-singularity p if and only if it satisfies condition $\Sigma(E)$ at p.*

The following theorem is proved in [7, 10] and a detailed proof clarifying some obscure points is presented.

Theorem 4

i) $Z \in \Omega^r$ is locally structurally stable at a hyperbolic fold–fold singularity p if and only if it satisfies condition $\Sigma(H)$ at p.
ii) $Z \in \Omega^r$ is locally structurally stable at a parabolic fold–fold singularity p if and only if it satisfies condition $\Sigma(P)$ at p.

Theorem 5 $\Sigma_0 = \Sigma(G) \cap \Sigma(R) \cap \Sigma(H) \cap \Sigma(P) \cap \Sigma(E)$.

Theorem 6 *Σ_0 is not residual in Ω^r.*

As a corollary of the characterization Theorem 5, we obtain:

Corollary 4.1

i) Σ_0 is an open dense set in $\Sigma(E)$. Moreover, $\Sigma(E)$ is maximal with respect to this property.

ii) If $Z \notin \Sigma(E)$, then Z has ∞-moduli of stability.

In addition, if Z has a T-singularity at p and ϕ_Z has complex eigenvalues, then a neighborhood $\mathscr{V}$ of Z in Ω^r is foliated by codimension one submanifolds of Ω^r corresponding to the value of the argument of the eigenvalues of the first return map. Moreover, the topological type along the corresponding leaf is locally constant.

We conclude that the local behavior around a T-singularity implies in the non-genericity of Σ_0 in Ω^r.

5 Fold–Fold Singularity

5.1 A Normal Form

In this section, we derive a normal form to study the fold–fold singularity and we present some consequences. This section is mainly motivated by the normal form of a fold point obtained by S. M. Vishik in [31] and some variants such as [7, 10, 11].

Proposition 2 *If $Z = (X, Y) \in \Omega^r$ is a nonsmooth vector field having a fold–fold point at p such that $S_X \pitchfork S_Y$ at p, then there exist coordinates (x, y, z) around p such that $f(x, y, z) = z$ and Z is given by:*

$$X(x,y,z) = \begin{pmatrix} \alpha \\ 1 \\ \delta y \end{pmatrix} \text{ and } Y(x,y,z) = \begin{pmatrix} \gamma + \mathscr{O}(|(x,y,z)|) \\ \beta + \mathscr{O}(|(x,y,z)|) \\ x + \mathscr{O}(|(x,y,z)|^2) \end{pmatrix}, \tag{4}$$

where $\delta = sgn(X^2 f(p))$, $sgn(\gamma) = sgn(Y^2 f(p))$, $\alpha, \beta, \gamma \in \mathbb{R}$.

Proof (Outline) Use the coordinates (x, y, z) of Theorem 2 from [31] to put X in the form $X(x, y, z) = (0, 1, \delta y)$ and $f(x, y, z) = z$. Now, consider the Taylor expansion of Y in this coordinate system and perform changes to put $Yf(x, y, z) = x + \mathscr{O}(|(x, y, z)|^2)$.

Definition 11 If $Z \in \Omega^r$ has a fold–fold singularity at p, then the coordinate system of Proposition 2 will be called **normal coordinates** of Z at p and the parameters of Z in the normal coordinates will be referred as **normal parameters** of Z at p. Denote $Z = Z(\alpha, \beta, \gamma)$.

Remark 4 If $\gamma = \pm 1$, $\alpha = V^+$, and $\beta = V^-$, then this normal form and the model used in [7, 8, 11] have the same semi-linear part. Geometrically, V^+ (V^-) measures the cotangent of the angle θ^+ (θ^-) between $X(0)$ $(Y(0))$ and the fold line S_X (S_Y). See [8] for more details.

Corollary 5.1 *If $Z = (X, Y) \in \Omega^r$ is a nonsmooth vector field having a fold–fold point at p such that $S_X \pitchfork S_Y$ at p, then there exist coordinates (x, y, z) around p defined in a neighborhood U of p in M, such that:*

1. *$f(x, y, z) = z$;*
2. *$S_X \cap U = \{(x, 0, 0);\ x \in (-\varepsilon, \varepsilon)\}$, for $\varepsilon > 0$ sufficiently small;*
3. *$S_Y \cap U = \{(g(y), y, 0);\ y \in (-\varepsilon, \varepsilon)\}$, for $\varepsilon > 0$ sufficiently small, where g is a $\mathscr{C}^r$ function such that $g(y) = \mathscr{O}(y^2)$, i.e., S_{Y_0} is locally a smooth curve tangent to the y-axis.*

Proof (Outline) It follows directly from Proposition 2 and the Implicit Function Theorem.

Proposition 3 *Let $Z = (X, Y) \in \Omega^r$ be a nonsmooth vector field having a fold–fold point at p such that $S_X \pitchfork S_Y$ at p. Then, the normalized sliding vector field of Z has a singularity at p and it is given by*

$$F_Z^N(x, y) = \begin{pmatrix} \alpha & -\delta\gamma \\ 1 & -\delta\beta \end{pmatrix} \cdot \begin{pmatrix} x \\ y \end{pmatrix} + \mathscr{O}(|(x, y)|^2), \tag{5}$$

in the normal coordinates of Z at p, where $\delta = sgn(X^2 f(p))$, $sgn(\gamma) = sgn(Y^2 f(p))$, $\alpha, \beta, \gamma \in \mathbb{R}$.

Proof (Outline) It follows directly from the expression of Z in this coordinate system.

Finally, we can classify a fold–fold singularity in four topologically distinct classes:

Definition 12 A fold–fold point p of $Z = (X, Y) \in \Omega^r$ is said to be:

- a **visible fold–fold** if $X^2 f(p) > 0$ and $Y^2 f(p) < 0$;
- an **invisible–visible fold–fold** if $X^2 f(p) < 0$ and $Y^2 f(p) < 0$;
- a **visible–invisible fold–fold** if $X^2 f(p) > 0$ and $Y^2 f(p) > 0$;
- an **invisible fold–fold** if $X^2 f(p) < 0$ and $Y^2 f(p) > 0$, in this case, p is also called a **T-singularity**.

Remark 5 Notice that the visible–invisible case can be obtained from the invisible–visible one by performing an orientation reversing change of coordinates. Also, we refer to a visible, invisible–visible/visible–invisible, invisible fold–fold point as a hyperbolic, parabolic, elliptic fold–fold point, respectively (Fig. 2).

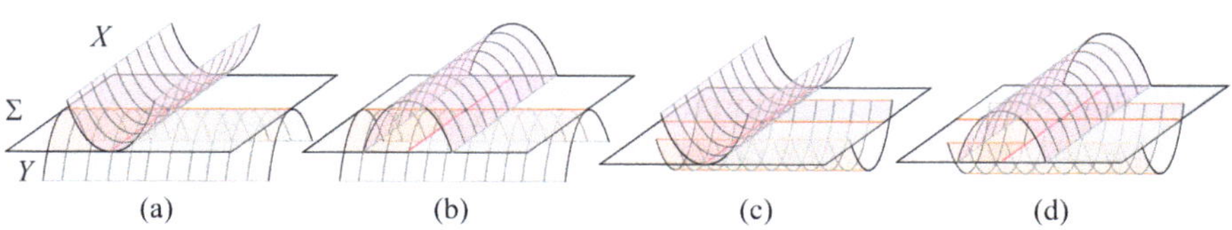

Fig. 2 Fold–fold singularity: (**a**) hyperbolic, (**b**) and (**c**) parabolic, and (**d**) elliptic

5.2 Sliding Dynamics

In this subsection, we discuss the sliding dynamics around a fold–fold singularity. This is a matured topic which has been well developed in [8, 10, 28].

From Proposition 2 and Lemma 3.1, we already know the behavior of the sliding vector field near a fold–fold singularity in a generic scenario (not only for the truncated system).

Let $Z = Z(\alpha, \beta, \gamma) \in \Omega^r$ having a fold–fold singularity at p, and consider its normalized sliding vector field F_Z^N in normal coordinates.

Consider:

$$
\begin{aligned}
R_E^1 &= \{(\alpha, \beta, \gamma) \in \mathbb{R}^2 \times \mathbb{R}^+;\ \alpha\beta > \gamma \text{ and } \alpha < 0,\ \beta < 0\} \\
R_E^2 &= \mathbb{R}^2 \times \mathbb{R}^+ \setminus \overline{R_I^1} \\
R_H^1 &= \{(\alpha, \beta, \gamma) \in \mathbb{R}^2 \times \mathbb{R}^-;\ \alpha\beta < \gamma \text{ and } \alpha > 0,\ \beta < 0\} \\
R_H^2 &= \mathbb{R}^2 \times \mathbb{R}^- \setminus \overline{R_V^1} \\
R_P^1 &= \{(\alpha, \beta, \gamma) \in \mathbb{R}^2 \times \mathbb{R}^-;\ \alpha\beta < \gamma \text{ and } \beta - \alpha > -2\sqrt{-\gamma}\} \\
R_P^2 &= \{(\alpha, \beta, \gamma) \in \mathbb{R}^2 \times \mathbb{R}^-;\ \alpha\beta < \gamma \text{ and } \alpha > 0\} \\
R_P^3 &= \{(\alpha, \beta, \gamma) \in \mathbb{R}^2 \times \mathbb{R}^-;\ \alpha\beta > \gamma,\ \beta + \alpha > 0 \text{ and } \beta - \alpha < -2\sqrt{\gamma}\} \\
R_P^4 &= \{(\alpha, \beta, \gamma) \in \mathbb{R}^2 \times \mathbb{R}^-;\ \alpha\beta > \gamma,\ \beta + \alpha < 0 \text{ and } \beta - \alpha < -2\sqrt{-\gamma}\}
\end{aligned}
$$

We claim that:

Claim 1 If p is an elliptic fold–fold singularity and $(\alpha, \beta, \gamma) \in R_E^1$, then F_Z has an invariant manifold W in Σ^s passing through p and each orbit of F_Z is transverse to S_Z and reaches p asymptotically to W (for a finite positive time in Σ^{ss} and negative time in Σ^{us}).

Claim 2 If p is an elliptic fold–fold singularity and $(\alpha, \beta, \gamma) \in R_E^2$, then F_Z has an invariant manifold W in Σ^s passing through p and each orbit is transverse to S_Z and does not reach p, with exception of W.

Claim 3 If p is a hyperbolic fold–fold singularity and $(\alpha, \beta, \gamma) \in R_H^1$ (resp. $(\alpha, \beta, \gamma) \in R_H^2$), then F_Z is of the same type of claim 1 (resp. claim 2) for reverse time.

Claim 4 If p is a parabolic fold–fold singularity and $(\alpha, \beta, \gamma) \in R_P^1$, then each orbit in Σ^{ss} (resp. Σ^{us}) is transverse to S_X (resp. S_Y) and reaches S_Y (resp. S_X) transversally for a positive finite time. In this case, we say that F_Z has transient behavior in Σ^s.

Claim 5 If p is a parabolic fold–fold singularity and $(\alpha, \beta, \gamma) \in R_P^2$, then there exist two invariant manifolds W_1 and W_2 in Σ^s passing through p which divide Σ^{ss} (and Σ^{us}) in three sectors. The intermediate sector is of hyperbolic type and in the other sectors the orbits are transversal to S_Z and go away from p (the orientation of the orbits is given in Fig. 3).

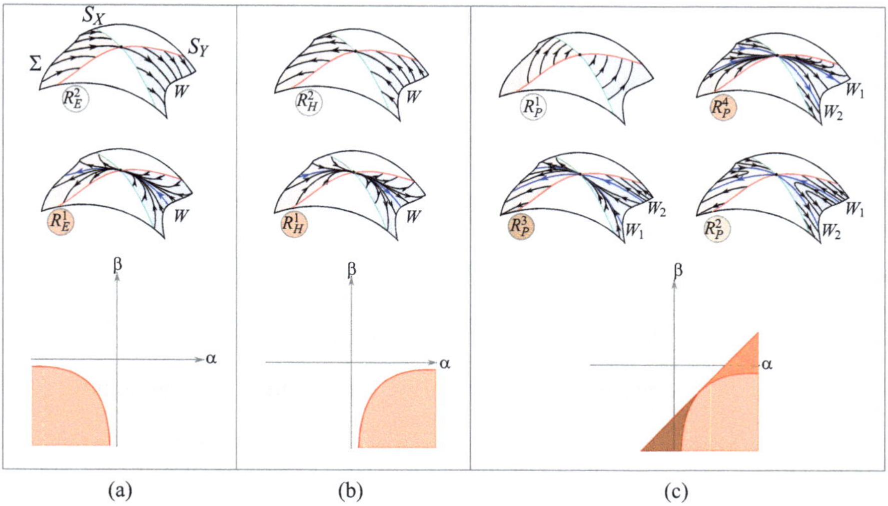

Fig. 3 Sliding dynamics near a fold–fold singularity of type elliptic (**a**), hyperbolic (**b**), and parabolic (**c**). In each case, the regions above are outlined in the (α, β)-parameter space for a fixed value of γ

Claim 6 If p is a parabolic fold–fold singularity and $(\alpha, \beta, \gamma) \in R_P^3$ then there exist two invariant manifolds W_1 and W_2 in Σ^s passing through p which divide Σ^{ss} in three sectors. In the intermediate sector, each orbit reaches p for a finite positive time asymptotically to W_1. In the left one, each orbit is transverse to S_Y and reaches p for a finite positive time asymptotically to W_1. In the right one, each orbit is transverse to S_X and goes away from p. The behavior in Σ^{us} is similar and can be seen in Fig. 3.

Claim 7 If p is a parabolic fold–fold singularity and $(\alpha, \beta, \gamma) \in R_P^4$, then F_Z has the same behavior as in claim 6 for reverse time and changing the role of W_1 and W_2, S_X and S_Y, right and left.

Claim 8 If (α, β, γ) is not in any of these regions, then F_Z presents bifurcations in Σ^s.

All these claims can be straightforwardly verified by analyzing the linear part of the normalized sliding vector field F_Z^N. We omitted the proofs due to the limitation of space.

6 Proofs of Theorems 3 and 4

This section is devoted to prove Theorems 3 and 4. In the sequel, we develop some Lemmas and Propositions which will lead us to the proof of the Theorems.

Assume that $Z \in \Omega^r$ has a T-singularity at p. Therefore, we have a first return map ϕ of Z defined around p. In order to study the local structural stability of Z, it will be crucial to study the dynamics of ϕ. Now, we derive the existence and some properties of φ.

Lemma 6.1 *Let $Z = (X, Y) \in \Omega^r$ be a nonsmooth vector field having a T-singularity at p such that $S_X \pitchfork S_Y$ at p. There exist two involutions $\phi_X : (\Sigma, p) \to (\Sigma, p)$ and $\phi_Y : (\Sigma, p) \to (\Sigma, p)$ associated to the folds X and Y such that:*

- *$Fix(\phi_X) = S_X$;*
- *$Fix(\phi_Y) = S_Y$;*
- *$\phi = \phi_X \circ \phi_Y$ is a first return map of Z such that $\phi(p) = p$.*

The proof of Lemma 6.1 can be found in [5] (Lemma 1). A straightforward verification shows the following results.

Lemma 6.2 *If $\phi = \varphi \circ \psi$, where φ and ψ are involutions of $\mathbb{R}^2$ at 0, then $\phi^n \circ \varphi = \varphi \circ \phi^{-n}$ and $\psi \circ \phi^n = \phi^{-n} \circ \psi$, for each $n \in \mathbb{Z}$.*

Proposition 4 *If $\phi = \varphi \circ \psi$, where φ and ψ are involutions of Σ at p, then the invariant manifolds W^s and W^u of ϕ at p are interchanged by φ and ψ in the following way:*

$$\psi(W^s) \subset W^u \text{ and } \varphi(W^u) \subset W^s.$$

Now, using the normal coordinates of $Z = (X, Y)$ at an elliptic fold–fold singularity, we obtain the following expressions for the associated involutions.

Lemma 6.3 *Let $Z = (X, Y) \in \Omega^r$ be a nonsmooth vector field having a T-singularity at p such that $S_X \pitchfork S_Y$ at p. Consider the normal coordinates (x, y, z) of Z at p. Then, the involutions ϕ_X and ϕ_Y are given by*

$$\phi_X(x, y) = (x - 2\alpha y, -y) \text{ and } \phi_Y(x, y) = \left(-x, -\frac{2\beta}{\gamma}x + y\right) + h.o.t., \tag{6}$$

in these coordinates, where α, β, γ are the normal parameters of Z at p.

Finally, we associate the local structural stability of Z at an elliptic fold–fold singularity with the local structural stability of the pair of involutions associated to Z.

Lemma 6.4 *Let $Z_0 = (X_0, Y_0) \in \Omega^r$ such that p is a T-singularity for Z_0. If Z_0 is locally structurally stable at p in Ω^r, then the pair of involutions (ϕ_{X_0}, ϕ_{Y_0}) associated to Z_0 is locally and simultaneously structurally stable at 0 in W^r.*

Proof In fact, since p is a T-singularity of Z_0, there exist neighborhoods $\mathscr{V}$ of Z_0 in Ω^r and V of p in M, such that each $Z \in \mathscr{V}$ has a unique Teixeira singularity at $q(Z) \in V \cap \Sigma$.

Consider the map $F : \mathscr{V} \to W^r$ given by:

$$F(X, Y) = (\phi_X, \phi_Y), \tag{7}$$

where ϕ_X and ϕ_Y are the involutions at $(0, 0)$ of $\mathbb{R}^2$ associated to X and Y, respectively.

From the continuous dependence of solutions with respect to initial conditions and parameters, it follows that F is a continuous map.

Moreover, there exists a neighborhood $\mathscr{U}$ of (ϕ_{X_0}, ϕ_{Y_0}) in W^r, such that, for each $(\tau, \psi) \in \mathscr{U}$, there exists a vector field $Z = (X, Y) \in \mathscr{V}$ such that $\tau = \phi_X$ and $\psi = \phi_Y$, and it can be done in a continuous fashion.

Then, reducing $\mathscr{V}$ if necessary, it follows that $F : \mathscr{V} \to W^r$ is an open continuous map.

Since Z_0 is locally structurally stable at p in Ω^r, $\mathscr{V}$ can be reduced such that every $Z \in \mathscr{V}$ is topologically equivalent to Z_0.

Thus, if $Z \in \mathscr{V}$, there exist a fold–fold singularity $q(Z) \in \Sigma$ of Z (with the same type of p) and a topological equivalence $h : (V_1, p) \to (V_2, q(Z))$ between Z_0 and Z, where V_1 and V_2 are neighborhoods of p in M, such that $q(Z) \in V_2$.

In particular, it induces a homeomorphism $h : \Sigma \cap V_1 \to \Sigma \cap V_2$ such that $h(p) = q(Z)$. Using coordinates, (x, y, z) around p and (u, v, w) around $q(Z)$ such that $f(x, y, z) = z$ and $f(u, v, w) = w$, the induced homeomorphism h can be seen as $h : U_1 \to U_2$, where U_1 and U_2 are neighborhoods of $(0, 0)$ in $\mathbb{R}^2$ and $h(0, 0) = (0, 0)$.

Now, given $(x, y) \in \Sigma - S_{X_0}$ (sufficiently near from $(0, 0)$), it follows from the definition of the involution ϕ_{X_0} that the points (x, y) and $\phi_{X_0}(x, y)$ are connected by an orbit γ_0 of X_0 contained in M^+. Analogously, the points $h(x, y)$ and $\phi_X(h(x, y))$ are connected by an orbit γ of X contained in M^+.

Since h is a topological equivalence such that $h(\Sigma) \subset \Sigma$, it follows that $h(\gamma_0) = \gamma$ and

$$h(\phi_{X_0}(x, y)) = \phi_X(h(x, y)). \tag{8}$$

It is trivial to see that (8) is also true when $(x, y) \in S_{X_0}$, by observing that $h(S_{X_0}) = S_X$. Hence, h is an equivalence between the germs of involution ϕ_{X_0} and ϕ_X.

Analogously, by changing the roles of X and Y, it can be shown that h is also an equivalence between the involutions ϕ_{Y_0} and ϕ_Y.

We conclude that h is a (simultaneous) topological equivalence between the pairs of involutions (ϕ_{X_0}, ϕ_{Y_0}) and (ϕ_X, ϕ_Y).

Since Z is arbitrary in $\mathscr{V}$, it follows that every pair of involutions in $\mathscr{U}$ is topologically equivalent to (ϕ_{X_0}, ϕ_{Y_0}), and since $\mathscr{U}$ is open in W^r, it follows that (ϕ_{X_0}, ϕ_{Y_0}) is local and simultaneous structurally stable in W^r.

The following result is obtained by combining Theorem 1 and Lemma 6.4.

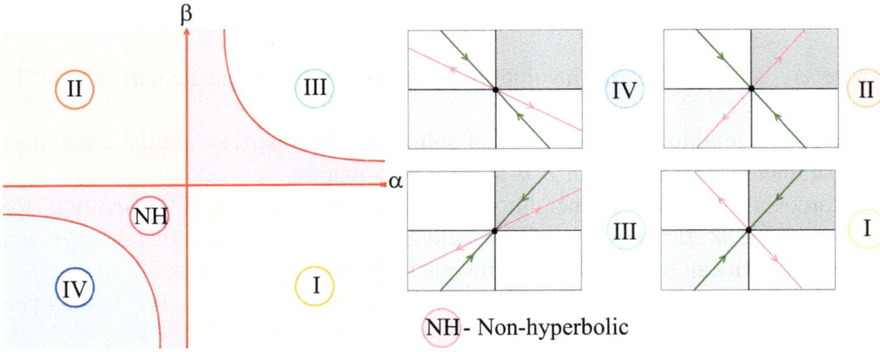

Fig. 4 Regions of the (α, β)-parameter space with the corresponding behavior of the first return map φ, for a fixed value of $\gamma > 0$

Proposition 5 *Let $Z_0 \in \Omega^r$ having a T-singularity at p, and let (ϕ_{X_0}, ϕ_{Y_0}) be the pair of involutions of $\mathbb{R}^2$ at $(0, 0)$ associated to Z_0. If 0 is not a hyperbolic fixed point of $\phi_{Y_0} \circ \phi_{X_0}$, then Z_0 is locally structurally unstable at p.*

A simple computation of eigenvalues and eigenvectors allows us to study the fixed point p of the first return map ϕ:

Lemma 6.5 *Let $Z = (X, Y) \in \Omega^r$ be a nonsmooth vector field having a T-singularity at p such that $S_X \pitchfork S_Y$ at p. Let (α, β, γ) be the normal parameters of Z at p.*

1. *If $\alpha\beta(\alpha\beta - \gamma) \leq 0$, then 0 is not a hyperbolic fixed point of ϕ. In addition, if $\alpha\beta(\alpha\beta - \gamma) < 0$, then ϕ has complex eigenvalues.*
2. *If $\alpha\beta(\alpha\beta - \gamma) > 0$, then 0 is a saddle point of ϕ (Fig. 4). In addition, if λ, μ are the eigenvalues of ϕ such that $|\mu| < 1 < |\lambda|$, and v_μ, v_λ are the correspondent eigenvectors, then:*

   a. *If $\alpha > 0$ and $\beta > 0$, then $v_\mu, v_\lambda \in \Sigma^s$.*
   b. *If $\alpha > 0$ and $\beta < 0$, then $v_\mu \in \Sigma^c$ and $v_\lambda \in \Sigma^s$.*
   c. *If $\alpha < 0$ and $\beta > 0$, then $v_\mu \in \Sigma^s$ and $v_\lambda \in \Sigma^c$.*
   d. *If $\alpha < 0$ and $\beta < 0$ then $v_\mu, v_\lambda \in \Sigma^c$.*

Proposition 6 *Let $Z_0 = (X_0, Y_0) \in \Omega^r$ be a germ of nonsmooth vector field having a T-singularity at p. Let (α, β, γ) be the normal parameters of Z_0 at p. If $\alpha\beta(\alpha\beta - \gamma) \leq 0$, then Z_0 is locally structurally unstable at p.*

Proof It follows directly from Proposition 5 and the fact that p is not a hyperbolic fixed point of the first return map $\phi_0 = \phi_{X_0} \circ \phi_{Y_0}$ associated to Z_0. In the sequel, we present an explicit argument for the local structural instability of Z_0. It is mainly based on [4] and the Blow-up procedure (see [2]).

Let $\phi_0 : (\Sigma, p) \to (\Sigma, p)$ be the (germ of) first return map associated to Z_0 at p. From the conditions assumed in the Theorem, it follows that ϕ_0 has eigenvalues

$\lambda_{\pm} = a \pm ib$, where $a^2 + b^2 = 1$. Using the normal form of Z_0 and basic linear algebra, it is easy to find coordinates (x, y) of Σ at p, such that:

$$\phi_0(x, y) = (ax - by, bx + ay) + \mathcal{O}(|(x, y)|^2).$$

Consider the germs of functions $h_1, h_2 : (\mathbb{R}^2, 0) \to (\mathbb{R}^2, 0)$, given by:

$$h_1(x, y) = (x, y) \text{ and } h_2(x, y) = \sqrt{x^2 + y^2}(x, y).$$

Notice that h_1, h_2 are germs of homeomorphisms if we exclude the origin in their domains.

If $(x, y) \neq (0, 0)$, a straightforward computation shows that:

$$\psi_0(x, y) = h_2^{-1} \circ \phi_0 \circ h_1(x, y) = \frac{1}{\sqrt{x^2 + y^2}}\phi_0(x, y).$$

Therefore, ϕ_0 and ψ_0 are topologically equivalent. Using the polar change of coordinates $\zeta(r, \theta) = (r\cos(\theta), r\sin(\theta))$, where $r > 0$ and $\theta \in \mathbb{R}/2\pi\mathbb{Z}$, we write $\psi_0 \circ \zeta$ as

$$\psi_0 \circ \zeta(r, \theta) = \begin{pmatrix} \cos(\theta + \tau) \\ \sin(\theta + \tau) \end{pmatrix} + \mathcal{O}(r),$$

where $a + ib = e^{i\tau}$.

If $r \to 0$, ζ blows up the singularity $r = 0$ into the circle $S^1 = \mathbb{R}/2\pi\mathbb{Z}$, and the map $\zeta^{-1} \circ \psi_0 \circ \zeta$ induces a dynamics in S^1 given by (Fig. 5)

$$\overline{\psi_0}([\theta]) = [\theta + \tau].$$

Let Z be a small perturbation of Z_0, take it small enough such that the normal parameters $(\tilde{\alpha}, \tilde{\beta}, \tilde{\gamma})$ of Z are close enough to (α, β, γ).

If ϕ is the first return map associated to Z at the fold–fold point $q(Z) \approx p$, then it has eigenvalues $\tilde{\lambda}_{\pm} = \tilde{a} \pm i\tilde{b}$.

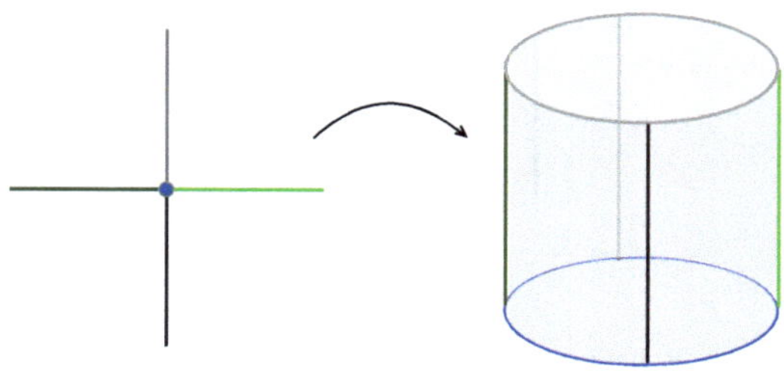

Fig. 5 Blowup of p into S^1

Applying the same procedure to ϕ, we can blow up its singularity $q(Z)$ into S^1, and the dynamics in S^1 is induced by $\overline{\psi} : S^1 \rightarrow S^1$, given by $\overline{\psi}(\theta) = \theta + \tilde{\tau}$, where $\tilde{a} + i\tilde{b} = e^{i\tilde{\tau}}$.

Now, if $h : V(p) \rightarrow V(q(Z))$ is an equivalence between Z_0 and Z, then $h(S_{X_0}) = S_X$. In adequate coordinates, it means that $h(x, 0) = (f(x), 0)$, where f is a homeomorphism of the real line such that $f(0) = 0$.

Notice that the motion of $S_{X_0} \cap \{x \geq 0\}$ (resp. $S_X \cap \{x \geq 0\}$) around the origin through ϕ_0 (resp. ϕ) is given by the orbit $\gamma_0 = \{\overline{\psi_0}^n(0), n \in \mathbb{Z}\}$ (resp. $\gamma = \{\overline{\psi}^n(0), n \in \mathbb{Z}\}$).

Since h is an equivalence, it follows that the orbits γ_0 and γ have the same topology. Nevertheless, if $\tau \in \mathbb{Q}$ (resp. $\tau \notin \mathbb{Q}$), we can take Z (sufficiently near of Z_0) such that $\tilde{\tau} \notin \mathbb{Q}$ (resp. $\tilde{\tau} \in \mathbb{Q}$). Therefore, γ_0 is a periodic orbit and γ is dense in S^1 (resp. γ_0 is dense in S^1 and γ is a periodic orbit).

It means that, when $\tau \in \mathbb{Q}$ (and γ_0 is periodic), the curves $\phi^n(S_X)$ are tangent to a finite number of directions at p, i.e., there exist m vectors $v_1, \ldots, v_m$ in $T_p\Sigma$ such that $T_p\phi^n(S_X) = \text{span}\{v_{i(n)}\}$, for some $i(n) \in \{1, \ldots, m\}$, for each $n \in \mathbb{N}$. Hence, we conclude that $\bigcup \phi^n(S_X)$ has zero measure in Σ.

On the other hand, if $\tau \notin \mathbb{Q}$ (and γ_0 is dense), we have that for each $v \in T_p\Sigma$, there exist a sequence $\phi^{n_k}(S_X)$, such that $T_p\phi^{n_k}(S_X) = \text{span}\{v_k\}$, and $v_k \rightarrow v$ when $k \rightarrow \infty$. We conclude that $\bigcup \phi^n(S_X)$ has full measure in Σ.

From these facts, we can see that the orbits $\phi_0^n(S_{X_0})$ and $\phi^n(S_X)$ do not have the same topology (Fig. 6).

Now, a Σ-equivalence between Z_0 and Z has to satisfy $h(S_{X_0}) = S_X$ and $h \circ \phi_0 = \phi \circ h$. Since $\phi_0^n(S_{X_0})$ and $\phi^n(S_X)$ have different topological type, it follows that there is no Σ-equivalence between Z_0 and Z.

We conclude that, in any neighborhood of Z_0 in Ω^r we can find a nonsmooth vector field Z such that Z_0 is not topologically equivalent to Z at p. Therefore, Z_0 is locally structurally unstable at p.

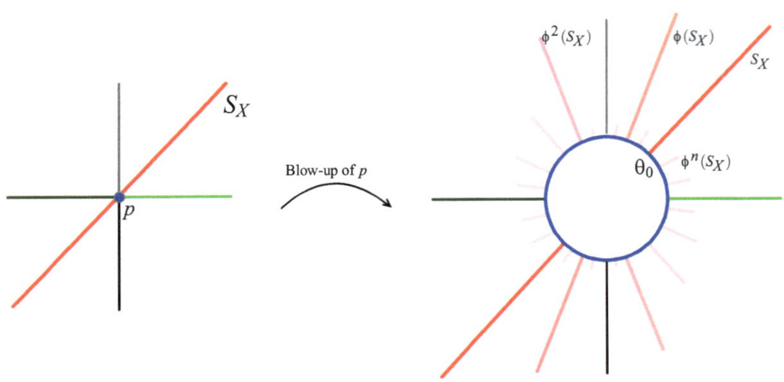

Fig. 6 Behavior of S_X when $\theta \notin \mathbb{Q}$

Remark 6 Let τ_Z be the argument of the eigenvalues $a \pm ib$ of the first return map ϕ associated to Z.

If Z_0 is a nonsmooth vector field satisfying the hypotheses of Proposition 6, then a neighborhood $\mathcal{V}_0$ of Z_0 in Ω^r is foliated by codimension one submanifolds of Ω^r corresponding to the value of τ_Z, i.e., $Z_1 \in \mathcal{V}_0$ and $Z_2 \in \mathcal{V}_0$ lie on the same leaf if and only if $\tau_{Z_1} = \tau_{Z_2}$.

The topological type of the first return map is locally constant along each leaf. Moreover, if Z_1 and Z_2 are elements of $\mathcal{V}_0$ lying on different leaves of the foliation, then they are not topologically equivalent.

We conclude that Z_0 has ∞-moduli of stability. (See [4, 16, 17] for more details.)

Now, we can prove Theorem D.

Theorem 7 *Σ_0 is not residual in Ω^r.*

Proof (Proof of Theorem D) It follows directly from Theorem 6. In fact, let $Z_0 \in \Omega^r$ and let $(\alpha_0, \beta_0, \gamma_0)$ be the normal parameters of Z_0 at p, they satisfy $\alpha_0\beta_0(\alpha_0\beta_0 - \gamma_0) < 0$.

From continuity (and Implicit Function Theorem), there exist neighborhoods $\mathcal{V}$ of Z_0 in Ω^r and V of p in M, such that each Z has a T-singularity at $q(Z) \in V$.

Moreover, if we apply Proposition 2 to Z at $q(Z)$, the normal parameters (α, β, γ) of Z at $q(Z)$ also satisfy $\alpha\beta(\alpha\beta - \gamma) < 0$.

From Theorem 6, each $Z \in \mathcal{V}$ is locally structurally unstable at the fold–fold singularity $q(Z) \in V \cap \Sigma$. It means that each $Z \in \mathcal{V}$ is locally structurally unstable at a point $q(Z) \in \Sigma$, hence each $Z \in \mathcal{V}$ is Σ-locally structurally unstable. Thus, $\mathcal{V} \subset \Omega^r \setminus \Sigma_0$ and Σ_0 is not residual in Ω^r.

Notice that the results obtained until this point are mainly concerned with the foliation $\mathcal{F}$ generated by a nonsmooth vector field near a T-singularity. The sliding dynamics does not have influence on these results. Nevertheless, the existence of sliding vector fields will be extremely important in the classification of the structural stability of a T-singularity having a first return map with hyperbolic fixed point.

Proposition 7 *Let $Z_0 = (X_0, Y_0) \in \Omega^r$ be a germ of nonsmooth vector field having a T-singularity at p. Let (α, β, γ) be the normal parameters of Z_0 at p. If either $\alpha\beta \geq \gamma$ and $\alpha, \beta > 0$ or $\alpha\beta < 0$, then Z_0 is locally structurally unstable at p.*

Proof In the conditions of the theorem, we can use Lemma 6.5 to conclude that the first return map ϕ_0 of Z_0 has a local invariant manifold of the saddle contained in Σ^s.

Without loss of generality, assume that $W^s \subset \Sigma^s$. Notice that the map ϕ_0^2 has the same invariant manifolds of ϕ_0, but it has both positive eigenvalues $0 < \lambda < 1 < \mu$.

Generically (i.e., $W^s \pitchfork W$ at p, where W is the invariant manifold of claim 2 in Sect. 5.2), we have that the sliding vector field F_0 of Z_0 is transverse to $W^s \cap \Sigma^{ss}$ for a small neighborhood of p. Let $V = U \cap \Sigma^s$, where U is a neighborhood of p such that F_0 is transverse to $W^s \cap V$ (Fig. 7).

Since $\lambda > 0$, we have that $\phi_0^2(W^s) \subset \phi_0^2(V) \cap V$. Moreover,

Fig. 7 Vector field F_0 near W^s

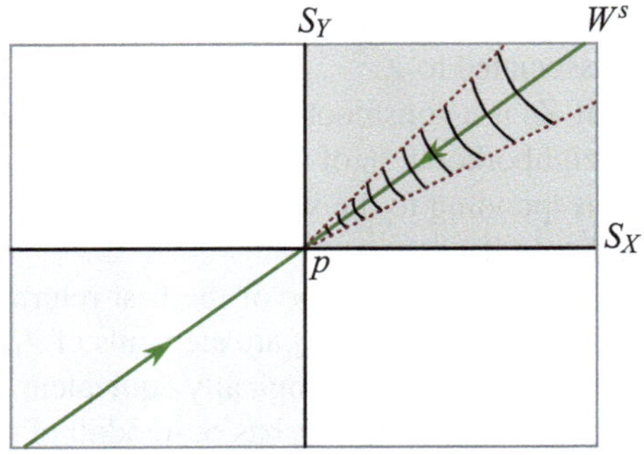

$$\phi_0^{2n}(W^s) \subset \phi_0^{2n}(V) \cap \phi_0^{2(n-1)}(V) \cap \cdots \cap \phi_0^2(V) \cap V,$$

for each $n \in \mathbb{N}$.

Let R_n be the open set $\phi_0^{2n}(V) \cap \phi_0^{2(n-1)}(V) \cap \cdots \cap \phi_0^2(V) \cap V$. Notice that, in each region $\phi_0^{2i}(V)$, we have a (push-forwarded) vector field

$$F_i = (\phi_0^{2i})^*(F_0),$$

defined on it. Therefore, there are $n+1$ vector fields defined on R_n. Moreover, we can reduce R_n such that F_i and F_j are transversal at each point of R_n, for $i \neq j$, generically. In fact, consider the expressions of ϕ_X, ϕ_Y, and F_Z^N in the normal coordinates. Consider the curves $\gamma_\pm(t) = tv_\pm$, where $v_\pm$ are the eigenvectors associated to the eigenvalues $\lambda_\pm$ of $d\phi_0^2$. A simple computation shows that:

$$F_{ij}^{\pm}(t) = \det(F_i(\gamma_\pm(t)), F_j(\gamma_\pm(t))) = A_{ij}^{\pm}(\alpha, \beta, \gamma)t^2 + \mathscr{O}(t^3),$$

where $A_{ij}^{\pm}$ is a rational function depending on α, β, and γ.

Clearly, if $A_{ij}^{\pm} \neq 0$, then F_i and F_j are transversal in a neighborhood of $\gamma_\pm$. In particular, they are transversal in a neighborhood of W^s.

Since $A_{ij}^{\pm} = 0$, for each $i, j = 0, 1, 2$, defines a zero measure set in the parameter space (α, β, γ), we achieved our goal.

Notice that each vector field F_i in R_n defines a codimension one foliation $\mathscr{F}_i$ of R_n (R_n is foliated by the integral curves of the vector field F_i). Moreover, $(\mathscr{F}_0, \ldots, \mathscr{F}_n)$ is in general position (by the reduction of R_n). In particular, for $n = 2$, we obtain 3 foliations $(\mathscr{F}_0, \mathscr{F}_1, \mathscr{F}_2)$ of R_2. This is called a 3-**web** in R_2 (see [3] and [20]) (Fig. 8).

Since R_2 is a 2-dimensional manifold, it follows that these foliations are structurally unstable in the following sense. If $(\widetilde{\mathscr{F}}_0, \widetilde{\mathscr{F}}_1, \widetilde{\mathscr{F}}_2)$ are the foliations correspondent to a nonsmooth vector field $\widetilde{Z} \approx Z_0$, then there exists at least one $\widetilde{Z}$ such that there is no homeomorphism $h : R_2 \to \widetilde{R_2}$ satisfying $h(\mathscr{F}_i) = \widetilde{\mathscr{F}}_i$, for every $i = 0, 1, 2$, preserving the leaves of each foliation.

Clearly, the property above has to be preserved by a Σ-equivalence, hence there exists a Z sufficiently near of Z_0 which is topologically different from Z_0 near p.

The instability of Z_0 at p follows directly from these facts.

Remark 7 In general, the Theory of Webs used in the last Theorem is developed for foliations on $\mathbb{C}^n$. Nevertheless, we can identify Σ with $\mathbb{C}$ at p (since Σ is 2-dimensional) and apply the results of this theory for this case.

Now, let $Z_0 = (X_0, Y_0) \in \Omega^r$ be a germ of nonsmooth vector field having a Teixeira singularity at p. Let (α, β, γ) be the normal parameters of Z_0 at p and assume that $\alpha\beta \geq \gamma$ and $\alpha, \beta < 0$.

Let $Z \in \Omega^r$ be any small perturbation of Z_0 and denote their first return maps by ϕ and ϕ_0, respectively. Our goal is to construct a topological equivalence between Z and Z_0.

Using the Implicit Function Theorem and the continuous dependence between Z_0 and its normal parameters, we can deduce the following result.

Lemma 6.6 *There exists a neighborhood $\mathcal{V}$ of Z_0 such that, for each $Z \in \mathcal{V}$, F_Z^N and $F_{Z_0}^N$ have the same topological type and the first return map ϕ of Z has a saddle at the origin with both local invariant manifolds in Σ^c.*

Remark 8 In what follows, $\mathcal{V}$ will denote the neighborhood of Lemma 6.6.

Now, we prove the existence of an invariant nonsmooth diabolo in an analytic way, this result was achieved by M. Jeffrey and A. Colombo for the semi-linear case (see [7]).

Proposition 8 *Let $Z_0 = (X_0, Y_0) \in \Omega^r$ be a nonsmooth vector field having a T-singularity at p such that the normal parameters (α, β, γ) of Z_0 at p satisfy $\alpha\beta \geq \gamma$ and $\alpha, \beta < 0$. Then, Z_0 has an invariant nonsmooth diabolo D_0 which prevents connections between points of Σ^{us} and Σ^{ss} through orbits of Z.*

Proof From Lemma 6.6, it follows that the first return map $\phi_0 = \phi_{X_0} \circ \phi_{Y_0}$ associated to Z_0 has a hyperbolic saddle at p with both eigenvectors in Σ^c.

Notice that the local stable manifold of the saddle W^s is tangent to the eigenvector v_- correspondent to the eigenvalue λ and the local unstable manifold of the saddle W^u is tangent to the eigenvector v_- correspondent to the eigenvalue μ, where $|\lambda| < 1 < |\mu|$.

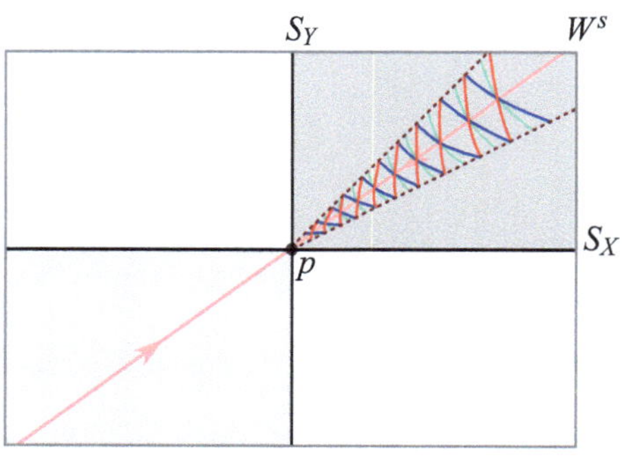

Fig. 8 Foliations $\mathscr{F}_0$, $\mathscr{F}_1$, and $\mathscr{F}_2$ originated from the vector fields F_0, F_1, and F_2, respectively, near W^s

Moreover, W^s and W^u are curves on Σ passing through p transverse to $S_X \cup S_Y$ at p and $W^s \pitchfork W^u$ at p (p is hyperbolic). Using coordinates (x, y) at p (which put Z_0 in the normal form (4)), we can see that $S_{X_0} = \text{Fix}(\phi_{X_0})$ is the x-axis, $S_{Y_0} = \text{Fix}(\phi_{Y_0})$ is a curve tangent to the y-axis at 0, and W^s and W^u are curves passing through 0 contained in the second and the fourth quadrants which are transverse to $S_{X_0} \cup S_{Y_0}$ at 0.

Therefore, we have the following situation:

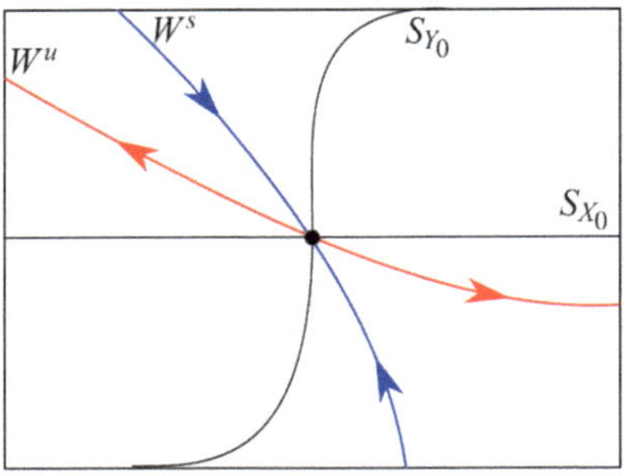

From Proposition 4, it follows that $\phi_{X_0}(W^u) \subset W^s$. Now, the image of a point in the semi-plane $\{y > 0\}$ through ϕ_{X_0} is a point in the semi-plane $\{y < 0\}$ by the construction of ϕ_{X_0}. It means that the branch of W^u in the second quadrant has to be taken into the branch of W^s in the fourth quadrant.

Also, $\phi_{Y_0}(W^s) \subset W^u$. Notice that S_{Y_0} splits $\mathbb{R}^2$ in two connected components, C_- and C_+. From the construction of ϕ_{Y_0}, the image of a point in C_- through ϕ_{Y_0} is a point in C_+. It means that the branch of W^s in the fourth quadrant is taken into the branch of W^u in the second quadrant.

These connections produce an invariant (nonsmooth) cone with vertex at the fold–fold point which contains Σ^{us} in its interior. Analogously, we prove that there exists an invariant (nonsmooth) cone with vertex at the fold–fold point which contains Σ^{ss} in its interior. These two cones produce the required nonsmooth diabolo (see Fig. 9).

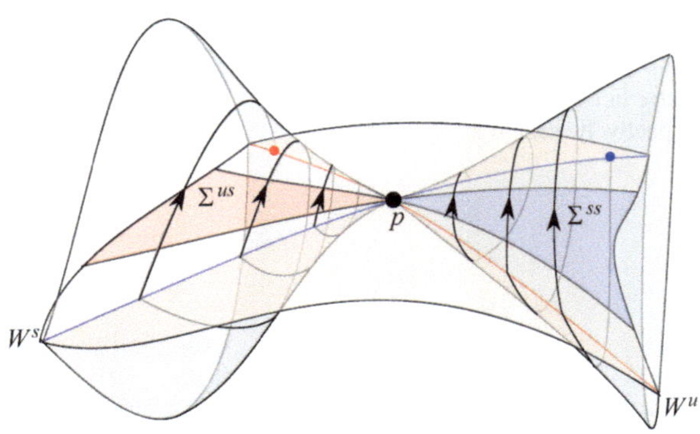

Fig. 9 A nonsmooth diabolo D_0 of Z_0

Remark 9 In other words, there is no communication between Σ^{us} and Σ^{ss} in this case.

Remark 10 Notice that the existence of the invariant diabolo D_0 implies that the T-singularity p_0 has stable and unstable invariant manifolds of dimension 2, and this is a phenomena which has no counterpart in smooth vector fields of dimension 3.

Now, we proceed by constructing a homeomorphism between $Z \in \mathcal{V}$ and Z_0.

Lemma 6.7 *If $Z \in \mathcal{V}$, there exists an order-preserving homeomorphism h : $\Sigma^s(Z_0) \to \Sigma^s(Z)$ which carries orbits of F_{Z_0} onto orbits of F_Z.*

The proof of this lemma follows straightforwardly from Lemmas 3.1 and 6.6.

Definition 13 If $\phi : (\mathbb{R}^2, 0) \to (\mathbb{R}^2, 0)$ is a germ of diffeomorphism at 0 having a saddle at 0, then the **deMelo–Palis invariant** of ϕ is defined as:

$$P(\phi) = \frac{\log(|\lambda|)}{\log(|\mu|)},$$

where λ, μ are the eigenvalues of $d\phi(0)$ such that $|\lambda| < 1 < |\mu|$.

Remark 11 In fact, the deMelo–Palis invariant P is a moduli of stability for ϕ. (See [16, 17].)

Proposition 9 *If $Z \in \mathcal{V}$, there exists a homeomorphism $h : \Sigma \to \Sigma$ which is a continuous extension of the homeomorphism $h : \Sigma^s(Z_0) \to \Sigma^s(Z)$ given by Lemma 6.7, such that $\phi \circ h = h \circ \phi_0$, i.e., it is a topological equivalence between ϕ and ϕ_0.*

Proof The proof of this proposition is divided into steps.

Let $h : \Sigma^s(Z_0) \to \Sigma^s(Z)$ be the homeomorphism obtained in Lemma 6.7.

Notice that Z has a T-singularity at $q(Z) \approx p$. Since $F^N_{Z_0}$ and F^N_Z are transversal to $S_{Z_0} \setminus \{p\}$ and $S_Z \setminus \{q(Z)\}$, respectively, we can easily continuously extend h on $\overline{\Sigma^s(Z_0)}$ via limit to obtain

$$h : \overline{\Sigma^s(Z_0)} \to \overline{\Sigma^s(Z)}.$$

Step 1: The first task is to define a **fundamental domain** for the first return maps, ϕ and ϕ_0.

We will detail it for ϕ_0. The process to construct the fundamental domain of ϕ is completely analogous.

By the Linearization Theorem (see [13]), we may assume that ϕ_0 is linear. Moreover, we can consider coordinates (x, y) of Σ at p such that:

$$\phi_0(x, y) = (\lambda_0 x, \mu_0 y),$$

where λ_0, μ_0 are the eigenvalues of ϕ_0 such that $|\mu_0| < 1 < |\lambda_0|$.

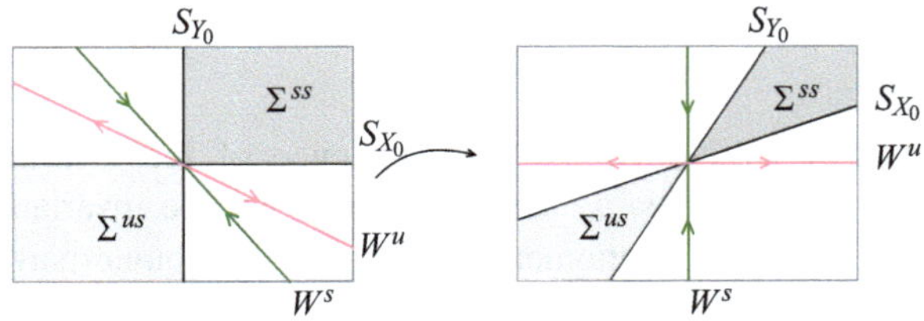

Fig. 10 Change of coordinates

By the position of S_{X_0}, S_{Y_0}, and the invariant manifolds of the saddle, obtained in Proposition 8, it follows that:

- S_{X_0} is a curve passing through 0, with one branch in the first quadrant and another in the fourth;
- S_{Y_0} is a curve passing through 0, with one branch in the first quadrant and another in the fourth;
- S_{X_0} is tangent to the line $y = k_0 x$;
- S_{Y_0} is tangent to the line $y = K_0 x$;
- $0 < k_0 < K_0$.

See Figure 10.

Without loss of generality, consider that $S_{X_0} = \{y = k_0 x\}$ and $S_{Y_0} = \{y = K_0 x\}$ and assume that these lines are the fixed points of ϕ_{X_0} and ϕ_{Y_0}, respectively. It will reduce our work, nevertheless it generates no loss of generality, since the same can be done with the original sets.

From the existence of the invariant diabolo in Proposition 8, it follows that $\phi_0^{-1}(S_{X_0})$ is a line in the same quadrants containing S_{X_0}; moreover, its inclination is greater than K_0.

Define:

$$\omega_0 = \{(x, y);\ k_0 x \leq y \leq K_0 x\} \text{ and } \widetilde{\omega}_0 = \phi_{Y_0}(\omega_0).$$

Notice that $R_0 = \omega_0 \cup \widetilde{\omega}_0$ is the region delimited by the lines S_{X_0} and $\phi_0^{-1}(S_{X_0})$.

Now, it is immediate that $\phi_0^n(S_{X_0}) \to W^u$ when $n \to \infty$ and $\phi_0^n(S_{X_0}) \to W^s$ when $n \to -\infty$. Therefore, the first and the third quadrants are partitioned by $\phi_0^n(R_0)$, $n \in \mathbb{Z}$.

In other words, if $Q = \{(x, y);\ xy > 0\}$, then

$$Q = \bigcup_{n \in \mathbb{Z}} \phi_0^n(R_0).$$

Therefore, we say that R_0 is the fundamental domain of ϕ_0 (Fig. 11).

Similarly, we can consider coordinates (x, y) of Σ at p such that:

$$\phi(x, y) = (\lambda x, \mu y),$$

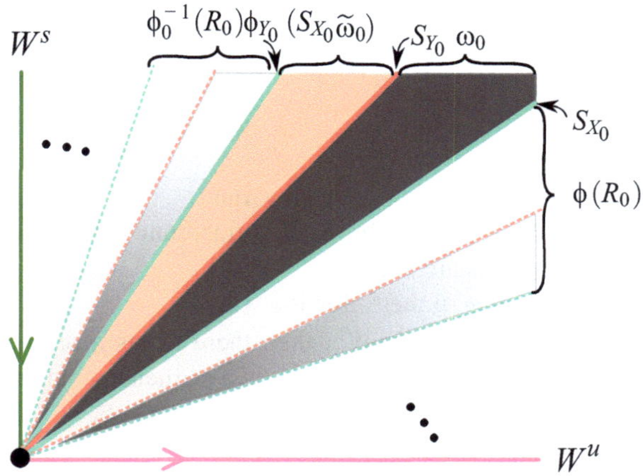

Fig. 11 Fundamental domain $R_0 = \omega_0 \cup \widetilde{\omega}_0$ in the first quadrant

where λ, μ are the eigenvalues of ϕ such that $|\mu| < 1 < |\lambda|$. Therefore, there exists $R = \omega \cup \widetilde{\omega}$, where ω is the region delimited by S_X and S_Y and $\widetilde{\omega} = \phi_Y(\omega)$.

Also, $Q = \bigcup_{n\in\mathbb{Z}} \phi^n(R)$, and R is the region delimited by S_X and $\phi^{-1}(S_X)$.

In both cases, each orbit of ϕ_0 (and ϕ) passes a unique time in each sector of the partition of Q.

Step 2: Extending the domain of h into $h : Q \to Q$.

Notice that $h : \omega_0 \to \omega$ is already defined (it is the homeomorphism $h : \overline{\Sigma^s(Z_0)} \to \overline{\Sigma^s(Z)}$ in these coordinates).

If $q \in \widetilde{\omega}_0$, then $q = \phi_{Y_0}(\tilde{q})$, for some $\tilde{q} \in \omega_0$, therefore, define:

$$h(q) = \phi_Y(h(\tilde{q})).$$

Clearly, it is a continuous extension of h from ω_0 into R_0. Now, we have defined a homeomorphism $h : R_0 \to R$.

The extension to Q follows in a natural way (since it is defined in a fundamental domain).

In fact, if $q \in Q$, there exist a unique $\tilde{q} \in R_0$ and a unique $n \in \mathbb{Z}$, such that $q = \phi_0^n(\tilde{q})$. Define:

$$h(q) = \phi^n(h(\tilde{q})).$$

Clearly, $h : Q \to Q$ is a homeomorphism satisfying:

$$h(\phi_0(q)) = \phi(h(q)),$$

for each $q \in Q$.

Step 3: Extending h on both W^u and W^s in a continuous fashion.

This is the most delicate part of the proof. Consider an arbitrary continuous extension of h on W^s.

Now, the difficult task is to continuously extend it to W^u, and it will be only possible because

$$P(\phi_0) = -1 = P(\phi),$$

where P is the deMelo–Palis invariant.

Only the extension in the first quadrant will be detailed. The extensions in the other quadrants are similar.

We extend ϕ in the following way.

Fix $w = (d, 0) \in W^u$, then there exists a sequence $w_i = \phi_0^{N_i}(y_i)$ such that $N_i \to \infty$ when $i \to \infty$ and y_i is a sequence contained in $S_{X_0} \cap \{x, y > 0\}$ such that $y_i \to 0$ when $i \to \infty$, which satisfies:

$$\lim_{i\to\infty} \phi_0^{N_i}(y_i) = w.$$

Notice that the homeomorphism h is already defined for the sequence w_i. Since we want a continuous extension and an equivalence, we must define:

$$h(w) = \lim_{i\to\infty} h(\phi_0^{N_i}(y_i)) = \lim_{i\to\infty} \phi^{N_i}(h(y_i)).$$

Our work is to prove that the limit above exists. In this case, h will be extended on W^u by doing this process for every $q \in [w, \phi_0(w)]$ and then extend it through the images of this fundamental domain by ϕ_0.

Now, we prove the existence of the limit.

Since $h(S_{X_0}) = S_X$ and $\phi^n(S_X) \to W^u$ as $n \to \infty$, it follows directly that:

$$\lim_{i\to\infty} \pi_2(\phi^{N_i}(h(y_i))) = 0.$$

Therefore, $\pi_2(h(w)) = 0$ and it is well-defined. The problem happens for the first coordinate. Consider:

1. $w = (d, 0)$;
2. $y_i \to 0$, $y_i \in S_{X_0}$, for every i;
3. $N_i \to$ such that $\phi_0^{N_i}(y_i) = w_i \to w$;
4. $t_i \to \infty$, $x_i \to x \in W^s$ such that $y_i = \phi_0^{t_i}(x_i)$.

See Figure 12.

Now, denote $\widetilde{y_i} = h(y_i)$, $\widetilde{x_i} = h(x_i)$, $\widetilde{w_i} = \phi^{N_i}(\widetilde{y_i})$, $d_i = \pi_1(w_i)$, $\widetilde{d_i} = \pi_1(\widetilde{w_i})$, $a_i = \pi_2(x_i)$, and $\widetilde{a_i} = \pi_2(h(x_i))$. Hence, we must prove that $\widetilde{d_i}$ converges.

Notice that since h is continuously extended for W^s and the sequence x_i converges to $x \in W^s$, it follows that $\widetilde{a_i}$ is a convergent sequence. Denote $\widetilde{a} = \lim \widetilde{a_i}$, and notice that

$$\widetilde{d_i} = \pi_1(\phi^{N_i}(h(y_i))) = \lambda^{N_i}\pi_1(\widetilde{y_i}).$$

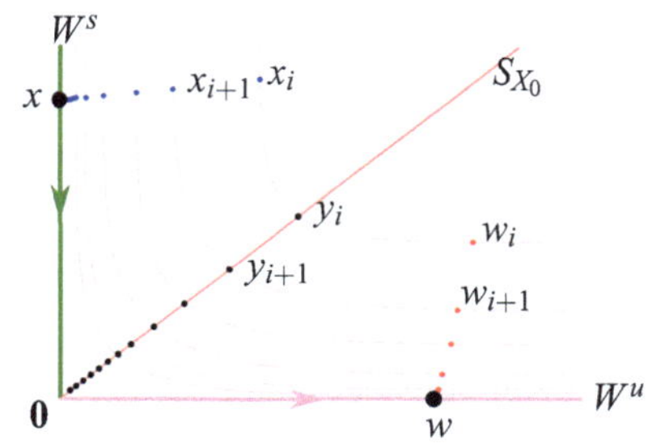

Fig. 12 Sequences (x_i), (y_i), and (w_i)

Now, observe that:

$$\widetilde{y_i} = h(y_i) = h(\phi_0^{t_i}(x_i)) = \phi^{t_i}(\widetilde{x_i}) = (\lambda^{t_i}\pi_1(\widetilde{x_i}), \mu^{t_i}\pi_2(\widetilde{x_i})).$$

Since $\widetilde{y_i} \in S_X = \{y = kx\}$, it follows that:

$$\pi_1(\widetilde{y_i}) = \frac{1}{k}\pi_2(\widetilde{y_i}) = \frac{1}{k}\mu^{t_i}\pi_2(\widetilde{x_i}).$$

Hence:

$$\widetilde{d_i} = \frac{1}{k}\lambda^{N_i}\mu^{t_i}\pi_2(\widetilde{x_i}) = \frac{1}{k}\lambda^{N_i}\mu^{t_i}\widetilde{a_i},$$

and applying the logarithm, we obtain:

$$\log(\widetilde{d_i}k) = N_i \log(\lambda) + t_i \log(\mu) + \log(\widetilde{a_i}).$$

With the same process, we also obtain:

$$\log(d_i k_0) = N_i \log(\lambda_0) + t_i \log(\mu_0) + \log(a_i).$$

Since $\log(d_i k_0)$ and $\log(a_i)$ converge, it follows that $N_i \log(\lambda_0) + t_i \log(\mu_0)$ converges.

Now, using that $P(\phi_0) = P(\phi)$, it is immediate that $N_i \log(\lambda) + t_i \log(\mu)$ converges.

Since $\widetilde{a_i} \to \widetilde{a}$, it follows that $\widetilde{d_i}$ converges and the proof is complete.

Remark 12 Notice that both ϕ and ϕ_0 are composition of elements of W^r; therefore, a perturbation of the first return map ϕ_0 still is a composition of two involutions. Hence, the diffeomorphism ϕ_0 is perturbed only over the codimension one submanifold $P^{-1}(-1)$ of $\text{Diff}(\mathbb{R}^2, 0)$ (space of germs of diffeomorphisms at 0).

It follows straightforwardly from the previous results:

Proposition 10 *Let $Z_0 = (X_0, Y_0) \in \Omega^r$ be a germ of nonsmooth vector field having a Teixeira singularity at p. Let (α, β, γ) be the normal parameters of Z_0 at p. If $\alpha\beta \geq \gamma$ and $\alpha, \beta < 0$, then Z_0 is locally structurally stable at p.*

Finally, we conclude the proof of Theorem 3:

Proof (Proof of Theorem 3) Notice that Z satisfies condition $\Sigma(E)$ at p if, and only if, the normal parameters (α, β, γ) of Z at p satisfy $\alpha\beta \geq \gamma$ and $\alpha, \beta < 0$.

The result follows directly from Propositions 6, 7, and 10,

7 Proofs of Theorems 4, 5 and Corollary 4.1

In this section, we intend to discuss the hyperbolic and the parabolic case of the fold–fold singularity in order to complete the characterization of Σ_0.

7.1 *Hyperbolic Fold–Fold*

Let $Z = (X, Y) \in \Omega^r$ be a nonsmooth vector field having a hyperbolic fold–fold point at p such that $S_X \pitchfork S_Y$ at p. Consider the normal coordinates (x, y, z) of Z at p and let (α, β, γ) be the normal parameters of Z at p. In this case, we do not have any orbit of X or Y connecting points of Σ, therefore the local structural stability of Z at p depends only on the sliding dynamics which is generically characterized in Sect. 5.2.

Proposition 11 *Let $Z_0 = (X_0, Y_0) \in \Omega^r$ be a nonsmooth vector field having a visible fold–fold point at p such that $S_{X_0} \pitchfork S_{Y_0}$ at p. Let $(\alpha_0, \beta_0, \gamma_0)$ be the normal parameters of Z_0 at p. Then, Z_0 is locally structurally stable at p if and only if $(\alpha_0, \beta_0, \gamma_0) \in R_H^1 \cup R_H^2$.*

Proof (Outline) The first implication is obvious since F_{Z_0} presents bifurcations in Σ^s. To prove the converse, let $(\alpha_0, \beta_0, \gamma_0)$ be the normal parameters of Z_0 at p. Using Implicit Function Theorem, we can find a neighborhood $\mathcal{V}$ of Z_0 in Ω^r such that every $Z \in \mathcal{V}$ has a hyperbolic fold–fold point $q(Z)$ near p and the normal parameters of Z at $q(Z)$ are close to $(\alpha_0, \beta_0, \gamma_0)$.

Now, it is easy to construct a homeomorphism $h : \Sigma \to \Sigma$ carrying sliding orbits of F_{Z_0} onto sliding orbits of F_Z. Extend it to a germ of homeomorphism $h : (M, p) \to (M, q(Z))$ using the flows in the same way of [10] (Lemma 3, page 271).

7.2 Parabolic Fold–Fold

Let $Z = (X, Y) \in \Omega^r$ be a nonsmooth vector field having an invisible–visible fold–fold point at p such that $S_X \pitchfork S_Y$ at p. Consider the normal coordinates (x, y, z) of Z at p, and let (α, β, γ) be the normal parameters of Z at p.

Proceeding as in the elliptic case, Z has an involution ϕ_X associated to the invisible fold of X, and recall that it is given by

$$\phi_X(x, y) = (x - 2\alpha y, -y),$$

in normal coordinates. Now, we use it to study the connections between sliding orbits, when they exist.

Lemma 7.1 *Let $Z = (X, Y) \in \Omega^r$ be a nonsmooth vector field having an invisible–visible fold–fold point at p such that $S_X \pitchfork S_Y$ at p. Let (α, β, γ) be the normal parameters of Z at p. Then, $\phi_X(S_Y) \pitchfork S_Y$ at p if and only if $\alpha \neq 0$.*

Proof From Corollary 5.1, we have that $S_Y = \{(g(y), y);\ y \in (-\varepsilon, \varepsilon)\}$, for some $\varepsilon > 0$, where g is a smooth function with $g(y) = \mathcal{O}(y^2)$. Therefore, $T_0 S_Y = \text{span}\{(0, 1)\}$.

On the other hand, $\phi_X(S_Y) = \{(g(y) - 2\alpha y, -y);\ y \in (-\varepsilon, \varepsilon)\}$. Then, $T_0\phi_X(S_Y) = \text{span}\{(-2\alpha, -1)\}$. The result follows from these expressions.

Lemma 7.2 *Let $Z = (X, Y) \in \Omega^r$ be a nonsmooth vector field having an invisible–visible fold–fold point at p such that $S_X \pitchfork S_Y$ at p. Let (α, β, γ) be the normal parameters of Z at p. Then, $\phi_X(\Sigma^{us}) \cap \Sigma^{ss} = \emptyset$ if and only if $\alpha > 0$.*

Proof In fact, in these coordinates, $S_Y = \{(g(y), y);\ y \in (-\varepsilon, \varepsilon)\}$, and $\phi_X(S_Y) = \{(g(y) - 2\alpha y, -y);\ y \in (-\varepsilon, \varepsilon)\}$, for some $\varepsilon > 0$, where g is a smooth function with $g(y) = \mathcal{O}(y^2)$.

Therefore, $T_0\phi_X(S_Y) = \text{span}\{(-2\alpha, -1)\}$. The sliding region Σ^s is the region delimited by S_X and S_Y.

Since $T_0 S_Y = \text{span}\{(0, 1)\}$ and $T_0 S_X = \text{span}\{(1, 0)\}$, it follows that $\phi_X(S_Y) \subset \Sigma^s$ if and only if $\alpha > 0$.

We conclude the proof by noticing that if $\phi_X(S_Y) \subset \Sigma^c$, then $\phi_X(\Sigma^{us}) \subset \Sigma^c$. Nevertheless, if $\phi_X(S_Y) \subset \Sigma^s$, then the region delimited by S_Y and $\phi_X(S_Y)$ in Σ^{us} is carried into the region delimited by S_Y and $\phi_X(S_Y)$ in Σ^{ss}.

Remark 13 In other words, there exist orbits of X in M^+ connecting distinct points in the sliding region Σ^s if and only if $\alpha > 0$.

Definition 14 If $\phi : \Sigma \to \Sigma$ is a diffeomorphism and F is a vector field in Σ, then define the **reflected vector field of** F **by** ϕ as $\phi^* F$.

Remark 14 The reflected vector field of F by ϕ can also be referred as **transport of** F **by** ϕ.

Lemma 7.3 *Let $Z = (X, Y) \in \Omega^r$ be a nonsmooth vector field having an invisible–visible fold–fold point at p such that $S_X \pitchfork S_Y$ at p. Let (α, β, γ) be the normal parameters of Z at p.*

Assume that there exists a region $S \subset \Sigma^{us}$ such that $\widetilde{S} = \phi_X(S) \subset \Sigma^{ss}$, and suppose that S is maximal with respect to this property. If $2(\alpha + \beta)(\alpha\beta - \gamma) \neq 0$, then F_Z^N and the transport of F_Z^N by ϕ_X are transversal vector fields defined in $\tilde{S}$.

Proof Consider $F_0 = F_Z^N$ and $F_1 = \phi^* F_Z^N$, where ϕ_X is the involution associated to X.

Clearly, F_0 and F_1 are transversal at $q \in \Sigma$ if and only if $F_0(q)$ and $F_1(q)$ are linearly independent vectors.

Considering the normal coordinates (x, y, z) at p. Define

$$D(x, y) = \det \begin{pmatrix} F_0(x, y) \\ F_1(x, y) \end{pmatrix}. \tag{9}$$

Notice that $D(x, y) \neq 0$ if and only if F_0 and F_1 are transversal at (x, y). Now, we use the expressions of the vector field in these coordinates to derive an approximation for the function D.

Since ϕ_X is a linear involution, it follows that $\phi_X^{-1} = \phi_X$ and $d\phi_X = \phi_X$, therefore:

$$\begin{aligned} F_1(x, y) &= d\phi_X(F_Z^N(\phi_X^{-1}(x, y))) \\ &= \phi_X(F_Z^N(\phi_X(x, y))) \end{aligned} \tag{10}$$

In order to compute D, we must analyze the influence of the higher order terms in the computation of F_Z^N. From Proposition 2, we have that:

$$X(x, y, z) = \begin{pmatrix} \alpha \\ 1 \\ -y \end{pmatrix} \text{ and } Y(x, y, z) = \begin{pmatrix} \gamma + \widetilde{F}(x, y, z) \\ \beta + \widetilde{G}(x, y, z) \\ x + \widetilde{H}(x, y, z) \end{pmatrix}, \tag{11}$$

where $\widetilde{F}(x, y, z) = \mathcal{O}(|(x, y, z)|)$, $\widetilde{G}(x, y, z) = \mathcal{O}(|(x, y, z)|)$, and $\widetilde{H}(x, y, z) = \mathcal{O}(|(x, y, z)|^2)$.

Hence, the sliding vector field is given by:

$$F_Z^N(x, y) = \begin{pmatrix} \alpha & \gamma \\ 1 & \beta \end{pmatrix} \cdot \begin{pmatrix} x \\ y \end{pmatrix} + \begin{pmatrix} \alpha H(x, y) + yF(x, y) \\ H(x, y) + yG(x, y) \end{pmatrix},$$

where $F(x, y) = \widetilde{F}(x, y, 0) = \mathcal{O}(|(x, y)|)$, $G(x, y) = \widetilde{G}(x, y, 0) = \mathcal{O}(|(x, y)|)$, and $H(x, y) = \widetilde{H}(x, y, 0) = \mathcal{O}(|(x, y)|^2)$.

Using the expressions of F_Z^N and $\phi_X(x, y) = (x - 2\alpha y, -y)$, we obtain:

$$D(x, y) = y^2[-2(\alpha + \beta)(\alpha\beta - \gamma) + P_1(x, y)].$$

where $P_1(x, y) = \mathcal{O}(|(x, y)|)$.

Now, if $(\alpha+\beta)(\alpha\beta-\gamma) \neq 0$, then the x-axis is the only solution of $D(x, y) = 0$, near the origin. Therefore, the vector fields F_0 and F_1 are transversal in the region $S \cup \widetilde{S}$, since it does not contain points of the x-axis.

Remark 15 Notice that in the curves $\alpha+\beta = 0$ and $\alpha\beta = \gamma$, the higher order terms may produce curves in $S \cup \widetilde{S}$ where the vector fields are not transversal, and they can be broken by small perturbations (making $\alpha + \beta \neq 0$ or $\alpha\beta \neq \gamma$). Clearly, this situation implies the instability of the system.

Lemma 7.4 *Let $Z = (X, Y) \in \Omega^r$ be a nonsmooth vector field having an invisible–visible fold–fold point at p such that $S_X \pitchfork S_Y$ at p. Let (α, β, γ) be the parameters given by Proposition 2 associated to Z at p. If $2\alpha(\alpha + \beta) - \gamma \neq 0$, then F_Z^N is transversal to $\phi_X(S_Y)$ in Σ^s.*

Proof In the coordinates of Proposition 2, we have that $S_Y = \{(g(y), y, 0);\ y \in (-\varepsilon, \varepsilon)\}$, for $\varepsilon > 0$ sufficiently small, where g is a $\mathscr{C}^r$ function such that $g(y) = \mathcal{O}(y^2)$.

Therefore, $\phi_X(S_Y) = \{(g(y)-2\alpha y, -y);\ y \in (-\varepsilon, \varepsilon)\}$. Since $\phi_X(S_Y)$ is tangent to the curve $\gamma(y) = (-2\alpha y, -y)$ at the origin, it is sufficient to prove that F_Z^N is transversal to γ.

Clearly, F_Z^N is transversal to γ at $\gamma(y)$ if and only if

$$T(y) = F_Z^N(\gamma(y)) \cdot (\gamma'(y))^{\perp} \neq 0. \tag{12}$$

Now, we use the expression of F_Z^N in these coordinates to obtain an approximation of T. In fact,

$$F_Z^N(\gamma(y)) = F_Z^N(-2\alpha y, -y) = (-2\alpha^2 y - \gamma y, -2\alpha y - \beta y) + \mathcal{O}(y^2)$$

and

$$(\gamma'(y))^{\perp} = (-2\alpha, -1)^{\perp} = (1, -2\alpha).$$

Substituting these expressions in (12), we obtain:

$$T(y) = [2\alpha(\alpha + \beta) - \gamma]y + \mathcal{O}(y^2)$$

Therefore, if the condition $2\alpha(\alpha + \beta) - \gamma \neq 0$ is assumed and $y \neq 0$, then F_Z^N is transversal to $\phi_X(S_Y)$. Since Σ^s does not contain points where $y \neq 0$ (because they belong to S_X), the result follows.

Remark 16 In the curve $2\alpha(\alpha + \beta) - \gamma = 0$, the higher order terms can be used to produce a curve such that F_Z^N is tangent to $\varphi_X(S_Y)$ in every point. Such structurally unstable phenomena is avoided.

Proposition 12 *Let $Z_0 = (X_0, Y_0) \in \Omega^r$ be a nonsmooth vector field having an invisible–visible fold–fold point at p such that $S_{X_0} \pitchfork S_{Y_0}$ at p. Let $(\alpha_0, \beta_0, \gamma_0)$ be the normal parameters of Z_0 at p. Then, Z_0 is locally structurally stable at p if and only if the following statements hold:*

1. $(\alpha_0, \beta_0, \gamma_0) \in \cup_{i=1}^4 R_P^i$;
2. $\alpha_0 \neq 0$;
3. $2\alpha_0(\alpha_0 + \beta_0) - \gamma_0 \neq 0$;
4. $\alpha_0 + \beta_0 \neq 0$, *if* $\alpha_0 > 0$.

Moreover, there exist only eleven topologically distinct classes of local structural stable systems at invisible–visible fold–fold points.

Proof (Outline) Proceeding as is the proof of Theorem 11. Consider the neighborhood $\mathscr{V}$ of Z_0 such that the correspondent parameters (α, β, γ) of any $Z \in \mathscr{V}$ are in the same region of $(\alpha_0, \beta_0, \gamma_0)$.

Let $Z = (X, Y) \in \mathscr{V}$. If there are no orbits of X connecting points of Σ^{ss} and Σ^{us}, then the proof can be done in the following way. We omit some details in this case, since it is very similar to the visible case.

- Construct $h : \Sigma^s(Z_0) \to \Sigma^s(Z)$ carrying orbits of F_0 onto orbits of F_Z. In addition, extend it to $S_{X_0} \cup S_{Y_0}$ via limit. Hence, $h(S_{X_0}) = S_X$ and $h(S_{Y_0}) = S_Y$;
- For each $p \in \Sigma \setminus S_{X_0}$, there exists $t_0(p) \neq 0$ such that $\varphi_{X_0}(t_0(p), p) \in \Sigma$. Similarly, there exists an analogous time $t(p) \neq 0$ for the vector field X;
- If $p \in \Sigma^s$, then $h(p)$ is already defined. Assume that $p \in \Sigma^c$. If $\varphi_{X_0}(t_0(p), p) \in \Sigma^s$, then define:

$$h(p) = \varphi_X(-t(\varphi_{X_0}(t_0(p), p)), h(\varphi_{X_0}(t_0(p), p))).$$

- Using Tietze Extension Theorem, we can extend h over Σ^c;
- Now, using the same idea of the third item, we can extend it to the whole Σ;
- Extend it to M^+ using the flow of X_0, X, and $h : \Sigma \to \Sigma$;
- Following the same idea of the hyperbolic case, extend it to M^-;
- Hence, we construct a germ of homeomorphism $h : M \to M$ at p, with $h(p) = q(Z)$, which is an equivalence between Z_0 and Z. Then, Z_0 is locally structurally stable at p.

Suppose that there exists a connection between Σ^{ss} and Σ^{us} for Z_0 and Z. Denote by S_0 and S, the regions of Σ^s presenting connections.

From the previous Lemmas of this subsection, it is possible to say that F_0 and $\phi_{X_0}^* F_0$ are transversal in each point of S_0, and the same holds for F_Z and $\phi_X^* F_Z$ in S.

Therefore, the orbits of F_0 and $\phi_{X_0}^* F_0$ define a coordinate system in S_0, such as the orbits of F_Z and $\phi_X^* F_Z$ in S.

Hence, let h be a function carrying S_{Y_0} onto S_Y, and $h(0) = 0$. Now, we can use these coordinate systems to construct $h : S_0 \to S$ satisfying

$$h \circ \phi_{X_0} = \phi_X \circ h.$$

By the transversality of F_0 to $\phi_{X_0}(S_{Y_0})$ (resp. F_Z to $\phi_X(S_Y)$), it is possible to extend h on $\Sigma^s(Z_0)$ using the sliding orbits. Then, we have a homeomorphism $h : \Sigma^s(Z_0) \to \Sigma^s(Z)$ carrying sliding orbits onto sliding orbits.

By construction, if $x \in S$, then $\phi_X(h(x)) = h(\phi_{X_0}(x))$. With this, we can use the same idea from the previous case without connections to extend such map to a germ of homeomorphism $h : M \to M$ at p, with $h(p) = q(Z)$, which is a topological equivalence between Z_0 and Z at p.

7.3 *Proof of Theorem 4*

Notice that Z satisfies condition $\Sigma(H)$ at p if, and only if, the normal parameters (α, β, γ) of Z at p satisfy the hypotheses of Proposition 11.

Moreover, Z satisfies condition $\Sigma(P)$ at p if, and only if, the normal parameters (α, β, γ) of Z at p satisfy the hypotheses of Proposition 12.

The result follows directly from Propositions 11 and 12.

7.4 *Proof of Theorem 5*

From Proposition 1, it follows that $\Sigma_0 \subset \Sigma(G)$.

The result follows from Theorems 2, 3, and 4.

7.5 *Proof of Corollary 4.1*

From the characterization of Σ_0, we can see that $\Sigma(G)$, $\Sigma(R)$, $\Sigma(H)$, and $\Sigma(P)$ are open dense sets in Ω^r.

Nevertheless, we also prove that $\Sigma(E)$ is not residual in Ω^r. Therefore, it follows that $\Sigma_0 \cap \Sigma(E)$ is open dense in $\Sigma(E)$ and $\Sigma(E)$ is the biggest set with this property.

References

1. A. A. Andronov, E. A. Leontovich, I. I. Gordon, A. G. Maier, *Theory of bifurcations of dynamic systems on a plane*. John Wiley and Sons, 1971.
2. D. K. Arrowsmith, C. M. Place, *An introduction to dynamical systems*. Cambridge University Press, 1990.
3. W. Blaschke, G. Bol, *Geometrie der gewebe: Topologische fragen der Differential Geometrie*. Springer, 1938.

4. C. Bonatti, M. A. Teixeira, *Topological equivalence of diffeomorphisms and curves*. Journal of Differential Equations, vol. 118, 371–379, 1995.
5. C. A. Buzzi, J. C. R. Medrado, M. A. Teixeira, *Generic bifurcation of refracted systems*. Advances in Mathematics, vol. 234, 653–666, 2013.
6. I. Ekeland, *Discontinuité des champs Hamiltoniens er existence de solutions optimales en calcul des variations*. Pub. IHES, 47, 5–32, 1977.
7. A. Colombo, M. R. Jeffrey, *The two-fold singularity of discontinuous vector fields*. SIAM J. Applied Dynamical Systems, **8**, 2, 624–640, 2009.
8. A. Colombo, M. R. Jeffrey, *Nondeterministic Chaos, and the Two-fold Singularity in Piecewise Smooth Flows*. SIAM J. Applied Dynamical Systems, **10**, 2, 423–451, 2011.
9. A. Colombo, M. R. Jeffrey, *The two-fold singularity of nonsmooth flows: leading order dynamics in n-dimensions*. Physica D: Nonlinear Phenomena, **263**, 1–10, 2013.
10. A. F. Filippov, *Differential equations with discontinuous righthand sides*. Kluwer, 1988.
11. S. Fernández-García, D. Angulo García, G. Olivar Tost, M. di Bernardo, M. R. Jeffrey, *Structural stability of the two-fold singularity*, SIAM Journal on Applied Dynamical Systems, 2012, **11**, 4, 1215–1230, 2012.
12. M. Guardia, T. M. Seara, M. A. Teixeira, *Generic bifurcations of low codimension of planar Filippov systems*. J. Differential Equations, **250** (2011).
13. P. Hartman, *On local homeomorphisms of Euclidean spaces*. Bol. Soc. Mat. Mexicana (2) 5, 1960.
14. V. S. Kozlova, *Roughness of a discontinuous system*. Vestinik Moskovskogo Universiteta, Matematika 5, 16–20, 1984.
15. Y. A. Kuznetsov, S. Rinaldi, A. Gragnani, *One-parameter bifurcations in planar Filippov systems*. International Journal of Bifurcation and Chaos, vol. 13, 8, 2157–2188, 2003.
16. W. Melo, *Moduli of stability of two-dimensional diffeomorphisms*. Topology, 19 (1), 9–21, 1980.
17. J. Palis, *A differentiable invariant of topological conjugacies and moduli of stability*. Astérisque, 51, 1978.
18. M. C. Peixoto, M. M. Peixoto, *Structural stability in the plane with enlarged boundary conditions*, An. Acad. Bras. Ciencias, 31, 1959.
19. P. B. Percell, *Structural stability on manifolds with boundary*, Topology, 12, 123–144, 1973.
20. J. V. Pereira, L. Pirio, *An introduction to web geometry* , IMPA Monographs, Springer, 2015.
21. E. Ponce, R. Cristiano, D. Pagano, E. Freire, *The Teixeira singularity degeneracy and its bifurcation in Piecewise Linear systems* , Fourth Symposium on Planar Vector Fields, 2016.
22. J. Sotomayor, M. A. Teixeira, *Vector fields near the boundary of a 3-manifold*, Dynamical systems, vol 1331, 169–195, 1988.
23. M. A. Teixeira, *Generic bifurcation in manifolds with boundary*. J. Differential Equations, **25** (1977).
24. M. A. Teixeira, *Local and simultaneous structural stability of certain diffeomorphisms*. Dynamical Systems and Turbulence, Warwick 1980, pp.382–390, 1981.
25. M. A. Teixeira, *Generic singularities of discontinuous vector fields*. Anais da Academia Brasileira de Ciências, **53**(2), 1981.
26. M. A. Teixeira, *On topological stability of divergent diagrams of folds*. Mathematische Zeitschrift, 180 (2), 361–371, 1982.
27. M. A. Teixeira, *Stability conditions for discontinuous vector fields*. J. Differential Equations, **88** (1990).
28. M. A. Teixeira, *Generic bifurcation of sliding vector fields*. Journal of Mathematical Analysis and Application, **176** (1993).
29. M. A. Teixeira, *Divergent diagrams of folds and simultaneous conjugacy of involutions*. Discrete and Continuous Dynamical Systems, 12 (4), 657–674, 2005.
30. M. A. Teixeira, *Perturbation Theory for Non-smooth Systems*. Encyclopedia of Complexity and Systems Science, Springer New York, 6697–6709, 2009.
31. S. M. Vishik, *Vector fields near the boundary of a manifold*. Vestnik Moskovskogo Universiteta Mathematika, **27(1)**, 21–28, 1972.

Appendix A
Non-smooth Dynamical Systems (NSDS): Reflections and Guidelines

Marco Antonio Teixeira

A.1 Introduction

In this note, I intend to discuss in very general terms what is currently occurring with an emerging structure theory of geometric and qualitative nature in non-smooth dynamical systems (NSDS) theory. Currently, the great interest in such theory is displayed by the rapid growth in the number of publications and specialized meetings in the area in recent decades. I believe that the main challenge in the study of NSDS is to understand and clarify some of the often very complicated dynamical behaviors and establish a precise mathematical framework for the problems encountered therein. This subject has been tainted with the reputation of being lax, mainly because there is an endless list of experimental research in genuine applied sciences. Due to the explosion of scientific developments in the smooth theory in the last century, NSDS has not attracted an expressive number of theoretical mathematicians. In my own research, I have experienced many challenges and extreme mathematical difficulties to elucidate a lot of problems in NSDS. On the other hand, I may say that some technical difficulties in general NSDS are rather formidable. My first motivation to study this field was the theoretical paper of Ekeland [1] where the main problem in Calculus of Variations was discussed via piecewise smooth systems. Personal discussions with V. Arnold, H. Sussman, I. Kupka, J. Sotomayor, and D. Anosov were also stimulating. In what follows, I briefly indicate directions in which the field can be developed as well as two very natural questions that are raised in this context: "How does the dynamical mathematical community react to these developments? Does the study of non-smooth systems have to be motivated by real-world applications?"

M. A. Teixeira (✉)
Institute of Mathematics, Statistics and Scientific Computing, University of Campinas, Campinas, São Paulo, Brazil
e-mail: teixeira@ime.unicamp.br

C. Lavor, F. A. M. Gomes (eds.), *Advances in Mathematics and Applications*,
https://doi.org/10.1007/978-3-319-94015-1

A.1.1 Some Words from Mauricio Peixoto

In the early 1970s, I was a young PhD student at USP-São Paulo and had been invited by J. Sotomayor to give a talk at IMPA on my master's thesis "On Stratified Sets." Mauricio Peixoto was a professor at USP-São Paulo and was constantly using the Rio-São Paulo air shuttle. That day, we traveled back together to São Paulo and of course being considered him (together with Leopoldo Nachbin) as the greatest Brazilian mathematician, I was anxious to have his intake on mathematical research. Inside the plane, at one point I naively asked him about a subject that intrigued me. I said: "Professor Maurício, something I find curious in Mathematics is an area that studies Differential Equations with Complex Time, since as I see it does not exist in the real world." He replied, "In Mathematics what matters in any problem is to have mathematically consistent statements that are, non trivial and to have correct proofs of what is asserted. Keep this advice always and be careful in saying whether something in mathematics is important or not." His philosophical view on mathematical development made quite an impression on me and I have treasured it and always follow the words of the Master.

A.2 Some of Non-smooth Dynamical Systems

(a) Why Filippov?
    It is clear that in studying the phase portrait of a differential equation the first object that comes to mind is the behavior of the solutions. I emphasize that the concept of a solution of a differential equation with second discontinuous member is not universal. Usually, it obeys a certain convention that is, a priori, stipulated or it is defined ad hoc according to the problem. The lack of uniqueness of solutions requires of course extra attention. For simplicity, we have chosen as the basis of our theoretical study the Filippov Convention due to its essentially geometric character although I understand that the Caratheodory Convention would be the most natural choice. Nowadays, the book of Filippov seems to be unanimously accepted as an important contribution to dynamical systems theory. On the other hand, some recent results have motivated me to better understand the Utkin sliding mode convention (see [2–4] and references therein).
(b) Take any heuristically proved result and try to give a rigorous mathematical proof.
    Obviously, there may be controversy over what is actually a heuristically proved result. Observe that it is common to find heuristic results for specific models and build upon them for generalization. In my point of view and in the strictly mathematical world, the above procedure has, philosophically speaking, a high scientific value. In short, our task would be to give formal mathematical justifications to the conclusions.

(c) Try to give non-smooth versions to classical results of the smooth world.
It is evident that this scenario is purely theoretical and in each case the first step would be to detect whether such a procedure is trivial or not. Hypotheses and new statements must be highly clear and new techniques and/or methods are welcome. To exemplify, in the literature we find the results of Classical Averaging Theory very well settled when one tries to detect limit cycles in NSDS (see [1] and references therein).

(d) Look for problems, for which there are no counterparts in the smooth universe.
In this item, the best argument is to cite the existence of the elliptical fold–fold singularity in high dimensions. There is no phenomenon in the smooth universe that is a counterpart of this singularity. Moreover, the proof of its stability/instability is indeed very complicated and uses several nontrivial techniques. Finding objects without smooth counterparts is a hard task, perhaps with arduous abstraction power. It seems that an analysis of the robustness of typical minimal sets would be highly encouraging.

(e) Approximating an $NSDS$ by smooth systems (regularization).
The regularization process of a non-smooth system Z_0, known as Sotomayor–Teixeira regularization, consists in approximating this system by m-parameter families Z_k ($k = (k_1, k_2, \ldots .k_m)$) of smooth systems. It is worthwhile to cite two directions: (1) depending on the characteristic of Z_0, each Z_k can have a very interesting intrinsic behavior under which deserves a deep analysis. (2) Sometimes, information about the behavior of Z_k can contribute to the understanding of the dynamics of Z_0. This can be observed in works involving averaging theory and also in the bridge established between NSDS and Singular Perturbation Theory. It would seem to me absolutely essential to reflect on the scope of general regularizations.

A.3 Miscellaneous in Geometric and Qualitative Theory in Non-smooth Dynamical Systems

1. Perturbative results are inherent to the methodology of structural stability and bifurcation theory. So, it is very important to specify the topological space to which the systems belong.
2. Some titles may be borrowed from the smooth universe whose contents can be successfully exploited: generic bifurcation (Local and Global), stability theorems, normal forms, ergodic theory (in certain classes, Lyapunov exponents could be, rigorously, extended to non-smooth systems), new trends in hyperbolic theory, generalization of Melnikov's method, Conley index, piecewise continuous mappings, synchronization, singular perturbation theory, integrability (piecewise Hamiltonian theory), $NSDS$ tangent to continuous foliations, symmetries in $NSDS$ (including Refractive Systems), stochastic differential equations, etc.…

3. Another point that also deserves reflection is one in which a smooth vector field is approximated by non-smooth systems. This assertion comes from the fact that many continuous phenomena, for purely technical reasons, are modeled by differential equations with discontinuous second member. In this direction, I recall the phenomenon named "Pinching" (see [5] and references therein).
4. Interesting problems appear when the switching set is not a differentiable manifold (see [6–8] and references therein).
5. I confess that I tried to understand the substance of the most practical models presented in various congress by engineers and physicists and I was unsuccessful. I would like to really understand something of the machinery but unfortunately cannot recognize a "good mathematical model."
6. *Why study piecewise smooth systems?* One finds in real life and in various branches of science distinguished phenomena whose mathematical models are expressed by discontinuous systems and deserve a systematic analysis. However, sometimes the treatment of such objects is far from the usual techniques or methodologies found in the smooth universe. This might be a good time to reflect on the role of the discussions found in [9–11]. In what follows, we reproduce the abstract of [9]:

   "Abstract - Despite the aphorism "Nature does not make jumps" (attributed to Newton, Leibniz, Linnaeus,. . .!!) it is frequently useful to work, either descriptively or prescriptively, with simplified models which involve switching between different modes of evolution. We describe a variety of examples of such modeling with particular attention to some situations in which the interpretation of the reduced model is a matter of concern."

   In [10, 11], one finds arguments that invite us to a discussion about the (non) use of a complete presentation of cases and subcases that appear in the general theory of $NSDS$.

   The following paragraphs were selected from a P. Glendining's talk:

   "To paraphrase Mike Field:

   a. Although most of dynamical systems theory for the last fifty years has been smooth, the real technological innovations and applications (computers, control, mechanics, some biological models) are not smooth.
   b. If you want a car to stop when the brakes are applied, don't choose an analytic or smooth system!!!"

A.4 Conclusion

The present work looks primarily to the future. It is mainly concerned with the intrinsic significance of the classification results in NSDS and to its range of applicability in other areas of science. Moreover, the importance of providing readily accessible proofs of the statements is assessed. Finally, I hope this text will help young researchers to face challenges in developing and performing consistent research projects in $NSDS$.

References

1. I. Ekeland, *Discontinuité des champs Hamiltoniens er existence de solutions optimales en calcul des variations*. Pub. IHES, 47, 5–32, 1977.
2. T. M. Seara, C. Bonet, M. R. Jeffrey, *A unified approach to explain contrary effects of hysteresis and smoothing in non-smooth systems*, Communications in Nonlinear Science and Numerical Simulation, Volume 50, 142–168, 2017.
3. M. R. Jeffrey, *The ghosts of departed quantities in switches and transitions*, arXiv:1508.04344, 2015.
4. V. I. Utkin, *Sliding modes in control and optimization*. Springer-Verlag, 1992.
5. M. Desroches, M. R. Jeffrey, *Canards and curvature: non smooth approximation by pinching*. Nonlinearity, 24, 1655–1682, 2011.
6. C. A. Buzzi, P. R. da Silva, M. A. Teixeira, *Slow -fast systems on algebraic varieties bordering piecewise smooth dynamical systems*. Bull. Sci. math., 136, 444–462 2012.
7. L. Dieci, L. Lopez, *Sliding motion on discontinuity surfaces of high co-dimension. a construction for selecting a Filippov vector field*. Numer Math, 117, 779–811, 2011.
8. J. Llibre, P. R. da Silva, M. A. Teixeira, *Sliding vector fields for non-smooth dynamical systems having intersecting switching manifolds*. Nonlinearity (Bristol. Print), v. 28, 493–507, 2015.
9. T. I. Seidman, *Some aspects of modeling with discontinuities*, 2007 - https://pdfs.semanticscholar.org/91ee/48cc825cb520fcdf79d4d3fb2fe7e6dab426.pdf
10. P. Glendinning, *Less is more I: a pessimistic view of piecewise smooth bifurcation theory*, - http://personalpages.manchester.ac.uk/staff/paul.glendinning/preprints/eca-glendinning1.pdf.
11. P. Glendinning, *Less is More II: an optimistic view of piecewise smooth bifurcation theory*, - http://personalpages.manchester.ac.uk/staff/paul.glendinning/preprints/eca-glendinning2.pdf.

MIX
Papier aus verantwortungsvollen Quellen
Paper from responsible sources
FSC® C105338

If you have any concerns about our products,
you can contact us on
ProductSafety@springernature.com

In case Publisher is established outside the EU,
the EU authorized representative is:
Springer Nature Customer Service Center GmbH
Europaplatz 3, 69115 Heidelberg, Germany

Printed by Libri Plureos GmbH
in Hamburg, Germany